Earth Science in the City: A Reader

Grant Heiken
Robert Fakundiny
John Sutter
Editors

American Geophysical Union
Washington, DC

Published under the aegis of the AGU Books Board

Library of Congress Cataloging-in-Publication Data
Earth science in the city: a reader / Grant Heiken, Robert Fakundiny, John Sutter, editors.
p.cm.
Includes bibliographical references.
ISBN 0-87590-299-5
1. Urban geology. I. Heiken, Grant. II. Fakundiny, Robert H. III. Sutter, John F.

QE39.5.U7E18 20003
624.1'51'091732--dc21

2003048070

ISBN 0-87590-299-5

Back Cover: Washington, DC, photographed on June 1, 2000, with the Advanced Spaceborne Thermal Emission and Reflection Radiometer (ASTER) on NASA's Terra satellite. NASA GSFC, MITI, ERSDAC, JAROS, and U.S./Japan ASTER Science Team

Printed in the United States of America.

CONTENTS

Section III. Urban Hydrology

Section IV. The Remotely Sensed City

Section V. Integrated Earth Sciences and Urban Development and Sustainability

Preface

Today nearly half of the Earth's peoples live in cities, and projected trends indicate a rise to five billion city dwellers by the year 2025. All cities become increasingly coupled with and vulnerable to their environment as they grow. For cities to be safe and sustainable we must be ever aware of the inter-relationships between natural processes and the urban environment, effects on the population, and—in turn—the effects of population on the environment. Many of these relationships, which become issues or problems of public note usually when we are faced with immediate conditions of risk—including water use sustainability or hazard mitigation from natural disasters—must be addressed via the geosciences. And geoscience researchers should be aware that the results of their work are vital to solving urban problems both in the practical and theoretical domains, and for immediate and future needs. The present book speaks to such concerns. We hope it also serves to stimulate discussion of, and research on, urban geoscience for professionals in the field or associated with it, and for students entering the field.

The current book derives from presentations given at "Earth Sciences in the Cities," a Union Session of the spring 2000 meeting of the American Geophysical Union (AGU). The 22 presentations also represented at least seven of the eleven sections that comprise the AGU, including: atmospheric sciences, hydrology, geodesy, ocean sciences, tectonophysics, seismology, and volcanology. The session highlighted current geosciences contributions to significant urban issues from disaster mitigation to environmental degradation, planning, and more.

Organization of the session was enthusiastically supported by the meeting program chair Carol Simpson and by AGU staff. We also thank the many reviewers of chapters in this book, our AGU acquisitions editor, Allan Graubard, and our production editor, Terence Mulligan.

The Editors

Introduction

Grant Heiken

Robert K. Leggett, the foremost expert on urban geology in the 20th Century, emphasized that the natural setting of a city is its foundation. In the past, most urban planning decisions were made with little or no regard for the role of the natural setting in the city's long-term health and stability. In the last several years, the number of cities with populations of more than a million people has topped 400 (Brinkerhoff, 2002), and the relevance of Leggett's philosophy has become more apparent. As the Earth's population shifts from predominately rural to urban settings and changing conditions make themselves felt in natural disasters and resource shortages, it is clear that we must adopt a new way of understanding and managing cities—by understanding their intimately linked man-made and environmental systems.

To date, few geoscience monographs and texts have focused on the application of Earth sciences to urban problems. Although Leggett's textbook was published in 1973 (yes, more than a quarter of a century ago), only one general text (McCall, et al. 1996) and several proceedings from UNESCO-sponsored conferences on urban geoscience for specific regions or cities (ESCAP, 1989) have been published. *Earth Science in the City: A Reader* is intended to be both an introduction to this emerging area of scientific study and a response to growing concern, expressed at national and international levels, about the lack of geoscientific analysis in urban development.

Earth Science in the City: A Reader

10.1029/056SP01

THE FUTURE OF EARTH SCIENCES IN THE CITIES

When the Union session, "Earth Sciences in the Cities," was organized for the American Geophysical Union meeting of spring 2000, from which this volume derives, two of the authors had prepared an editorial for the journal *Environmental Science and Policy* (Valentine and Heiken, 2000). Part of that editorial is quoted here to re-emphasize the desperate need for growth in the new field before us.

The Problem

According to the United Nations Population Division, the world population will undergo a major transition in about 2005 *(this transition occurred in 2002)* when the majority of humans will live in cities. There are multiple reasons for this transition. In many developing nations the change is driven by a perception of increased economic opportunities in cities. Rural populations can only grow to a point where subsistence agriculture cannot be sustained. In developed nations, urban population growth is accelerating as rural migrants move into the cities for access to employment, goods and services. In the world today there are 405 cities with populations of over one million and 28 cities with populations over eight million, the so-called "megacities." There are 47 U.S. metropolitan areas with over a million people, and 198 U. S. cities with populations of 100,000-500,000. As cities increase in size and complexity, so do the issues of economic and political stability that are responsible for our quality of life. Increasingly, cities face problems that may have technical solutions—if the technical solutions are integrated with the more traditional approaches to municipal infrastructure maintenance, emergency response, public health, and planning.

Large cities are places where infrastructure elements such as telecommunications, transportation, and electricity, as well as economic activity converge. This convergence makes cities more vulnerable to natural and human-made disasters as well as poor long-term planning decisions. Disruption of a city's infrastructure can have far-reaching effects. Some authors write of the "footprint" of a city as the region from which a city pulls its resources, that receives the city's waste, or that depends to first order on the city's economy. City footprints usually extend far beyond a city's limits. For example, Los Angeles draws electricity that is generated in Washington, relies on lumber from the Pacific Northwest, and plays a role in the global economy. Most western states and, to some extent, the nation would be damaged by physical and economic collapse in Los Angeles.

To understand and better manage the cities of the world there is a need for changes in the underlying thinking of governments and the scientific community. The scientific community needs to further embrace urban systems as an important and credible field of research. For their part, university departments should collaborate to promote interdisciplinary curricula to train a new generation of science-based urban planners. Physical and biological scientists should increase their collaborations with social scientists, economists, and infrastructure engineers. Government laboratories should use their interdisciplinary horsepower and experience in applying science and technology to large societal problems for improving the urban condition. Given the inertia that has built up over the last century, we recognize that none of these changes will be easy to secure. But new thinking is mandatory as we move into the urban era of the 21st century.

Innovative Approaches to Urban Issues

According to G. S. Cheema, of the United Nations Development Program, "The urban research agenda ...should focus on the identification of innovative approaches to deal with the complex issues in urban management and on strengthening national capacities to plan and implement urban development programs."

Integrated approaches must be used within the governing bodies of individual metropolitan areas as well. The managers of systems that are the lifeblood of the world's cities are mostly isolated from one another. For example, the mutual dependence of water, electrical, and sewage systems and the economy and environment are not evident until there is a disaster such as an earthquake or hurricane. We need to understand these connections by running cities without the management barriers that currently block our understanding of all natural and man-made systems. Cities need integrated teams to collect data and make observations from a holistic viewpoint that then can be wrapped into quantitative models to use in alleviating vulnerabilities to natural disasters, terrorist attacks, and bad planning decisions. From our perspective, key players now missing as city employees come from the sciences, especially systems, earth, and atmospheric sciences. At this time, such professions are mostly housed in universities, and state and federal agencies. Where can they help? The list is long and includes, to name a few: water—sources and quality; air quality; energy resources and building materials; the fate and transport of chemical species from pollution or chemical/biological attacks; natural hazards mitigation—by hazards mapping, zoning, and hazards scenarios; public health after a natural disaster; greenbelts and urban agriculture; the effects of sea-level rise; thermal extremes and health effects; and understanding urban microenvironments as incubators of disease.

A focus on integrated urban systems also presents a new challenge for the scientific community, especially those of us in the natural sciences, who, like many (but not all) city managers and infrastructure managers, "work in a box." Integrated science is becoming a common approach to understanding natural (and man-made) systems and can be seen in the reorganization of some traditional, discipline-oriented university departments into interdisciplinary institutes or divisions. Students trained in this way must be willing to leave the academic fold to work for the cities, and the cities must be convinced that they will gain by hiring these pioneers and looking at new ways of managing a city.

The scientific community must recognize the increasing effect of cities on the Earth. The complexity of cities demands that integrated approaches be taken to achieve an understanding of the urban "system of systems" required to identify weaknesses and to enhance sustainability. The use of integrated studies should revolutionize the way governments address urban issues. Now is the time to prepare for humanity's transition from a rural to an urban human environment.

This book is but a step along the way toward a time when earth scientists realize that they are needed by the world's cities and when cities realize that, for sustainable growth, the sciences and especially the earth sciences are necessary.

REFERENCES

Brinkerhoff, T., 2002. Principal Agglomerations and Cities of the World, 11.05.2002. http://www.citypopulation.de/

ESCAP (United Nations Economic and Social Commission for Asia and the Pacific), 1989. *Geology and Urban Development, Atlas of Urban Geology, Vol. 4*—Hong Kong, Malaysia, The Netherlands, Thailand, 207 pp.

Leggett, R. F., *Cities and Geology.* McGraw-Hill, New York, 1973.

McCall, G. J. H., De Mulder, E. F. J., and Marker, B. R., 1996. *Urban Geoscience.* A. A. Balkema, Rotterdam, 273 pp.

Valentine, G. A. and Heiken, G., 2000. The need for a new look at cities. *Environmental Science and Policy,* 3: 231-234.

SECTION I

BACKGROUND

EARTH SCIENCE IN THE CITIES

Cities grow on a geologic framework that supplies renewable and nonrenewable resources, serves as defensive high ground, and provides foundations for dwellings, both private and public. Five thousand years ago the first urban centers were catalysts for trade, protection, and the evolution of societies, spurring the development of organized economies, transportation systems, and infrastructures for water and energy. Cities have also become foci for the most intense anthropogenic change on the Earth's surface yet known, including the devastation wrought by modern war and terrorism.

Engineering geology was crucial to the construction of New York's World Trade Center in the 1970s and, along with remote sensing, served to safely guide the excavation and removal of debris after the collapse of the twin towers on September 11, 2001, due to terrorist attack.

The public has little concept of the value of the Earth sciences to the health and stability of their cities. Help in remedying that deficiency can come from "geotopes" (geological parks) as research and public education sites established throughout a city. Geotopes will not only supply windows to a city's past but illuminate the effects of the anthropogenic change that accompanies urban development.

1

Large Urban Concentrations: A New Phenomenon

George Bugliarello

INTRODUCTION

Today's worldwide rapid rate of urbanization and the development of very large urban concentrations—the so-called megacities—are a new phenomenon that impacts in unprecedented ways human society and the Earth. Very large urban concentrations absorb an ever larger portion of Earth's resources. Their actual and virtual footprints and their globe-encircling emissions are changing environment and ecology, making even the atmosphere an artifact. Their design and internal dynamics impact not only watersheds and climate, but also human biology and the human psyche. From the energy, materials and water they consume to the halos from their night illumination, megacities have become the single most concentrated and accelerating source of anthropogenic change on the surface of the Earth. That change is the product basically of two factors, population and affluence. If the urban concentrations in the developed world are few but have greater affluence, the emerging and rapidly multiplying megacities in the developing world, being more populous and numerous, are bound to have an even greater comparable impact, that will increase as living standards increase.

Unfortunately, many facets of the phenomenon of large urban concentrations still elude us, and so do the actions needed to make an urbanized world more livable and sustainable.

THE PHENOMENON

Cities emerged some ten thousand years ago as the result of the invention of agriculture, which freed human populations from the nomadic existence of

Earth Science in the City: A Reader

10.1029/056SP02

hunter-gatherers. Major urban concentration arose already some five thousand years ago in the fertile crescent of the Middle East, in Egypt, and later in China. Later yet, Athens and Rome became the epitome of urbanization in the period of classical antiquity in Europe; large urban concentrations also occurred in Mexico, Central America and Peru. Rome, with a larger population than Athens, and with sophisticated public works, became a prototype of integrated urbanization—of the integration of production, trade and habitation that are the most fundamental functions of a city. That integration became a guiding concept for cities throughout the Roman world, but vanished in the medieval decline that followed dissolution of that world, to emerge again, imperfectly, in the cities of the late medieval period. It continues to be a universal aspiration for today's cities (Saalman).

The technological explosion ushered in by the Industrial Revolution, and the improvement of public health in the last century spurred a growth of cities that continues unabated today. Thus since the earliest beginnings, technology—at first agricultural technology, but also construction technology and civil engineering—has been the determining factor in the genesis and evolution of the cities. Among the many technological revolutions that have influenced that evolution, at the turn of the nineteenth century water supply and purification works made possible the widespread development of healthy large urban concentrations and the elevator, together with steel and concrete, made possible the vertical city; in the twentieth century, the automobile created the extended suburb and urban sprawl, and later in the century, aviation gave direct international access to landlocked cities in the interior of continents. Today telecommunications are affecting cities in still unfathomable ways.

Statistics about the concentration of populations in cities have led to a widespread characterization of the urbanization phenomenon as an explosion. A few statistics will suffice. The percentage of world population living in cities greater than one hundred thousand rose from five in 1900 to forty-five in 1995 (2.5 billion people). It is projected by the U.N. to reach sixty-one percent in 2025 (United Nations, 1996, 1998). In China, for example, the number of those cities increased from 193 in 1978 to 663 in 1999, and the urban population from 172 million to 388 million (Yongxiang). In the developed world, cities of one million increased from forty-nine in 1950 to 112 in 1995, and in the developing world, from twenty-four to 213. In 1975, cities with population greater than ten million, currently defined by the United Nations as megacities, were only two in the developed world and three in the developing world; in 2015, that number is expected to double in the developed world to four, but to increase to twenty-two in the developing world. Thus very large urban concentrations are primarily a phenomenon of the developing world. This is underscored by the changes in the eleven largest urban agglomerates. In 1980 New York City (the New York City metropolitan area) with a population of 15.6 million, ranked second among the eleven largest urban agglomerates in the world. It remained so in 1994, but by 2015 it is projected to fall to eleventh place. At the same time, Mumbai and

Jakarta, which did not rank among the eleven largest urban agglomerations in 1980, rose respectively to the sixth and the eleventh place in 1994, and are expected to further rise to the second and the fifth place in 2015. In the year 2000, about 4.3 percent of world population lived in megacities. By the year 2015, that figure is expected to exceed five percent (Brennan-Galvin; Population Institute; United Nations, 1996; United Nations, 1998; World Bank).

Pragmatically, large urban concentrations are unique instruments of creativity, ideas, and psychic energy; instruments of wealth creation and globalization because of the connection that they establish with each other; instruments of enhanced social development because of the institutions that are housed in them; and also powerful instruments of birth rate reduction. Cities, and particularly the large urban concentrations, attract because of expectations of a higher quality of life—jobs, less hardship than in the countryside, education, better health care, and a higher level of social interactions. The ability of a city to support many elements that contribute to the quality of life, such as theaters, sports arenas and universities, is generally correlated with its size. A city of thirty to fifty thousand inhabitants often cannot support a stable orchestra, which typically may require aggregations of some 250,000 in population; a large sports arena requires larger populations and a large international airport even larger ones.

At the same time, large urban concentrations are in most cases dysfunctional. They are concentrated sources of pollution; they are harbors of poverty; they are congested (for instance, China has ten times the number of persons per room as the United States); they have, particularly in the developing world, large infrastructure deficits in water supply, in sanitation, in transportation and in telecommunications (World Bank). They are difficult to manage and they are also risky. In their expansion large urban concentrations are increasingly exposed to natural hazards, from earthquakes to floods, and to the spread of communicable disease among a very concentrated human population, as well as to the interruption or destruction of the many systems on which their life depends. In a more subtle way, life in highly artificial environments also may make it easier for the inhabitants of the city to lose awareness of global ecological issues.

By and large, the increase in large population concentrations occurs not only because a city attracts, but also because in its growth a city engulfs populated areas that surround them. However, population statistics about urbanization must be taken with a degree of caution because of lack of uniformity as to how they are gathered and reported, starting with the question of how a city is defined. For instance, if New York City is defined as a municipality in the state of New York, its population would be about 7.5 million people. If it is defined as Greater New York, spanning also its northern suburbs, New Jersey and the southern part of Connecticut, its population would be much larger. Also problematic are projections into the future, starting with those of the U.N., based, as most demographic projections are, on interpolation of past trends rather than on models that take into account the major economic and social factors that will affect the dynamics of the population (Brennan-Galvin, 2000). Furthermore, there are still very few metrics about large urban concentrations that would make it possible to gain a better understanding of the interrelations among these factors and to measure and benchmark the performance of urban concentrations across the globe. Typically, each city has its own way of making projections, and each department within the city has its own data base, seldom integrated with other data bases. Thus, projections can be way off the mark for a city and for the ensemble of all cities.

A description of the phenomenon represented by these large urban concentrations requires, however, far more than just population statistics. Data are needed on social and environmental costs and benefits of the economic activities in the city, because increasingly, urban regions are the principal basis of the global economy. There is, for instance, even in the most advanced cities, little systematic information about the economic impact that projects in a city have on households, particularly by income level, on national and local government, on individual industries and on the urban region as a whole. Neither are there clear data for the gross urban product, that is, the gross economic product of an urban region, and for money flows between production, consumption, savings and investment, data about the share of income that goes to workers, to taxes and profits, about the urban area's balance of payment, or data about the consumption and production of energy (Shore). (In general, it appears that higher population densities lead to lower *per capita* energy consumption.)

Environmental accounts are needed to establish a monetary value of the degradation and enhancement of environmental assets in order to calculate, for instance, the increased value of clean-up activities *versus* their cost or to add to the cost of urban travel the cost of the environmental damages that travel creates.

Another set of needed data has to do with the quality of life. The assessment of the quality of life is somewhat arbitrary, although a number of attempts have been made to identify criteria, such as availability of recreational and transport facilities, crime statistics, education, jobs, etc., that, taken together, give a possible indication of the quality of life. The cost of waiting, an ubiquitous phenomenon in our large population concentrations, should also be taken into account. It can be said that, in most cases, cities of the developing world have poorer quality of life than the developed world cities. However, regardless of whether they are developing or developed, some cities are characterized by higher quality in certain parameters, for instance sports and leisure, and other cities in other parameters, such as efficiency of transportation and quality of health care.

Collection and integration of the information about the geographical size of the cities, mortality, water and land use, health, education, income distribution, etc., are today only episodical, but urgently needed in order to understand urban complexity and the global impact of these human habitats. For instance, we need to better understand the causes and possible remedies for the persistent urban poverty and disease that characterize the explosive urban growth in developing countries and lead to dangerous imbalances with the developed world. That world, however, is not immune either to the influences of poverty and disease within its own cities. In the American cities, twenty-five percent of AIDS occurs in African Americans, who represent only fourteen percent of the population.

From the purely geographical viewpoint, physical features of the landscape or the dangers presented by natural hazards such as floods, earthquakes, or volcanoes, are increasingly less of a deterrent to urban expansion. Throughout the world, that expansion continues undeterred by obstacles or potential dangers, engulfing also ever-greater portions of coastlines. The changes to the Earth's surface caused by the presence of a city are dramatic. For instance, over eighty percent of the surface of Tokyo is occupied by buildings, concrete and asphalt and has thus become excluded from the normal hydrological cycle. In the typical American city with a high proportion of individual dwellings that percentage is less, but not much less in Manhattan. Much of the groundwater removed from under a dense city footprint cannot be replenished when precipitation cannot penetrate the surface because of the large extension of paving. In several cities this can lead to inordinate manifestations of subsistence, as in the case of Mexico City (World Resources Institute).

The footprint of a large urban concentration has multiple dimensions. It encompasses not only the area physically occupied by the city, but also the area that contributes resources to the city and the area that, in turn, is affected by the outflows from the city, from waste to air and water pollution. This extended footprint is that much greater, the greater the population of the city and its affluence. (Hence the enormous impact of megacities and other large urban concentrations.) Although some studies have been performed to determine the size of that footprint, information is still very scanty. It has been reported, for instance, that a Baltic city of one square kilometer uses the resources of eighteen square kilometers of forest, fifty square kilometers of arable land, and thirty-three square kilometers of marine surface (Rowland). An affluent city may use daily some 0.6 tons of water per inhabitant, most of it transformed into waste water, and may absorb daily some five pounds of food per inhabitant, virtually all of it becoming waste (most of it dispersed not too far, within two to three hundred miles). Again, reliable information as to this balance is very limited and episodical. It has been roughly estimated that in a modern city in a developed country, if all the materials that flow into it, from stone to wood to metals to plastics to carbon-based fuels, were to be spread evenly over its surface, the ground would increase in height by five centimeters per year (Graedel). About three centimeters of that height are removed as waste every year, so that the net material growth of the city would be two centimeters per year, or twenty meters in a millennium. Thus the city is a great accumulator of materials embedded, e.g., in the cement of houses and bridges, the asphalt of roads, and the metals of machines. Rather than eventually being left as detritus or waste, those materials could be mined ("city mining") and remanufactured to provide part of the resources consumed by the city, thereby reducing the city's resource footprint.

The urban atmosphere contains two main pollutants, ozone and particulate matter. Ozone is formed by photochemical processes and arises from the interaction of CO and NOx; typically the concentration of NOx is now more than double its background values. In general, the higher the temperature, for instance in the carburetor of an automobile, the higher the NOx production. Hence, in large urban areas with intense automobile traffic that production is very high and very concentrated. CO is due to the carbon content of the typical fuel. In cities without strong measures to reduce traffic pollution it can reach twenty to thirty times its background concentration. In general, pollution plumes from the affluent and energy intensive northern hemisphere travel a very long distance. They reach from the Asian continent to the American continent (Wilkening et al.); new pollution plumes generated there can reach to Europe, where again the pollution generated there reaches back to Asia. In effect everyone is downwind of someone else. Within a city, topography has a significant effect on pollution levels. Thus the average permanence of air over New York City is one half day, but in Mexico City air stays over the city much longer, one and a half days, because of the configuration of the valley in which the city is located (Rowland).

THE CHALLENGES

Beyond the environmental challenges, a number of social and socio-technological challenges affect the fitness of a large city as a human habitat and hence its future dynamics and configuration. Major among those challenges are jobs and education, health, infrastructure, and management. In the developing world, the challenge of jobs is extremely serious. There are large segments of the population of large cities working in an informal sector, devoid of health care assistance and other benefits, being made rapidly redundant, as in the case of artisans, by new technologies and large enterprises, and having limited mobility and hence access to jobs in other parts of a city because of lack of transportation. This perpetuates poverty and the existence of slums and of barrios and favelas, typically at the margins of the city. As to health challenges, two major causes of concern, beyond air and water pollution, and the disposal of waste, are disease and violence. The exposure to unfamiliar pathogens can lead to a high rate of infection, exacerbated when poor nutrition weakens resistance, as well as by the fact that many cities, particularly in the developing world, have a low level of immunization and an inadequate public health infrastructure. The danger of contagion is high and can spread worldwide. Violence is an extreme challenge to human survival in cities. The most extreme case of violence, barring wars, is a terrorist attack, for which large urban concentrations can be a prime target. Also, the ubiquitous frustration of the daily life in a congested city and the opportunities that may be denied to segments of the population can lead to violence.

Infrastructure challenges are universal, but, again, particularly acute in the rapidly growing cities of the developing world, not only because of major capital shortages, but also because of deficits in knowledge, both in its generation through research, and in its dissemination and utilization.

The management challenges of a large urban concentration are extremely complex and crucial. The first challenge is growth *versus* stability, that is, how to find a viable balance between social equity and economy efficiency—between jobs and good living standards for all citizens on the one hand and the ability of the city on the other hand to compete in worldwide markets with other cities. This requires at times some preferential treatment that conflicts with an equitable distribution of resources to all the citizens (what can be called "the Mayor's dilemma" (Bugliarello, 1999)). In the long run, only by placing itself in a position to successfully trade and compete can a city acquire the resources it wishes to have for its inhabitants.

Subsidies are a second major management challenge. One of the problems of many cities, particularly in the developing world, is that subsidies preclude new investments in urban services. For instance, when subsidized water is distributed to everybody, even to those who are willing to pay for it, rather than just to those who cannot, revenues are insufficient for maintenance and for bringing the water supply system up to date, thus creating a spiral of increasing inadequacies and decay.

A third management challenge is how to avoid the vicious circle that starts with the attraction a large urban concentration may hold for people from the outside. That attraction leads to growth that brings with it high real estate costs, slums, health care problems, shortages of water and energy and environmental problems. These counterproductive consequences end up by reducing the attractiveness of the urban concentration to those people who came to it to seek opportunities, a better environment and a better life. For example, Bangalore, in India, became a good base for growth because of climate, skilled population, and transportation, attracting business and jobs, but now begins to suffer from many of these negatives (Niath). A related dilemma associated with large urban concentrations—a national dilemma—is their relation to the rest of the country. In virtue of the magnitude of their population and the concentration of economic activities, large urban concentrations exert an overwhelming influence on the rest of the country. For instance, Karachi, in Pakistan, represents twenty percent of Pakistan's gross domestic product and generates fifty percent of the government revenues.

The last, and ultimately the most basic challenge in the management of large urban concentrations—indeed of all cities—is how to involve the citizens in the decisions that affect their lives and determine the nature and configuration of a city. When that involvement and active participation are deficient, cities suffer and plans are unrealistic, as in the design of Brasilia.

Increasingly available to the management of a large city are a number of powerful new tools that can help to address these issues. They range from geographical information systems to simulators, enhanced communications systems, city-wide area networks, and data banks. With these tools, management can, for the first time in history, obtain more precise data about the city, project those data into the future, develop effective mechanisms for community participation, improve the possibility of developing synergies with other large urban centers that face the same problems and, by joining forces with those centers, find the resources and create a market for needed urban innovations. The tools also include new technologies, such as environmental bio-technologies and technologies for rapid excavation and construction, that reduce the upheaval in the streets and make it possible to build rapidly new elements of the infrastructure.

BUT WHAT IS A CITY?

Views of what a city is are more than purely philosophical speculations with no practical impact. They can influence powerfully the development of large urban concentrations. Le Corbusier saw the city as the grip of man over nature (Le Corbusier). Others may see the city as part of a continuum of natural systems that start at the cellular level and lead all the way to the city and the biosphere. The President of Chinese Academy of Science says that, "unlike biological communities . . . [the city is] a kind of artificial ecosystem dominated by technology, sustained by natural life support systems and motivated by social behavior. It is a socio-economic natural complex ecosystem" (Yongxiang). The present author views the city as a bio-socio-machine ("biosoma") entity in which the advantages, balances and trade-offs among as well as within its three inextricably interwoven components affect the design and function of the city. The biological component, constituted by the inhabitants, is the realm of emotion, feelings, self-replication (Bugliarello, 1998, 2000). Other living organisms within the city are at the base of many natural processes and of recycling. The machine component—that is, all the artifacts in the city, from bridges, roads, buildings, to machines, automobiles and power lines—provide reliability, precision and power. The social component—the society in which the lives of humans are embedded in the city—has characteristics between those of the biological and the machines realms; it exhibits precision in its bureaucracies, emotions in its collective moods and self-replication in the continuity and regenerative power of its organizations. Biosoma balances are exemplified by those in the biological domain between humans and other species—plants and animals—with impact on bioremediation and city vegetation, by those between the individual and society (e.g., issues of employment, privacy, health care), by those among disparate social organizations and activities, by those among a multitude of machines and technologies (e.g., the automobile and the streetcar), and by those between biological organisms and machines (e.g., between vegetation and structures). Examples of trade-offs and substitutions are those between materials and information (e.g., expediting traffic by electronic controls *versus* building more roads), between energy and information (e.g., the use of telecommunications to reduce the need for physical travel), between materials and energy (e.g., insulation *versus* heating), between biological energy and machine energy (as between walking and transportation), as well as between biological information—carried and manipulated by humans—and machine information manipulated and processed by computers. However, in a large city it is easy to lose sight that the social and the machine components are projections of the individual and that the individual component of the biosoma—the human—is the ultimate *raison d'être* of the city.

The societal component of the city changes continuously and so do the city's machines. But eventually the human component of the bio—the individual—might change also and some machines and biological organisms may combine in new bio-machines. This may still be far in the future, but the rapidly expanding cities in the developing world have a better opportunity than the well established cities of the developed world to rethink fundamentally the balance among the three components of the biosoma and their relation to the environment. They can more easily make changes in that balance and avoid the creation of impersonal and alienating environments.

A city acquires different characteristics according to what major themes within the domains of biology, society and machines are emphasized. The *leit motivs* of traditional industrial cities are materials and energy. Those of the eco-industrial cities which are beginning to emerge, for instance, in Scandinavia, are the balance between biology and machines. The knowledge city is an example of a biosomic city in which information is the *leit motif*—biological information, as in biotechnology, social information, as in education and in other human services, and machine information, as in computers and other telecommunications devices. Its manifestations include the knowledge parks now beginning to emerge (Bugliarello, 1996).

PRAGMATIC IMPERATIVES AND THE FUTURE

The future of any large urban concentration—and hence its impact on the surface of the Earth—depends on its ability to respond to pragmatic imperatives, reducing potential hazards to its inhabitants, improving livability in its multiple aspects, and being sustainable. These imperatives can only be satisfied if a city is intelligent, ecological and emotionally satisfying. To be intelligent, a city needs to be self-adapting, that is, able to respond and adjust rapidly and adequately to the challenges and opportunities it faces, both internally and on the outside; it needs also to be efficient in the use of resources, in the flexible scheduling of its operations, and in traffic control. An intelligent city strives to eliminate poverty, with its associated impacts on physical and social health, and pursues the providing of education at all levels as a fundamental tool of efficiency. In effect, being intelligent means that a city is able to address its challenges with new organization and services, by deciding on an appropriate balance between local activities and centralized activities, and by controlling technologies such as the automobile that otherwise can lead to undesirable results, from pollution to congestion to uncontrollable development, as in the case of urban sprawl.

To be sustainable and ecological a city needs, in the first place, to contain its geographical footprint so as to avoid environmentally destructive urban sprawl—a task extremely difficult in well established cities, but still possible in rapidly developing ones. The city needs also to reduce its resources footprint by reducing the pollution and the waste material it generates and by being able to mine its own resources, extracting from within its territory by mining or recycling those materials that have accumulated there in various forms. Being ecological for a city also means reliance on natural means, such as bioremediation, alternative energy sources, and on new concepts in organizing the city, such as balances and trade-offs among the elements of the biosoma, and development of urban environments that are knowledge-driven(the knowledge city), or driven by the development of more ecological industry operations (the eco-industrial city).

A city is a system of systems in which synergies have to be developed among different goals. For instance, the goal of elimination of slums requires the city to be a system that is caring and emotionally satisfying, as well as efficient and the goal of reducing consumption requires a city to be a system that is efficient and manageable.

TECHNOLOGICAL CHALLENGES

Technology is a key factor in the future trajectory of large urban concentrations, giving them form, purpose and vitality. Technology presents today major new challenges (Bugliarello, 1990; Moss; OECD; Tarr). Several key questions arise in this context, both in the developed and the developing world, but particularly in the latter. For instance, to what extent do totally new systems need to be developed, *versus* bringing to the cities systems that are only locally new? To what extent should new and older technologies coexist? The older technologies, though less sophisticated, offer at times the large cities of the developing world simpler and more affordable solutions, as in the case of streetcars. The newer technologies, as in the case of cell phones, make it possible to bypass cumbersome and inefficient older systems. Or, to what extent should a large city rely on the locally produced—to which, given the city's scale, it can offer a large local market—*versus* imported technologies? Also, what kind of standards would be required to facilitate low cost and low maintenance construction, ease of repair, good-enough technologies to enhance local content, to respond to different labor/machines equations than in the high labor cost economies of the highly developed world, and to create products potentially exportable to other urban concentrations, while being socially and environmentally acceptable? Examples of needed technologies in both developed and developing world cities include simpler and cheaper people movers, vehicles with smaller street footprint to alleviate the congestion and parking problem, local energy transformers, and flexible multi-modal systems for transportation, water supply and waste removal. Needed technologies for the developing world also encompass simpler sanitation systems, the creation of materials, methods and supplies for self-help, as well as the development of pay-per-use systems, e.g., for energy, water and highway usage, that reduce waste and help financing maintenance and expansion. Major engineering challenges for all cities include relating the built environment to the natural landscape. Another challenge is to make manageable sub-units of a large city (Bugliarello, 2001).

Many large urban concentrations, as they expand, must reach with their services marginal, peripheral areas without inhibiting their eventual transformation into new, more affluent centers of economic development. There is a need for systems that do not rely completely on rigid trunks, such as a metropolitan railroad or a sewage system, but that extend them at the periphery with flexible, less permanent and cheaper devices that can be replaced eventually with more permanent systems. Especially in the cities of the developing world, infrastructural systems built on the model of those of more affluent and highly industrialized countries are often prohibitively costly.

Addressing these technological challenges can only be successful if the fact is accepted that the city cannot be totally planned, because it is not a machine but a complex bio-socio-machine entity. It can, however, be encouraged to develop in certain directions. New bio-socio-technological conceptions and policies are needed to help guide realistically a large city and avoid freezing its future in patterns that are unsustainable economically, demographically and environmentally and lead to the neglect of areas and populations within the city because of the inability to serve them. Often, today, urban infrastructures are designed without much thought of how cities will evolve. Rare is the case when a city is planned so as to consider its future growth. Past projections have frequently been faulty, also because they have been based on extrapolation of past data rather than on a systematic analysis of the variables that affect growth.

CONCLUSIONS

In the foreseeable future, large urban concentrations will become more numerous, and probably even larger.

The ultimate question is whether they and the extreme urbanization that the globe is experiencing are in the long run good or bad for the species. This multifaceted question implies issues of both fact and human values in a world increasingly artificial and removed from nature. Is the world sustainable if today's rates of consumption of natural resources are reduced? Is it fatally vulnerable? Does it destroy essential human values? At this moment, these questions cannot be answered. But, for that very reason, a better understanding of the phenomenon and of the mechanisms for ameliorating and changing living conditions in cities is that more urgent and important.

The questions subsume a slew of other questions, such as: How do large urban concentrations affect poverty? Will the developing telecommunications systems, from the Internet to satellites to wireless, lessen the need for concentrated human habitats? In an era of exploding telecommunications, will the big urban infrastructural component—highways, bridges, theaters, hospitals, schools, airports—continue to be the glue that binds together a community? Are we irreversibly locked in growing cities? Will expanding cities that sit astride environmental corridors be able to mitigate the great environmental threat that they represent for those corridors? Can the elimination of poverty trigger environmental disasters by enhancing the demands on the environment by a population that has become more affluent? Today, we do not have answers to any of these questions.

REFERENCES

Brennan-Galvin, Ellen, Presentation, Megacities Workshop, National Research Council, Washington, DC, September 26, 2000.

Bugliarello, G., "Rethinking today's cities: designing tomorrow's urban centers," *The Bridge*, Vol. 31, Spring 2001.

Bugliarello, G., "The biosoma: the synthesis of biology, machines and society," *Bulletin of Science, Technology & Society*, Vol. 20, No. 6, pp. 452-464, December 2000.

Bugliarello, G., "Technology and the city," in *Megacity Growth and the Future*, R.J. Hooks, B. Brennan, J. Chamie, F. Lowe, J.I. Uitto, eds., Tokyo, United Nations University Press, 1990.

Bugliarello, G., "Urban knowledge parks and economic and social development strategies," *Journal of Urban Planning and Development*, ASCE, New York, June 1996.

Bugliarello, G., "Biology, society and machines," *American Scientist*, Vol 86, May-June 1998.

Bugliarello, G., "Megacities and the developing world," *The Bridge*, Vol. 29, No. 4, Winter 1999.

Graedel, Thomas E., "Industrial ecology and the ecocity," *The Bridge*, Vol. 29, No. 4, Winter 1999.

Le Corbusier, *The City of Tomorrow and Its Planning*, (reprint) Dover, New York, 1987.

Moss, Mitchell L., "Technology and cities," *Citiscape*, Vol. 3, No. 3, pp. 107-127, 1998.

Niath, I., "Urbanization in India—challenges and some solutions," paper presented at Inter-Academy Forum Meeting, United Nations Habitat II Conference, Istanbul, June, 1996.

Organization for Economic Cooperation and Development (OECD), *Cities and New Technologies*, OECD, Paris, 1992.

Population Institute, *World Population Overview 2000: Population and the Urban Future*, No. 1, Washington, DC, 2001.

Rowland, F. Sherwood, Presentation, Megacities Workshop, National Research Council, Washington, DC, September 26, 2000.

Saalman, Howard, *Medieval Cities*, George Braziller, New York, 1968.

Shore, William, "Project: to prepare regional economic accounts for the tri-state New York metropolitan region integrated with environmental and social accounts," Regional Research Consortium, Institute of Public Administration, New York, January 2001.

Tarr, Joy A. and Gabriel Dupuy, *Technology and the Rise of the Network City in Europe and America*, Temple University Press, Philadelphia, 1988.

United Nations, "Trends in urbanization and the components of urban growth," in *Proceedings of the Symposium on Internal Migration and Urbanization in Developing Countries*, New York, United Nations Population Fund, 22-24 January, 1996.

United Nations, *World Urbanization Prospects*, United Nations, New York, 1998.

Wilkening, Kenneth E., Leonard X. Barrie and Marilyn Engle, "Trans-Pacific air pollution," *Science* Vol. 290, pp. 65-67, October 6, 2000.

World Bank, *World Development Report 2000/2001: Attacking Poverty*, World Bank, New York, 2000.

World Resources Institute, *World Resources 1996-97: The Urban Environment*, Oxford University Press, 1996.

Yongxiang, Lu, *Eco-Integration: A New Approach in Dealing with the Challenge of Cities' Expansion in Developing Countries*, Chinese Academy of Science, Beijing, China, 2001.

2

Geoantiquities in the Urban Landscape: Earth History Records in the Cities

Marjorie A. Chan, Donald R. Currey,
Andrea N. Dion, and Holly S. Godsey

INTRODUCTION

During late Cenozoic time, the surface of every region was impacted by a succession of Earth system processes involving the atmosphere, hydrosphere, cryosphere, lithosphere, and biosphere. However, in recent centuries and decades, the Earth has been transitioning at an accelerating rate to an anthropogenically dominated system.

It is clear that humans can be agents of erosion and irreversible change [e.g., Hooke, 1994, 2000]. Rates, volumes, and intensities of human action in shaping the landscape have been well-documented [Nir, 1983; Panizza, 1996]. An important question in adapting to change and achieving sustainable urban systems is how humans interact with the natural, physical environment [Brown, 1997]. Where the human and natural domains intersect (Figure 1), "...we find ourselves in an increasingly unstable and disorienting region.... The intersection, where most real-world problems exist, is the one in which fundamental analysis is most needed. Yet virtually no maps exist; few concepts and words serve to guide us…" [Wilson, 1998]. An opportunity for Earth science and the community interaction lies within the intersection of the natural and urban environments. Geoconservation provides the concepts and tools to help communities wisely manage their Earth science resources.

We herein define a geoantiquity as a natural record of Earth history that documents environmental change at local, regional, and global scales. These records of Earth history commonly are natural landscapes that preserve material evidence of geologically recent surface processes and environments. However,

Earth Science in the City: A Reader

10.1029/056SP03

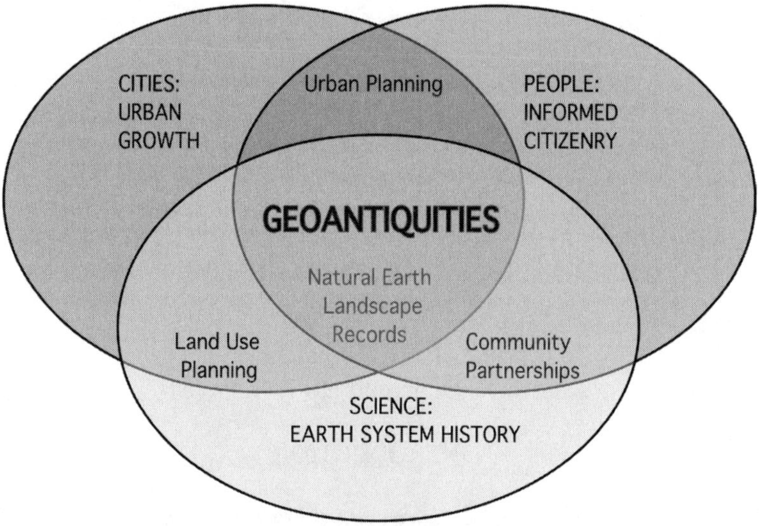

Figure 1. Intersecting regions of Earth science, people, and cities. The focus of this paper is the central intersection region of geoantiquities, where cities and people impact the natural landscape. Furthermore, this paper discusses the concept of geoantiquities as natural records of Earth history, and their importance, scientific value, and why some should be conserved and preserved by society.

especially in regions undergoing rapid rural-to-urban land conversion, many geoantiquities are in danger of being permanently lost.

The purpose of this paper is to introduce and apply the concept of geoantiquities for the wise use of valuable geological and geomorphological landscapes. This concept is consistent with international trends in geoconservation, and parallels the well-established cultural antiquities management (cultural resource management) for preservation of prehistoric human artifacts.

Late Cenozoic geoantiquities of the Bonneville Basin are preserved in the sedimentary record of deltas, spits and beaches, glacial moraines, alluvial fans and debris flows, fluvial terraces, fault scarps, modern lakeshores, playas, and salt flats. A large variety of geoantiquities is exposed along Utah's Wasatch Front, where long-term population growth rates exceed the national average. Most of these geoantiquities are unconsolidated surficial deposits that are extremely vulnerable to destruction, corruption, removal, and burial.

Geoantiquities provide educational and scientific opportunities, open space, and enhanced environmental quality and safety. The Stockton Bar, a large barrier beach in the Bonneville Basin, serves as an exemplary geoantiquity and is documented as a case study in this paper. Education and information transfer are key to protecting geoantiquities and prioritizing those landscapes that should be left intact. Partnerships with government agencies, educational institutions, non-governmental organizations, public interest groups, and committed individuals provide pathways to raise awareness and produce broad involvement in planning. Ideally, the ultimate goal and outgrowth would be implementation of geoantiquities resource management. Earth Science in the cities requires credible scientific investigations, coupled with public outreach and informational tools, to help guide community decision making.

GEOCONSERVATION

Much of the development of natural resource conservation is from a utilitarian standpoint, which focuses on the economic or productive uses of resources. For instance, soil conservation is fueled mainly by concern for loss of agricultural productivity. In contrast, the preservationist concept emphasizes the intrinsic value of nature, rather than its potential use to humans. Few conservation movements in the United States have considered long-lasting impacts of human action on the physical landscape, and even fewer have recognized the value of geologic/geomorphic features beyond their aesthetics.

Geoconservation seeks to protect the intrinsic qualities of Earth resources in order to sustain their non-consumptive uses in Earth science research and education. It integrates the two often-opposing concepts in conservation theory: that of the intrinsic value of landscapes, and that of the utility of land for human benefit. By developing and integrating geology and geomorphology into conservation theory, geoconservation can contribute to more holistic planning and policies.

With the possible exception of national and state parks and monuments, the concept of geoconservation has not been fully articulated in the United States and is thus reflected in the lack of relevant policies. However, concepts similar to geoantiquities have been adopted in many other countries worldwide, and these efforts may provide valuable examples for geoconservation application in the United States.

European countries have recognized the importance of geoconservation by adopting means to help protect Earth resources. In 1969, a small working group of Dutch scientists began developing concepts and goals of Earth science conservation, and over the years this has evolved into an international organization. In 1991, representatives from over 30 countries attended the 1st International Symposium on the Conservation of Our Geological Heritage, held in Digne, France. The meeting culminated in the signing of the International Declaration of the Rights of the Memory of the Earth; a document which advocates the international consideration and protection of geological heritage [Krieg, 1996]. The declaration stresses the idea of geological heritage being an "archive" of Earth history: "Just as an old tree keeps all the records of its growth and life, the Earth retains memories of its past- a record inscribed both in its depth and on the surface, in the rocks and in the landscapes, a record which can be read and translated" [Martini and Pages, 1991].

Throughout all of the international geoconservation efforts, a uniform term for important Earth features and landforms has not yet emerged. The popular term *geotope* was coined and defined as a "geologically or geomorphically valuable or sensitive part of the countryside" by the Swiss geologist Stürm [1994]. Geotope is meant as a parallel to the word *biotope*: *geo* meaning Earth, and *tope* meaning place. Stürm [1994] further defines the actions that threaten the existence of geotopes, and called for the inclusion of geotopes in the local planning process as well as an international convention on geological heritage.

Australia also has developed extensive methods in *geoheritage assessment* and management. The Australian Heritage Commission seeks to promote awareness of geological, geomorphological, and soil features by listing them in the Register of the National Estate. This list includes "those places...that have aesthetic, historic, scientific or social significance or other special value for future generations as well as the present community" [Eberhard, 1997].

In response to numerous national geoconservation efforts, UNESCO (United Nations Educational, Scientific and Cultural Organization) recently implemented the first program aimed at international recognition of important geological sites. The UNESCO Division of Earth Sciences, along with the International Union of Geological Sciences (IUGS), developed the GeoPark Program, which coordinates international efforts dealing with "geoconservation, geotopes, geosites, or general geological heritage" [UNESCO, 2000]. The goal of the program is to promote the conservation of geological heritage through the initiation of a worldwide network of geoparks under the existing program, "Man and the Biosphere."

Ultimately, it will be important to develop, model, and implement strategies for managing the interactions between geoantiquities and areas of rapid urban growth. Wise stewardship of geoantiquities requires something comparable to the cultural resources management (CRM) ethic that has, for several decades, been in place for managing archaeological antiquities and historic places. The research presented here focuses on the genesis, inherent values, and urban lessons to be learned from geoantiquities, utilizing science-based evaluations and models that can be applied to similar areas where urban growth intersects areas of valuable natural archives.

GEOANTIQUITIES

Definition and Significance

A geoantiquity is a record of Earth history in which natural landscape preserves material evidence of geologically recent surface processes and environments. These records are classics of Earth history, written in nature's handwrit-

Table 1. Geoantiquities in the Bonneville Basin (keyed to Fig. 2B)

DEPOSITS FORMED IN HYDROCLIMATIC ZONES	
PG	Periglacial: rock glaciers
G	Glacial: lateral moraines, end moraines, cirque features
F	Fluvial: piedmont alluvial fans, valley-floor alluvial systems (channels, floodplains, terraces)
FL	Fluviolacustrine: deltas (fine-grained, coarse-grained), estuaries, Lagoons
W	Wetlands
LACUSTRINE DEPOSITS	
S	Shorezone
	Fringing beaches: beach ridges
	Barrier beaches: cuspate and baymouth barriers, barrier islands
	Projecting beaches: spits, tombolos
OS	Offshore
	Ice-rafted debris, marls
PL	Playa
	Salt Flats
	Saline mud flats
	Non-saline mud flats
DEPOSITS FORMED BY CROSS-CUTTING, AZONAL PROCESSES	
M	Mass-wastings: landslides, debris flows
E	Eolian: dunes (silica, gypsum, ooid)
O	Organic: packrat middens, peat bogs, algal bioherms (tufa)
V	Volcanic: lava flows, cinder cones, volcanic ash
T	Tectonic: fault scarps
GA	Geoarcheological sites

ing. Geoantiquities form in a wide range of environmental zones, and are shaped by a host of surficial processes (Table 1). We herein restrict our use of geoantiquities to features of the late Cenozoic because present landscapes for the most part date from late Cenozoic time.

Important geoantiquities are somewhat like a geologic type section [North American Stratigraphic Code, 1983], but in this case, the section is in danger of being lost because it is unlithified and easily removed or corrupted in comparison to most lithostratigraphic type sections. The vulnerability of a physical environment commonly shows an inverse relationship to its geologic age. Geologically young deposits are the features most susceptible to change in the urban landscape because they comprise the uppermost, most exposed and most accessible layers, and are typically unlithified. The lack of lithification means that young deposits can be readily excavated and removed or rearranged in the urbanization process. Moreover, many of these unlithified deposits (e.g., sand and gravel) are important sources of construction materials. Thus, in the dynamics of urban development, geoantiquities are likely to be removed from the natural physical environment and added to the built physical (urban) environment.

Geoantiquities can provide valuable scientific information (Table 2) of the recent geologic past that may be relevant to prediction of climatic trends and/or future geologic events. The sedimentary record in geoantiquities can provide scientific proxies such as: 1) climatic signatures that can be used for interpreting temperature, extreme conditions, wind directions and strengths; 2) hydrologic changes in the lake history and river discharge; 3) geodynamic events such earthquake and volcanic recurrence intervals and markers; and 4) other environmental parameters such as fire occurrences, biotic communities, and rare or unusual depositional events. Scientific studies also have applications that relate to science and policy issues, such as hazard mitigation and environmental quality.

Occurrences

Geoantiquities can occur in nearly any location and setting, but this study focuses on near-surface, well-exposed examples from the basin of Pleistocene Lake Bonneville, which can serve as a model for other regions. The natural envi-

Table 2. Scientific and Educational Values of Geoantiquities

CLIMATIC INFORMATION
- Temperatures (isotope geochemistry, amino acid analysis, borehole temperature gradients)
- Extreme cold (thermal contraction wedges, gelifluction deposits, ice-rafted dropstones)
- Extreme dustiness (wind-blown silt, i.e., loess)
- Wind directions and strengths (aeolian bedforms and lacustrine wave-shaped features)

HYDROLOGIC INFORMATION
- Drought recurrence (lake levels, soil salinization, aeolian features, oxidized layers)
- Flood recurrence (lake levels, floodplain deposits, spring deposits, reduced layers)
- Extreme discharge events (channel geometry, bedload sediments)

GEODYNAMIC INFORMATION
- Earthquake recurrence (surface fault rupture, liquefaction, rockfall)
- Volcanic eruption recurrence (tephra layers, lahars)

OTHER ENVIRONMENTAL INFORMATION
- Biotic communities of pre-urban ecosystems (microfossils, macrofossils)
- Wildfire frequency and magnitude (charcoal particle abundance)
- Mudflows, debris flows, debris avalanches, and snow avalanche recurrence

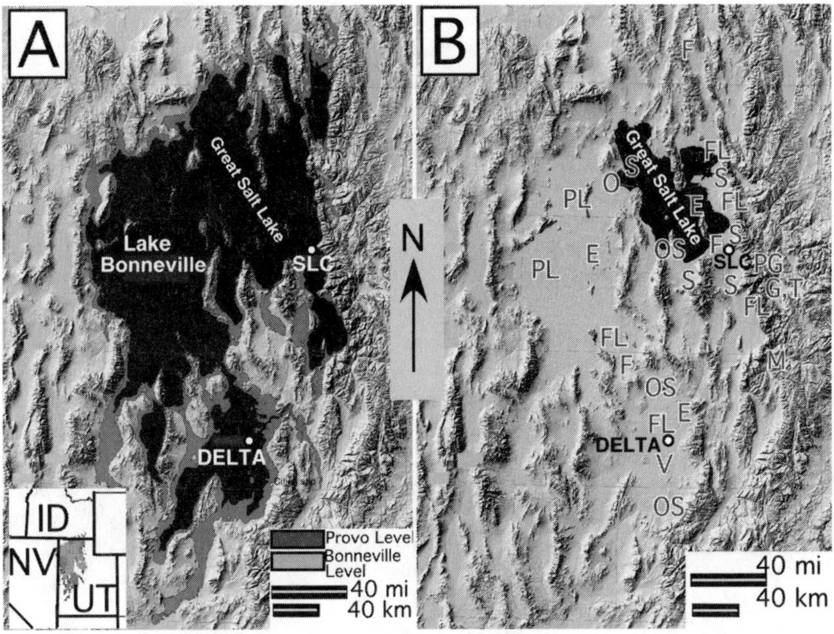

Figure 2. Bonneville Basin of western Utah. SLC=Salt Lake City. A. Lake Bonneville was the largest Pleistocene lake within the Basin and Range physiographic province of western North America. The lake formed during oxygen isotope stage 2, and was over 350 m deep at its highest stage (Bonneville shoreline). The Provo shoreline forms a prominent bench, due to stillstand following a rapid lake-level drop of over 100 m caused by catastrophic flooding through drainages emptying into the Pacific Ocean 14,500 radiocarbon years or 17,300 calendar years ago. B. Great Salt Lake is the Holocene descendent of Pleistocene Lake Bonneville. Few areas rival the geoantiquities of Lake Bonneville in terms of size, form, definition, and scientific value for paleoclimatology, sedimentology, tectonics, and geomorphology. Examples (not inclusive) of geoantiquities are shown by letters, keyed to Table 1.

ronment of this area is dominated by Quaternary landscape history, including evidence of lakes, marshes, glaciers, dunes, alluvial fans, deltas, floodplains and extensional tectonics. During the last major ice age, Lake Bonneville covered much of western Utah and extended into southern Idaho and eastern Nevada [Currey et al., 1984] (Figure 2a). The significant natural records left by Lake Bonneville (24,000-12,500 radiocarbon years or 28,000-15,000 calendar years ago) are prominent shorezone deposits including beaches and spits (Table 1, Figure 2b).

Within the Bonneville Basin, the goal is to conserve geoantiquites that can yield valuable long-term benefits as community natural landscapes and open space (Table 3). The objectives of geoantiquities implementation are to: 1) study

Table 3. Importance of geoantiquities with examples from the Bonneville Basin

COMMUNITY AESTHETICS: Geoantiquities form a picturesque landscape and natural open space (for example, Bonneville Salt Flats, Bonneville Shoreline Trail of the Wasatch Front, Stockton Bar of Tooele County, and glacial valleys and moraines of Little Cottonwood Canyon and Bells Canyon).

COMMUNITY ETHICS: The community can endow to future generations a window on Earth history, and a landscape preserved to enhance the quality of life.

BASIC SCIENCE: Scientists use sediment records to learn what physical, chemical, and biological processes have acted in the geologically recent past. For example, studies on Lake Bonneville geoantiquities tell us about global change, past climatic conditions, and how wind and water influence sediment transport.

APPLIED SCIENCE: Geoantiquities help us understand geologic processes and allow us to better predict rates of change, and assess local natural hazards.

COMMUNITY AND ENVIRONMENTAL EDUCATION: Geoantiquities provide people of all ages with a natural outdoor laboratory of Earth-surface history (for example, Great Salt Lake, Antelope Island, the Stockton Bar, and shorelines of the Wasatch Front).

and inventory geoantiquities in selected regions, 2) identify those that warrant community recognition, and 3) foster community appreciation of geoantiquities.

It is recognized that conservation is successful in the U.S. where federal lands can be managed, conserved, and preserved with existing laws, legislation, and guidelines (e.g., National Parks and Monuments, Bureau of Land Management-Area of Critical Concern, National Landmarks, etc.). Cultural Resource Management for cultural antiquities has evolved over many years and now provides a mechanism for evaluation and possible protection in exceptional cases, even if the resources are on private land. Legislative measures to protect geoantiquities represent a distant goal, but one that must begin at the local level, involving community interest and scientific vision.

GEOANTIQUITIES HERITAGE AREAS

Geoantiquities are important to the community for many reasons: aesthetics, ethics, basic and applied science, and environmental education (Table 3). Many geoantiquities have high value as intact landscapes. For example, they may serve as tourist destinations, cultural landmarks, or as crucial habitat for a particular species. These may prove to be very important considerations when land use decisions are made. Although aesthetic, ethical, and historical considerations are important, they do not provide criteria for distinguishing which geoantiquities are especially unique and worthy of protection. Some sites may be extraordinary geoantiquities but have no open space or particular aesthetic value. Other sites may provide open space, but may be a very common geoantiquity that has no special scientific or educational value. It must be emphasized that geoantiquities should be defined by their scientific and educational values (Table 2) that represent significant opportunities for research and learning. These values derive from geological or geomorphological qualities, which are used to identify the most scientifically valuable sites. Both basic and applied Earth-science research can benefit from geoantiquities, and are often contingent upon them.

There is a unique memory that the Earth retains in geoantiquities. One way to illustrate this concept is to compare a particularly important geoantiquity to a unique, rare and complete book that has all of its pages of history. Although it may not look particularly valuable from the outside cover, the inside tells its true story. There may be many other common books that cover similar topical material but are incomplete, and therefore not as important or as valuable. All geoantiquities are not equal; thus those that have greater intrinsic or scientific value must be given the highest priority.

Prioritization is important in any resource management scheme, and is essential in the Bonneville Basin, where there are hundreds of geoantiquities. A local natural landscape of highest management priority is termed a *geoantiquities heritage area.* Geoantiquities heritage areas have three essential characteristics: 1) they include intact remnants of distinctive natural landscapes, 2) they contain scientifically important records of geologically recent environmental history, and 3) they are at great risk of damage or destruction by urbanization. Thus, geoantiquities heritage areas are world-class natural archives of Earth system history, which, if managed properly, will provide outstanding visitor experiences, teaching resources and research opportunities.

GEOANTIQUITIES IN CHANGING URBAN LANDSCAPES

This paper reflects the development of a conservation concept that is applicable to any region where urban growth is impinging on geoantiquities. A typical example is one where urbanization spreads from lowlands to more physically challenging, adjacent uplands. Lowlands are easily developed and contain abundant resources, including water, arable lands with fertile soils, and natural transportation corridors. As urban centers grow they tend to expand into more challenging areas, where the terrain is steeper, wetter, drier, or not as easily developed. Although populations are spread unevenly, there are physical linkages such as highways and roads between the urban cores (densely populated areas) and the urban fringes (outlying areas increasingly impacted by urban cores). Geoantiquities need to be identified, inventoried, and described, within both urban cores and urban fringes. From this information, several management options may be considered (Table 4). This information then needs to be disseminated to the public as an essential foundation for informed public policy and land-use planning.

Table 4. Management Options for Geoantiquities in Urban Regions

CONSUMPTIVE USES (permanent loss of the geoantiquities resource):
- Removal by urban development (excavation)
- Mutilation by urban development (severe mechanical disturbance)
- Corruption by urban land use (severe chemical disturbance)

NON-CONSUMPTIVE, SCIENTIFICALLY UNPRODUCTIVE USES:
- Burial by urban development (non-permanent loss)
- Idle land (short-term protection, long-term uncertainty)
- Park land (long-term protection, can be scientifically productive)

NON-CONSUMPTIVE, SCIENTIFICALLY PRODUCTIVE USES:
- Urban outdoor science museum (visitor-based education)
- Urban outdoor science classroom (school-based education)
- Urban outdoor science laboratory (ongoing research)

Within the Bonneville Basin is Utah's 10-county metropolitan area known as the Wasatch Front. The Wasatch Front includes 98 municipalities [QGET, 1998], and is the main urban zone between Denver, Colorado, and San Francisco, California. Utah has a population growth rate that exceeds the national average [QGET, 1998] (Figure 3). The current population of 1.6 million is projected to increase to 5 million by 2050, making the Wasatch Front a metropolitan region the size of present-day Philadelphia [QGET, 1998]. In 1990, Utah ranked as the sixth most urban state in the country, with 87% of its population classified as urban (more than 1000 people or jobs per square mile) [Donner, 1997]. The projected spatial urban expansion is depicted in Figure 3.

The area's growth is an issue that is conspicuous and highly politicized. Governmental and other organizations are striving to manage or direct growth in ways they deem best. Consideration of the landscape is only starting to be acknowledged through efforts such as this geoantiquities paper and its scientific outgrowth, where science and the cities can interact and try to work together.

PARTNERSHIPS

Partnerships with local and national governing bodies, special interest groups, community organizations, scientists, and educational leaders are vital for raising awareness of geoantiquities, and for establishing actions that can lead to protection and integration of geoantiquities heritage areas into urban planning. Partnering includes aspects of community education, information transfer, and organizational cooperation. Implementation and evaluation of partnerships provide a model that can be transferred to other projects in the region and nationally.

Because there is no legal framework with which to defend geoantiquities, it is essential to work with existing programs and agencies. For example, the local

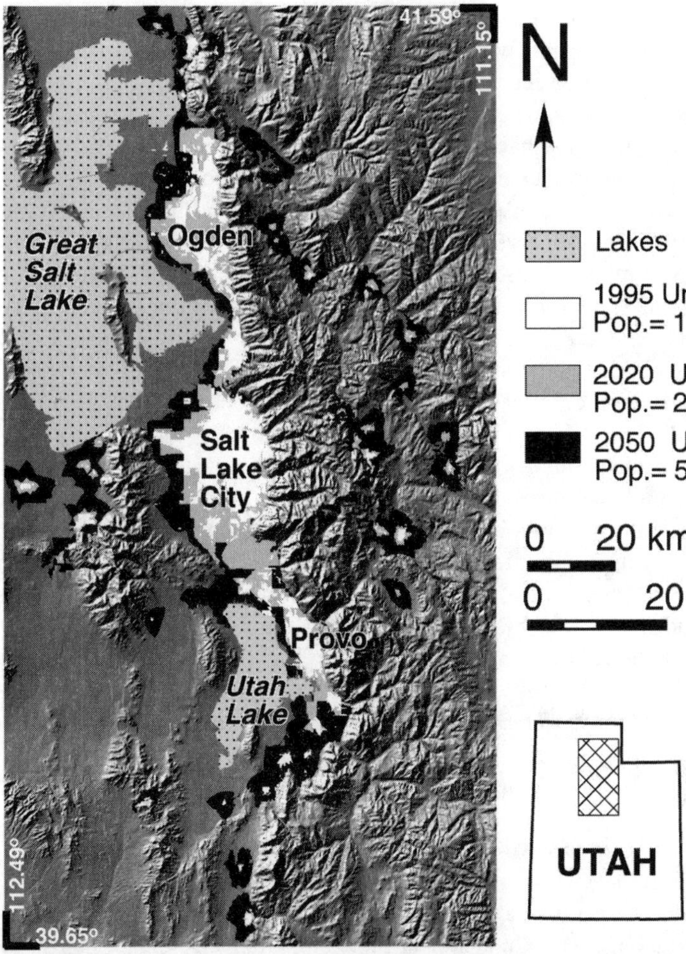

Figure 3. Population projections (for years 2020, 2050) in an area of high concentration of geoantiquities. Shaded relief map area covers ~ 40 x 80 km area on the east side of Great Salt Lake. Spatial patterns of growth show the highest concentration against the mountain front (Wasatch Front) at the east, with projected urban expansion north and south of current cities, and projection growth to the west. Irregular thin lines represent fault traces. Map adapted from QGET.

advocacy group "Coalition for Utah's Future" received a legislative mandate to facilitate quality growth through governmental, public, and private cooperation. The result of this was the establishment of the QGET (Quality Growth Efficiency Tools) committee which is designed to help local government understand, manage, and plan for growth. The outgrowth of QGET is Envision Utah, an organization committed to partnerships for quality growth. These and other similar organizations can serve as valuable information resources and springboards for developing additional liaisons with potential partners. Many of the partnerships cited in this study build on the existing commitment of the state to thoroughly examine the issues of dynamic change in the urban landscape.

Both formal and informal liaisons with local, regional and national advocacy groups and governmental agencies help accomplish common goals toward community awareness and the conservation of geoantiquities heritage areas. These partnerships bring together diverse yet allied groups and individuals who can help accomplish common goals for preserving both open space and Earth history records in the urban setting. Without these partnerships, geoantiquity resources would be at much greater risk of loss.

There is no doubt that geoantiquity efforts bring about a new level of community awareness, if even on a local scale. One of the dividends of geoantiquities research is that individuals and groups within the community increasingly look to the Earth science community to help understand and evaluate their local landscape. The case study described below illustrates how information transfer to the community can influence policy making with regard to potential geoantiquities heritage areas. Even if in particular instances geoantiquities are not ultimately preserved as open space, developed partnerships can still help pave the way for future joint endeavors, and may raise community awareness to affect other potentially important future decisions.

CASE STUDY - THE STOCKTON BAR

Scientific Value - Geologic History

The Stockton Bar, a massive body of Pleistocene sand and gravel in Tooele County, Utah (Figure 4), was chosen for specific case study as a potential geoantiquities heritage area. Over a century ago, this landform was documented, described, and named "The Great Bar at Stockton, Utah," by U.S. Geological Survey geologist G. K. Gilbert [1890]. The Stockton Bar is a shoreline remnant of Lake Bonneville, the largest and most recent deep-lake cycle that occurred in the Bonneville Basin. Consequently, it contains an exceptionally well-developed sequence of littoral deposits. Components of a full study (not presented here) include documentation of the sedimentologic, geomorphic and climatologic information contained in the deposits of the Stockton Bar. A short documentation of geologic processes interpreted from the Stockton Bar geoantiquities heritage area are outlined here to show the developmental history of the Bar, and the sedimentary response to lake level change and other environmental parameters during the Late Pleistocene.

The Stockton Bar is unique in that no other feature in the Bonneville Basin contains such a complete and detailed record of the history of Lake Bonneville. It is almost 3 km long and is one of the two largest lacustrine cross-valley barriers on the North American continent (the other is the Humboldt Bar of western Nevada). Deposition of the Bar began when the lake transgressed to a brim-full condition and paused long enough to allow sediments to prograde across the strait between Tooele and Rush valleys (Figures 4 and 5).

Two large (2.5 and 1.2 km long) spits were attached to the Stockton Bar during subsequent highstand oscillations. At an elevation of 1595 m, the smaller of the two spits is thought to represent the very highest stage of Lake Bonneville and can be correlated to similar, but less detailed, shoreline features elsewhere in the basin. The distinct geomorphic expression of the Bar and spit complex and exposures of well-preserved stratigraphy indicate southward longshore transport at the time of the Bonneville highstand.

Sedimentary structures, including ripple marks and cross-bedding, are present in a sandy facies at the junction of the Stockton Bar and attached spits. The stratigraphic position of this facies along with abundant freshwater gastropods are used to infer paleoenvironmental conditions and the relative timing of this deposit in Lake Bonneville history. These sediments mark the rapid lake-level drop to the Provo shoreline, known as the Bonneville Flood.

The Stockton Bar is in close proximity to major urban areas and is easily accessible by road, making it an ideal location for educational field trips. The multiple large-scale shorelines exposed in this area provide a tangible illustration of the concepts of Lake Bonneville history and environmental change. The geologic and educational aspects of the Stockton Bar are complemented by its environmental and aesthetic qualities. The site serves as valuable habitat for a variety of desert plants and animals and provides green-space in a rapidly developing urban area. The Stockton Bar is also a popular site for recreational activities such as horseback riding, hiking, off-road vehicle usage and hang gliding.

Major threats to the Stockton Bar due to urbanization are several-fold.

(1) The young, unconsolidated deposits of the Stockton Bar are at risk of destruction by rapidly growing demands for aggregate materials. Permits for the removal of 400,000 tons of sand and gravel from the Stockton Bar were issued in 1999 [Meli, 1999]. Sand and gravel quarries increasingly threaten permanent loss of the opportunity to study geomorphic and sedimentologic expressions of the Stockton Bar. Comparative aerial photos illustrate the change in the landscape from quarry operations due to the increased urbanization pressures in the last few decades (Figure 5).

(2) Areas just to the south of the Stockton Bar have been subdivided into 2-acre ranchettes, and it is expected there will be increased demand and desire for home construction on top of the Stockton Bar. Housing development would regrade the surface of the Bar, forever altering its original form, and render it inaccessible to visitors and unavailable as open space.

(3) Environmental and aesthetic aspects of the Stockton Bar are also threatened by the presence of a nearby solid waste reduction center and a toxic waste pond (Figure 5b).

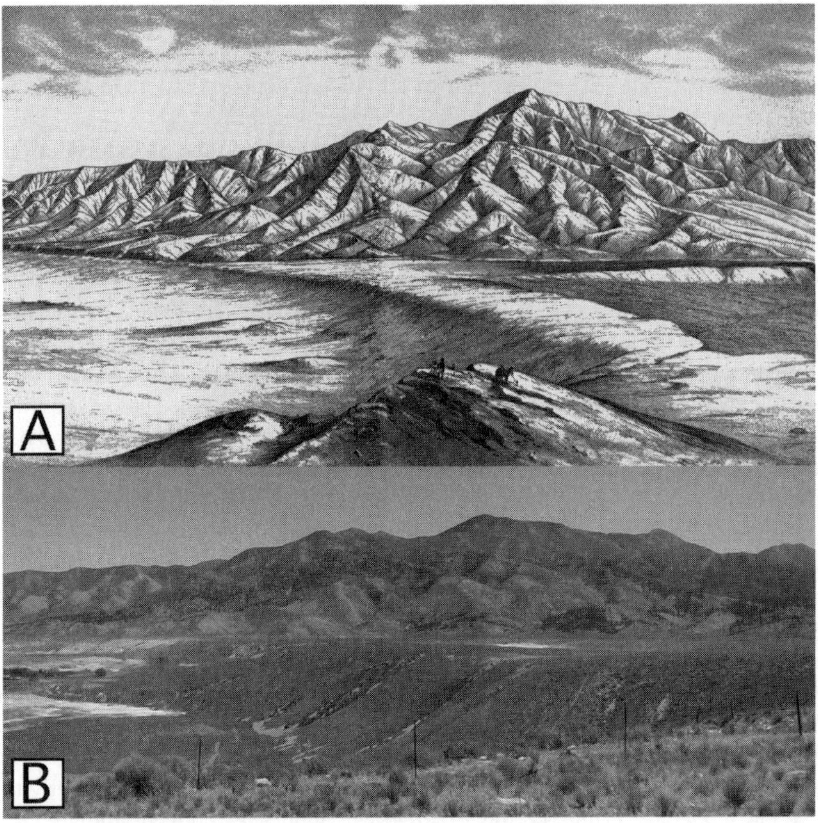

Figure 4. Case study of a geoantiquities heritage site at the Stockton Bar, Utah. A. Stockton Bar (view to the northeast) depicted in Gilbert [1890]. B. Stockton Bar today (same view looking northeast), still relatively intact, with excellent potential to be a geoantiquities heritage area worth saving and preserving.

Partnership Outcome

Preservation of the Stockton Bar is dependent on partnerships among various community groups, local government bodies and national advocacy organizations. In order for these partnerships to be successful, it is critical that there is a fundamental understanding of the attributes of the Stockton Bar that make it a unique feature. Information about the Stockton Bar was made available to the general public by meeting with various organizations such as Utah Open Lands, The Nature Conservancy, the Bureau of Land Management, Friends of Great Salt Lake, and the National Energy Foundation. As an outgrowth of these meetings, field trips were offered to the residents of neighboring communities and advertised in monthly utility bills.

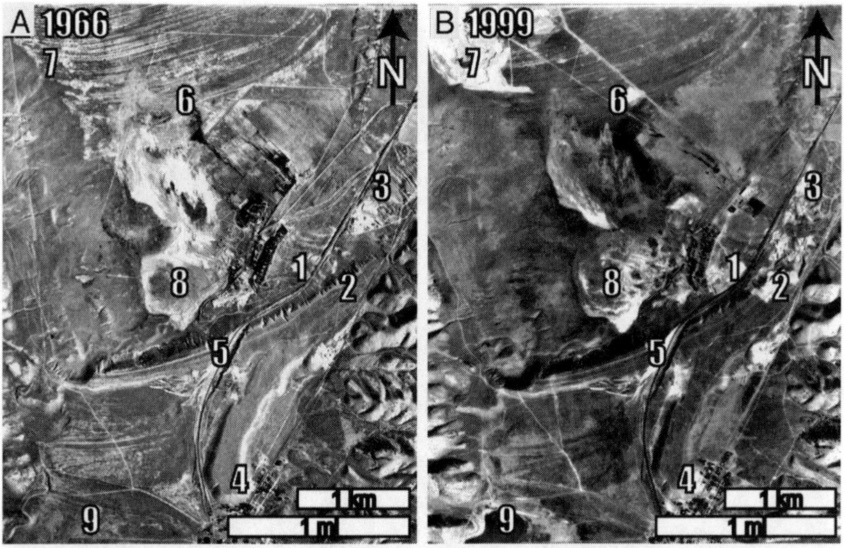

Figure 5. Air photo comparison of the Stockton Bar geoantiquities heritage site and surrounding region. Numbers depict areas of rapid change over the past 30 years. A. 1966 view showing relatively little human impact. Numbers 1-4 = sites of early-stage gravel pit development on the Stockton Bar. Number 5 = railroad cut. Numbers 6 and 7 = pristine Lake Bonneville beach ridges. Number 8 = tailings/toxic waste pond. Number 9 = Rush Valley. B. 1999 view showing significant gravel pit development, particularly in the vicinity of numbers 1-4 and 6-7. Note also the appearance of Rush Lake (number 9), an ephemeral lake that fluctuates with changes in local climate.

In an effort to promote the educational aspects of the Stockton Bar, a geoantiquities module was created for use in the local school district's curriculum. Information was also disseminated through various newspaper articles, the creation of a website designed to inform, display and link research data, and direct mailings of informative brochures. Additionally, the Tooele County planning commission requested that scientific and geologic information about the Stockton Bar be presented to the public preceding a vote to re-zone the Bar for sand and gravel mining. Educating the public about the significance of the Stockton Bar has facilitated communication among interested parties and resulted in a campaign to halt further excavation of the site. Within a year of initial contacts and cooperation with the local community, the Tooele County planning commission denied a request to re-zone the eastern portion of the Stockton Bar for aggregate extraction. This action spurred a widespread interest in the future of the Stockton Bar and many solutions have been proposed, including preservation as open space and the development of a multi-use park.

The concept of multi-use landscapes plays a significant role in the development of geoantiquities heritage areas by providing alternatives for various pri-

Figure 6. Two years of change in Gilbert delta exposures at the mouth of Big Cottonwood Canyon, Salt Lake City, Utah [Chan and Milligan, 1985]: A. 1993 view (top) of large Provo-level delta foresets, dipping to the west, where sand and gravel was actively quarried. B. 1994 view (bottom) of the same site, where the delta foresets were remolded and regraded for conversion to a county golf course. Presently the area is covered with manicured grass, and houses are rapidly being built along the available shorelines just behind and overlooking the golf course.

vate and public landowners. In the case of the Stockton Bar, landowners who have received revenue in the past from aggregate mining have now sought alternate ways to generate returns. For example, a local engineering company has received a conditional land-use permit to build wind-powered turbines on top of the Stockton Bar at the same site where only a few months earlier a request for an aggregate extraction permit had been denied. Construction of the windmills will adversely impact small areas of the Bar, but most of the Bar will remain intact. Lease of the land to the engineering firm will provide income to the landowners for several decades, as opposed to a one-time profit from sale of the land for aggregate. Although perhaps not the ideal future for a geoantiquity as important as the Stockton Bar, this progression of events demonstrates how it is possible for geoantiquities to coexist with the needs of land owners, community members and developers. Presently, this site is a rare island of relatively intact Bonneville deposits in one of the most rapidly urbanizing counties in Utah. Designation of this site as a geoantiquities heritage area increases the probability that the integrity of its scientific, educational, environmental and aesthetic values might be maintained.

URBAN IMPACTS ON GEOANTIQUITIES

Human impact and change in geoantiquities can be dramatic over time periods of just a few years (Figs. 6-8), to longer decadal (Figure 9) and centennial changes (Figure 10). These loose, unconsolidated sediments can easily be disturbed and are vulnerable to removal and burial, particularly in areas with high growth rates, such as the Wasatch Front (Figure 3). These changes are graphically illustrated when comparing the same localities over different years.

In just the space of a few years, an active gravel quarry that was formerly on the outer fringes of the city was encroached upon by high-end new homes and housing tracts. As a gravel quarry, it displayed fine examples of large Gilbert-delta foresets (Figure 6a), but with the encroaching urbanization it was rapidly

Figure 7. In a southern suburb of Salt Lake City, the Draper Spit (center) is enveloped in housing. The only part of the spit that does not have houses is the sandy face that is too steep for development.

Figure 8. Only a few miles east of downtown Salt Lake City, the Bonneville shoreline (bench just below the "U" on the hill) is now obscured by expensive homes.

regraded, seeded with grass, and turned into a golf course (Figure 6b). Consequently, gravel operations have relocated to other previously pristine geoantiquities (area of Figure 9). This demonstrates the extremely rapid change that can occur to geoantiquities in just the space of two years.

Other geoantiquities of Bonneville spits (Figure 7) and shorelines (Figs. 8 and 9) show decadal change. Most of these examples were relatively pristine only 10 to 20 years ago, and now have been engulfed or covered by homes or actively quarried.

A small post-Bonneville alluvial fan studied and drawn by G. K. Gilbert's team [Gilbert, 1890] was deposited over approximately ten thousand years of post-Bonneville time (Figure 10a). A photograph taken from the same viewpoint a century later shows that the fan has been removed and the area has been developed as an industrial park (Figure 10b). Thus, the anthropogenic rate of removal is approximately two to three orders of magnitude greater than the natural rate of deposition.

SUMMARY

Geoantiquities are geologically young, near-surface, natural records of Earth history that document environmental change at local, regional, and global scales. Geoantiquities comprise important landscapes that are vulnerable to destruction and removal, particularly in rapidly growing cities and urban corridors. The concept of geoantiquities parallels existing concepts in cultural resource management, and is consistent with international trends in geoconservation.

The impact of humans can be permanent, and the loss of geoantiquities is an impending reality, particularly near cities where land is privately owned and urban growth is rapid. Earth science studies can provide a mechanism to educate society on the long-lasting values of landscapes that were previously seen only as expendable aggregate resources.

Case study sedimentary and geomorphic examples from Late Cenozoic of the Bonneville Basin illustrate the scientific and societal values of geoantiquities that could be lost without a commitment to geoconservation. The Stockton Bar is considered a geoantiquities heritage area that can serve as an example of how science, information transfer, and partnerships can work at a community level, even if the land is privately owned.

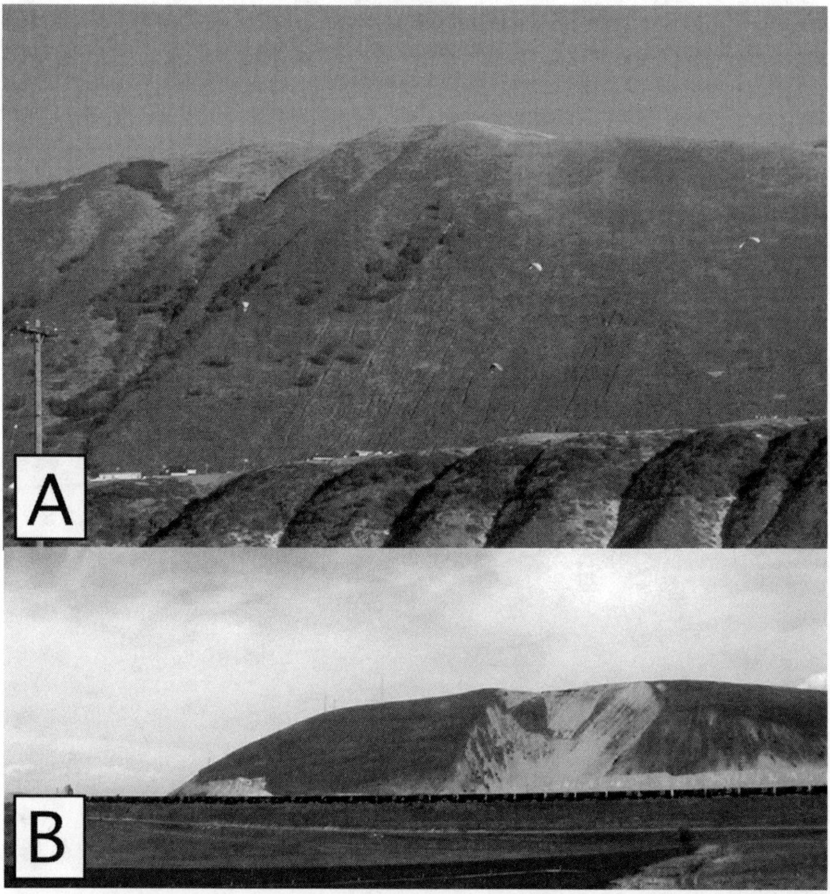

Figure 9. Competing uses of the Point of the Mountain spit, at the south end of the Salt Lake Valley. A. This geoantiquity is a world class paragliding and hang gliding area, that is being altered to high density residential housing (left), which will soon cover this whole shoreline bench. Rapid growth is encroaching upon the Utah State Prison, which was originally outside the populated areas. B. The south side of the Point of the Mountain is rapidly being removed by sand and gravel quarrying. Large tonnages of gravel are hauled away by railway.

Figure 10. A. Engraving "Fault scarp crossing alluvial cone, near Salt Lake City", drawn by W. H. Holmes, Plate XLIV, p. 348 in Gilbert [1890]. This geoantiquity illustrates Lake Bonneville shoreline deposits, alluvial fan development, and fault displacement. B. In the 20th century, this fan has been entirely removed for its values as a sand and gravel resource. The site is now an industrial area of North Salt Lake City, Utah [Chan and Milligan, 1985].

The conservation of geoantiquities is dependent upon informed citizenry and community vision. This paper underscores the importance of science and community cooperation in promoting wise management of geoantiquities in the urban environment.

Acknowledgments. This material is based on work supported by the National Science Foundation under grants SBR-9817777 and EAR-9809241. We thank the following reviewers Ramon Arrowsmith and Brian Marker for their constructive input on the manuscript. The content of this paper remains solely the responsibility of the authors. We gratefully acknowledge the research contributions of Genevieve Atwood, Tammy Wambeam, and Elliot Lips. Additional information is available at the Utah Geoantiquities Heritage Program Web Site: http://www.geog.utah.edu/geoantiquities/.

REFERENCES

Brown, D. M., Understanding urban interactions: Summary of a research workshop (National Science Foundation executive summary): http://www.nsf.gov/pubs/1998/sbe981/sbe981.htm, 1997.

Chan, M. A., and Milligan, M. R., Gilbert's vanishing deltas: A century of change in Pleistocene Lake Bonneville deposits of northern Utah, edited by W. Lund, Environmental and engineering geology for the Wasatch Front region: *Utah Geological Association Publication* 24, 521-532, 1995.

Currey, D. R., G. Atwood, and D. R. Mabey, Major levels of Great Salt Lake and Lake Bonneville, *Utah Geological and Mineral Survey* Map 73, scale 1:750,000, 1984.

Donner, P., Population estimates for Utah: Demographic and Economic Analysis Section, Governor's Office of Planning and Budget, 17, 1997.

Eberhard, R., ed., Pattern and Process: Towards a Regional Approach to National Estate Assessment of Geodiversity, *Australian Heritage Commission and Environment Forest Taskforce Technical Series* no. 2, Environment Australia, Canberra, 102 pp., 1997.

Gilbert, G. K., Lake Bonneville: *U. S. Geological Survey Monograph* 1, 438 pp., 1890.

Krieg, W., Progress in management for Conservation of Geotopes in Europe. *Geologica Balcanica* 26, 1, Sofia, Mart., 13-14, 1996.

Hooke, R. LeB., On the efficacy of humans as geomorphic agents: *Geological Society of America Today*, Sept. 1994.

Hooke, R. LeB., On the history of humans as geomorphic agents: *Geology*, v. 28, 843-846, 2000.

Martini, G. (Secretary of the symposium) and Pages, J.-S. (Editor of Texts), Memoires de la Societe Geologique de France, no. 165, *Actes du Premier Symposium International sur la Protection du Patrimoine Geologique*, Digne-le-bains, European Working Group on Earth Science Conservation, 276 p., 11-16 juin 1991.

Meli, N., State of Utah Department of Environmental Quality, Division of Air Quality Report, Approval number DAQE-683-99, Aug 9, 1999.

Nir, Dov. *Man, a Geomorphological Agent: and Introduction to Anthropic Geomorphology*. Jerusalem, Israel: Keter Publishing House, 165 pp, 1983.

North American Stratigraphic Code, North American commission on stratigraphic nomenclature: *American Association of Petroleum Geologists Bulletin* v. 67, 841-875, 1983.

Panizza, M., *Environmental Geomorphology*., New York: Elsevier, 268 p, 1996.

QGET (Quality, Growth Efficiency Tools), Utah Governor's Office of Planning and Budget, State of Utah, Technical Committee which seeks to improve the quality of information available to plan for Utah's future: http://www.governor.state.ut.us/dea/qget/1.htm.

Stürm, B., Integration de la protection du patrimoine geologique dans l'amenagement du territoire en Suisse. *Mem. Soc. Geol. Fr.,* vol. 165, 93-97, 1994.

UNESCO (United Nations Educational, Scientific and Cultural Organization) United Nations for education, science and culture, Hundred sixtieth session, Paris: *Report of the General Manager on the feasibility study of the installation of a program Geosites/Geoparks of UNESCO*, http://www.unesco.org/science/earthsciences/geoparks/geoparks.htm, Aug 2000.

Wilson, E. O., Back from chaos: *Atlantic Monthly*, March 1998, 45-59 (quote from p. 46), 1998.

3

Engineering Geology of New York City: Continuing Value of Geologic Data

Charles A. Baskerville and Robert H. Fakundiny

INTRODUCTION

New York City owes its high-rise buildings and unique skyline to the underlying geological materials, which allow for its diverse architecture. Solid, metamorphic crystalline rock anchors the famous skyscrapers that rise in midtown Manhattan and the lower end of the island, while soft sediments and artificial fill restrict the construction of tall buildings in between. The scarcity of groundwater forced the City to install an aqueduct system that draws water from the Catskill and Taconic Mountains to the north into reservoirs and thence into a vast tunnel system that is hailed as one of the world's engineering marvels. Engineering projects, such as bridges, wharves, and transportation tunnels along and under the Hudson and East Rivers, relied on well-documented geologic studies of the City's surface and subsurface foundation materials. Most of the geological information used by engineers came from the study of drill-hole core, chips, and soil. Outcrops are rare, and those that remain are being covered by construction projects or isolated by walls for privacy. Engineering geologists have to use all opportunities available to study surface and subsurface exposures, such as the walls of the new Water Tunnel No. 3 that is being installed currently. This report is a condensation of one of those efforts to collect engineering geology data and provide advice based on the insights gained from their analysis.

New York City is composed of five counties, called Boroughs by the original Dutch settlers, a name still retained. These boroughs are complete counties, unlike most cities, such as the City of Los Angeles, which occupies only a part of the County of the same name. The five boroughs, alphabetically listed, are The Bronx, Borough of Brooklyn, Borough of Manhattan, Borough of Queens,

Earth Science in the City: A Reader

10.1029/056SP04

Borough of Staten Island (see Figure 1); together they encompass an area of 950 square kilometers.

This paper discusses the stratigraphic relations of Earth materials found in the New York City area and their structural makeup, as well as where their names originated. We also review how these units are studied, the use of their physical properties for the engineering geology characteristics that are necessary for designing structures, such as buildings, bridges, and tunnels. One major example, provided here, is how engineering geology played a part in the design and construction of the World Trade Center (WTC), which was subsequently destroyed on September 11, 2001, almost 4 decades after its construction, by terrorism. In addition, we discuss how the same archived information was utilized in the rescue and cleanup efforts, which were accomplished ahead of schedule and under budget. The ideas presented here may possibly be a model for geologic mapping that leads to similar detailed planning for other large urban areas.

STRATIGRAPHY

New York City is at the southern end of the New England Upland physiographic province, which is part of the Appalachian mountain chain. The oldest and stratigraphically lowest rocks in this region are Late Middle Proterozoic (Grenvillian) Fordham Gneiss. The immediate overlying rock is the Yonkers Gneiss of Late Proterozoic age. The Lowerre Quartzite is stratigraphically next highest above the Yonkers, and in unconformable contact with the Fordham. The Lowerre is considered to be Lower Cambrian and is overlain by the Lower Cambrian to Lower Ordovician Inwood Marble. The Middle Ordovician Walloomsac Formation, mostly interbedded marbles and schists, unconformably overlies the Inwood in this region (Figure 1). All of these units form an autochthonous sequence of early continental shield rocks that are overlain by deep-ocean shelf deposits. Subsequently, thrust sheets of allocthonous units were placed above these shield-and-shelf units. The thrust units consist of the Manhattan Schist and units of the Hartland Formation, all of which extend in age from Early Cambrian to Middle Ordovician.

Structurally these units range from rather open folds, with ptygmatic folds on their eastern limbs, to tight to isoclinal folds on their western limbs. This fold regime occurs on the east side of a generally NE-striking belt. The larger folds are crossed by SE-striking, smaller, open folds that give a "porpoising" appearance to the limbs of the larger folds. The New York City region is a reentrant of the folded Appalachian Mountain Belt, in contrast to a section across the Appalachian Mountain Belt along the latitude of Maryland and Virginia, which is a salient. The tighter folds here indicate folding and thrusting forces that were more intense at the latitude of New York City than to the south.

The bedrock geology of New York City has been tangled and reorganized by the thrusting of large parts of foreign basement that once lay to the east over the

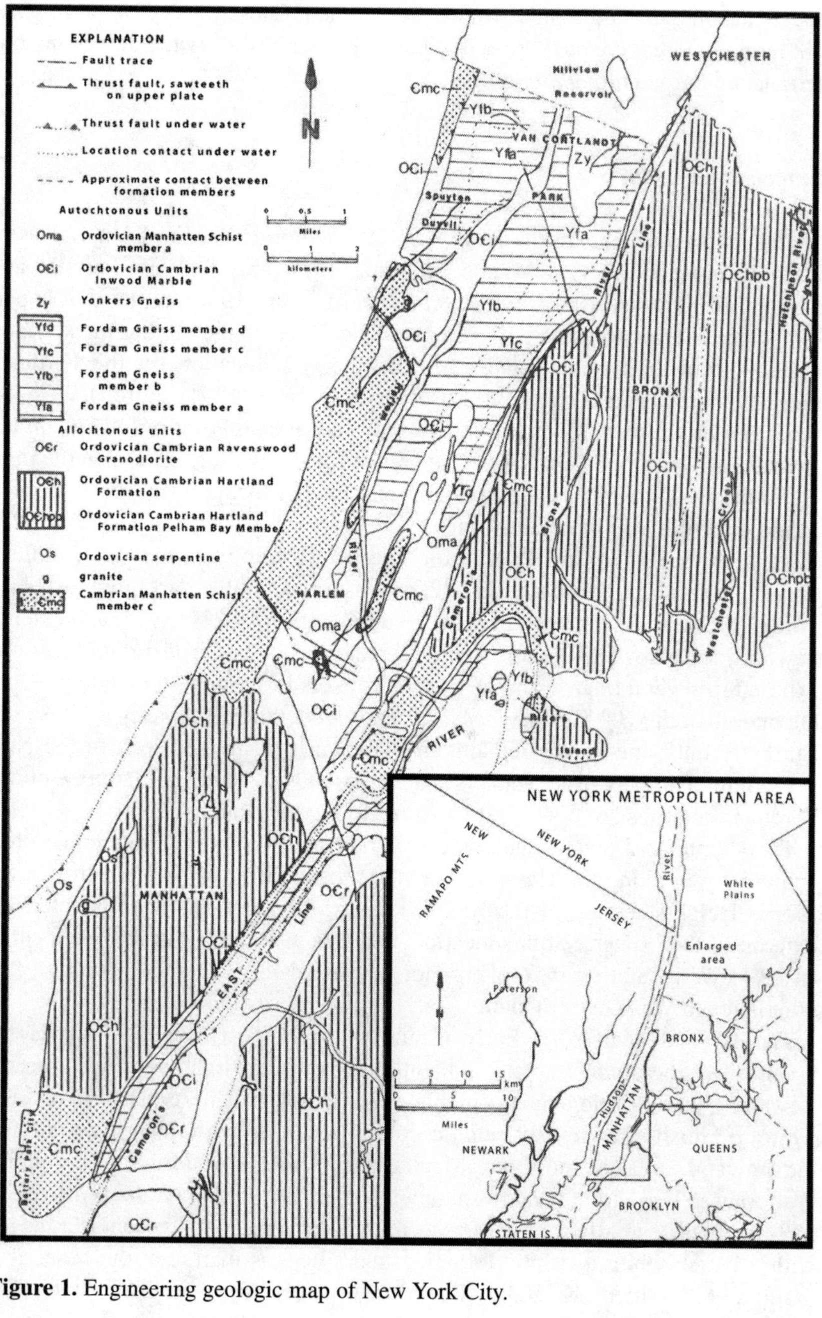

Figure 1. Engineering geologic map of New York City.

rocks that are still in the place where they formed. Those rocks that are in place are termed *autochthonous*; those that have been shoved over the autochthonous terrane are termed *allochthonous*.

Autochthonous Units

The oldest rock unit in the New York City region is the Fordham Gneiss of Grenvillian age (1.3 Ga, Middle Proterozoic; Mose, 1982; Baskerville and Mose, 1989; Baskerville, 1992; Ratcliffe and Burton, 1990), a unit that is part of the North American shield. Outcrops can be found on the northernmost ridges of the Bronx at Riverdale and Fordham, the type location for the Fordham Gneiss (Merrill and others, 1902; Baskerville, 1992, 1994). Fordham Gneiss crops out in the south Bronx in the vicinity of the Oak Point rail yards, on the Brother Islands in the East River, and on Roosevelt (Welfare) Island in the middle of the East River; field observations at the latter location showed Roosevelt island to be the axis of an antiform.

Yonkers Gneiss of late Proterozoic age overlies and intrudes the Fordham Gneiss. Bands of the pink granitic gneiss intrude the dark-gray to black Fordham Gneiss within the aureole surrounding the Yonkers Gneiss and extending outward a few kilometers from its main body. The Yonkers Gneiss was exposed in several quarries in the City of Yonkers in Westchester County before the middle of the 20th Century; these quarries were the type locality for this unit. Currently only one quarry remains in Yonkers; it is situated west of the New York State Thruway. The Yonkers Gneiss extends 1.5 km into Bronx County. Fordham outcrops are not found south of Roosevelt Island.

The Cambrian Lowerre Quartzite unconformably overlies Proterozoic units and crops out along the Harlem River next to the Metro-North rail line in the Morris Heights section of the Bronx. Urbanization projects have covered the outcrop of the Lowerre at this location, and it is visible only in its type locality at the Lowerre Station in Yonkers, not far from Morris Heights, and at a few other sites in Westchester County.

The Inwood Marble of Early Cambrian to Early Ordovician age is the youngest of the autochthonous units and lies with apparent conformity on the Lowerre Quartzite and unconformably on the Fordham Gneiss. The Inwood is a white to bluish gray, mostly calcitic dolomite marble with its type location in the Inwood area of northern Manhattan Island. The Middle Ordovician Walloomsac Formation unconformably overlies the Inwood (Baskerville, 1992, 1994), and is generally a garnetiferous muscovite-biotite-quartz schist at the top with a bluish-white to white dolomitic and siliceous marble at the base. This basal unit is called the Balmville Formation elsewhere in N.Y. State. The Wallomsac Formation is overlain in thrust contact by the Lower Cambrian Manhattan Schist.

Allochthonous Units

The allochthonous units are generally of Early Cambrian to Middle Ordovician age and correlate with the Hartland Formation (Hartland Terrane of Rogers, 1985) of western Connecticut (Baskerville, 1992, 1994). Middle Ordovician units are overlain by Silurian (?) metaigneous granitic and amphibolitic intrusives and, possibly, some Permian igneous rhyodacites. The Hartland Formation crops out in The Bronx and Manhattan. It is encountered in drill holes in Kings, Queens, and Staten Island, and exposed in tunnels. Ravenswood Granodiorite, first mapped and described by Ziegler (1911), was named after the Ravenswood neighborhood in Queens where it crops out. This Lower Paleozoic unit intrudes Hartland rocks; its Early Ordovician age is based on isotope dating (Baskerville and Mose, 1989; Baskerville, 1992; unpublished data, Mose and Baskerville). The Ravenswood straddles the East River and is exposed in Queens County south of the Triborough Bridge; it extends southwest to the Lower East Side of Manhattan and the Brooklyn Heights section of northern Kings County (Baskerville, 1992, 1994).

Serpentinites, greenish rocks containing asbestiform crysotile, are associated with Hartland rocks. The largest mass within the city limits is found in outcrop on Staten Island in the Todt Hill area. Another serpentinite mass is found at slight depth, centered at W. 57th Street and 10th Avenue in Manhattan (Baskerville, 1994). Several other serpentinite bodies occur at depth in tunnels near the City. Many serpentinite and some associated amphibolite specimens have chemical analyses that are similar to those of Kilauea basalts (Baskerville, 1994), and therefore, may be related to mid-ocean-ridge basalts, meteors, or Taconic basalts because they contain the trace elements iridium and chromium (Ratcliffe, 1987). The chemical compositions of the serpentinites suggest that their protolith was an ophiolite that probably came from the floor of Iapetus (Proto Atlantic). All of these ophiolite units have been found to border the approximate leading edge of the Hartland-type rocks, which define the eastern side of a major ancient thrust fault called Cameron's Line.

The Manhattan Schist is generally in thrust contact with units above and below. It is a gray, medium- to coarse-grained, layered sillimanite-muscovite-biotite-quartz schist and gneiss interlayered with other garnetiferous schists, gneisses, and black amphibolites. The Manhattan Schist crops out in The Bronx and Manhattan, and overlies the younger Walloomsac and Inwood units. The Manhattan Schist was named by Merrill and others (1902) for the type location where it was first found.

Mesozoic rocks are found at depth along the west shore of Staten Island. Intrusive Jurassic diabase in the form of a dike—a southward extension of the New Jersey Palisades—crops out in Graniteville at the northwest corner of the Island. Outcrops of variegated and cross-bedded Cretaceous Raritan Formation sediments are exposed south of Latourette Park in the west-central part of the island (Baskerville, 1965; Cousminer and others, 1981). These sediments are unconsolidated sands, silts, and clays that cover most of Staten Island east of the serpentinite and the entire southern end of the island; they also occur at depth along the southern parts of Kings and Queens Counties. In Queens County, the Magothy Formation (Matawan Group Undifferentiated) lies above the Raritan Clay (Soren, 1978). The Magothy has a continental origin and disconformably overlies the Raritan; it has sparse exposures in Brooklyn and Staten Island (Lyttle and Epstein, 1987).

Tertiary rocks are not found in New York City. Pleistocene glacial deposits extend throughout all five counties. Glacial grooves and striations on crystalline-rock outcrops indicate a generally southeastward movement of the ice across the City.

STRUCTURE

The structural complexity of Manhattan's bedrock results from the emplacement of basement rock from eastern foreign sources along thrust faults, which were subsequently folded during at least five different episodes.

Thrust Faults

Cameron's Line is a regional northeast-striking thrust fault with a southeast dip that extends across all of the five boroughs of New York City. Hall (1968a) and Spinek and Hall (1985) indicate that evidence for thrusting along this line is the truncation of rock units above and below this fault and the stratigraphic succession of older rocks over younger units or of similar age in Connecticut and White Plains, NY. This fault is also the suture that ties North American rocks to those of Africa and Europe.

Thrust faults on Manhattan Island and Bronx County dip gently to the east or are nearly horizontal, and commonly contain enough decomposed gouge to hinder measuring a direction of movement (Baskerville, 1992). Several low-angle to nearly horizontal thrust faults fail to reach the ground surface and, therefore, can be depicted only in cross-sectional views.

High-angle Faults

Many more faults occur in the Metropolitan area than can be seen in outcrops, as revealed by their exposure in tunnels (Figure 2); they are depicted on archival geologic maps. Urban development and sedimentary cover prevent tracing most of these faults for any distance underground or to the surface. Most faults in New York City appear to have displacements of only a few cms; the exceptions are low-angle thrust faults. Most of these faults are high-angle, normal or reverse, faults with NW strikes, and many have a movement sense of right-lateral, as well as vertical movement. Most exposed faults in the Bronx have NW trends. The

Figure 2. Fault traces on the north wall of the Brooklyn Tunnel of the City Water Tunnel Number 3 project. Bolt plate is approximately 10 cm on a side.

age of faulting in the New York City area may range from late Middle Ordovician Taconic deformation (Hall, 1968b) to, and possibly through, the Triassic (Rodgers, 1967). Normal and reverse faults in the region commonly cut across Paleozoic folds. No evidence of movement since Mesozoic time has been observed on any of the faults. The 125th Street fault, however, is currently being monitored by seismologists at Lamont-Doherty Earth Observatory of Columbia University with portable seismographs to establish whether recent small earthquakes in northern Manhattan occurred on it.

Folds

At least five phases of folding have been observed throughout New York City (Taterka, 1987; Langer and Bowes, 1969). First-phase folds are sparse and infrequently observed. Evidence for this phase is regional foliation that is subparallel to bedding, and is marked by parallel alignment of micas and segregation layering of light and dark minerals (Taterka, 1987). Second- and third-phase folds are the most abundant and the easiest to recognize. Second-phase folds are generally isoclinal, with their axial planes overturned to recumbent to the west. These folds plunge southward from 5° to 40°. The most numerous third-phase folds have upright axial planes with northeast trends and are open structures with southwestward plunging axes. Only small versions of fourth- and fifth-phase folds are exposed. Fourth-phase folds are crenulations or kinks that have a northerly trend. Fifth-phase folds are broad, open warps of the third-phase folds that are seen as "porpoising" axial plunge reversals.

ENGINEERING GEOLOGY

The physical and chemical properties of rock and soil determine their capability to function as foundation materials for engineering-construction purposes. The engineering characteristics of materials at a proposed site, such as confined- and unconfined-compressive strength, grain size, whole-rock chemical analyses, and ease or difficulty of excavation, must be measured. Surface-water and groundwater samples should be tested for corrosivity. Groundwater-level-monitoring wells should be placed at the site. The resulting laboratory and field data, although not exhaustive, indicate the material properties that must be factored into the design of New York City's construction projects.

Excavation of Rock

Most of the white to blue-white Inwood Marble and weathered, near-surface schists of the Walloomsac Formation are rippable and can be excavated without blasting, or can be shattered with pneumatic concrete breakers. Rippability refers to the ability of heavy machinery, adapted to the task, being able to rip apart friable rock and thus save the cost and inconvenience of blasting (Goodman, 1980). Excavation becomes increasingly difficult at depth where the rock must be blasted. Blasting is required in most places for excavation of Manhattan Schist or of sound, unweathered Hartland Formation rocks such as interbedded amphibolites, gneisses, and granites.

Schists and gneisses commonly are moderately difficult to cut. Rapid wear can be expected on cutting tools where very hard quartzose layers or granitic intrusives are encountered; this will increase cutting time. Severe wear can also be expected on cutter discs of tunnel-boring machines (TBMs) by these rocks, as well as on drill rods and diamond- and tungsten-carbide-drill bits. Schist and granulite rocks in the Hartland Formation are generally fine-textured and can cause excessive wear on tungsten-carbide percussion-drill bits. Roller bits used with rotary drills are more satisfactory in drilling holes for placement of blasting charges in these medium-hard to hard rocks (Hemphill, 1981).

Bedrock beneath the soil is deeply decomposed in some places. These highly weathered zones, known as saprolite, can be excavated with hand tools or power shovels and backhoes; but excavation becomes increasingly difficult with depth and the consequent increase in the rock's integrity. Light-green serpentinite is generally extremely weathered from 2 to 13 m beneath the surface. Asbestiform rock can be found in association with the serpentinites. Weathering makes excavation comparatively easy near the surface; tight joints (cracks in the rock) that are sealed by mineralization make the rock less easy to excavate. Heavy minerals give the rock a dark greenish-to-black appearance where weathering is minimal or absent. Drilling with diamond bits into fresh, unweathered serpentinite is slow and causes a great amount of wear on the diamonds.

Excavation Around Shear Zones and Joints

Crush and shear zones seen in tunnels, shafts, and underground chambers are fault zones with many close-spaced fault surfaces between crushed rock. Loose rock material in these zones of weakness may collapse from the tunnel walls or roof. Additionally, many of these zones have been observed to carry large volumes of water. Associated with these major faults are many closely spaced joint sets. The number and spacing of these joints can affect the engineering characteristics of the rocks in which they occur (Farmer, 1983). Joints may be spaced as close as 2 to 12 cm apart near the fault, but the spacing may increase to 2 m or more away from the fault. A companion joint set commonly intersects the first set at some large angle and with similar spacing, thus creating a joint system that produces loose rock blocks. Blocky rock can present production delays to the operation of TBMs. Poor tunnel-wall and roof stability may hinder tunnel completion and operation after the excavation has been made. Most of these joints have NW or NE trends and near-vertical dip angles to both the NW and NE. An additional prominent joint system has a lower dip angle ($\geq 55^{\circ}$) in all four compass quadrants. This latter joint set appears related to folding rather than faulting (Hobbs and others, 1976, p. 294 and 299) and recalls the porpoising of fold structures referred to previously. The orientation of joint sets at large angles to each other is common in massive rock and controls the ultimate bulk strength of these rocks. Intact blocks of rock may have test properties that falsely indicate greater strength than actually present in the bulk rock (Farmer, 1983). Rock with many joint sets at close spacing will exert a greater load on retaining structures than intact rock, and thereby lead to potential collapse from the forces of gravity or other deformational processes (Goodman, 1980; Farmer, 1983).

Types of Soils

The most abundant surficial material in New York City is till, which consists of a mixture of clay, silt, sand, gravel, and boulders; although extensive deposits of stratified drift abound also. The glacial material in Manhattan and parts of The Bronx is derived from the Triassic and Jurassic red beds and diabase of New Jersey. Many of the glacial erratics found in Manhattan are diabase from the Palisades sill, which is exposed on the west side of the Hudson River. Large amounts of miscellaneous fill can be found in former building lots. Miscellaneous (artificial) fill is considered to be a type of soil and contains mixtures of glacial deposits, riprap (large blocks of rubble), and building-demolition rubble composed of glass, wood, brick, and concrete. Miscellaneous fill is readily excavated with power equipment such as backhoes, front-end loaders, and power shovels.

Freshwater rhythmites (systematically layered fine-grained sediments) and tidal-marsh sediments consisting of organic silts and clays of varying thickness may be found overlying sand, silt, clay, gravel or combinations of these, which in turn overlie bedrock. The most common deposits are organic silts and clays, known as "sensitive soils" and locally as "bull's liver" (Parsons, 1976). Varved silts and clays are found at sites within the geographical limits of ancient glacial lakes Hudson and Flushing, and represent glaciolacustrine organic silts and clays. Post-Pleistocene erosion, engineering removal, and nondeposition interrupt the continuity of these lake-bottom sediments across these former lakebeds. Glacial lakes Hudson and Flushing covered most of Manhattan Island except for the area around Central Park and the parks along the upper west side of the island. These lakes covered at least three-quarters of Kings and Queens Counties northwest of the Harbor Hill terminal moraine, which strikes northeast through Brooklyn and Queens Counties, generally parallel to the orientation of Long Island, and about 15 km inland from its north shore. Engineering problems related to "sensitive soils" in New York City, such as differential settlement during construction of tall buildings, have been thoroughly discussed by Parsons (1976). Soil thickness ranges from 0 to more than 35 m in Manhattan and The Bronx, and to more than 100 m in Brooklyn and Queens Counties.

Water

The five Boroughs contain streams and lakes as well as groundwater. Artificial fill derived from urbanization has been used to eliminate many of the drainage channels that existed in the 1600s. Many of the old, buried creek beds exposed in excavations still maintain underground flow. Old tidal-marsh shoreline areas have been filled to provide land for development. Groundwater levels under these reclaimed lands are still affected by tides, although the time of the highest groundwater level lags the time of highest tide water; the amount of lag time depends on the distance inland from the ocean. Flooding from escaping groundwater occurs in excavations that expose these old stream channels or buried marshes.

Potable water for the city is obtained from upstate reservoirs through deep tunnels, except for a part of Queens County that depends on groundwater. Groundwater was used for industrial purposes through much of the 20th Century, but has become contaminated by urban runoff. Some of the pollutants that have entered the ground from surface runoff are: (1) street refuse, including animal droppings, leaves, and dust from natural erosion; (2) industrial spills; and (3) road-salt stockpiles exposed to rainfall.

Raw sewage, which can enter stream channels during periods of excessive storm flow, feeds bacteria that multiply and use up great amounts of dissolved oxygen. These bacteria produce carbon dioxide (Hem, 1970), which in solution attacks concrete by converting calcium hydroxide, the normal product of hydration of portland cement, to calcium bicarbonate with a consequent increase in volume (McConnell and others, 1950). Naturally acidic water, which is produced by organic acids derived from decaying vegetation, weakens portland-cement concrete by dissolving the calcium hydroxide and calcium carbonate out of the concrete (Mielenz, 1962). Concrete made from certain cements tends to deteriorate in sulfate-bearing soils. The sulfate is produced by dissolution of sulfur-bearing minerals such as pyrite, and tends to increase in concentration with depth in clay soils.

Dissolved impurities and subsequent acidity in surface and subsurface waters can corrode metal and concrete. Buried iron or steel structures are commonly corroded by bacterially derived sulfate. Chloride, which can be a product of sewage, contributes to the electrical conductivity of water, which increases its potential to corrode metal; dissolved metals in turn contribute to the water's acidity.

VALUE OF ENGINEERING GEOLOGY TO THE WORLD TRADE CENTER CONSTRUCTION AND CLEANUP

Engineering geology data, analyses, and interpretations, once obtained, can be valuable over long periods for the design, construction, and demolition of structures, especially those data sets that are obtained with difficulty from highly urbanized areas, such as Manhattan. The preliminary engineering geology studies of subsurface geology for the World Trade Center (WTC) are an excellent, if not also timely, example.

Planning for the WTC was underway in 1963, when the original diamond-drill core borings were made into buried bedrock along the sides and through the site planned for the high towers (Figure 3). Soil samples and core was analyzed by the engineering staff of the Port Authority of New York and New Jersey. These data were reviewed and integrated with older geologic data sets into a final set of conclusions by the head consulting engineering geologist on the project, Thomas W. Fluhr. The following narrative is based upon his reports, maps, and cross sections (Fluhr, 1963, 1964a, b).

Figure 3 is a map of the preliminary plans for the footprints of the North and South Towers of the WTC. This map depicts some of the borehole locations that were used in Fluhr's analyses. Figure 4 is cross section CC' (which he labeled C-C), one of the 5 cross sections Fluhr made from borehole data. Cross section C-C' is located along the southern part of the planned location of the North Tower, extending from borehole 145 to east of borehole 138A. The eastern side of the wedge of organic silt marks the approximate location of the original Hudson River shore. The organic silt and overlying artificial fill extended the western shore of Manhattan Island westward into the Hudson River. The glacial sand and

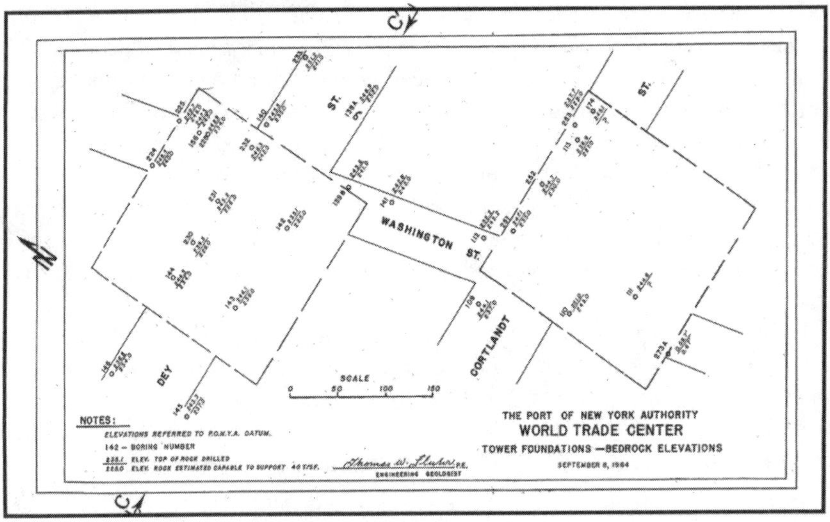

Figure 3. Index map of the footprint of the North and South Towers of the World Trade Center, New York City, NY. North arrow and section-line arrows were added by the authors to the outside of the original map. Figure is reduced from a scale of 1 in = 50 ft to 1 in = 200 ft (From Fluhr, 1963, 1964a, b).

gravel forms a saturated aquifer that probably is hydraulically connected to Hudson River water and presents the probability of massive water infiltration to the excavated foundation area.

Engineering geology analyses and conclusions by Fluhr about the bedrock included the following: (1) the basement consists of Manhattan Schist, which has variable thickness and varying degrees of weathering at its top surface, which lies 20 to 23 m below sea level; (2) some of the schist is jointed and would require removal of loose blocks by "hand methods" rather than by blasting; (3) laboratory analyses of the schist indicated that the modulus of elasticity (the ratio of stress to strain under a given load for elastically-deformable materials) varied from 1.17 to 2.27 x 10^6, and the lowest shear strength is 6,000 pounds per square inch; (4) the bedrock could be loaded to a force of 75 tons per square foot, much more than required for the weight of the planned towers; and (5) the surface and subsurface geologic conditions would permit construction of two high towers.

Soils included artificial fill, organic silt, glacial sand and gravel, glacial lakebed sediment, till in the form of ground moraine, and *in situ* decayed rock. Laboratory testing and field observations indicated that the soils at the site would not support the tower structures and, thus, should be excavated. The stratigraphy of the surficial deposits was interpreted from boreholes to be generally a sequence of mostly stratified units, starting at the bottom with a continuous layer of ground moraine overlying the bedrock surface and varying in thickness from

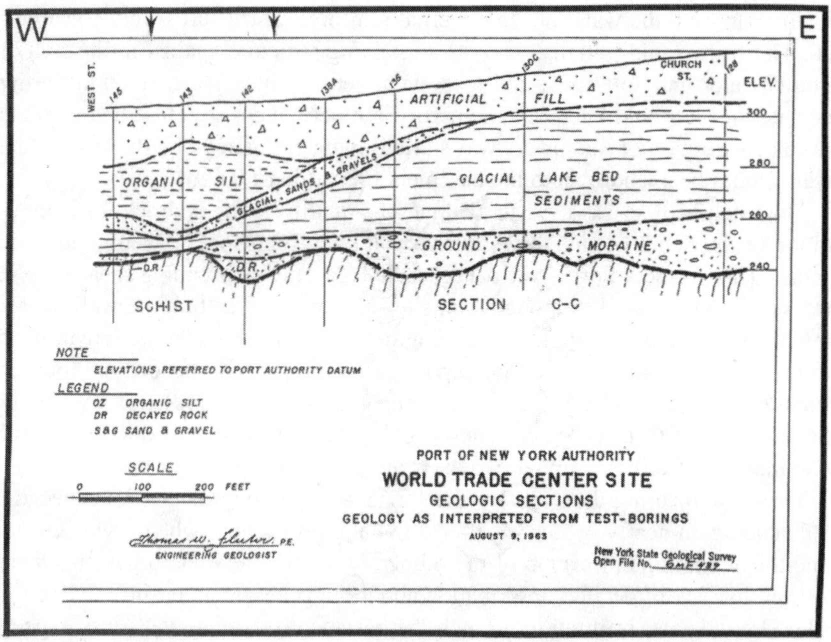

Figure 4. Cross-section C-C' (C-C on Fluhr's original cross section) through the planned area of the North Tower of the World Trade Center, New York City, NY. The letters "W" and "E" and the arrows were added by the authors to the original drawing. The arrows delineate the planned position of the west and east walls of the North Tower. Figure is reduced from 1 in = 80 ft to 1 in = 300 ft (From Fluhr, 1963, 1964a, b).

2 to 7 m. Glacial sand and gravel unconformably overly the partly eroded surface of the varved glacial lake-bed sediments ("bull's liver") and ground moraine, with a gentle dip to the west and southwest and thickness varying from 0 m., east of the site, to about 20 m, on the southwest. From here these materials continue under the Hudson River as a 3-m-thick layer. A wedge of organic silt overlies the glacial sand and gravel; its thickness ranges from 0 m at the center of the site to greater than 20 m on the west and southwest. Artificial fill, 3- to 7-m thick, caps the sequence across the entire site.

Water could enter the site either by flowing in from the sides through the glacial sand and gravel unit, or upward from joints within the bedrock. A suggestion to use a 1-m-thick gravel bed at the base of the lowest basement floor to allow drainage to a sump pump and prevent floor buckling was not accepted by Fluhr; rather he advised that the basement be sealed off before its excavation by a retaining concrete wall that would be emplaced through the surficial materials into bedrock by injecting concrete into closely spaced holes that joined each other and contained reinforcing rods. Outside tie rods would be grouted to bedrock to hold the wall until the internal reinforcing structures were constructed. The wall would initially serve as a retaining structure against infiltration of groundwater and soft soils, and to prevent settling of adjacent ground during dewatering; it would later become part of the permanent structure below grade once construction of the towers was started. Fluhr estimated that the base of the wall should be anchored into the basement rock at least 0.6 m.

The final drilling was done to evaluate the quality of the bedrock for computation of the depth needed to which the wall's footing should penetrate into the schist. This last set of drill-hole data revealed a northwest-trending shear zone in the schist that crossed the foundation sites and required setting the wall over it into the bedrock from 2 to 3 m. Thus, engineering geology methods were utilized to measure the rock and surficial deposit properties that influenced the foundation design of the WTC's lower structure. The WTC was built and stood until September 11, 2001, when it was destroyed by the crashing and exploding of two commercial jet airliners into the higher part of each tower.

These engineering data sets for the foundation geology of the WTC became valuable again nearly 4 decades after original construction, when rescue efforts and the beginning of cleanup of the rubble overlying the WTC basements were slowed by lack of detailed information about the geologic units surrounding the site. Groundwater infiltration through the basement retaining walls was expected, and the engineers feared it might halt demolition and recovery of bodies. Engineers and rescue personnel needed to know precisely what paths infiltrating groundwater and surface-water would take to flood the basement. The 38-year old maps, cross sections, and borehole-log descriptions were retrieved from the New York State Geological Survey Open File, where they had been stored for future use, and were provided to the engineers working at the site. The map and cross-section depictions of the precise location of the water-bearing units were indispensable to the planning and design of the clean up. This illustration of the continued use of archived engineering geology data portrays the value of geological studies at major construction sites for all cities, and the value of collecting and carefully protecting archived geologic data.

COLLECTION, COMPILATION, AND AVAILABILITY OF ENGINEERING GEOLOGY DATA

Engineering geology studies in large cities are required for proper planning and construction of new facilities and the repair or dismantling of old or damaged structures. Engineering geology studies are difficult to perform in large cities where exposures are few and commonly inaccessible. This work must continue, however, so that structures in our cities of the future can be as safe as possible.

The compilation of New York City's data was based on many years of fieldwork. Virtually detective-like approaches are commonly required to "sniff out" the location of rock outcrops. Inspection behind rows of apartment buildings and backyards, in addition to the obvious places, such as parks, is mandatory for finding outcrops. Many basements of older structures, such as those built before the 20th Century, are carved out of the bedrock and expose the original foundation rock, which serves as the "basement walls."

Another source of subsurface information is municipal building-permit agencies and public works departments, which can provide data such as boring logs for large structures, including bridges, tunnels, and highways. This information can be plotted on topographic maps, which are then used to develop engineering geology maps. These maps in turn are used in the design of engineering structures to provide the future inhabitants with a healthy and structurally sound environment in which to live and work.

Acknowledgments. Thanks are extended to Jill Schneiderman and James W. Skehan, S.J. for their reviews and comments on the paper. Robert H. Fickies managed the storage and retrieval of the original Fluhr documents and provided them to the engineers working at the World Trade Center. We thank Anne J. Finch for her review and editorial suggestions.

REFERENCES

Baskerville, C.A., A micropaleontological study of Cretaceous sediments on Staten Island, New York, New York University, New York, Ph.D. Dissertation, 65 p., 1965.

Baskerville, C.A. and Mose, D.G., The separation of the Hartland Formation and Ravenswood Granodiorite from the Fordham Gneiss at Cameron's Line in the New York City area, *Northeastern Geology* 11(1): 22-28, 1989.

Baskerville, C.A., Bedrock and engineering geologic maps of Bronx County and parts of New York and Queens Counties, New York, U.S. Geological Survey Miscellaneous Investigations Series Map I-2003, 2 sheets, scale 1:24,000, 1992.

Baskerville, C.A., Bedrock and engineering geologic maps of New York County and parts of Kings and Queens Counties, New York, and parts of Bergen and Hudson Counties, New Jersey, U.S. Geological Survey Miscellaneous Investigations Series Map I-2306, 2 sheets, scale 1:24,000, 1994.

Cousminer, H.L., Conners, S.D., and Loring, A.P., 1981. Stratigraphy and sedimentology of Upper Cretaceous (Raritan) sediments of Staten Island, N.Y, Atlantic Margin Energy Conference, Atlantic City, New Jersey [Poster Session], 1981.

Farmer, I. W., *Engineering Behavior of Rocks (second ed.),* Chapman and Hall, London and New York, 208 p., 1983.

Fluhr, T. W., Memorandum No. 1 on geologic features of the World Trade Center Site, Lower Manhattan, City of New York, The Port of New York Authority, *New York State Geologic Survey Open File* No. 6mF372 1 of 4, 14p., 1963.

Fluhr, T. W., Memorandum No. 2 on geologic features of the World Trade Center Site, The Port of New York Authority, *New York State Geological Survey Open File* No. 6mF372, 2 of 4, 5p., 1964a.

Fluhr, T. W., 1964b. Memorandum No. 3 on geologic features of the World Trade Center site, The Port of New York Authority, *New York State Geologic Survey Open File* No. 6mF372, 3 of 4, 8p.

Goodman, R.E., *Introduction to Rock Mechanics*, John Wiley and Sons, New York, 478 p., 1980.

Hall, L. M., Geology of the Glenville area, southwesternmost Connecticut and southeastern New York, Trip D-6, in Orville, P. M., (Ed.), Guidebook for field trips in Connecticut-New England Intercollegiate Geological Conference, 60th annual meeting, New Haven, Conn., Oct. 25-27, 1968, *Connecticut Geological and Natural History Survey Guidebook* 2, Section D-6, 12 p., 1968a.

Hall, L. M., Times of origin and deformation of bedrock in the Manhattan Prong, *in* Zen, E-an, White, W. S., Hadley, J. B. and Thompson, J. B., Jr., (Eds.), *Studies of Appalachian Geology—Northern and Maritime* (Billings volume), New York, Interscience Publishers: 117-127, 1968b.

Hem, J.D., Study and interpretation of the chemical characteristics of natural water (second edition), *U.S. Geological Survey Water-Supply Paper* 1473, 363 p., 1970.

Hemphill, G.B., *Blasting operations,* McGraw-Hill, New York, 258 p., 1981.

Hobbs, B.E., Means, W.D., and Williams, P.F., *An outline of structural geology*, John Wiley and Sons, New York, 571 p., 1976.

Langer, A.M. and Bowes, D.R., Polyphase deformation in the Manhattan Formation (sic) of Manhattan Island, New York City, *Geological Society of America Memoir* 115: 361-377, 1969.

Lyttle, P.T. and Epstein, J.B., Geologic map of the Newark 1° x 2° quadrangle, New Jersey, Pennsylvania, and New York, *U.S. Geological Survey Miscellaneous Investigations Series* I-1715, 2 sheets, scale 1: 250,000, 1987.

Merrill, F.J.H., Darton, N.H., Hollick, Arthur, Salisbury, R.D., Dodge, R.E., Willis, Bailey, and Pressey, H.A., Description of the New York City district, *in Folio 83 of Geologic Atlas of the United States*, U.S. Geological Survey, 1902.

McConnell, Duncan, Mielenz, R.C., Holland, W.Y., and Greene, K.T., Petrology of concrete affected by cement-aggregate reaction, *in* Paige, S., (Ch.), Application of geology to engineering practice, *Geological Society of America, Engineering Geology (Berkey) Volume*: 225-250, 1950.

Mielenz, R.C., Petrography applied to portland-cement concrete, *in* Fluhr, T.W. and Legget, R.F., (Eds.), *Geological Society of America, Reviews in Engineering Geology* 1: 1-38, 1962.

Mose, D. G., 1,300-million-year-old rocks in the Appalachians, *Geological Society of America Bulletin* 93: 391-399, 1982.

Parsons, J.D., New York's glacial lake formation of varved silt and clay, Journal of the Geotechnical Engineering Division, *Proceedings of the American Society of Civil Engineers* 102(GT6): 605-538, 1976.

Ratcliffe, N.M., Basaltic rocks in the Rensselaer Plateau and Chatham slices of the Taconic allochthon, chemistry and tectonic setting, *Geological Society of America Bulletin* 99(4): 511-528, 1987.

Ratcliffe, N.M. and Burton, W.C., Bedrock geologic map of the Poughquag quadrangle, New York, *U.S. Geological Survey Geologic Quadrangle Map* GQ-1662, scale 1: 24.000, 1990.

Rodgers, J., Chronology of tectonic movements in the Appalachian region of eastern North America, *in* Symposium on the chronology of tectonic movements in the United States, *American Journal of Science* 265(5): 408-427, 1967.

Rodgers, J., *Bedrock geologic map of Connecticut*, Connecticut Geological and Natural History Survey, scale 1:250,000, 1985.

Soren, J., Subsurface geology and paleogeography of Queens County, Long Island, New York, *U.S. Geological Survey Water-Resources Investigations Open-File Report* 77-34: 1-7, 1978.

Spinek, T.W. and Hall, L.M., Stratigraphy and structural geology in the Bethel area, southwestern Connecticut, *in* Tracy, R.C. (Ed.), Guidebook for field trips in Connecticut and adjacent areas of New York and Rhode Island, New England Intercollegiate Geological Conference, 57th annual meeting, New Haven, Conn., Oct. 4-6, 1985, *Connecticut Geological and Natural History Survey Guidebook* 6: 219-240, 1985.

Taterka, B.D., Bedrock geologic map of Central Park, New York City, Amherst, Mass., University of Massachusetts, Department of Geology and Geography, Masters Thesis, Contribution No. 62, 84 p., 1987.

Ziegler, Victor, The Ravenswood Granodiorite, *New York Academy of Sciences Annals* 21: 1-10, 1911.

SECTION II

NATURAL HAZARDS AND THE CITY

The public is aware of the power of the dynamic Earth when a natural event occurs, such as an earthquake or tsunami. Past earthquakes and tsunamis are evident in both the historical and recent geologic records. These data can be used to determine a city's susceptibility to such events and for careful planning and zoning to prevent undue damage and loss of life.

In many ancient cities natural hazards over centuries or millennia have affected the course of history. The Neapolitan region of Italy has seen the eruptions of Vesuvius and the volcanoes of the Phlegrean Fields and the island of Ischia. The best-preserved records of Greek and Roman history in Campania lie under volcanic ash deposits.

Gravity is the bane of the developer who builds home sites with scenic views. For with gravity comes landslides, which have ranged from a nuisance to a catastrophe in cities worldwide. To mitigate landslide dangers requires that scientists interweave data from the geologic and atmospheric sciences then make the public aware of what they may face.

4

Towards Integrated Natural Hazard Reduction in Urban Areas

Greg A. Valentine

INTRODUCTION

The last century has seen increasing migration of people to urban areas around the world, and as a result humankind is now undergoing a sort of phase transition to being dominantly based in cities rather than rural settings [Fuchs, 1994; World Resources Institute, 1996]. Many of these cities are very large and are growing at a pace faster than careful planning and infrastructure engineering can accommodate. Even in developed nations where urban growth may be more deliberate, sudden perturbations by events such as earthquakes, severe weather, flooding, and volcanic eruptions can quickly bring a city to its knees. In a sense, most of our cities can be considered to be in a metastable state—currently operating (how well depends on the individual city) but vulnerable to small perturbations. As Heiken et al. [2000] and Valentine and Heiken [2000] point out, the vulnerability of large cities to natural hazards needs to be viewed as a problem of national and international significance, rather than just a municipal one. This stems from the fact that cities are nodes for infrastructure and regional economic networks; urban vulnerability implies regional vulnerability, which may not be contained by political boundaries such as national borders. Regional and global stability increasingly depend on the stability of cities in our increasingly interconnected world.

Cities are exposed to nearly every natural hazard imaginable:

Meteorological—Hurricanes, tornadoes, high winds, heavy snow and rain.

Hydrologic—Floods, drought, water-induced slope failures.

Earth Science in the City: A Reader

10.1029/056SP05

Seismic—Strong ground motion, surface ruptures, earthquake-induced slope failures, tsunamis.
Volcanic—Lava flows, pyroclastic density currents, fallout, debris flows and avalanches, ground deformation, volcanic seismicity, tsunamis.
Ecological—Wildfires, slope instability induced by vegetation changes.

One of the key features that is common for most of these hazards is that they happen relatively quickly, releasing tremendous amounts of energy. In some cases, the underlying processes that cause the hazards can be accentuated or actually caused by human development—for example slope failures or surface flooding in highly urbanized areas.

This introductory paper focuses on volcanic hazards because they encompass many features of the other types of hazards listed above. If a city is unfortunate enough to be situated near a large explosive eruption there is a good chance it will experience earthquakes, fallout, lateral dynamic pressure loads from pyroclastic currents (similar to wind), fires, debris flows and avalanches, and disruption of the hydrologic system. Other papers within this section touch upon individual types of hazards in more detail.

Due to the toll on human life and the huge financial costs to individual nations and by the international community to recover from natural disasters to urban areas a focused, comprehensive and multi-agency approach is proposed as a sound investment. At the end of this paper I briefly suggest how such a program could be formulated, patterned after a major defense-related program that has been quite successful in the United States during the past several years. A major goal of this suggestion is to strengthen ongoing dialogs on how the U.S. and international organizations should address natural hazards.

URBAN VOLCANIC HAZARDS, PAST AND PRESENT

There are several historical examples of eruptions that have had effects, usually devastating, on cities and large towns. Akrotiri, on the Greek island of Thera (a.k.a. Santorini) represented a thriving city in the Late Bronze Age, but was destroyed and inundated by pyroclastic fallout and density currents during the caldera forming eruption of 1630 B.C. [McCoy and Heiken, 2000]. Pompeii and surrounding towns such as Herculaneum were similarly destroyed by pyroclastic phenomena associated with the AD 79 eruption of Vesuvius [e.g., Sigurdsson et al., 1985]. The 1902 eruptions of Mt. Pelee, on the island of Martinique, produced pyroclastic density currents by dome collapse. These currents overwhelmed the city of St. Pierre, killing 29,000 and nearly completely devastating the city [Lacroix, 1904; Fisher et al., 1997]. Three other examples occurred within the past two decades, highlighting the vulnerability of increased urbanization near volcanoes. The 1980 and 1981 eruptions of Mount St. Helens (Washington,

U.S.A.) produced widespread fallout that affected infrastructure and daily life in cities of the region [see Lipman and Mullineaux, 1981]. Over 20,000 people were killed in the small city of Armero (Colombia) by debris flows that were triggered by small scale explosive eruptions far downstream of Nevado del Ruiz volcano [e.g., Voight, 1990]. Finally in the mid to late 1990s the city of Plymouth (island of Montserrat) was damaged by fallout, and its outlying areas by pyroclastic density currents, associated with eruptions of nearby Soufriere volcano [e.g., Young et al., 1998].

Table 1, though not exhaustive, lists several cities that currently have varying degrees and types of volcanic risk. For example, volcanic risk in Tokyo is mainly a result of potential eruptions at volcanoes like Mount Fuji that are tens to hundreds of kilometers away, but that can produce significant eruption plumes, fallout from which could cause major disruption to the city. Because of the economic links that other countries have to Tokyo, such a disruption would have global impacts. Other cities, like Naples, Mexico City, and Auckland, sit directly on volcanic fields that have been historically (or nearly historically) active, as well as being near recognized major composite volcanoes. These cities are at risk to direct disruption as a new volcano could form right in their metropolitan areas. The cities listed in Table 1 also represent a range of economic and infrastructure development and therefore a range of abilities to respond to and recover from a volcanic event. Overlaid on the physical setting is cultural complexity. Even if the cities were identical in their size and their hazards, mitigation of the hazards would be different due to cultural variations.

INTEGRATION

One of the key areas where volcanology needs to improve its ability to address hazards, particularly in complex urban areas, is in integration between geologic monitoring, numerical modeling, and probabilistic assessment subdisciplines. Lack of integration causes problems on two fronts: first, if we do not integrate across subdisciplines with a common focus, the effectiveness of our science is reduced. Second, lack of coordination causes confusion among decision makers and the local population, again reducing our ability to save lives.

Integration within one of the subdisciplines identified above is already known to increase effectiveness in hazard prediction; for example it is most powerful to combine monitoring data such as deformation, seismology, and gas chemistry rather than to base predictions on one isolated technique. Here we focus on another type of integration, which brings a variety of techniques together to feed into a common end result, as illustrated in Figure 1 [Valentine and Keating, 2000]. Although the discussion here focuses on volcanic hazards, the framework can be applied to work on all types of natural hazards [e.g., Heiken et al., 2000, applied parts of this approach to seismic hazards in the Los Angeles area].

Table 1. Selected Cities and Their Volcanic Hazards.

City, Country	Approximate population (metropolitan area) and other aspects	Volcanic setting	Types of volcanic hazards present
Naples, Italy	3,000,000. Shipping, agriculture, NATO facilities.	Vesuvius composite volcano at eastern edge of metro area. City of Naples and suburbs growing rapidly within historically active Campi Flegrei.	Pyroclastic fallout, pyroclastic density currents, lava flows, cone formation, ground deformation, seismicity, tsunamis.
Mexico City, Mexico	20,000,000.National economic and federal government center.	Popocatepetl composite volcano 40 km SE, other large composite volcanoes in region. City is built on historically active basaltic volcanic field.	Pyroclastic fallout, pyroclastic density currents, lava flows, cone formation, debris flows, ground deformation, seismicity.
Quito, Ecuador	1,200,000. Capital of Ecuador	Guagua Pichincha composite volcano near western margin of city. Other large composite volcanoes in the region.	Pyroclastic fallout and density currents, debris avalanches, debris flows, seismicity.
Manila, Philippines	10,000,000. Capital of Philippines, major shipping harbor.	Near Taal Caldera (within which is Taal volcano), composite volcanoes in the region.	Pyroclastic fallout, seismicity, tsunamis.
Kagoshima, Japan	500,000. Prefecture capital.	City is on the edge of Aira caldera, within several km of Sakurajima composite volcano, and several composite volcanoes are in the region.	Pyroclastic fallout, seismicity, local tsunamis.
Tokyo, Japan	12,000,000. Capital of Japan, major global economic and commerce center.	Composite volcanoes in surrounding regions.	Pyroclastic fallout, tsunamis.
Auckland, New Zealand	1,000,000. Shipping center. Major city in New Zealand.	City is built on a field of basaltic scoria cones, maars, tuff rings and cones. Calderas and composite volcanoes are present in the region.	Lava flows, fallout from proximal and distant eruptions, pyroclastic density currents within a few km of basaltic eruptions, seismicity, ground deformation, tsunamis.

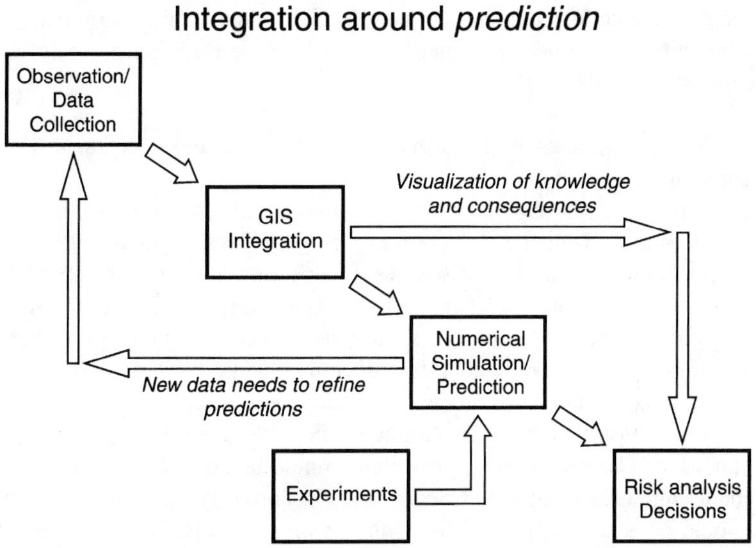

Figure 1. Illustration of the integration of techniques so that they all contribute towards predictions that can be used for risk assessment and decision making.

Observational data are the first step and foundation of an integrated predictive approach; however, it can be difficult to combine observational data of different types in a quantitative way. In large part, this is due to the range of techniques that contribute these data. For example fumarole temperature/chemistry measurements provide data at specific points on a volcano, while synthetic aperture radar provides spatial data over the whole of a volcano. Seismic and petrologic data provide data that pertain to processes at depth, and for which processes that occur between the source and the sample site or seismometer may need to be filtered out. The volcanological community needs to make much more use of Geographic Information Systems (GIS), which can provide important mechanisms for integrating such a wide range of data. In addition, GIS provides tools for visualization of complex, multiple features in a manner that helps both scientific interpretation and communication with the public and decision makers. Another value is that as numerical modeling grows in its ability to capture the real complexities of volcanoes, data residing in a GIS can be used directly to set up model geometry, boundary, and initial conditions. GIS is more than a set of "off-the-shelf" software tools. GIS is one of the most rapidly advancing research areas in geography and the environmental sciences. Developing GIS capabilities that integrate the types of data sets that are specific to volcanology, that manage and visualize these data sets, and that integrate with numerical modeling, is an area where volcanologists need to actively collaborate with experts in the fields of applied mathematics, informatics, and computer science.

Once data are collected and integrated into GIS, they need to form part of the basis for numerical models. Numerical models serve three purposes [e.g., see Valentine et al., 2002].

1) They are basic research tools that help us explore and understand the volcanic processes.
2) Numerical models should be used iteratively with observations and experiments to help focus the collection of new data. This is especially important where there are limited resources (e.g., time, instruments, people) for addressing a potential volcanic hazard. It is much more effective to use models to help identify what new data will most reduce uncertainty in predictions, and then to target those data rather than to collect data in a more haphazard way. It is often stated that modeling should never be done "in a vacuum" with respect to observations and experiments. In modern volcanology the opposite should also be true: Field observations should be conducted iteratively with modeling studies. To larger degree than is currently the case, field studies should be aimed at specific issues that improve predictive capabilities. This is also a major role for experimental work (as shown in Figure 1), which should be increasingly aimed at developing parameters, constitutive relations, and validation tests for numerical models [e.g., Valentine, 1994].
3) Numerical models are the mechanism for making specific predictions. In order to optimize the application of the predictive capability, simulations of different volcanic phenomena (e.g., lava flows, debris flows, pyroclastic fallout) should be overlaid on GIS-based economic, land use, demographic, and infrastructure data. If fragilities of these features to the range of conditions that could be generated by eruptions are known, consequences of the scenarios can be predicted and mitigation/response planned more efficiently. Some examples of this approach are already being undertaken in volcanology [Iverson et al. 1998]. A critical link here is the vulnerability information [e.g., Blong, 1984; Valentine, 1998], which requires collaboration with civil engineers, infrastructure experts, economists, and a variety of other disciplines with which volcanologists have had only limited interactions with until recently.

A key in the integrated approach being discussed here is the central roles of GIS and numerical modeling. In order to greatly advance our ability to minimize and respond to volcanic crises we must view GIS and modeling as much more than tools for convenience or for specialized applications. Modern volcanic hazards efforts should have these as key integrating pieces. This requires a higher level of awareness and communication across all of the boxes in Figure 1 than is currently often the case. The need for these integrating pieces has been well articulated by F. Dobran and coworkers [Dobran 1993, 1995] and is gradually taking root in parts of the volcanology community.

Finally, because of the inherent uncertainties in volcanology, it is necessary to incorporate all of the above into a probabilistic framework. This is another area where interdisciplinary work is needed, in this case with statisticians and risk assessment experts. Developing approaches that account for uncertainty, or ranges of possibilities, can be very complicated for volcanic hazards because many parameters are not independent. Therefore simple "Monte Carlo" approaches that calculate probabilistic risk by sampling from independent probability distributions of parameters (e.g., probability that a pyroclastic density current will move down a certain path, probability of a certain level of damage to a building from a certain range of pyroclastic density current conditions) may not be adequate. One ongoing program where probabilistic volcanic risk assessment is being done is the Yucca Mountain Project in the U.S.A. This program is exploring the possibility of permanently disposing of spent nuclear fuel and other high level radioactive waste underground in southern Nevada. The presence of several small-volume, Quaternary basaltic volcanoes in the region, combined with the long time frame over which a potential repository must isolate waste (10,000 years or greater), necessitates assessment of the probability and consequences (i.e., risk) from future volcanism in the area [CRWMS M&O 2000a-d]. Similar work has begun for potential geological repositories in Japan [Perry et al., 2001]. While the types of volcanic hazards for this potential underground facility are different from those that are of interest to an urban area, much can be learned from the general probabilistic approach. Probabilistic assessment is a key in the integrated predictive approach. Such assessments make the results of all the complex science in previous steps (Figure 1) useable by decision makers.

The flow of research and information shown in Figure 1 is already well-utilized by other fields of natural science where there are direct impacts on humans, and the power of such an approach is simply demonstrated by its acceptance in the lay public and by decision makers. Perhaps the best example is meteorology, where a range of data (e.g., satellite-based remote sensing data, ground-based data) are routinely collected and fused in spatial databases. These data are used directly as boundary and initial conditions for physics-based numerical models to make predictions on a variety of time and space scales. Experimental and theoretical studies, combined with observations, constantly hone the predictive capability. Finally these results are interpreted in a probabilistic framework that is used to make decisions. When one considers how the general public commonly sees the results of complex weather observations, numerical predictions, and even probability maps (e.g., contours of severe thunderstorm probability) and how readily the public uses this information, it is easy to see what the value of a similar integrated approach would be before and during a volcanic crisis. The seismic hazards community is also making strides towards such integrated approaches. Other parallels can be found in the exploitation of petroleum reservoirs and in environmental management.

PROPOSAL: A PROGRAM ON SCIENCE BASED NATURAL HAZARD REDUCTION

Given the degree of human and economic vulnerability to the impact of natural hazards on urban centers, it makes sense for the geophysics community to try to drive national and international policy towards implementation of programs that integrate across observation, monitoring, GIS, experimental, modeling, and probabilistic assessment subdisciplines as described above. While our current vulnerability is the motivation for such a program, other factors key in formulating a large program include:

1) Prediction and probabilistic assessment of future events is the ultimate goal.
2) Observations of natural hazards at a specific city in terms of the response of the whole urban system are limited to past behavior and likely different conditions than the future.
3) A wide range of data on "pieces" of an urban hazard system must be fused together.
4) Full-scale experiments of urban hazards cannot be conducted; therefore there will need to be a heavy reliance on numerical modeling for understanding the behavior of a whole system. At the same time, experimental work on key pieces of a hazard problem are necessary, with the goal of testing and improving numerical models.

There is a precedent for a program that was developed around the same factors, but for a national defense issue in the U.S.A. This program is called Science-Based Stockpile Stewardship (SBSS). The ultimate goal of SBSS is to certify the performance of nuclear weapons in the U.S.A.'s stockpile in the absence of full-scale testing of the weapons; some of the key scientific issues in this program were recently reviewed by Jeanloz [2001]. Like natural hazards problems, nuclear weapons involve a wide range of complex, coupled processes (e.g., hydrodynamics, nuclear reactions, radiative transport). Under the best of conditions, there has always been some uncertainty in the performance of these coupled processes. The coupled processes are further complicated by the aging of components beyond their original design lifetime. Individual components may be tested under the current program, but the full integrated system may not be tested due to the nuclear testing moratorium. Therefore, similar to natural hazards, a range of data on individual subsystems in the weapons must be fused together and numerical models must be used in both a predictive mode for the whole systems, and to help guide experimental and theoretical work on individual components. Furthermore, the only experimental data for the performance of the whole system are historical and represent different conditions compared to the present ones.

National governments and international agencies should look at SBSS as an analog program for how urban natural hazards might be reduced. To a large degree, global stability will be tied to urban stability in the future. The cost in human suffering and capital outlays in response to natural disasters in urban areas is already extensive. A Science-Based Natural Hazard Reduction program that is wholly or partly patterned after SBSS (but with an additional factor of integrating across many agencies) would bring together a myriad of agencies, institutions, and disciplines with a common focus and a framework that would effectively guide work aimed at improving our ability to mitigate hazards before they happen, and to respond to them afterwards. A program in the U.S.A. that is funded at about the same order of magnitude as the cost of SBSS (about five billion dollars per year) would serve as a huge catalyst to integrate organizations that are now loosely coordinated, if at all, and could facilitate the exchange of techniques between large-scale disciplines like meteorology and seismology. The cost savings compared to the investment could potentially be quite large. The promotion of policies that would result in such a program is a pursuit that needs the advocacy of influential organizations like the American Geophysical Union and the U.S. National Academy of Sciences. The newly developing U.S. Department of Homeland Security potentially could be a nucleus for such a program.

COLLABORATION WITH OTHER URBAN DISCIPLINES

The need for Earth and atmospheric scientists to integrate with other urban disciplines has already been mentioned, but it is worth reinforcing this point. In order for our science to be most effective, we need to work closely with people whose expertise lies in the actual workings of cities. An example is power and transportation infrastructure. Recent studies by Heiken et al. [2000] provide an example of such collaboration. In this work, the authors incorporated the complex three-dimensional geology of the Los Angeles Basin into a first-principles seismic wave propagation numerical model. They used this model to predict ground motion over the basin for a variety of potential fault rupture scenarios. The results where overlaid on GIS-based data on the distribution of electrical power substations in the basin. Probabilistic ground motion fragility curves were obtained for all the substations. The predicted ground motions and the fragilities were convolved in a Monte Carlo type of approach to produce probabilistic predictions of the response of the power grid to scenario earthquakes. These predictions showed areas where varying degrees of power outage (brownouts and blackouts) are likely for each scenario. Such results can be used to help utilities prioritize their upgrades of substations and, in the event that an earthquake did occur, to know where to focus recovery efforts. In the end it is necessary for all natural hazards work to integrate with urban engineering in this manner.

CONCLUSION: THE HUMAN ELEMENT

This paper attempts to touch upon the importance of urban "natural" hazards research, drawing on volcanic hazards as a specific example. A framework for a more integrated approach to urban hazards is discussed, and a large-scale program is proposed that could be partly patterned after the defense-related Science-Based Stockpile Stewardship program. The need for integration with urban engineers is also highlighted.

In the end, reducing urban hazards is all about people – reducing the danger to their lives and livelihoods. While this discussion has focused on science and engineering approaches, the technical community must be open to the human element in such work. The human element is complex in ways that may be confusing to those of us who practice the scientific method. For example, a comprehensive natural hazards program may produce results that the people in a city do not want to hear. We may bring our best science to bear on a problem, integrated with urban engineers, and come up with a technically strong solution, only to find that cultural reasons prevent its implementation. These issues will never go away completely, and scientists and engineers need to be prepared to deal with them in constructive ways. The human side and the technical side can reduce misunderstanding and surprises between them by involving urban populations in the scientific work, even to the point of helping to prioritize it, at an early stage. This may seem an insurmountable problem for a city with several million people, but it still needs to be pursued. One mechanism that can facilitate the interaction of practically unlimited numbers of people in problem solving involves novel uses of the internet. Heiken et al. [2000] describe preliminary studies on the use of the internet as a mechanism for generating collective intelligence from large numbers of people in an interactive way that allows incorporation of scientific results into the "discussion." The use of such techniques, along with strong teaming with the social sciences, will be crucial in the future of reducing natural hazards to cities.

Acknowledgments. The thoughts presented in this paper have been aided in their development over the years from interactions with a number of people, especially Grant Heiken, Frank Perry, and Steen Rasmussen of Los Alamos National Laboratory, all of whom I would like to thank for many stimulating collaborations and discussions. Grant Heiken, Cathy Hickson, and Fouad Bendimerad all provided useful reviews, and Paula Geisik aided in preparing the manuscript.

REFERENCES

CRWMS M&O, (Civilian Radioactive Waste Management Systems Management and Operating Contractor), *Characterize Framework for Igneous Activity at Yucca Mountain, Nevada.* Document ANL-MGR-GS-000001 Rev 00, ICN 01, U.S. Department of Energy, Office of Civilian Radioactive Waste Management, 2000a.

CRWMS M&O, *Characterize Eruptive Processes at Yucca Mountain, NV. Document* ANL-MGR-GS-000002 Rev 00, U.S. Department of Energy, Office of Civilian Radioactive Waste Management, 2000b.

CRWMS M&O, *Dike Propagation Near Drifts*. Document ANL-WIS-MD-000015 Rev 00 ICN 01, U.S. Department of Energy, Office of Civilian Radioactive Waste Management, 2000c.

CRWMS M&O, *Igneous Consequences Modeling for the TSPA-SR*. Document ANL-WIS-MD-000017 Rev 00 ICN 01, U.S. Department of Energy, Office of Civilian Radioactive Waste Management, 2000d.

Fisher, R.V., Heiken, G., Hulen, J.B., *Volcanoes, Crucibles of Change*. Princeton University Press, Princeton, N.J.: 317, 1997.

Fuchs, R. J., Brennan, E., Chamie, J., Lo, F., and Uitto, J.I. (eds), *Mega-City Growth and the Future*. United Nations University Press, Tokyo, 428, 1994.

Heiken, G, Valentine, G.A., Brown, M., Rasmussen, S., George, D.C., Greene, R.K., Jones, E., Olsen, K., Andersson, C., Modeling Cities: The Los Alamos urban security initiative. *Public Works Manag. and Pol.*, vol. 4, 198-212, 2000.

Iverson, R.M., Schilling, S.P., Vallance, J.W., Objective delineation of lahar inundation hazard zones. *Geol. Soc. Am. Bull.* 110: 972-984, 1998.

Jeanloz, R., Science-based stockpile stewardship. *Phys. Today* 53: 44-50, 2000.

Lacroix, A., *La Montagne Pelee et ses Eruptions*. Masson et Cie, Paris, 662, 1904.

Lipman, P.W., Mullineaux, D.R., eds, The 1980 eruptions of mount St. Helens. U.S. Geological Survey Professional Paper 1250, 1981.

McCoy, F.W., Heiken, G., The late-bronze age explosive eruption of Thera (Santorini), Greece: regional and local effects. In *Volcanic Hazards and Disasters in Human Antiquity*, McCoy, F.W. and Heiken, G., editors. Geological Society of America Special Paper 345: 43-70, 2000.

Perry, F.V., Valentine, G.A., Desmarais, E., WoldeGabriel, G., Probabilistic assessment of volcanic hazard to radioactive waste repositories in Japan: intersection by a dike from a nearby composite volcano. *Geology*, 29: 255-258, 2001.

Valentine, G.A., Multifield governing equations for magma dynamics. *Geophys. Astrophys. Fluid Dyn.*, 78: 193-210, 1994.

Valentine, G.A., Damage to structures by pyroclastic flows and surges, inferred from nuclear weapons effects. *J. Volcanol. Geotherm. Res.*, 87: 117-140, 1998.

Valentine, G.A., Heiken, G., The need for a new look at cities. *Env. Sci. Pol.* 3: 231-234, 2000.

Valentine, G.A., Keating, G.N., Integration between data, simulation, and risk assessment: examples from other disciplines and implications for volcanology. *Eos Trans.* 81: F1254, 2000.

Valentine, G.A., Zhang, D.X., Robinson, B.A., Modeling complex, nonlinear geological processes. *Annual Rev. Earth Planet. Sci.*, 30: 35-64, 2002.

Voight, B., The 1985 Nevado-del-Ruiz volcano catastrophe: anatomy and retrospective. *J. Volcanol. Geotherm. Res.* 42: 151-188, 1990.

World Resources Institute, *World Resources—1996-1997* (special issue on the urban environment). Oxford University Press, Oxford, 1-156, 1996.

Young, S.R., Sparks, R.S.J., Aspinall, W.P., Lynch, L.L., Miller, A.D., Robertson, R.E.A., Shepherd, J.B., Overview of the eruption of Soufriere hills volcano, Montserrat, 18 July 1995 to December 1997. *Geophys. Res. Lett.* 25: 3389-3392,1998.

5

Seismic-Risk Evaluation in Cities of New York and Surrounding Regions: Issues Related to all Intraplate Cities

Robert H. Fakundiny

INTRODUCTION

Large earthquakes occasionally occur in the interiors of tectonic plates even though these regions are considered to be tectonically stable (Johnston, 1989; Kanter, 1994). New York State and surrounding regions (NYSSR) lie within one such tectonically stable region in the central part of the North American tectonic plate (Sbar and Sykes, 1973; Muehlberger, 1992) and, like many other intraplate regions, have experienced moderately sized earthquakes. The seismic hazard of plate interiors, consequently, has been estimated to be at least half as large as that at some plate-edge areas, such as Oregon and Washington (FEMA, 2001b). Discussions of potential damage to humans, constructed facilities, and economic endeavors commonly use the terms *seismic hazard* and *seismic risk*, but rarely define these two terms. Definitions in the literature vary, yet commonly follow the general concept that *hazard* refers to the danger, and *risk* is the probability of loss if that danger is manifested. In the following discussion, *seismic hazard* represents an estimate of the probability that a site or area will be affected by the energy and seismic-wave properties of potential earthquakes generated at given distances and during future time intervals, as quantified by either deterministic or probabilistic analyses, whereas *seismic risk* represents an estimate of the cost of property damage, economic recovery, and human injury resulting from the probable earthquake that was estimated in the hazard calculation.

Seismic hazard and seismic risk in intraplate areas are difficult to evaluate because the calculations have large implicit uncertainties (Basham and Adams, 1989; Algermissen, 1997; Donovan and Bornstein, 1997). Risk assessments for

Earth Science in the City: A Reader

10.1029/056SP06

cities that lie within the central regions of tectonic plates, where data on earthquake-generation mechanisms are sparse and, thus, incompletely understood, commonly rely on conservative estimates that anticipate large seismic events—those of magnitudes greater than 6.0 (M[1] > 6.0). Some measure of seismic risk must, therefore, be factored into the construction and retrofitting of those facilities that must remain functional after an earthquake, referred to herein as *critical* facilities and *lifelines*[2] (Hall, 1980). Current calculations of seismic risk within the NYSSR rely on incomplete knowledge about earthquake generation, seismic-wave transmission through the lithosphere[3], local geologic conditions where the shaking occurs, and building-construction characteristics. Earthquakes throughout the NYSSR are unevenly distributed in space and time, and they vary widely in size (Ebel and Kafka, 1991; Johnston, 1994). A thorough knowledge of the geologic and tectonic framework of the NYSSR, which is not yet attained, would aid in the understanding of the mechanics of earthquake generation in this region and in quantifying the associated seismic risks.

The current incomplete understanding of earthquake mechanisms and of their controls on their geographic distribution requires that public policies be precautionary (Anderson, 2001) and assume that earthquakes with M > 6.0 could occur in the NYSSR at any time. Adequate mitigation and remediation programs require: (1) a refined understanding of the region's geologic framework, (2) installation and operation of an advanced, national seismic-monitoring system; (3) refined models of the propagation of seismic waves through the Earth's lithosphere; (4) implementation of appropriate seismic building-code provisions; and (5) public education in remediation procedures. Only when these conditions are met might the seismic risks within the NYSSR be minimized. A corollary is

---

[1] In this paper the letter M is the symbol for "generalized magnitude." Many magnitude scales are used in the quantification of earthquake size, as discussed further on. Generalized magnitudes may be derived from many sources and be given a value by averaging specific magnitudes, by comparison with magnitude-intensity relations, and by other techniques. Most magnitude values for earthquakes prior to the 1920s are probably converted from intensities. Many lists of magnitudes of historical earthquakes do not provide the calculation procedure for deriving the magnitudes given and, therefore, must be considered generalized. The magnitudes designated M herein have values close to or equivalent to what might be considered a "Richter magnitude."

[2] "Critical" or essential facilities include hospitals, schools, police and fire stations, and emergency-operation centers, nuclear and conventional powerplants, dams, military installations, chemical-processing plants, offshore petroleum facilities, and liquefied natural-gas terminals (Hall, 1980; FEMA, 2001b). "Lifelines" include highways, bridges, gas and water pipelines, and power-transmission lines (Whitman, 1989).

[3] "Lithosphere" is used here to mean that part of the upper layers of the Earth that has higher shear strength than the asthenosphere below. It includes all of the crust and the upper part of the mantle.

that seismic risks could be greatly reduced if the processes that create earthquakes were sufficiently understood to allow development of a prediction technique to give adequate warning of impending seismicity.

Factors and Definitions Used in Seismic-Risk Evaluations

The amount of damage that any specific earthquake causes to a structure is a function of several factors: (1) location of the earthquake; (2) the size of the earthquake; (3) the earthquake's *focal mechanism*—that is, the characteristics of the source fault's orientation and displacement; (4) the efficiency of transmission of the various types of seismic waves through the Earth from the source to the affected site; (5) the recurrence rates of variably sized earthquakes; (6) the effects of geologic-foundation conditions and local geology in modifying the amplitudes of impinging seismic waves; and (7) the effects of the structure's design and quality of the construction on its response to seismic-waves. The seismic hazard at a specific site is a measure of items 1 through 5 of this list and is, thus, a calculation of the seismic-wave energies and frequency distributions expected to arrive at a site from hypothetical earthquakes placed at distant possible epicentral locations and assigned the largest credible magnitudes. The seismic risk at a site is a function of the hazard multiplied by a factor that attempts to quantify the vulnerability of the constructed facilities and their contents (Jacob, 1993), including humans and, thus, is a measure of the total damage expected from hypothetical earthquakes with assigned hazard values. Indirect damages, such as economic losses to those sectors of the community that did not sustain direct damage (Cochrane, 1997) must also be factored into the risk evaluation. Seismic-risk evaluations provide a rationale for mitigation and remediation at either individual building sites or over larger developed areas, and they are needed in the NYSSR in preparation for future earthquakes.

An illustration of seismic risk in one part of the NYSSR has been provided by the U.S. Federal Emergency Management Agency (FEMA), which estimates that $84 million in building stock is lost on average each year in New York State from earthquakes alone (HAZUS earthquake-loss estimate cost by FEMA, 2001b). This estimate applies mainly to the New York City-Northern New Jersey metropolitan area because of the dense population and high concentration of buildings that lie within an *epicenter cluster*—an area of historic seismicity that exceeds the region's average. The estimates of earthquake-loss costs in other large cities in the NYSSR that are within clusters of historic seismicity can be expected to be proportional to the ratio of their population to that of New York City. Prediction of the occurrence of specific future earthquakes is not currently within the capability of science and engineering (Press, 1968), but the four steps toward resistant design of buildings—prevention, mitigation, remediation, and the development of real-time response to earthquakes—are feasible, as discussed below and by Jacob and Turkstra (1989).

Two terms that are ubiquitous in discussions of earthquakes and earthquake risk are *hypocenter* and *epicenter.* The *hypocenter,* or earthquake *focus,* is the location of breakage initiation in the rock at depth where strain energy is first converted to elastic-wave energy. The *epicenter* is the location at land surface directly above the hypocenter (Bates and Jackson, 1980).

Purpose and Scope

This paper briefly discusses some of the factors used in making seismic-risk evaluations and explains why these factors are difficult to quantify. It also reviews: (1) the geologic and tectonic framework of earthquakes in the NYSSR; (2) the temporal and spatial distribution of earthquakes in this region; (3) the difficulty in accurately estimating earthquake-epicenter locations; (4) methods of seismic-hazard evaluation; (5) seismic risks in the NYSSR; (6) methods of earthquake prediction and prevention; (7) aspects of damage mitigation; (8) aspects of damage remediation; and (9) political responses to risk estimates. This discussion assumes a rudimentary knowledge of the basic principles and vocabulary of geology. A glossary of earthquake terms can be found in von Hake (1975). The references herein, although not comprehensive, serve as a partial bibliography of seismic hazard- and seismic-risk evaluations in the NYSSR.

The area under discussion includes the northeastern United States and eastern Canada between the latitudes 40° - 46° N and longitudes 72° - 80° W. The locations of the major cities of the NYSSR are depicted in Figure 1. Epicenters of many of the earthquakes with $M > 2.5$ are depicted in Figure 2. Epicenters of some of the most notable earthquakes with $M \geq 3.0$ or $I > V$ for New York State and $M \geq 5.0$ for surrounding regions are listed in Table 1 and depicted as stars in Figure 2. The major facilities that are used in measuring seismic risk are the infrastructure, critical facilities, lifelines, and building stock within and among these cities. Nuclear facilities include the Pickering and Darlington nuclear powerplants on the north shore of Lake Ontario east of Toronto, the Nine Mile 1 and 2 and James A. Fitzpatrick nuclear powerplants on the southeast shore of Lake Ontario at Oswego, NY, the Robert E. Ginna nuclear powerplant on the south shore of Lake Ontario east of Rochester, NY, the Indian Point 1, 2 and 3 nuclear powerplants on the Hudson River 40 km north of New York City, and the West Valley Nuclear Service Center 60 km southeast of Buffalo, NY with its radioactive-waste burial grounds and high-level liquid radioactive-waste solidification plant. Several reactor-testing facilities also reside in New York State.

GEOLOGIC SETTING

The NYSSR lies within the North American tectonic plate, which extends from the Mid-Atlantic ridge westward as oceanic crust to the North American eastern continental seaboard, from where continental crust continues its exten-

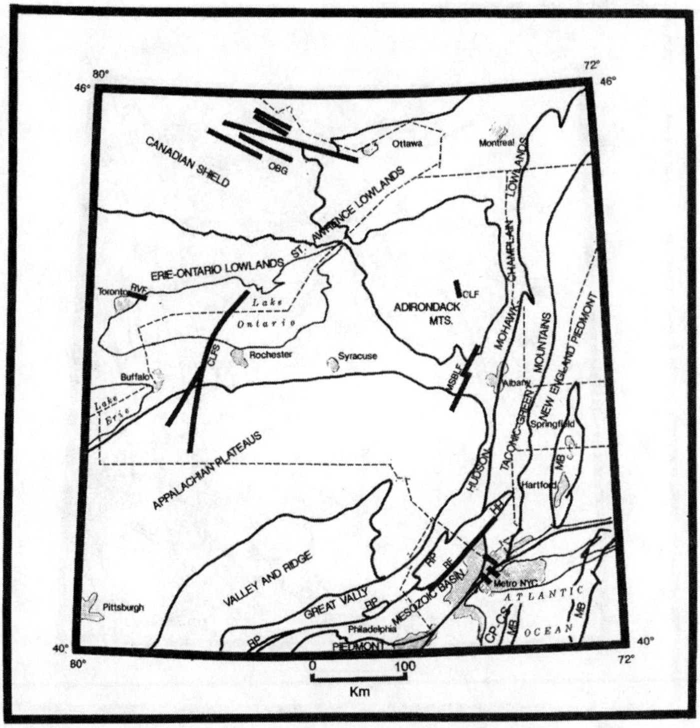

Figure 1. Tectonic map of New York State and surrounding regions with possible active faults. Heavy lines are possible seismogenic faults mentioned in the text: CLF=Catlin Lake fault; CLFS=Clarendon-Linden fault system; MF=Mosholu fault; MSBLF=McGregor-Saratoga-Ballston Lake fault system; 125F=125th Street fault; OBG= Ottawa-Bonnechure graben; DFF-Dobbs Ferry fault; RF=Ramapo fault system; RVF=faults of the Rouge River Valley. Tectonic province abbreviations: CP-CS=Coastal Plain-Continental Shelf; HH=Hudson Highlands; MB=Mesozoic Basins; RP=Reading Prong

sion west to the transform faults and subduction zones that separate it from the Pacific and others oceanic plates. Proterozoic and Phanerozoic geologic units form the upper part of the continental crust in the NYSSR. The plate-tectonic theory requires added data on the history of formation and structural framework of the lithosphere of eastern Canada and the United States before is can be augmented to explain intraplate-earthquake generation.

Plate Tectonics and Tectonic Provinces

The plate-tectonics concept is the scientific rationale currently used by most geologists for the study and evaluation of earthquakes and seismic hazard. Plate tectonics places all crustal geologic processes, including seismicity, in a context

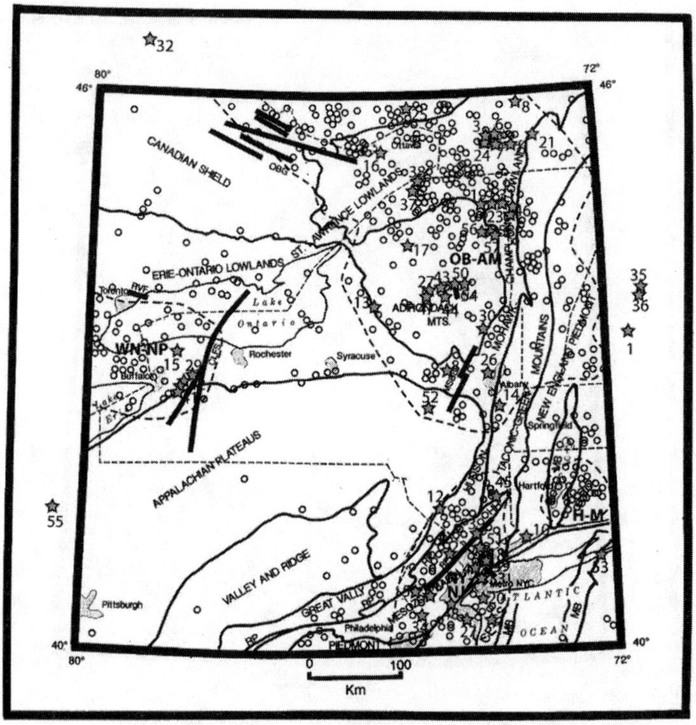

Figure 2. Map of some notable earthquakes in New York State and surrounding regions. Abbreviation of zones of earthquake clusters, which have a dashed line around them: H-M=Haddam-Moodus; NY-NJ=New York-New Jersey greater metropolitan area; OB-AM=Ottawa Bonnechere graben-Adirondack Mountains; WN-NP=Western New York-Niagara Peninsula. Large stars with numbers are those listed in Table 1. Small circles are earthquakes with M>2.5.

of semirigid lithospheric plates that move away from each other to form long rifts, toward each other at collision and subduction zones, or past each other along transform faults. Faults along these plate borders are frequently the sources of earthquakes, and their mechanisms of generation are fairly well understood. Mechanisms for earthquake generation within the central parts of plates, however, are not well understood and may be operating under a set of tectonic conditions different from those that exist at plate boundaries. Therefore, accurate seismic-hazard evaluations are more difficult to make within the plates, such as in the NYSSR and other similar intraplate areas, than at plate boundaries.

The concept of *tectonic provinces*—areas within a given plate that have similar structural styles and tectonic histories—is a common geological basis for the evaluation of seismic characteristics at a site. The tectonic-province concept is

Table 1. Notable Earthquakes in the New York and Surrounding Regions:
M > 5 (Canada); M ≥ 3 or I > V (New York)[1]

No.[2]	Place	Date	Size	Damage in NY
1.	Central NH	06/11/1638	6.5	Damage uncertain, wide spread shaking
2.	South of Granby, Que.	02/10/1661	VII	
3.	Montreal, Que.	09/16/1732	VIII	
4.	New York, NY	12/19/1737	5.2	Bells rang, several chimneys fell
5.	Morris Counrt, NJ	11/30/1783	VI	
6.	Montreal, Que.	09/09/1816	VII	
7.	Montreal, Que.	09/16/1816	VI	
8.	Herkimer, NY	01/16/1840	VI	No reports of damage
9.	Trois Rivieres, Que.	11/09/1842	VI	
10.	Bridgeport, CY	10/26/1845	VI	
11.	Offshore NY City	no date/1847	VI	No reference to reports of damage
12.	Rockland Lake, NY	09/09/1848	V	Felt by many
13.	Lowville, NY	03/12/1853	VI	Machinery knocked down
14.	Saugerties(?[3]), NY	02/07/1955	VI	Cryoseism?
15.	Buffalo (Lockport?), NY	10/23/1857	VI	Bells rang, crocks fell from shelves
16.	Ottawa, Ont.	07/13/1861	VII	
17.	Canton, NY	12/18/1867	VI	Sleepers awakened
18.	Terrytown, NY	12/11/1874	VI	
19.	Lyon Mt. (?), NY	11/04/1877	VII	Chimneys down, walls cracked, windows damaged
20.	New York Bight	08/10/1884	5.2	Chimneys and bricks fell, walls cracked
21.	East of Montreal, Que.	11/27/1893	VII	
22.	High Bridge, NJ	09/01/1895	VI	
23.	Dannemora, NY	05/28/1897	VI	No damage reported
24.	Montreal, Que.	03/23/1897	VII	
25.	NE of Adele, Que.	02/10/1914	VII	Water pipes broken in Canton, NY, objects thrown from shelves in Syracuse, NY
26.	Schenectady, NY	02/03/1916	VI	Windows broken, people thrown out of bed
27.	Fairhaven, NY	06/01/1927	3.9	
28.	Saranac Lake, NY	03/18/1928	VI	No damage reported
29.	Attica, NY	08/12/1929	5.2	250 chimneys fell, brick buildings damaged, Attica Prison wall cracked, wells went dry
30.	Warrensburg, NY	04/20/1931	VII	Chimneys down, church spire twisted
31.	Dannemora, NY	04/15/1934	VI	House shifted

[1]Sources: Smith, 1962, 1966; Nottis, 1983 and sources therein; Stover and Coffman, 1993; Wheeler and others, 2001; Won-Young Kim, Lamont-Doherty Earth Observatory, written communication, 2002.
[2]Number refers to numbered epicenters on Figure 2.
[3]Question mark indicates that the location is unclear.

Table 1. (Continued) Notable Earthquakes in the New York and Surrounding Regions: M > 5 (Canada); M ≥ 3 or I > V (New York)

No.	Place	Date	Size	Damage in NY
32.	Timiskiming, Que.	11/01/1935	5.9	Heavy damage, chimneys damaged Cortland, NY
33.	Brooklyn (?), NY	07/19/1937	3.5	No damage reported
34.	Trenton, NJ	08/23/1938	3.8	
35.	Ossipee, NH	12/20/1940	5.5	
36.	Ossipee, NH	12/24/1940	5.5	
37.	Massena, NY	09/05/1944	5.8	Nearly all chimneys fell, buildings damaged, $2 million damage
38.	Massena, NY	09/05/1944	4.5	Chimneys destroyed, houses damaged
39.	Rockland County, NY	09/03/1951	3.6	No damage reported
40.	Schooly Mt., NJ	03/23/1957	3.5	
41.	Attica, NY	01/01/1966	4.6	Chimneys and walls damaged
42.	Attica, NY	06/13/1967	3.0	Chimneys and walls damaged
43.	Blue Mt. Lake, NY	05/23/1971	3.6	No damage reported, numerous small aftershocks
44.	Blue Mt. Lake, NY	05/23/1971	3.4	No damage reported
45.	Wappingers Falls, NY	06/07/1974	3.8	Windows broken
46.	Plattsburgh (Altona), NY	06/09/1975	4,2	Chimneys and fireplaces cracked
47.	Raquette Lake, NY	11/03/1975	3.9	No damage reported
48.	Barnardsville, NJ	03/10/1979	3.2	
49.	Scarsdale, NY	02/26/1983	3.0	Chimneys cracked
50.	Newcomb, NY	10/07/1983	5.1	Tombstones rotated, chimneys cracked
51.	Ardsley, NY	10/19/1985	4.0	Windows broken, walls damaged
52.	Richmondville, NY	06/17/1991	4.0	No damage reported
53.	East Hampton, NY	03/10/1992	4.1	No damage reported?
54.	Newcomb, NY	04/20/2000	3.8	Aftershock of 1983 event?, no damage reported
55.	Pymatuning Res., PA	09/25/1998	5.2	Some chimneys fell, walls damaged
56.	Au Sable Forks, NY	04/20/2002	5.3	Cracked walls, chimneys fell, road collapse, power outages, landslides
57.	Au Sable Forks, NY	04/20/2002	4.0	Aftershock of Au Sable Forks event 14 minutes later
58.	Au Sable Forks, NY	05/24/2002	3.1	Aftershock of Au Sable Forks events of 04/24/2002, no damage reported

based on the geographic distribution of tectonic terrains, each with a different set of regional structures and geologic histories. For the eastern United States and eastern Canada these terrains formed during the collisional and extensional episodes of the Proterozoic and Phanerozoic. One delineation of possible tectonic provinces within the NYSSR is depicted in Figure 1. Comparison of the earthquake epicenter patterns (Figure 2) with the tectonic provinces (Figure 1) reveals little obvious spatial correlation, except perhaps for the seismicity along the Ottawa-Bonnechere graben (Kay, 1942; Adams and Basham, 1991). Earthquakes north of the Ottawa-Bonnechere graben do not appear to be associated with any recognized crustal structure, however (Adams and Basham, 1991). Each tectonic province is considered by many geologists to have its own unique seismic and tectonic character and, thus, requires individual consideration in seismic-hazard evaluations (Yang and Aggarwal, 1981). The clusters of seismicity in Figure 2 appear either to be restricted to certain parts of a tectonic province or to straddle tectonic-province boundaries; this implies that the tectonic-province approach may not lead to reliable seismic-hazard evaluations (Hall, 1980). Analysis of the spatial relations between hypocenters and brittle faults in the basement, however, may explain the distribution of seismicity in the NYSSR better than the traditional tectonic-province approach (Fakundiny and Pomeroy, 2002).

Geology of the NYSSR

Geologic units of the NYSSR can be divided into six major types that were emplaced during five distinct geologic episodes: (1) Proterozoic (Grenville) high-grade metamorphosed crystalline rock that forms the basement; (2) Paleozoic unmetamorphosed sedimentary rock and igneous intrusions; (3) Paleozoic mildly to strongly metamorphosed sedimentary and igneous rock; (4) Mesozoic sedimentary and igneous rock; (5) Mesozoic and Tertiary sediments, and (6) Quaternary glacial (Pleistocene) and postglacial (Holocene) deposits. Each geologic episode had its own tectonic history, and each type of material responds uniquely to regionally imposed stresses.

The regional geology of the NYSSR can be summarized by geologic province and geologic age. The Grenvillian-aged Canadian Shield and Proterozoic-Paleozoic-aged Piedmont Provinces have high-grade metamorphic rocks, which are exposed north of Lake Ontario, in the Adirondack Mountains of New York, in the Green Mountains of Vermont, and within the Piedmont and Reading Prong of Pennsylvania, New Jersey, and the Hudson Highlands of New York. Unmetamorphosed Paleozoic sedimentary strata lie on the basement in the three interior lowlands, and in the Appalachian Plateau, the Appalachian Valley and Ridge Province, and the Great Valley Province. Unmetamorphosed sedimentary and igneous rocks fill faulted Mesozoic basins that formed during the opening of the modern Atlantic Ocean. The Atlantic Coastal Plain and Continental Shelf host Mesozoic to Holocene sediments deposited on older metamorphic and sedimentary rocks and oceanic basalts. Quaternary glacial deposits cover northern Pennsylvania and most of the region to the north.

Basement Structure of the NYSSR

The ~10- to 20-km depths of most hypocenters (Acharya, 1980) suggests that seismically active faults of the NYSSR are primarily within the Proterozoic crystalline basement and are mostly obscured by overlying Paleozoic strata or glacial deposits, although some may extend upward into Phanerozoic rocks. These faults and fault systems are thought by some researchers to lie along major anomalies in the regional aeromagnetic and gravity fields (Forsyth and others, 1994a, b). Areas with differing regional geophysical characteristics have been proposed to represent different basement geologic terranes (Fakundiny, 1981; Sanford and others, 1985; Forsyth and others, 1994a, b; Fakundiny and Pomeroy, 2002). Possible seismically active faults (Table 2) border and cut Paleozoic strata in Ontario, within and around the Adirondack dome and the Piedmont of southeastern New York (Dawers and Seeber, 1991), as well as strata of the Mesozoic rift basins (Ratcliffe, 1971). Faults without precise spatial correlation with historic earthquakes (within the last 5 centuries) are assumed by most seismologists to be inactive.

Detection of Possible Active Faults

A primary geologic concern in earthquake-risk evaluations is the history of seismic activity along specific faults (Muir Wood and Mallard, 1992). Faults within the NYSSR have been seismically active over a wide range of periods that may exceed 1 billion years (Fakundiny and others, 1978a). The term *active fault* refers to faults with a history of Holocene seismicity and the potential to host future earthquakes; but applying the concept of active fault to seismic-hazard analysis in the NYSSR may be inappropriate in areas where the age of past activity cannot be measured. The term *modern history* of seismicity refers to the time since Europeans started to colonize the continent in the 16th Century. The common opinion of most seismologists is that active faults pose seismic hazards, whereas inactive faults probably present no future danger.

The instruments of most seismic-station networks in the NYSSR are placed at such large distances from one another that the calculated locations of earthquakes may be more than 5 km from the actual epicenter (Mohajer, 1993; Seeber and Armbruster, 1993). One exception is the Southern Ontario Seismic Network, which was established in 1991 and can locate epicenters within ±2 km and can measure depths to hypocenters that are within 25 km of a station with accuracies of generally ±7 km (Mereu and others, 2000).

The inability of most current seismic networks to locate epicenters or hypocenters with enough accuracy to spatially correlate them with faults makes the process

Table 2. Possible Active Faults in the NYSSR

Fault or Fault System	Symbol	Reference
125th Street Fault (Manhattan, NY)	125F	Baskerville (1994)
Mosholu Fault (Manhattan, NY)	MF	Baskerville (1994); Merguerian (1996); Merguerian and Sanders (1997)
Dobbs Ferry Fault	DFF	Dawers and Seeber (1991)
Ramapo Fault System (Hudson Highlands)	RF	Page and others (1968); Ratcliffe (1971); Aggarwal and Sykes (1978)
McGregor-Saratoga-Balston Lake Fault System	MSBLF	Isachsen and McKendree (1977)
Ottawa-Bonnechere Gragen (Ontario-Quebec border)	OBG	Kumarapeli and Saull (1966)
Rouge Valley fault (Toronto, Ont.)	RVF	Mohajer and others (1992)
Clarendon-Linden fault system (Western New York	CLFS	Chadwick (1920) Van Tyne (1975)

of identifying active faults in the NYSSR difficult to impossible (Mohajer, 1993; Eyles and others, 1993; Stevens, 1995). The exception is where portable seismographs are installed in the field above a seismically active fault that generates aftershocks or regular microseismicity (Dawers and Seeber, 1991). The alignment of aftershock hypocenters of the 1983 Newcomb (Goodnow) earthquake with the Catlin Lake lineament (CLF) in the central Adirondack Mountains (Figure 1) (Dawers and Seeber, 1991) strongly indicates that the lineament is associated with an active fault. Field studies have provided data about the history of movement on a fault, such as study of past sediment disruption near faults (Tuttle and Seeber, 1991; Tuttle, 1996; Tuttle and others, 1996). Spatial correlation of an earthquake with a fault does not necessarily prove that the fault is active, however. Establishing which faults are active requires either direct observation of movement, indirect evidence of movement during the Holocene, or plotting of hypocenters that are unequivocally aligned in three dimensions with the fault surface and with focal mechanisms that are consistent with possible movement on the fault plane. Several structures in the NYSSR that are visible at land surface or are inferred at depth from regional geophysical-field data have been nominated for active status, although none meet the first of the two requirements mentioned above, direct observation of movement. The Clarendon-Linden fault system (CLFS) in western New York (Chadwick, 1920; Van Tyne, 1975; Fletcher and Sykes, 1977; Fakundiny and others, 1978b; Hutchinson and others, 1979; Pomeroy and others, 1978; Jacobi and Fountain, 1993; Fakundiny and Pomeroy, 2002) is an example of a tectonic structure that has a long history of spatially associated activity that may have continued for the last 1.3 billion years (Fakundiny and others, 1978a). Other faults or fault systems (Figure 1) that have been suggested to be currently active are listed in Table 2. Modern mapping of these structures in New York began with the work of Oliver and others (1970), who confirmed the need for continued field studies and mapping to identify areas of potential seismic risk (Davis and others, 1979). Some seismologists think, however, that no seismically active structure has been unequivocally identified in the NYSSR (Wheeler and Johnson, 1992; Stevens, 1995).

Seismically Active Areas

Steady and apparently randomly dispersed, low-level seismicity has been recorded in the NYSSR since the first colonists arrived in the 16th century. Paleoseismic studies suggest that this general constant seismicity holds for the entire Holocene. Some areas in the NYSSR have experienced higher seismic activity than average for the region; the epicenters in these areas are distributed in geographic clusters (Smith, 1962, 1966; Nottis, 1983; Ebel and Kafka, 1991; Gordon and Dewey, 1999). Some examples are: (1) along the Ottawa-Bonnechere graben and northern Adirondack Mountains (OB-AM) and north of that line (Adams and Basham, 1991); (2) western New York and the Niagara Peninsula of Ontario (WN-NP) (Seeber and Armbruster, 1993); (3) southeastern New York and northern New Jersey (NY-NJ) (Ebel and Kafka, 1991); and (4) Haddam-Moodus in southeastern Connecticut (H-M) (Nottis, 1983) (Figure 2).

Broad linear zones of seismicity have been proposed also. Although not apparent in Figure 2, these include: (1) the St. Lawrence River linear zone in the St. Lawrence Lowlands (Kumarapeli and Saull, 1966); (2) the Niagara-Pickering linear zone and Hamilton-Lake Erie linear zone, each trending north-northwest through eastern Lake Erie and western Lake Ontario (Thomas and others, 1993; Wallach and others, 1998); and (3) the Georgian Bay linear zone between Georgian Bay and western Lake Ontario (Wallach, 1990; Wallach and Mohajer, 1990; Eyles and others, 1993).

Locations of Some Notable Earthquakes

Some notable earthquakes recorded within the NYSSR in modern time ($M > 5.0$ for Canada, $M \geq 3.0$ for New York State, or $I > V$) are listed in Table 1. The largest seismic event within the NYSSR was the 1944 earthquake at Cornwall, Ont. and Massena, NY, with M = 5.8; the second largest is the 2002 event (M = 5.3) at Au Sable Forks in the northwestern Adirondack Mountains. Five other large earthquakes that have occurred within 100 km of the region, as depicted in Figure 2, are the 1638 (M = 6.5) earthquake in central New Hampshire, the 1935 (M = 6.3) event near Timiskaming, Que.; the two 1940 (M = 5.5) shocks at Ossipee, NH; and the 1998 (M = 5.2) event at Pymatuning Reservoir, PA (Smith, 1962, 1966; Nottis, 1983 and sources therein; Wheeler and others, 2001). None of these surrounding earthquakes have been spatially associated with faults accurately enough to assign those faults to the active category.

Associated Geologic Hazards

The amount of shaking at any given site during an earthquake depends on the mechanical properties of both the bedrock and any overlying unconsolidated sediments (McGuire and Toro, 1989), as well as the characteristics of the earthquake and attenuation of seismic waves. Geologic foundation materials with contrasting engineering properties, such as hard bedrock or soft saturated sand, will respond to the different seismic waves in different ways. For example unconsolidated sediments and thixotropic clays, especially those that are saturated at a site, might amplify some incoming seismic waves much more than solid rock.

Saturated sediment may take on fluidlike properties, or liquefy, when shaken, much like saturated sand on beaches when stamped on; this can result in damage to structures built on them when shaken by an earthquake (Dobry, 1989). A catastrophic example of damage to buildings over liquefied sediments was the destruction of the Turnagain Heights suburb of Anchorage, Alaska, during the Good Friday earthquake of March 27, 1964 (Bolt, 1978).

Other associated or *collateral* hazards that may accompany the effects of shaking include landslides, changes in groundwater, release of subterranean gasses, ground collapse, fires, and floods. Landslides can present a hazard where seismic shaking dislodges unstable surficial material on steep slopes or loose rock on cliffs. Motion-picture depictions of great chasms opening along faults during earthquakes and swallowing men and machines before closing are erroneous, however; such chasms do not commonly appear along faults, even during great earthquakes. The concept probably originated from observations of cracks and larger openings forming at the heads of earthquake-triggered landslides. Earthquakes also can muddy well water and disrupt groundwater levels, release gasses along fault lines, initiate ground collapse over cavities or loosely compacted soils, start fires at broken powerlines or damaged gas pipelines, cause floods from failed dams and levees, and break water mains, among other hazards (Bolt, 1978). The tsunami, or seismic sea wave, is a geologic hazard associated with near-coast and undersea earthquakes that can devastate entire coastal cities (FEMA, 1995; Werner and others, 1997).

EARTHQUAKE CHARACTERISTICS

Earthquakes can be characterized in terms of several properties; those that are most commonly considered in seismic-hazard evaluation are earthquake focal mechanisms, size (magnitude or intensity), seismic-wave propagation characteristics, and recurrence rates. The parameters of these earthquake characteristics are factored into the seismic-hazard measurement; thus, the seismic-hazard evaluation can be only as accurate as the values assigned to these parameters.

Focal Mechanisms

Earthquakes (or more precisely "tectonic earthquakes") originate where differential stresses within the Earth overcome the rock's internal friction or strength sufficiently to create a break (Bolt, 1978; Mitronovas, 2000) and converts strain energy to elastic energy (Bolt, 1978). The shaking and tremors that are perceived during an earthquake are the vibrations produced as the rock breaks, which are transmitted as seismic waves through the Earth and along the land surface. Slippage along the fault can occur in one of four ways: (1) horizontally as "strike-slip" motion; (2) vertically as "reverse-slip" motion where the block above the fault moves upward; (3) vertically as "normal-slip" motion where the block above the fault moves downward; or (4) a combination of these. Each type of slip motion broadcasts its own distinctive pattern of seismic waves that will affect distant sites uniquely. The basic generalized concepts of faulting and seismology are given by Bolt (1978) and Smith and others (2001).

Earthquakes can originate deep within the Earth (at depths of 600 km or greater) at plate boundaries, but within the NYSRR they generally occur at depths of <20 km (Acharya, 1980; Ebel and Kafka, 1991). Earthquakes far from subduction zones or transform faults can be triggered in many ways, but always occur in response to differential stress within brittle rock. The favored current concept is that most earthquakes in the NYSSR are triggered by the buildup of differential stress on faults in response to a regionally imposed stress system that is oriented ENE in most of the area (Sbar and Sykes, 1973, 1977; Lo, 1978; Adams and Basham, 1991; Adams and Bell, 1991; Zoback and Zoback, 1989, 1991). Other possible causes include: (1) unnaturally high hydrostatic pressure produced by water-reservoir loading that overcomes the frictional stress along faults and, thus, allowing them to slip (Talwani and Acree, 1984/1985; Costain and others, 1987); (2) the industrial injection of fluids under high pressure into a differentially stressed region of rock (Mereu and others, 1986; Wesson and Nicholson, 1987: Nicholson and others, 1988; Seeber and Armbruster, 1993), as occurred at Dale, NY in 1992 (Fletcher and Sykes, 1977) and possibly at Avoca, NY in 2001 (Price, 2001), or in some producing gas fields, such as in Uzbekistan (Plotnikova and others, 1992); (3) glacial rebound that disturbs the regional stress field (Quinlin, 1984; James and Morgan, 1990); and (4) the unloading of rock in the excavation of deep quarries (Pomeroy and others, 1976; Seeber and others, 1998).

Most earthquakes in the NYSSR occur on faults that have reverse-slip movement, strike-slip movement, or combinations of the two. Controversy and debate continue as to whether each earthquake results from the formation of a new fault or occurs on old faults that have become reactivated through time (Prucha, 1992). One hypothesis, among many, for the western part of the NYSSR is that earthquakes occur at the intersections of faults, within the regional stress field, that have orientations that accommodate the formation of stress-concentration points (Barosh, 1990; Talwani, 1998). Many earthquake catalogues list historic seismic events and evaluate whether they are real or "pseudo" earthquakes[4] (Smith, 1962, 1966; Nottis, 1983; Seeber, 1991; Stover and Coffman, 1993).

Magnitude

Earthquake size can be measured by the energy that is released, the *magnitude (M)*. Magnitude is a logarithmic measure of released earthquake energy that is calculated from the seismic-wave amplitudes recorded on a seismogram. The concept of earthquake magnitude was introduced by C. F. Richter and is commonly called the "Richter scale of magnitude." Richter's calculation does not apply well to the eastern United States, however, because the crust in the East has different seismic-wave-energy attenuation than the west. Thus, the magnitudes calculated from eastern data are commonly assigned some designation other than Richter Magnitude, such as moment magnitude (M_m), which is independent of the type of seismograph used or the regional geology. The amplitudes of the body and surface waves (explained further on) at large distances from the epicenter can be used to calculate the size of earthquakes; this magnitude is commonly reported as m_b or M_s, respectively (Chung and Bernreuter, 1980). Many seismic parameters are not incorporated into the magnitude calculation, such as the nature of the source mechanism or variations in attenuation because of the long distances from source to measuring site and the incomplete understanding of the lithosphere's structure (Duda, 1978).

Intensity

An earthquake's size can also be measured by an interpretation of the effect it has on people, structures, and the natural environment, the *intensity* (I). Intensity is a descriptive classification of ground shaking or damage caused by earthquakes at a site or as felt by individuals. The most commonly used intensity scale is the

[4] Pseudo earthquakes are shocks that are created by means other than the breakage of rock under regional differential stresses; some examples are quarry or construction dynamite blasts (Scharnberger, 1991; Brady, 1992; Mitronovas, 1997), acoustic waves through the air, such as sonic booms, rock bursts in tunnels or on quarry floors, breakage of lake or river ice, or cryoseisms (frost quakes), which are shocks from breaking of frozen ground during freeze-thaw conditions (Mitronovas, 1991; Barosh, 2000).

Table 3. Rough Comparison of Earthquake Rating Scales in Terms of Energy Released[1]

MMI[2]	Mag	V	A	TNT	Witnessed Observations
I	1-2	-	-	6 oz.	Not felt accept by very few in exceptional circumstances.
II	2-3	-	-	<400 lbs	Felt by only a few persons, especially in upper floors of buildings; delicately suspended objects may swing.
III	3-4	-	-	<6 tons	Felt quite noticeably indoors, especially on upper floors, but many people do not recognize it as an earthquake; standing motor cars may rock slightly; vibration like a passing truck.
IV	4->4	1-2	0.015-<0.03g	<32 tons	Felt indoors during the day by many, outdoors by few; some awakened; dishes, windows, and doors disturbed; walls creak; sensation like a heavy truck striking the building; standing motor cars rocked noticeably.
V	>4-5	2-5	0.03-0.04g	<200 tons	Felt by nearly everyone; many awakened; some dishes, windows broken; cracked plaster; unstable objects overturned; trees and poles disturbed; pendulum clocks may stop.
VI	5-<6	5-8	>0.04-0.07g	<1000 tons	Felt by all; many frightened; some heavy furniture moves; damaged chimneys; buildings slight damaged.
VII	<6->6	8-20	>0.07-0.15g	<32,000 tons	Everyone runs outdoors; damage negligible in buildings of good design and constructions; some chimneys broken, noticed by persons driving.

[1]Adapted from Young (1975) and Bolt (1978).

[2]MMI is Modified Mercalli Intensity; Mag is generalized magnitude; V is average peak velocity in cm/s; A is average peak acceleration as a percentage of gravity (g), where g is 980 cm/s^2; TNT is energy released in terms of weight of TNT. Witnessed observations are the descriptions of intensity.

Table 3. (Continued) Rough Comparison of Earthquake Rating Scales in Terms of Energy Released[1]

MMI[2]	Mag	V	A	TNT	Witnessed Observations
VIII	>6-7	20-45	>0.15-0.40g	<200,00 tons	Damage slight in specially designed structures; considerable damage in ordinary buildings; chimneys fall; factory stacks fall; heavy furniture overturned; sand and mud ejected in small amounts; changes in well water; noticed by persons driving cars.
IX	7->7	45-60	>0.40-0.60g	<1,000,000 tons	Damage considerable; in well-designed structures; frame structures thrown out of plumb; partial collapse of ordinary buildings; buildings shifted on foundations; small cracks in ground are conspicuous; underground pipes broken.
X	>7-8	>60	>0.60g	<6,000,000 tons	Some well-built structures are destroyed; ground badly cracked; rails bent; landslides considerable from river banks and steep slopes; shifted sand and mud.
XI	8-9	-	-	<200,000,000 tons	Few, if any masonry buildings remain standing; bridges destroyed; broad fissures in ground (probably shifting ground and landslides); underground pipelines completely out of service; earth slumps and land slips in soft ground; rails bent greatly.
XII	9->9	-	-	>200,000,000 tons	Damage total; waves seen on ground; lines of sight and level distorted; objects thrown into the air.

Modified Mercalli scale of intensity (MMI), which uses Roman numeral I for the weakest earthquakes to XII for the strongest. Intensity depends on the amount of energy released by an earthquake, the distance from the epicenter, the attenuation of seismic-wave energy, the local geology at the site of measurement, the type of construction that is affected, and the population density at the site. A rough comparison of the MMI intensity scale with a magnitude scale is given in Table 3.

Seismic-wave Propagation and Resultant Shaking

The efficiency and speed of seismic-wave propagation through the Earth depends on the earthquake's focal mechanism, seismic-wave properties, and the variation in physical properties of the Earth materials along the path. Properties of seismic waves include: (1) wave form; (2) wave amplitudes; (3) direction of movement from source; (4) velocity through various geologic materials; and (5) duration. Seismic waves propagate in two ways: (1) as body waves that travel through the interior of the Earth either as compression waves (P waves) or shear waves (S waves); and (2) surface waves that travel over the land surface either as (1) Love waves, which displace the ground in a sideways manner, perpendicular to the direction of travel with no vertical movement (essentially a specific form of shear wave); and (2) Rayleigh waves, which move the ground vertically and horizontally along a vertical plane that is aligned in the direction of wave travel, like rolling ocean waves (Bolt, 1978), which are made of specific combinations of P and S waves. The P wave is the faster of the body waves, and the Love wave is the faster of the surface waves. The velocity of each type of seismic wave is directly related to the density and elastic modulus of the geologic material through which it is traveling, hence the need to have good models of the 3-dimensional character of the lithosphere. The difference in arrival times between the P wave and the first S wave at several widely spaced seismographs can be used to calculate the distance from the focus to the recorder. Locations of epicenters then can be inferred from the differences in the arrival times of various seismic waves at several seismographs (Long, 1974; Press and Siever, 1974; Klein, 2000).

The shaking at a site is the result of various seismic waves impinging on a facility.[5] The primary wave, or P wave, is a compression wave that arrives first and commonly gives the effect of a sonic boom, or a truck hitting the building, which rattles windows and makes the noise. Seconds later, the secondary wave, or S wave, arrives with a vertical- and horizontal-shearing motion that generally produces most of the damage. Love waves arrive next and cause damage by producing long-period horizontal shaking at the building's base. Finally, the Rayleigh waves arrive with vertical and horizontal motion. The surface waves are commonly the largest recorded on seismographs; the body waves (P and S) traveling through the Earth get refracted and reflected off deep geologic structures and, thus, may lose some of their energy before reaching the site. The decrease in average intensity of shaking with distance from the earthquake is termed *attenuation*; this is a measure of efficiency of seismic-wave transmission through the lithosphere that results from the damping effect of encountering rocks with contrasting physi-

[5] Much of the discussion in this paragraph is taken from Bolt (1978) and response to review comments by John Ebel. This description of seismic-wave propagation is generalized and does not present the total complexity of the seismic waves that reach a site.

cal properties and structures (Bolt, 1978). The lithosphere's varying composition across the North American plate causes attenuation to vary regionally; for example, attenuation of seismic-wave amplitudes in the eastern United States is less than in the West and earthquakes of equivalent size are, therefore, felt over a larger area in the East than in the West (Nuttli, 1978; Sighn and Herrmann, 1983; Stanton and others, 1988; McGuire and Toro, 1989; Somerville, 1989; Atkinson and Mereu, 1992). This heterogeneity of the lithosphere can focus, amplify, or impede seismic waves as they radiate from the earthquake. Attenuation can even vary across small regions such as the NYSSR as well (Stanton and others, 1985). Attenuation calculations rely on refined models of the crust. Deep seismic reflection profiling and other geophysical studies are adding new data for the construction of these models (Forsyth and others, 1994a, b). Therefore, attenuation must be factored into seismic-hazard evaluations in the East.

Recurrence Rates and Earthquake Sequences

Each region of continental plate has its individual rate of earthquake activity for given divisions of earthquake size. Plotting the logarithm of the frequency of occurrence of earthquakes as a function of earthquake size provides a linear relation for the middle range of magnitudes (M = 2.5 to 7.0) or intensities that describes the recurrence rate for a region. Smaller earthquakes are the most difficult to detect and the resulting incomplete data set from the historical record results in a deviation from the linear relation at that end of the plot. The shape of the end of the curve that represents the largest earthquakes is speculative because the recurrence of the largest possible events may take longer than the historic record. The slope of the line on the plot is called the *b-value* and is always near the value of -1 and within the range of -0.5 to -1.5 (Kissinger, 1975). Each earthquake-source zone commonly has its individual b-value. New York State can be divided into four or more regions with different recurrence rates or b-values, three of which may represent areas of high seismicity while the fourth encompasses the remaining part of the State (Mitronovas, 1982). These four regions are essentially the same as the four regions with different seismic characteristics, as depicted in Figure 2. Recurrence rates for earthquakes of any given size in the central and eastern United States are lower than those west of the Rocky Mountain front (McGuire, 1977; Seeber, 1991; FEMA, 2001b).

Seismicity on a fault commonly occurs as a series of shocks in a sequence over time with the largest called *the main shock.* Those events that occur before the main shock are called *foreshocks* and those occurring later are called *aftershocks.* The sequence may occur over a range of seconds to weeks, and even years for some fault systems. Aftershocks commonly decrease in frequency and magnitude with time. Events to be considered in the sequence must have focal mechanisms that are compatible with the main shock and occur near the focus of the main shock (Bates and Jackson, 1980).

Dating of Prehistoric and Modern Earthquakes

Dating of prehistoric (> 500 yr) seismic activity on faults for seismic-hazard calculations is based on analysis of materials found in Holocene sedimentary deposits that have been interpreted to be disrupted by strong seismic shaking or by surface faulting. Paleoseismic studies examine the structures within unconsolidated materials for evidence of disruption of beds or liquefaction from prehistoric earthquakes (McCalpin and Nelson, 1996). The dating of post-glacial disruptions is commonly based on ^{14}C measurements of strata-bound organic material. Dating of materials associated with a fault, such as *fault gouge*—clay-like material that results from the grinding of rock as the fault moves, or cross-cutting veins and dikes, can put age limits on the time its of origin or the time of its cessation of movement (for example, Ratcliffe, 1971).

Other techniques that have been used to estimate recent movement on faults are more inferential than the dating approaches listed above. Geodetic studies of modern movement of benchmarks, such as in the Adirondack Mountains (Isachsen, 1975), or of other presumably stable features, such as shorelines along the coast of Maine have been used to suggest uplift or sagging (Anderson and others, 1989). Features formed in response to modern stresses, such as pop-ups of exposed bedrock (Fakundiny and others, 1978b; Wallach and others, 1993), or rock displacements by aseismic creep, are studied to estimate the amount of stress that might exist in the rock. High-differential stress can be used to indicate earthquake potential and, consequently, high modern seismic hazards (Wallach and others, 1993). Geomorphic features, such as fault scarps and landslides, are commonly used in seismic-hazard studies to estimate potential sizes and recurrence rates of earthquakes. Damage to, or response of, modern constructed facilities may also give useful information about the history of frequency and relative displacement on active faults; for example, systematic rotation of grave monuments in cemeteries or chimney collapses near epicenters (Revetta and others, 1983). Even with all of these techniques the history of seismicity in eastern North America is still incompletely documented and, thus, presents challenges for seismic-hazard evaluations.

SEISMIC-HAZARD EVALUATION

Seismic-hazard evaluation makes use of the location, size, focal mechanism, recurrence rate of earthquakes, the attenuation of seismic-wave energy on its way through the lithosphere to the site of measurement, and local geology (Milne and Davenport, 1969; Kulhawy and Ninyo, 1977; McGuire and others, 1989). Conservative seismic-hazard evaluations may use, for design purposes, the largest historical earthquake that could occur at any given time and at any place within that seismic-source zone; many seismologists think that a larger earthquake than has been experienced could occur (Hall, 1980; Coppersmith, 1994). The estimated probabilities of an $M = 5$, $M = 6$, and $M = 7$ earthquake hitting southern Ontario in the next 50 years are about 57 percent, 6 percent, and <1 percent, respectively (Mohajar, 1997), and for New York City an M = 5, M = 6, M = 7 could strike once every 500 years at distances of 20, 50, and 130 km, respectively (Jacob, 1997). Engineers can design structures to withstand the corresponding ground motions once the degree of horizontal and vertical shaking has been estimated.

Seismic Hazards and Seismic Risks

The damage incurred from an earthquake at a site is a function of the nature of shaking, the geologic-foundation conditions, the structural stability of the building, and the susceptibility to secondary associated collateral geologic hazards, such as tsunamis, floods, or landslides, that may result from that shaking (Jacob, 1997). For example, a small earthquake that occurs near a structure or facility commonly presents a minimal hazard, but may present a high risk, if a substantial part of the facility is inadequately designed to withstand small shocks. Conversely, a large earthquake may present a large hazard, but a small risk, if the affected building can withstand the seismic-wave energy and if the surrounding area contains no other potential geologic hazards. Thus, a well-constructed facility near a large earthquake may present a smaller risk than a poorly constructed facility at some distance from a small earthquake. Therefore, identification of seismic risks entails making an inventory of structures in a given area that includes such factors as the age, design, construction materials, construction techniques, and geologic-foundation conditions of the structure, and their type of use. Old structures with less-stringent design features than required by modern building codes may present a higher risk than new structures.

A *seismic-source zone* is an area of seismicity that is thought to have a single, uniform earthquake-generating cause. Seismic-source zones are sometimes arrived at by a poll of experts who provide their best analyses or guesses (EPRI, 1986; Basham, 1989). Both the tectonic-province and source-zone concepts assume that the internal tectonic regime has remained unchanged throughout the last 10,000 years and that each tectonic province or source zone has a uniform distribution of seismic hazard; this second aspect has been contested (Coppersmith, 1994). A more useful determination than uniform distribution of earthquakes would be that the distribution of *seismicity* within a tectonic province or seismic-source zone, whether uniform or not, will remain unchanged in the foreseeable future. Any commonality of current seismicity with prehistoric earthquakes can only be interpreted from paleoseismic data—the "fossil" indications of signs of seismic shaking in Quaternary deposits.

A common parameter in the approach to regional seismic-hazard evaluation is the estimation of *the maximum credible earthquake* that could occur anywhere within a tectonic province or seismic-source zone (Veneziano and Chouinard, 1989). The maximum credible earthquake is a term for the earthquake that is used for engineering design and designates the highest possible magnitude for that source region, irrespective of time and assuming that the tectonic regime does not change (Coppersmith, 1994), and is commonly higher than the magnitude derived from seismic-hazard calculations.

Probabilistic and Deterministic Approaches to Seismic-Hazards Evaluation

Seismic-hazard calculations commonly use either *probabilistic* approaches—those that consider a statistical analysis of the earthquakes that might affect a site (See Chinnery, 1979; Kimbal and Pasedag, 1992; Algermissen, 1997; for reviews of the approach), or *deterministic* approaches—those that use seismic-history data to locate source faults and fault characteristics to estimate the maximum size of the earthquake that might affect the site (McGuire, 1977). Deterministic approaches to modeling seismic hazard can be further separated into *rational*—those derived from basic scientific principles or laws of nature, or *empirical*— those based on observations without any underlying theoretical basis. The deterministic approach to estimating building-design parameters has the disadvantage of relying on less-than-complete data for source-fault geometry, and provides estimates of only the largest credible earthquake (McGuire, 1977). Probabilistic approaches, in contrast, are synonymous with "random" or "stochastic" approaches (Haneberg, 2000). Earthquake-recurrence data from eastern Canada have been used to estimate probabilistic seismic hazard (Basham and others, 1979).

Seismic-Hazard Maps

Several techniques have been developed to map seismic hazards. One is to delineate seismic-source zones on maps and to assign seismic-activity rates and maximum-earthquake magnitudes to these zones on the basis of historic and paleoseismic data. These assessments of zones are then applied to the probability calculations of seismic hazards and the creation of seismic-hazard maps (Richter, 1959; Howell, 1974; Hadley and Devine, 1974; Algermissen and Perkins, 1976; Basham, 1983; Seeber and others, 1985; Frankel, 1995; Frankel and others, 1996). A common approach for mapping seismic-source zones is tectonic history. For example, some researchers postulate that earthquakes on the eastern seaboard occur along old faults created in the Mesozoic Era during the continental breakup of the supercontinent, Pangaea, to form the modern Atlantic Ocean (Adams and Basham, 1991; Adams and others, 1995); others postulate that the loci of modern earthquakes in NYSSR are on faults created during the late Proterozoic- early-Paleozoic breakup of the supercontinent, Rodinia (Wheeler, 1995). The tectonic history of a structure or an area can be clarified partly through the study of its history of regional uplift or collapse by techniques such as fission-track studies of apatite (Roden-Tice and others, 2000). Thus, some seismic-hazard maps are derivatives of tectonic maps.

An alternative approach is based on the hypothesis that "soft," ultramafic igneous plugs that were intruded into "stiffer" quartzo-feldspathic gneisses within the basement locally concentrate regional stresses to levels that exceed the strength of the basement host rock and create a catastrophic break (Kane, 1977; Campbell, 1978; McKeown, 1978); a map based on this hypothesis would be a derivative of maps that depict regional gravity and magnetics (for example, see Fakundiny and Pomeroy, 2002). Some seismologists see no consistent map trend in earthquake locations over the last century, especially within the NYSSR (Stevens, 1995).

Several other techniques and approaches have been used to generate seismic source-zone maps of much of the NYSSR; one is the epicenter map. The epicenter map for the NYSSR (Figure 2) provides a useful depiction of data for seismic-hazard calculations because it illustrates the geographic distribution of epicenters, some of which are more concentrated, or clustered, than the regional average. For example, Figure 2 depicts the northern Appalachian Plateaus as almost devoid of modern moderate-sized earthquakes, except for the 1998 Pymatuning Reservoir M = 5.2 event in westernmost Pennsylvania. The WN-NP region (Figure 2) which encompasses western New York, western Lake Ontario, and the Niagara Peninsula of Ontario, has greater than average seismicity than much of the NYSSR. Its largest modern earthquake was the 1944 M = 5.1 event at Attica, NY. The map depicts another region of modern seismic quiescence in the Canadian Shield between Lake Ontario and the Ottawa-Bonnechere graben (OB-AM on Figure 2). One of the most geographically expansive distributions of low- to intermediate-level seismicity in the eastern part of the North American continent is the region extending from the southeastern Adirondack Mountains northwestward into the Ottawa-Bonnechere graben; its hallmark earthquake is the 1935 M = 5.9 event at Timiskaming, Que.. Two other clusters of seismicity on the map are: (1) the New York City metropolitan area that includes the Hudson Highlands and Northern New Jersey, with its 1884 M = 5.2 earthquake ~28 km offshore in the New York bight (NY-NJ on Figure 2); and (2) the Haddam-Moodus area (H-M on Figure 2) of southeastern Connecticut, a region of persistent, but quite-minor earthquakes. Some seismologists do not consider the Hadam-Moodus area to have seismicity much above the regional background recurrence rates (John Ebel, written communication, 2002), but Nottis (1983) lists 48 events in that area, most of which are felt reports of small-sized events. Three large earthquakes that are not associated with epicenter clusters are the 1638 event (M = 6.5) in central New Hampshire and the two (M = 5.5) earthquakes that struck Ossipee, NH in 1940.

Another type of map, the isoseismal map, depicts the extent of felt effects of an earthquake distant from the epicenter by contouring areas of equal seismic intensity, as reported by citizens to police stations, to newspapers, and on questionnaires that are distributed by seismologists. Isoseismal maps roughly depict the attenuation of seismic-wave energy in terms of intensity from the epicenter and, thus, can be used to estimate the decrease in seismic-wave response that can be expected away from a source zone or fault.

Among the remaining approaches used for seismic-hazard evaluations that can be depicted on maps are: (1) paleoseismicity data, which provide prehistoric recurrence rates (Youd and Idress, 1997; Tuttle, 1996; Tuttle and others, 1996); (2) geodetic data, which can measure the current land movement that may be related to seismic activity or aseismic creep, such as the rapid uplift of the Adirondack Mountains (Isachsen, 1975); (3) paleostress measurements in rock, which may describe the variation in the stresses acting in the lithosphere, for example in northern New York (Engelder and Sbar, 1976); (4) geomorphologic features, which define the changes in landforms in response to stress buildup and release within the lithosphere, commonly in the form of fresh fault scarps (Press and Siever, 1974) or landslides; and (5) delineation of presumed or possible active seismic structures, such as faults (Wheeler, 1995).

Ground Motions and Response Spectra

The properties of ground shaking from an earthquake at a site are termed *ground motions*, and their graphic records are used for the design and construction of buildings. Ground motions are recorded by seismographs and include horizontal and vertical ground accelerations and velocities at various frequencies. Plotting ground motions with their natural periods (or frequencies) provides graphic curves that represent *response spectra* (Shannon & Wilson, 1979; Hall and McCabe, 1989). Seismic-response spectra are Fourier-analysis representations of the energy released during an earthquake; they include acceleration, frequency, and amplitude of shear waves at the source (McGuire and Toro, 1989). Many cities in the eastern United States have buildings that were designed with fundamental vibration periods that are comparable to the natural high frequencies of eastern earthquakes. These buildings may respond catastrophically to seismic waves and, therefore, present seismic risks comparable to many of those in the more seismically active parts of the western United States (Atkinson, 1989). An explanation of critical building characteristics is presented in FEMA (1995). Adequate seismic-risk evaluations, therefore, require seismic-response spectra, information on local geologic foundation conditions, building design, and construction conditions. Ground-motion data for eastern North America are sparse (McGuire and Toro, 1989), but the available response-spectra curves can be applied by engineers to engineering designs (Kircher, and others, 1997a, b). The response spectra for earthquakes in eastern North America differ from those for earthquakes of equivalent size in the western states (Atkinson, 1989), possibly because of the lower attenuation of energy during seismic-wave transmission (Somerville, 1989). Eastern earthquakes have more energy at high frequencies than western events at the same distance from the recording site, and ground motions decay more slowly with distance (Atkinson, 1989).

Earthquake Prediction

One of the primary long-term goals of seismologists is to predict earthquakes. A valid prediction of an earthquake must meet three specific criteria: (1) the expected time of occurrence within some specified confidence limits; (2) the location of the epicenter or hypocenter within some defined area with specified confidence limits; and (3) the magnitude of the event with specified confidence limits. Earthquake prediction would give emergency-management officials the opportunity to shut down and evacuate critical structures such as nuclear power-plants.

Seismologists were optimistic 2 decades ago that techniques for predicting earthquakes would be discovered (Kissinger, 1975). Some approaches that seemed promising were: (1) locating gaps in time and space of earthquakes along active faults, that is delineating parts of active faults that had not produced earthquakes for periods much longer than the average for the entire length of these faults; (2) relying on recurrence rates as measured by b-values; (3) studying changes in physical and chemical properties of the crust that might be premonitory phenomena or precursory events, such as elastic-wave velocities, electrical conductivity, magnetic properties, geochemical tracers (radon in groundwater, for instance), and geodetic measurements (such as elevations or geographic position of benchmarks, for instance). Even unusual behavior of animals before an earthquake has been a topic of prediction studies. Although these approaches have been partly successful under certain circumstances, none has stood the test of time over wide areas. Consequently, prediction is still considered by most seismologists to be impossible, because the three required conditions, listed above, to different degrees, are currently indeterminable. In the absence of ability to predict an earthquake, damage prevention and early warning are the most reliable approaches to minimizing earthquake damage.

EARTHQUAKE-RESISTANT DESIGN

Much can be done to "earthquake-proof," or protect structures from earthquake damage through earthquake-resistant design. Four steps toward minimizing damage from earthquake damage are: (1) prevention of human-induced earthquakes; (2) mitigation of potential damage through proper construction techniques; (3) remediation (retrofitting) of structures that have already been built; and (4) development of real-time response to earthquakes. These four lines of defense could have prevented much damage and loss of life throughout those parts of the world where cities destroyed by earthquakes were rebuilt on the rubble without adequate seismic design (Kircher and others, 1997a, b). The NYSSR can minimize future damage if proper prevention, remediation, and mitigation techniques are used.

Prevention

Prevention of natural earthquakes is beyond current technology, but the generation of human-induced earthquakes can be lessened or eliminated. For example, some large freshwater reservoirs have the potential to produce earthquakes (Simpson, 1976), yet the seismicity associated with reservoir filling may be largely prevented by: (1) good geologic studies to define faults that could be activated by high pressure; (2) correctly designing dams and overflow systems; and (3) carefully, monitoring filling. Another source of human-induced seismicity is the injection of fluids into bedrock (Fletcher and Sykes, 1977; Price, 2001). Some common reasons for injection are: (1) disposal of liquid wastes; (2) creation of underground cavities for natural-gas storage by dissolution of salt; and (3) production of brine. Such systems can be designed to minimize the potential for causing seismicity, however. A third type of human-induced earthquake can result from the removal of rock from deep quarries by eliminating the confining load over highly stressed rock (Pomeroy and others, 1976). All of these human-induced earthquakes result from mining or storage of fluids and can be reduced by stringent regulation and careful monitoring.

Mitigation

Earthquake-damage mitigation is the planning, design, and construction that ensures structures will withstand seismic shaking (Eguchi, 1997). Mitigation is currently the most effective way to address seismic risks and is attained by instituting adequate building-code provisions and inspecting structures as they are constructed. Mitigation provides seismic resistance for an added cost of as little as 4 percent (Whitman, 1989). A recommended approach to the engineering design of structures is to use a *design earthquake*—a hypothetical earthquake that, at a specified location, results in ground motions that have a 90-percent probability of not being exceeded in 50 years (FEMA, 1995). A similar criterion, the *safe-shutdown earthquake*, is used in the design of nuclear powerplants and is based on the evaluation of the maximum earthquake potential as indicated by the regional and local geology, seismology, and characteristics of the local subsurface material (CFR10, 1978; NRC, 1997).

Another critical factor in earthquake-damage mitigation is site selection. Finding an adequate geologic medium for the foundation, such as competent bedrock, requires geological studies to assess the potential for landslides, rockfalls, ground failure, flooding, tsunamis, and other associated collateral geologic hazards that might be triggered by earthquakes. A structural design that will permit easy and fast evacuation of occupants is desirable. Structures near large water bodies should be designed to withstand the inundation by floods or tsunamis.

Remediation

The third line of defense in earthquake-resistant design is *remediation* (Poland, 1989), which refers to the retrofitting of current building stock and infrastructure against future earthquakes. This can be accomplished through efforts similar to those now underway in the western coastal states (Davis and others, 1979; Eguchi, 1997; Moore, 1997). Engineering studies have produced numerous quick and inexpensive methods to minimize damage and injury in buildings, among which are bolting or strapping all loose standing objects, such as hot-water heaters, refrigerators, and bookcases, to stable walls.

Real-time response is the newest proposed approach to avoid damage and injury from an earthquake (Goltz and Hanksson, 2001). Although real-time response is not earthquake prevention, it can decrease the deleterious effects on society and possibly prevent serious damage to critical structures. The seismic waves produced by large earthquakes at medium to far distances from a given site travel much more slowly than radio waves or electric signals through transmission lines. Therefore, a real-time response system could be triggered to alert distant populations by telecommunication systems to the location and size of the event. The several seconds to minutes that might elapse from the time of the warning signal and the first arrival of seismic waves might allow local emergency-management personnel to mobilize rescue and fire equipment, close bridges, chemical facilities, and other critical structures, shut down nuclear powerplants, and call for the evacuation of buildings. Ideally, an advanced national seismic-monitoring system with a real-time response system could save lives and reduce property damage (Ward, 2001). Real-time response to data from the Advanced National Seismic System, discussed further on, could become the fourth line of defense against earthquake risk.

POLITICAL RESPONSE TO EARTHQUAKE RISKS[6]

Response to earthquake risks can be activated at the National, State, and local levels of government. Some of the U.S. Federal participants are:

1. The U.S. Federal Emergency Management Agency (FEMA), which has supported earthquake studies (FEMA, 1995; Moore, 1997), suggested seismic building-code regulations (FEMA, 1997), and developed the HAZUS program (Whitman and others, 1997), among other programs.

2. The U.S. Geological Survey, through its seismic-research programs, geologic mapping, and geologic-hazards studies, including the operation of the Advanced National Seismic System.

---

[6] This discussion emphasizes political response in the United States and New York State because of the author's familiarity with developments in these two political arenas.

3. The U.S. National Science Foundation, with its research-grants program.

4. The U.S. National Institute of Standards and Technology through its study of earthquake engineering.

5. The U.S. Nuclear Regulatory Commission, through its review of seismic hazards at nuclear facilities (for example, NRC, 1997).

(These first 5 agencies form the U.S. National Earthquake Hazards Reduction Program-NEHRP).

6. The U.S. Army Corps of Engineers, through its studies of critical structures such as dams, Veterans' Administration hospitals, military facilities, and others, and the maintenance of strong- motion seismographs at dam sites.

7. The U.S. Department of Energy, through its seismic studies at radioactive-waste facilities (for example, DOE, 1996, 2000).

8. The U.S. Department of Defense, through programs such as its monitoring of international nuclear-weapons testing.

9. The Atomic Energy Control Board of Canada, which assesses seismic hazards at nuclear facilities (Leblanc and Klimkiewicz, 1994), for example, in reviewing the seismic hazards at the Pickering and Darlington nuclear-power plants in Ontario (Wallach, 1990; Jacobi and others, 1997).

HAZUS (hazards in the United States) is a geographic-information-system analysis of seismic risk developed by the National Institute of Building Sciences for FEMA, which can make earthquake-loss estimates and, thus, forms the basis for the development of damage-mitigation procedures (Kircher and others, 1977a, b; Whitman and others, 1997; Moore, 1997; FEMA, 2001a). The HAZUS program is being expanded to include other natural hazards, such as hurricanes, tornadoes, winds, and floods. Emergency managers can use HAZUS in the response period immediately after damaging earthquakes to identify likely damaged areas, provide a rapid estimate of the damage and casualties, and estimate the type and amount of resources that should be sent to assist the damaged area. A preliminary trial run of the HAZUS estimation of probable losses throughout the United States calculated that New York State has the highest degree of risk after California, Washington, and Oregon. This high estimate results from the large number of buildings, the large population, and large number of lifelines in the New York City metropolitan area, which lie in a geographic cluster of seismicity that has as its largest earthquake the M = 5.2 event of 1884.

The Advanced National Seismic System (ANSS) was authorized by Congress in 1998 as part of the NEHRP. The purpose of the ANSS is to become a national real-time seismic-monitoring system that has modern strong-motions seismographs deployed throughout the United States. Once triggered by an earthquake the stations send the acquired recordings by satellite telemetry instantaneously to all facilities that require seismic data. As of the end of 2001, 273 instruments have been installed in Alaska and the coterminous United States, with 1 in Pennsylvania and 1 in Massachusetts.

The NRC has placed within its Code of Federal Regulations that apply to seismic and geologic siting criteria for nuclear powerplants a political definition of an active fault that may not be easily applied to the northeastern United States. This is the concept of the *capable fault*:

"...having had movement at or near the ground surface at least once within the past 10,000 years or recurring movement within the past 500,000 years, or macro-seismicity measured with sufficient precision to demonstrate a direct spatial relation with a fault, or a structural relation with a capable fault such that movement on one could be accompanied by movement on the other" (CFR10, 1978; Hall, 1980).

Movement at or near the ground surface during the last 10,000 years has not been unequivocally proven for any fault in the NYSSR.

States respond politically by establishing:

1. Seismic building codes and special laws, such as California's Alquist-Priolo Special Studies Act of 1972 (Davis and others, 1979).

2. The geology and seismology provisions of the powerplant-siting rules and regulations of the New York State Public Service Law (Davis and others, 1979).

3. The earthquake-load provisions of the Massachusetts State Building Code (Luft, 1989).

Multiple states, in cooperation with Federal agencies, have formed regional earthquake-mitigation programs such as the Western States Seismic Policy Council and the Central United States Seismic Consortium.

Many cities have instituted earthquake-response procedures, such as the seismic provisions of the New York City Building Code (Miele, 1997).

Building Codes, Geologic Studies, and Land-use Planning

All levels of government adopt building codes for public structures and emergency-response systems (MSBCC, 1979; BSSC, 1986; BOCAI, 1987; SBCCI, 1987; ICBO, 1988; NIBS, 1997; FEMA, 1997). Building codes formulate the fundamental hazard-mitigation strategies for the construction of buildings, lifelines, communication and transportation corridors, and critical structures. Local governments address seismic hazards through: (1) building codes, such as the New York City seismic building-code provisions (Jacob, 1993; Miele, 1997; Nordenson, 1997); (2) land-use planning (Fakundiny, 1975; Davis and others, 1978); and (3) emergency-response programs. Local governments control 75 percent of the enforcement of building regulations (FEMA, 1995). The Uniform Building Code (ICBO, 1988), with its seismic-hazard maps, is designed to be used at the national level, but can be adapted for use at the State and local levels of government. The U.S. Nuclear Regulatory Commission (NRC) has strict seismic provisions that govern the design and construction of nuclear facilities. For example, a protracted 3-year hearing by the NRC reviewed the design and location of the Indian Point 3 nuclear powerplant at Peekskill, NY, 40 km north of New York City and found the design adequate to withstand the greatest probable earthquake that could strike the site (Davis and others, 1979). The New York City model of earthquake damage (Scawthorn and Harris, 1989) predicted that an earthquake of M = 6.0 at the site of the 1884 earthquake (M = 5.2), 28 km southeast of City Hall, would cause damage of about $11 billion and that about 130 post-earthquake fires would occur. This type of scenario led to the adoption of an earthquake-building code by New York City in 1995 (Jacob and others, 1990; Miele, 1997). Newer modeling techniques have raised the estimate for an M = 6 earthquake striking New York City to $20 billion (Scawthorn and others, 1997). New York State has been provided with a suggested set of revisions to its building code for seismic-risk design, but the revisions have not as yet been accepted (Kelly, 1991; Gergely, 1993; Jacob and others, 1990; Jacob, 1993; Kelly, 1995). Massachusetts also has addressed the issue of seismic safety with its Earthquake Load section of the Massachusetts State Building Code (MSBCC, 1979; Luft, 1989), as has California, with its Alquist-Priolo Special Studies Act of 1972 to identify and map active faults (Davis and others, 1979). At least 37 other nations have instituted seismic building codes (Paz, 1994).

One of the most effective and least expensive methods for earthquake-hazard mitigation at any level of government is land-use planning. Proper land-use planning relies on modern geologic mapping and soils investigations to locate the least geologically hazardous sites for the siting of facilities and structures. Geologists and seismologists clearly must be involved in all decisions related to the establishment of seismic building codes and the estimation of seismic risk (Woodhouse, 1989).

APPLICATION OF EARTHQUAKE-RISK STUDIES TO THE NYSSR

Seismic risk can be calculated with sufficient certainty to identify regions where prevention, mitigation, and remediation should be applied, even though seismic hazards in the NYSSR are difficult to quantify. Citizens of the NYSSR can respond to the high earthquake risk they face, even though earthquake epicenters and hypocenters cannot be accurately located, seismic-wave-propagation models of transmission to a site are as, yet, poorly defined, and obvious surface expression of active faults is lacking.

Cities in Danger

The major cities in the eastern United States lack the ability to adequately quantify siesmic hazards. The high seismic risk for many of the major cities of the NYSSR, including Buffalo, Rochester, Toronto, Montreal, Ottawa, Hartford, the greater metropolitan area of New York City, Philadelphia, as well cities adjacent to the NYSSR, such as Cleveland, Baltimore, Boston, and Washington D.C., is a result of their high population densities and building stock, and because each is situated within or near a cluster of historic seismicity. New York City, by awareness of its vulnerability and by its incorporation of a seismic provision into its building code, sets an example of prudent action for all of these other major metropolitan areas. The current incomplete understanding of earthquake mechanisms and controls on their geographic distribution require that public policies must assume that a large earthquake with M > 6.0, could occur in the NYSSR at any time. The 2002 earthquake (M = 5.3) at Au Sable Forks in the northeastern Adirondack Mountains of New York State underscores the need for public awareness of local seismic hazards in other areas of the NYSSR.

Some recommendations for planning include: (1) careful evaluation of anticipated losses; (2) initiation of selected damage-mitigation efforts; (3) fostering public demand for earthquake-loss protection; (4) aiding the insurance industry to build the capacity to handle catastrophic loss; (5) adopting construction performance standards, such as building-code provisions; and (6) considering the broader economic consequences of earthquake disasters (Jones, 1997). A massive effort to educate the citizens of the major cities of the world is sorely needed, but will come about only through an increased effort by Earth scientists to communicate on how to reduce seismic risks (Buckle, 1989; Nufer and others, 1993). Reliable correlation between earthquakes and specific geologic structures is lacking for metropolitan areas throughout much of the world. An approach to make the cities of the world safer from earthquakes would be: (1) to incorporate seismic provisions into building codes; (2) to apply modern hazard-mitigation and damage-remediation techniques to strengthen building stock and critical structures; (3) to apply the HAZUS approach to earthquake-risk evaluations; and (4) to establish an advanced global seismic system, modeled on the proposed U.S. Advanced National Seismic System.

Earthquake Insurance

This paper has not addressed the issue of earthquake insurance, which is a topic unto itself (Lecomte, 1997; Tierney, 1997). Until a decade ago, earthquake-insurance problems had not been analyzed at the national level, nor by any groups outside of the insurance industry (Bernstein, 1989). The institution of a national earthquake-insurance program remains for the future in the United States.

CONCLUSIONS

1. Steady and apparently randomly dispersed, low-level seismicity has been recorded in the NYSSR since the 16th Century.

2. The largest earthquake recorded in the NYSSR was the M = 5.8 event in Cornwall, Ont.-Massena, NY in 1944. The largest earthquake located close to the NYSSR is the M = 6.5 event in central Massachusetts in 1638.

3. The mechanisms for earthquake generation in the NYSSR are not well understood but are assumed to be consistent with plate-tectonics models of the

lithosphere in the interior of the North American plate. Earthquakes within NYSSR probably reflect response to a regional horizontal stress system, and are produced by some set of mechanisms probably different from those that operate at plate boundaries.

4. A better understanding of the history of formation and structural framework of the lithosphere in the eastern United States and Canada than currently exists will be required to augment plate-tectonics theory such that it can explain intraplate earthquakes in the NYSSR.

5. The traditional approaches to seismic-hazard evaluations that use the tectonic-province concept, which is based on the collisional and extensional tectonic history of the Proterozoic and Phanerozoic, do not provide an adequate explanation for the distribution of modern seismicity in the NYSSR.

6. The history of seismicity in the eastern United States and Canada is till incompletely documented and, thus, presents challenges for seismic-hazard evalauations.

7. No correlation is apparent between the surface geology of the NYSSR and the distribution of seismicity, even though many geologic structures have been proposed to be seismically active.

8. Plots of historic epicenters reveal clusters of seismicity that are restricted to certain part of arbitrarily delineated tectonic provinces and, consequently, raise doubts about the effectiveness of the tectonic-province approach to seismic-hazard evaluations.

9. The inability to locate epicenters within less than 5 km of their actual occurrence in most of the NYSSR makes spatial association of single earthquakes with specific tectonic features difficult, except where portable seismometers are used to pinpoint aftershocks.

10. Seismic-risk calculations take into account: (1) the location of the earthquake; (2) the magnitude and seismic-wave characteristics of the earthquake; (3) the focal mechanism of the earthquake; (4) the variations in properties of the rock along the path of the seismic waves that affect attenuation; (5) the recurrence rates of various sized earthquakes; (6) the local geology of the site being evaluated; and (7) the design and construction of the buildings at the site. Consequently, thorough geological studies at the site of concern, regional geophysical studies, and a refined model of the Earth's crust that adequately characterizes the focusing, amplification, or impedance of seismic waves is required to calculate the seismic hazard. If scientists understood thoroughly the processes that create earthquakes in intraplate regions, they might be able to predict impending seismicity.

11. A large ($M > 6.0$) earthquake could strike any metropolitan area of the NYSSR at any time.

12. The probabilities of a $M = 7.0$ earthquake hitting the NYSSR in the next 50 years is less than 1 percent; New York City can expect to experience the shaking from a $M = 7.0$ earthquake at a distance of 130 km or less once every 500 years.

13. Prevention, hazard mitigation, and damage remediation are the three lines of defense against earthquake risk. A fourth, real-time response from data sent by the Advanced National Seismic System, could be in operation in the near future.

14. Adequate mitigation and remediation programs require: (1) refined understanding of the region's geologic framework; (2) installation and operation of an advanced national seismic-monitoring system; (3) refined models of the propagation of seismic waves through the lithosphere; (4) implementation of appropriate seismic building-code provisions; and (5) public education in remediation procedures.

15. Reduction of earthquake risk must address associated or collateral hazards, such as liquefaction of saturated sediments, landslides, changes in groundwater, release of subterranean gases, ground collapse, fires, and floods.

16. HAZUS is a FEMA national-planning tool that uses a geographic-information system to calculate seismic-risk. It is forming the basis for mitigation programs throughout the United States. A preliminary analysis by HAZUS calculated that New York State has the fourth-highest seismic risk in the United States.

17. Modern land-use planning that relies on good geologic studies and adequate seismic provisions in building codes is the most effective and least expensive method for earthquake-hazard mitigation.

18. Planning should include: (1) evaluation of anticipated losses; (2) initiation of damage-mitigation; (3) fostering public demand for earthquake-loss protection; (4) aiding the insurance industry to handle earthquake losses; (5) adopting building-code provisions; and (6) considering the broader economic consequences of earthquake damage.

19. All the major cities in the NYSSR lack the ability to adequately quantify seismic hazards, yet all are located within or near a cluster of historic seismicity.

20. A massive effort to educate the citizens of the major cities of intraplate regions is sorely needed.

21. Reliable correlation between earthquakes and specific geologic structures is lacking for metropolitan areas throughout much of the world. An approach to make the cities of the world safer from earthquakes would be: (1) to incorporate seismic provisions into building codes; (2) to apply modern hazard-mitigation and damage-remediation techniques to strengthen building stock and critical structures; (3) to apply the HAZUS approach to earthquake-risk evaluations; and (4) to establish an advanced global seismic system, modeled on the proposed U.S. Advanced National Seismic System.

Acknowledgments. Thanks are extended to the Technical Advisory Committee to the New York State Office of Naval Affairs and Disaster Preparedness Earthquake Hazards Study Advisory Panel, the West Valley Nuclear Service Center in western NY, the N.Y.S. Energy Research and Development Authority, the U.S. Geological Survey, the U.S. Nuclear Regulatory Commission, and the U.S. Environmental Protection Agency for support of research on seismicity in western New York State. Robert D. Jacobi and Joseph L. Wallach were co-authors with me on a precursor paper given in the symposium "Earth

Sciences in the Cities" at the American Geophysical Union 2000 Spring Meeting. M. James Aldrich, Jr., John E. Ebel, and Paul W. Pomeroy reviewed the manuscript and made many critical suggestions that greatly improved the paper. Any mistakes and all opinions are mine, however. Claudia Anderson provided library services.

REFERENCES

Acharya, H., Possible minimum depths of large historical earthquakes in eastern North America. *Geophysical Research Letters* 7(8): 619-620, 1980.

Adams, J. and Basham, D., The seismicity and seismotectonics of eastern Canada, *in* Slemmons, D. B., Engdahl, E. R., Zoback, M. D., and Blackwell (Eds.), *Neotectonics of North America.* Geological Society of America, *Geology of North America, Decade of North America,* Map 1, Chapter 14: 261-276, 1991.

Adams, J., Basham, P. W., and Halchuk, S., Northeastern North American earthquake potential—new challenges for seismic hazards mapping. *Geological Survey of Canada Research* 1995-D: 91-99, 1995.

Adams, J. and Bell, J. S., Crustal stresses in Canada, *in* Slemmons, D. B., Engldahl, E. R., Zoback, M. D., and Blackwell, D. D. (Eds.) *Neotectonics of North America.* Geological Society of America, *Geology of North America, Decade of North America* Map 1, Chapter 20: 367-386, 1991.

Aggarwal, Y. P. and Sykes, L. R., Earthquakes, faults, and nuclear power plants in southern New York and northern New Jersey. *Science* 200: 425-429, 1978.

Algermissen, S. T., Some problems in the assessment of earthquake hazard in Eastern United States, *in* Jones, B. G. (Ed.), *Economic consequences of earthquakes: preparing for the unexpected. National Center for Earthquake Engineering Research* NCEER-SP-000: 51-66, 1997.

Algermissen, S. T. and Perkins, D. M., A probabilistic estimate of maximum accelerations in rock in the contiguous United States. *U.S. Geological Survey Open-file Report* 76-416: 1-45, 1 map, 1976.

Anderson, J. G., Precautionary principle: applications to seismic hazard analysis. *Seismological Research Letters* 72: 319-322, 2001.

Anderson, W. A., Borns, H. B., Jr., Kelly, J. T., and Thompson, W. B., Neotectionic activity in coastal Maine, *in* Anderson, W. A., and Borns, H. W., Jr. (Eds.), Neotectonics of Maine. *Maine Geological Survey Bulletin* 40: 1- 10, 1989.

Atkinson, G. M., Implications of eastern ground-motion characteristics for seismic hazard assessment in eastern North America, *in* Jacob, K. H. and Turkstra, C. J. (Eds.), *Earthquake hazards and the design of constructed facilities in the eastern United States. New York Academy of Sciences Annals* 558: 128-135, 1989.

Atkinson. G. M. and Mereu, R. F., The shape of ground motion attenuation curves in southeastern Canada. *Seismological Society of America Bulletin* 82: 2014-2031, 1992.

Barosh, P. J., Neotectonic movement and earthquake assessment in the eastern United States. *Geological Society of America Reviews in Engineering Geology* XIII: 77-109, 1990.

Barosh, P. J., Frostquakes in New England. *Engineering Geology* 56: 389-394, 2000.

Basham, P. W., New seismic zoning maps for Canada. Energy, Mines, and Resources of Canada. *GEOS* 12(3): 10-12, 1983.

Basham, P. W., Problems with seismic hazard assessment on the eastern Canadian continental margin, *in* Gregersen, S. and Basham, P. W. (Eds.), *Earthquakes at the North-Atlantic passive margin: neotectonics and post-glacial rebound.* Kluwer Academic Publishers, Dordrect, The Netherlands: 679-686, 1989.

Basham, P. W. and Adams, J., Problems of seismic hazard estimation in regions with few large earthquakes: examples from eastern Canada. *Tectonophysics* 167: 187-199, 1989.

Basham, P. W., Weichert, D. H., and Berry, M. J., Regional assessment of seismic risk in Eastern Canada. *Seismological Society of America Bulletin* 69(5): 1567-1602, 1979.

Baskerville, C. A., Bedrock and engineering geologic maps of New York County and parts of Kings and Queens counties, New York, and parts of Bergen and Hudson counties, New Jersey. *U.S. Geological Survey Miscellaneous Investigation Series* Map I-2306 (Sheets 1 and 2), Bedrock, scale=1:24,000, 1994.

Bates, R. L. and Jackson, J. A., (Eds.), *Glossary of Geology* (2nd Edition). American Geological Institute, Falls Church, Virginia, 749 pp., 1980.

Bernstein, G. K. (Chair), Commentary and recommendations of the Expert Review Committee, 1987, Earthquake Insurance. *FEMA*-164: 65-69, 1989.

BOCAI (Building Officials and Code Administrators International), *BOCA Basic National Building Code*. Country Club Hills, IL, 1987.

BSSC (Building Seismic Safety Council), NEHRP recommended provisions for the development of seismic regulations for new buildings. *Earthquake Hazards Reduction Series* 19, FEMA 97, Federal Emergency Management Agency, Washington, D.C., 1986.

Bolt, B. A., *Earthquakes: A primer*. W. H. Freeman & Co., San Francisco, 241 pp., 1978.

Brady, B. H. G., Computational analysis of rock stress, structure and mine seismicity, *in* Knoll, P. (Ed.), *Induced Seismicity*. A. A. Balkema/Rotterdam/Broodfield: 403-417, 1992.

Buckle, I. G., The need for earthquake education, *in* Ross, E. K. (Ed.), *Proceedings from the Conference on Disaster Preparedness—the place of earthquake education in our schools. National Center for Earthquake Engineering Research,* NCEER-89-00178: 23, 1989.

Campbell, D. L., Investigation of the stress-concentration mechanisms for intraplate earthquakes. *Geophysical Research Letters* 5(6): 477-479, 1978.

CFR 10 (Code of Federal Regulations), *Title 10, Energy, Part 100 (10 CFR 100). Reactor site criteria: Appendix A, Seismic and geologic siting criteria for nuclear power plants.* U.S. Nuclear Regulatory Commission, Washington, D.C., U.S., 1978.

Chadwick, G. H., Large fault in western New York. *Geological Society of America Bulletin* 31: 117-120, 1920.

Chinnery, M. A., Investigations of seismological input to the safety design of nuclear power reactors in New England. *Nuclear Regulatory Commission*, Washington, D.C. NUREG/CR-0563: 1-67, 1979.

Chung, D. H. and Bernreuter, D. L., Regional relationships among earthquake magnitude scales. *U.S. Nuclear Regulatory Commission* NUREG/CR-1457-5274, 42 pp., 1980.

Cochrane, H. C., Forecasting the economic impact of a Midwest earthquake, *in* Jones, B. G. (ed.), *Economic consequences of earthquakes: preparing for the unexpected. National Center for Earthquake Engineering Research*, NCEER-SP-0001: 223-247, 1997.

Coppersmith, K. J., Conclusions regarding maximum earthquake assessment. *Electric Power Research Institute* EPRI TR-102261s-V1-V5: 6-1-6-24, 1994.

Costain, J. K., Bollinger, G. A., and Speer, J. A., Hydroseismicity: a hypothesis for the role of water in the generation of intraplate seismicity. *Seismological Research Letters* 58(3): 44-66, 1987.

Davis, J. F., Chazen, C. A., and Fakundiny, R. H., An evaluation of the influences of geologic and other natural resource information products on land use decision making (abs.).

Geological Society of America, Northeastern Section, Abstracts with Programs 10(2): 38, 1978.

Davis, J. F., Slosson, J. E., and Fakundiny, R. H., The State Federal partnership in the siting of nuclear power plants, *in* Hathaway, A. W. and McClure, C. R., Jr. (Eds.), *Reviews of Engineering Geology IV. Geological Society of America*: 47-64, 1979.

Dawers, N. H. and Seeber, L., Intraplate faults revealed in crystalline bedrock in the 1983 Goodnow and 1985 Ardsley epicentral areas, New York. *Tectonophysics* 186: 115-131, 1991.

DOE (U.S. Department of Energy), Natural phenomena hazards assessment criteria. *U. S. Department of Energy, Washington, D.C., DOE Standard* DOE-STE-1023-95, Change Notice #1: 1-38, A-1-A17, 1996.

DOE (U.S. Department of Energy), Guide for the mitigation of natural phenomena hazards for DOE nuclear facilities and nonnuclear facilities. *U.S. Department of Energy, Office of Environment, Safety, and Health, Washington, D.C., DOE* G 420.1-2: 1-21, A1-A2, B1-B6, C1-C2, 2000.

Dobry, R., Some basic aspects of soil liquefaction during earthquakes, *in* Jacob, K. H. and Turkstra, C. J. (Eds.), *Earthquake hazards and the design of constructed facilities in the eastern United States. New York Academy of Sciences Annals* 558: 172-182, 1989.

Donovan, N. C. and Bornstein, A. E., The problems of uncertainties in the use of seismic risk procedures. *Dames and Moore*, San Francisco, KEE77-4: 1-37, 1977.

Duda, S. J., Physical significance of the earthquake magnitude—the present state of interpretation of the concept. *Tectonphysics* 49: 119-130, 1978.

Ebel, J. E. and Kafka, A. L., Earthquake activity in the northeastern United States, *in* Slemmons, D. B., Engdahl, E. R., Zoback, M. D., and Blackwell, D. D. (Eds.), *Neotectonics of North America. Geological Society of America, Geology of North America, Decade of North America* Map, Chapter 15: 277-290, 1991.

Eguchi, R. T., Mitigating risks to lifeline systems through natural hazard reduction and design, *in* Jones, B. G. (Ed.), *Economic consequences of earthquakes: preparing for the unexpected. National Center for Earthquake Engineering Research,* NCEER-SP-0001: 111-123, 1997.

Engelder, T. and Sbar, M. L., Evidence for uniform strain orientation in the Potsdam Sandstone, northern New York, from *in situ* measurements. *Journal of Geophysical Research* 81: 3013-3017, 1976.

EPRI (Electric Power Research Institute), Seismic hazard methodology for the central and eastern United States. *EPRI Report* NP-4726, 10 volumes. Electric Power Research Institute, Palo Alto, California (reference from Frankel, 1995), 1986.

Eyles, N., Boyce, J., and Mohajer, A. A., The bedrock surface of the western Lake Ontario region: evidence of reactivated basement structures? *Géographie physique et Quaternaire* 47: 269-283, 1993.

Fakundiny, R. H., Geologic resources, *Genesee Finger Lakes Regional Planning Board Technical Study Series* 16, HUD Project No. B/FL RPB: 75-NYP-1039-TSR 16: 1-169, 1975.

Fakundiny, R. H., Basement tectonics and seismicity in New York State (abs.). *Geological Society of America, Northeastern Sections Annual Meetings, Abstracts with Programs*: 132, 1981.

Fakundiny, R. H. and Pomeroy, P. W., Seismic-reflection profiles of the central part of the Clarendon-Linden fault system of western New York in relation to regional seismicity.

Tectonophysics, 353: 175-213, 2002.

Fakundiny, R. H., Pferd, J. W., and Pomeroy, P. W., Clarendon-Linden fault system of western New York: longest (?) and oldest (?) active fault in eastern United States (abs.). *Geological Society of America, Northeast Section Annual Meeting, Abstracts with Programs* 10(2): 42, 1978a.

Fakundiny, R. H., Pomeroy, P. W., Pferd, J. W., and Nowak, T. A., Structural instability features in the vicinity of the Clarendon-Linden fault system, western New York and Lake Ontario, *in* Thompson, J. C. (Ed.), *Proceedings of the 12th Canadian Rock Mechanics Symposium, University of Waterloo Solid Mechanics Division Study* 13(4): 121-178, 1978b.

FEMA (Federal Emergency Management Agency), A non-technical explanation of the 1994 NEHRP (National Earthquake Hazard Reduction Program) Recommended Provisions. *Building Seismic Safety Council*, Washington, D.C., FEMA99: 1-82, 1995.

FEMA (Federal Emergency Management Agency), NEHRP Recommended Provisions for seismic regulations for new buildings and other structures. *Building Seismic Safety Council*, Washington, D.C., FEMA302, Part 1: 1-335, 1997.

FEMA (Federal Emergency Management Agency), Enhancing the Nation's risk assessment capabilities: The HAZUS strategic plan 2001-2005. *FEMA Mitigation Directorate*, Washington, D.C., 79 pp., 2001a.

FEMA (Federal Emergency Management Agency), HAZUS 99: estimated annualized earthquake losses for the United States. *FEMA Mitigation Directorate*, Washington, D.C.: 1-33, 2001b.

Fletcher, J. P. and Sykes, L. R., Earthquakes related to hydraulic mining and natural activity, western New York State. *Journal of Geophysical Research* 82: 3767-3780, 1977.

Forsyth, D. A., Milkereit, B., Davidson, A., Hamner, S., Hutchinson, D. R., Hinze, W. J., and Mereu, R. F., Seismic images of a tectonic subdivision of the Grenville orogen beneath Lakes Ontario and Erie. *Canadian Journal of Earth Sciences* 31: 229-242, 1994a.

Forsyth, D. A., Milkereit, B., Zett, C. A., White, D. J., Easton, R. M., and Hutchinson, D. R., Deep structure beneath Lake Ontario: crustal-scale Grenville subdivisions. *Canadian Journal of Earth Sciences* 31: 255-270, 1994b.

Frankel, A., Mapping seismic hazard in the central and eastern United States. *Seismological Research Letters* 66: 8-21, 1995.

Frankel, A., Mueller, C., Barnhard, T., Perkins, D., Leyendecker, E. V., Dickman, N., Hanson, S., and Hopper, M., National seismic hazard maps: documentation June, 1996. *U.S. Geological Survey Open-file Report* 96- 532: 1-63, 1996.

Gergely, P., Incorporation of seismic considerations in the New York State building code. *Multidisciplinary Center for Earthquake Engineering Bulletin* 7(2): 5, 1993.

Goltz, J. D. and Hanksson, E., TriNet, new tool for rapid earthquake response through interdisciplinary research. *Natural Hazards Observer* XXV(6): 1-2, 2001.

Gordon, D. W. and Dewey, J. W., Chapter 53: Earthquakes *in* Shultz, C. H. (Ed.), *The geology of Pennsylvania. Pennsylvania Geological Survey Special Publication* 1: 763-769, 1999.

Hadley, J. F. and Devine, J. F., Seismotectonics map of the eastern United States. *U.S. Geological Survey Field Study* MF-620, 3 maps, scale=1:5,000,000, 1974.

Hall, W. J. (Chair), Earthquake research for the safer siting of critical facilities. *National Academy of Sciences/National Research Council*, Washington, D.C.: 1-49. ISBN 0-309-03082-x, 1980.

Hall, W. J. and McCabe, S. L., Current design spectra: background and limitations, *in* Jacob, K. H. and Turkstra, C. J. (Eds.*), Earthquake hazards and the design of constructed facilities in the eastern United States. New York Academy of Sciences Annals* 558: 222-233, 1989.

Haneberg, W. C., Deterministic and probabilistic approaches to geologic hazard assessment. *Environmental and Engineering Geoscience* VI(3): 209-226, 2000.

Howell, B. F., Seismic regionalization in North America based on average regional seismic hazard index. *Seismological Society of America Bulletin* 64(5): 1509-1528, 1974.

Hutchinson, D. R., Pomeroy, P. W., Wold, R. J., and Halls, H. C., A geophysical investigation concerning the continuation of the Clarendon-Linden fault across Lake Ontario. *Geology* 7: 206-210, 1979.

ICBO (International Conference of Building Officials), *Uniform Building Code*. Whittier, CA., 1988.

Isachsen, Y. W., Possible evidence of contemporary doming of the Adirondack Mountains, New York and suggested implications for regional tectonics and seismicity. *Tectonophysics* 29: 169-181, 1975.

Isachsen, Y. W. and McKendree, W., Preliminary brittle structures map of New York: 1:250,000 and 1:500,000 generalized map of recorded joint systems in New York. *New York State Museum and Science Service Map and Chart Series* 31, 1977.

Jacob, K., Seismic vulnerability of New York State: code implications for buildings, bridges, and municipal facilities. *Multidisciplinary Center for Earthquake Engineering Research Bulletin* 7(2): 4-5. Or 2000, http:// mceer.buffalo.edu/infoservice/faqs/jacob.html., 1993.

Jacob, K. H., Scenario earthquakes for urban areas along the Atlantic seaboard of the United States, *in* Jones, B. G. (Ed.), *Economic consequences of earthquakes: preparing for the unexpected. National Center for Earthquake Engineering Research*, NCEER-SP-0001: 67-97, 1997.

Jacob, K. H., Gariel, J-C., Armbruster, J., Hough, S., Friberg, P., and Tuttle, M., Site-specific ground motion estimates for New York City. *4th U.S. National Conference on Earthquake Engineering*, Chicago v(iii): 1-10, 1990.

Jacob, K. H. and Turkstra, C. J., Preface *in* Jacob, K. H. and Turkstra, C. J. (Eds.), *Earthquake hazards and the design of constructed facilities in the eastern United States. New York Academy of Sciences Annals* 558: xi-xii, 1989.

Jacobi, R. and Fountain, J., The southern extension and reactivation of the Clarendon-Linden fault system. *Géographie physique et Quaternaire* 47: 285-302, 1993.

Jacobi, R. D., Price, R. A., Seeber, L., Stepp, J. C., and Thurston, P. C., Report of the AECB (Atomic Energy Control Board) Multi-Disciplinary Review Panel on probabilistic seismic hazard assessment for Pickering and Darlington sites. *Atomic Energy Control Board, Ottawa, Ont., Canada, Revised Final Draft*, dated 30 Sept. 97: 1- 158, 1997.

James, T. S. and Morgan, W. J., Horizontal motions due to post-glacial rebound. *Geophysical Research Letters* 17: 957-960, 1990.

Johnston, A. C., The seismicity of 'stable continental interiors' *in* Gregersen, S. and Basham, P. W. (Eds.), *Earthquakes at North-Atlantic passive margins: neotectonics and post-glacial rebound*. Kluwer Academic Publishers: 299-327, 1989.

Johnston, A. C., The earthquakes of stable continental regions: assessment of large earthquake potential. *Electric Power Research Institute Report*, EPRI TR-102261s-V1-V5: 4-1-4-103, 1994.

Jones, B. G., Recommendations for future action, *in* Jones, B. G. (Ed.), *Economic conse-*

quences of earthquakes: preparing for the unexpected. National Center for Earthquake Engineering Research, NCEER-SP-0001: 269-271, 1997.

Kane, M. F., Correlation of major eastern earthquake epicenters with mafic/ultramafic basement masses, *in* Rankin, D. W. (Ed.), *Studies related to the Charleston, South Carolina earthquake of 1886: a preliminary report.* U.S. Geological Survey Professional Paper 1028-O: 199-204, 1977.

Kanter, L. R., Tectonic interpretation of stable continental crust. *Electric Power Research Institute* EPRI TR-102261-V1-V5: 2-1-2-98, 1994.

Kay, G. M., Ottawa-Bonnechere graben and Lake Ontario homocline. *Geological Society of America Bulletin* 53: 585-646, 1942.

Kelly, J. (Chair), New York State Division of Housing and Community Renewal, New York State Seismic Provisions RS 24-11 (submitted to 1991 Uniform Building Code Sections 2710 (I)(k) and 2711 as modified. *New York State Division of Housing and Community Renewal,* Troy, NY: 1-47, 1991.

Kelly, J. (Chair), Proposed amendment to the State Uniform Fire Prevention and Building Code relating to structural requirements and seismic provision by the Structural Update Advisory Committee. *New York State Division of Housing and Community Renewal*, Troy, NY: 1-120, 1995.

Kimbal, J. and Pasedag, W. (Co-Chairs), Guidelines for use of probabilistic seismic hazard curves at Department of Energy sites. Kimbal, J. (Chair), Department of Energy Seismic Working Group. *Department of Energy Standard* DOE-STD- 102224-92: 1-35, 1992.

Kircher, C. A., Nasser, A. A., Kustu, O., and Holmes, W. R., Development of building damage functions for earthquake loss estimations. *Earthquake Spectra* 13: 663-682, 1997a.

Kircher, C. A., Reitherman, R. K., Whitman, R. V., and Arnold, C., Estimation of earthquake losses to buildings. *Earthquake Spectra* 13(4): 703-720, 1997b.

Kissinger, C., Earthquake prediction, *in* McKenzie, G. D. and Utgard, R. O. (Eds.), *Man and his physical environment.* Burgess Publishing Company, Minneapolis, Minnesota: 26-35. (Reprinted from Kissinger, C., 1974. Earthquake prediction. *Physics Today* 27:36-42), 1975.

Klein, F., Finding an earthquake's location with modern seismic networks. *U.S. Geological Survey Earthquake Hazards Program.* http://quake.wr.usgs.gov/info/eqlocation/index.html, 2000.

Kulhawy, F. H. and Ninyo, A., Earthquakes and earthquake zoning in New York State. *Association of Engineering Geologists Bulletin* XIV(2): 69-87, 1977.

Kumarapeli, P. S. and Saull, A. A., The St. Lawrence valley system: a North American equivalent to the East African rift valley system. *Canadian Journal of Earth Sciences* 3: 639-659, 1966.

Lecomte, E. L., Impact of catastrophic losses on the insurance industry, *in* Jones, B. G. (ed.), Economic consequences of earthquakes: preparing for the unexpected. *National Center for Earthquake Engineering*, NCEE-SP-0001: 249-257, 1977.

Leblanc, G. A. and Klimkiewicz, Seismological issues: history and examples of earthquake hazard assessment for Canadian nuclear generating stations. *Geological Survey of Canada Open File* 2929: 1-1-8-8, 1994.

Lo, K. Y., Regional distribution of *in situ* horizontal stresses in rocks of southern Ontario. *Canadian Geotechnical Journal* 15: 371-381, 1978.

Long, L. E, *Geology*. McGraw-Hill, New York: 287-291, 1974.

Luft, R. W., Experience with changes in design requirements in the Boston area, *in* Jacob, K.

W. and Turkstra, C. J. (Eds.), *Earthquake hazards and the design of constructed facilities in the eastern United States. New York Academy of Sciences Annals* 558: 408-420, 1989.

McCalpin, J. P. and Nelson, A. R., Introduction to paleoseismology, *in* McCalpin, J. P. (Ed.), *Paleoseismology.* Academic Press, Toronto: 1-32, 1996.

McGuire, R. K., Effects of uncertainty in seismicity of the East Coast of the United States. *Seismological Society of America Bulletin* 67: 827-845, 1977.

McGuire, R. K. and Toro, G. R., Issues in strong-motion information in eastern North America, *in* Jacob, K. H. and Turkstra, C. J. (Eds.), *Earthquake hazards and the design of constructed facilities in the eastern United States. New York Academy of Sciences Annals* 558: 136-147, 1989.

McGuire, R. K., McCann, M. W., Jr., Drake, L., Toro, G., Veneziano, D., Van Dyke, J., Boissonnade, A. C., Dong, W., and Hadidi-Tamjed, H., Seismic hazard methodology for the central and eastern United States, Volume 3: User's Manual (Revision 1*). Electric Power Research Institute*, EPRI NP-4726-CCML-A, 1989.

McKeown, F. A., Hypothesis: many earthquakes in the central and southern United States are causally related to mafic intrusive bodies. *Journal Research U.S. Geological Survey* 6: 41-50, 1978.

Mereu, R. F., Brunet J., Morrisey, K., Price, B., and Yapp, A., A study of the microearthquakes of the Gobles Oil Field area of southeastern Ontario. *Seismological Society of America Bulletin* 76(5): 1215-1223, 1986.

Mereu, R. F., Asmis, H. W., Dunn, B., Eaton, D., and Yapp, A., The seismicity of the western Lake Ontario area: results from the Southern Ontario Seismic Network (1991-2000) (abs.). *Canadian Geophysical Union Annual Meetings Abstracts with Program,* Banff, Alberta, Canada, 2000.

Merguerian, C., Stratigraphy, structural geology, and ductile- and brittle faults of New York City, *in* Benimoff, A. I. And Ohan, A. A. (Eds.*), Field trip guide for the 68th Annual Meeting of the New York State Geological Association.* New York State Museum, Albany, NY 12230: 53-77, 1996.

Merguerian, C. and Sanders, J. E., Bronx River Diversion—Neotectonic implications. *Journal of Rock Mechanics and Mining Sciences* 34(3-4), paper 198 on CD-ROM, 1997.

Miele, J. A., Adopting a seismic building code to protect the buildings of New York City, *in* Jones, B. G. (Ed.), *Economic consequences of earthquakes: preparing for the unexpected.* National Center for Earthquake Engineering Research, NCEER-SP-0001: 175-179, 1997.

Milne, W. G. and Davenport, A. G., Distribution of earthquake risk in Canada. *Seismological Society of America Bulletin* 59(2): 729-754, 1969.

Mitronovas, W., Earthquake statistics in New York State. *Earthquake Notes* 53: 5-22, 1982.

Mitronovas, W., Instrumental study of cryoseisms in east-central New York State (abs,). *American Geophysical Union EOS, Spring Meeting, 1991, Program and Abstracts*: 202, 1991.

Mitronovas, W., Seismic monitoring of the AKZO-Nobel Retsof salt mine collapse (March 94-Aprill 96)(abs.), *IRIS Consortium 9th Annual IRIS Workshop*, no pages given, 1997.

Mitronovas, W., Earthquakes! What, where, when, why? *In* Isachsen, Y. W., Landing, E., Lauber, J. M., Rickard, L. V., and Rogers, W. B. (Eds.), *Geology of New York: A simplified account.* New York State Museum Educational Leaflet 28: 231-238. Also: 2000. Do earthquakes occur in New York State? http://mceer.buffalo.edu/ infoservice/faqs/eqlist.html., 2000.

Mohajer, A. A., Relocation of seismic events in western Québec. *Multiple Agency Group for*

Neotectonics in Eastern Canada 90-02: 1-17, 1992.

Mohajer, A. A., Seismicity and seismotectonics of the western Lake Ontario region. *Géographie physique et Quaternaire* 47: 353-362, 1993.

Mohajer, A. A., Earthquake hazard in the greater Toronto area, *in* Eyles, N. (Ed.), *Environmental Geology of urban areas*. Geological Association of Canada GEOtext 3: 409-421, 1997.

Mohajer, A., Eyles, N., and Rogojina, C., Neotectonic faulting in metropolitan Toronto: implications for earthquake hazard assessment in the Lake Ontario region. *Geology* 20: 1003-1006, 1992.

Moore, R. T., A tale of two cities, *in* Jones, B. G. (Ed.), *Economic consequences of earthquakes: preparing for the unexpected.* National Center for Earthquake Engineering Research, NCEER-SP-0001: 99-109, 1997.

MSBCC (The Commonwealth of Massachusetts State Building Code Commission), *1979 Building Code, Third Edition*, Sections 716-720. Boston, MA., 1979.

Muehlberger, W. R., Tectonic map of North America, Southeast Sheet. *American Association of Petroleum Geologists*, Tulsa, OK 74101-0979, 1992.

Muir Wood, R. and Mallard, D. J., When is a fault extinct? *Geological Society, London, Journal* 149: 251-255, 1992.

NIBS (National Institute of Building Sciences), NEHRP recommended provisions for seismic regulations for new buildings and other structures, Part 1: Provisions. *FEMA* 302: 1-335, 1997.

Nicholson, C., Roeloffs, E., and Wesson, R. L., The northeastern Ohio earthquake of 31 January 1986: was it induced? *Seismological Society of America Bulletin* 78(1): 188-217, 1988.

Nordenson, G. P., Built value and earthquake risk, *in* Jones, B. G. (Ed.), *Economic consequences of earthquakes: preparing for the unexpected.* National Center for Earthquake Engineering Research, NCEER-SP-0001: 167-174, 1997.

Nottis, G. N., Epicenters of northeastern United States and southeastern Canada, onshore and offshore: time period 1534-1980. *New York State Museum Map and Chart Series* 38: 1-39, 2 plates, 1983.

NRC (U.S. Nuclear Regulatory Commission), Regulatory Guide 1.165: Identification and characterization of seismic sources and determination of safe shutdown earthquake ground motion. *U.S. Nuclear Regulatory Commission, Office of Nuclear Regulatory Research*, Draft DG-1032: 1-10, 1997.

Nuhfer, E. B., Proctor, R. J., and Moser, P. H., *The citizens guide to geologic hazards*. American Institute of Professional Geologists, Winchester, CO: 38-53, 1993.

Nuttli, O. W., Seismic wave attenuations and magnitude relations for eastern North America. *Journal of Geophysical Research* 78: 876-885, 1973.

Oliver, J., Johnson, T. and Dorman, J., Postglacial faulting and seismicity in New York and Quebec. *Canadian Journal of Earth Sciences* 7: 579-590, 1970.

Page, R. A., Molnar, P. H., and Oliver, J., Seismicity in the vicinity of the Ramapo fault, New Jersey-New York. *Seismological Society of America Bulletin* 58(2): 681-687, 1968.

Paz, M. (Ed.), *International handbook of earthquake engineering codes: codes, programs, and examples (IHEE).* Chapman & Hill, New York: 1-545, ISBN 0-412-98211-0, 1994.

Plotnikova, L. M., Flyonova, M. G., and Machmudova, V. I., Induced seismicity in the Gazly gas field region, *in* Knoll, P. (Ed.), *Induced Seismicity*. A. A. Balkema/ Rotterdam/Brookfield: 309-319, 1992.

Poland, C. D., Seismic rehabilitation techniques for buildings, *in* Jacob, K. H. and Turkstra, C. J. (Eds.), *Earthquake hazards and the design of constructed facilities in the eastern United States. New York Academy of Sciences Annals* 558: 378-391, 1989.

Pomeroy, P. W., Nowak, T. A., Jr., and Fakundiny, R. H., Clarendon-Linden fault system of western New York: a vibroseis seismic study. *New York State Geological Survey Open-File Report*, 68 pp., 1978.

Pomeroy, P. W., Simpson, D. W., and Sbar, M. L., Earthquakes triggered by surface quarrying—the Wappingers Falls, New York sequence of June, 1974. *Seismological Society of America Bulletin* 66(3): 685-700, 1976.

Press, F., A strategy for an earthquake prediction research program. *Tectonophysics* 6(1): 11-15, 1968.

Press, F. and Siever, R., *Earth.* Freeman and Company, San Francisco: 1-660, 1974.

Price, R., Earthquake in Avoka. *The Spectator, Elmira, NY*: C1 and C3, 2001.

Prucha, J. J., Zone of weakness concept: a review and evaluation, *in* Bartholomew, M. J., Hyndman, D. W., Mogk, D. W., and Mason, R. (Eds.). *Basement Tectonics 8: Characterization and comparison of ancient and Mesozoic Continental Margins. Proceedings of the 8th International Conference on Basement Tectonics.* Kluwer Academic Publications, Dordrecht, The Netherlands: 83-92, 1992.

Quinlan, G., Postglacial rebound and the focal mechanisms of eastern Canadian earthquakes. *Canadian Journal of Earth Sciences* 21: 1018-1023, 1984.

Ratcliffe, N. M., The Ramapo fault system in New York and adjacent northern New Jersey: a case of tectonic heredity. *Geological Society of America Bulletin* 82: 125-141, 1971.

Revetta, F. A., Harrison, W. P., Barstow, N., and Schlesinger-Miller, E., Seismology, tectonics and engineering geology in the St. Lawrence Valley and northwest Adirondacks. *New York State Geological Association 55th Annual Meeting Field Trip Guidebook*, New York State Museum, Albany, NY 12230: 1-5, 1983.

Richter, C. F., Seismic regionalization. *Seismological Society of America Bulletin* 49(2): 123-162, 1959.

Roden-Tice, M. K., Tice, S. J., and Schofield, I. S., Evidence for differential unroofing in the Adirondack Mountains, New York State, determined by apatite fission-track thermochronology. *Journal of Geology* 108: 155- 169, 2000.

Sanford, B. V., Thompson, F. J., and McFall, G. H., Plate-tectonics—possible controlling mechanism in the development of hydrocarbon traps in southwestern Ontario. *Bulletin of Canadian Petroleum Geology* 33(1): 52- 71, 1985.

Sbar, M. L. and Sykes, L. R., Contemporary compressive stress and seismicity in eastern North America: an example of intra-plate tectonics. *Geological Society of America Bulletin* 84: 1861-1882, 1973.

Sbar, M. L. and Sykes, L. R., Seismicity and lithospheric stress in New York and adjacent areas. *Journal of Geophysical Research* 82(36): 5771-5786, 1977.

SBCCI (Southern Building Code Congress International), *Standard Building Code.* Birmingham, AL., 1987.

Scawthorn, C. and Harris, S. K., Estimation of earthquake losses for a large eastern urban center: scenario events for New York City, *in* Jacob, K. H. and Turkstra, C. J. (Eds.), *Earthquake hazards and the design of constructed facilities in the eastern United States. New York Academy of Sciences Annals* 558: 435-451, 1989.

Scawthorn, C., Lashkari, B., and Naseer, A., What happened in Kobe and what if it happened here: *in* Jones, B. G. (Ed.), *Economic consequences of earthquakes: preparing for the unexpected. National Center for Earthquake Engineering Research*, NCEER-SP-0001: 15-49, 1997.

Scharnberger, C. K., The pseudoearthquakes of 21 and 23 February, 1954, in Wilkes-Barre, Pennsylvania. *Seismological Society of America Bulletin* 62(2): 135-138, 1991.

Seeber, L., The NCEER-91 earthquake catalog: improved intensity-based magnitudes and recurrence relations for U.S. earthquakes east of New Madrid. *National Center for Earthquake Engineering, Technical Report* 91- 0021: 1-98, 1991.

Seeber, L. and Armbruster, J. C., Natural and induced seismicity in the Lake Erie-Lake Ontario region: reactivation of ancient faults with little neotectonic displacement. *Géographie physique et Quaternaire* 47(3): 363-378, 1993.

Seeber, L., Armbruster, J. C., Kim, W.-Y., Barstow, N., and Scharnberger, C., The 1994 Cacoosing Valley earthquake near Reading, Pennsylvania: a shallow rupture triggered by quarry unloading. *Journal of Geophysical Research* 103(B10): 24,505-24,521, 1998.

Seeber, L., Statton, T., Wu, F., Pomeroy, P., Bhatia, S., Ishibashir, I., Zandt, G., Kozin, F., Revetta, F., Thurber, C. H., Fakundiny, R., Shinozaka, M., Nordenson, G., Papageorgeo, A., and Jacob, K., Earthquake hazards identification, New York State. *New York State Emergency Management Office, Report* FY1985: 1-63, 1985.

Shannon & Wilson, Statistical analysis of earthquake ground motion parameters. *U.S. Nuclear Regulatory Commission* NUREG/CR-1175, 188 pp., 1979.

Simpson, D. W., Seismicity changes associated with reservoir loading. *Engineering Geology* 10: 123-150, 1976.

Smith, W. E. T., Earthquakes of eastern Canada and adjacent areas 1534 to 1927. *Publication of the Dominion Observatory*, Ottawa 26(5): 271-301, 1962.

Smith, W. E. T., Earthquakes of eastern Canada and adjacent areas 1928 to 1959. *Publication of the Dominion Observatory*, Ottawa 32: 87-121, 1966.

Smith, M. J., Southard, J. B., and Remery, R., EarthCom-Earth system science in the community. It's About Time, Inc., Armonk, NY, Earthquakes:G-120-G-172, *American Geological Institute*, Alexandria, VA 22302. ISSBN 1-5859-160-0, 180 pp., 2001.

Singh, S. and Herrmann, R. B., Regionalization of crustal coda Q in the continental United States. *Journal of Geophysical Research* 88: 527-538, 1983.

Somerville, P., Earthquake source and ground-motion characteristics in eastern North America, *in* Jacob, K. H. and Turkstra, C. J. (Eds.), *Earthquake hazards and the design of constructed facilities in the eastern United States. New York Academy of Sciences Annals* 558: 105-121, 1989.

Statton, C. T., Quittmeyer, R. C., and Davison, F., Attenuation of seismic energy in New York State: Final Report. *Empire State Electric Energy Research Corporation Report* EP 83-14: 1-35, 1985.

Stevens, A. E., Earthquakes in the Lake Ontario region: intermittent scattered activity, not persistent trends. *Geoscience Canada* 21(3): 105-111, 1995.

Stover, C. W. and Coffman, J. L., Seismicity of the United States, 1568-1989 (revised). *U.S. Geological Survey Professional Paper* 1527: 1-17, 313-319, 400-418, 1993.

Talwani, C. W., The intersection model for intraplate earthquakes. *Seismological Research Letters* 59: 305- 310, 1998.

Talwani, P. and Acree, S., Pore pressure diffusion and the mechanism of reservoir-induced seismicity. *Journal of Pure and Applied Geophysics* 122: 947-965, 1984/1985.

Thomas, R. L., Wallach, J. L., McMillan, R. K., Bowlby, J. R., Frape, S., Keyes, D., and Mohajer, A. A., Recent deformation in the bottom sediments of western and southeastern

Lake Ontario and its association with major structures and seismicity. *Géographie physique et Quaternaire* 47: 325-335, 1993.

Tierney, K. J., Impacts of recent disasters on businesses: the 1993 Midwest floods and the 1994 Northridge earthquake, *in* Jones, B. G. (Ed.), *Economic consequences of earthquakes: preparing for the unexpected. National Center for Earthquake Engineering Research*, NCEER-SP-0001: 189-202, 1997.

Tuttle, M. P., Case study of liquefaction induced by the 1944 Massena, New York-Cornwall, Ontario earthquake. *U.S. Nuclear Regulatory Commission* NUREG/CR-6495: 1-23, 1996.

Tuttle, M., Dyer-Williams, K., and Barstow, N., Seismic hazard implications of paleoliquefaction study along the Clarendon-Linden fault system in western New York State (abs.). *Geological Society of America Abstracts with Programs* 28(3): 106, 1996.

Tuttle, M. P. and Seeber, L., Historic and prehistoric earthquakes-induced liquefaction in Newbury, Massachusetts. Geology 19: 594-597, 1991.

Van Tyne, A. M., Clarendon-Linden structure, western New York. *New York State Geological Survey Open- File Report*: 1-10, 6 plates, 1975.

Veneziano, P. and Chouinard, L., Combination of seismic-source and historic estimates of earthquake hazards, *in* Jacob, K. H. and Turkstra, C. J. (Eds.), *Earthquake hazards and the design of constructed facilities in the eastern United States. New York Academy of Sciences Annals* 558: 148-161, 1989.

von Hake, C. A., Earthquake history of New York. *Earthquake Information Bulletin* 7(4), 1975.

Wallach, J. L., Newly discovered geological features and their potential impact on Darlington and Pickering. *Atomic energy Control Board of Canada Report* INFO-0342: 1-17, 1990.

Wallach, J. L. and Mohajer, A. A., Integrated geoscientific data related to assessing seismic hazard in the vicinity of the Darlington and Pickering nuclear power plants, *in Prediction and performance in géotechnique. Canadian Geotechnical Conference*, Quebec, Que., October 10-12, 1990, 2: 679-686, 1990.

Wallach, J. L., Mohajer, A. A., McFall, G. H., Bowlby, J. R., Pearce, M., and McKay, D. A., Pop-ups as geological indicators of earthquake-prone areas of intraplate eastern North America. *Quaternary Proceedings* 3: 67-83, 1993.

Wallach, J. L., Mohajer, A. A., and Thomas, R. L., Linear zones, seismicity, and the possibility of a major earthquake in the intraplate western Lake Ontario area of eastern North America. *Canadian Journal of Earth Sciences* 35: 762-786, 1998.

Ward, P. L., Effective disaster warnings, a national tragedy. *National Hazards Observer* XXV(6): 3-4, 2001.

Werner, S. D., Dickenson, S. E., and Taylor, C. E., Seismic performance of ports: lessons from Kobe, *in* Jones B. G. (Ed.), *Consequences of earthquakes: preparing for the unexpected. National Center for Earthquake Engineering Research*, NCEER-SP-0001: 125-153, 1997.

Wesson, R. L. and Nicholson, C., Earthquake hazard associated with deep well injection: a report to the U.S. Environmental Protection Agency. *U.S. Geological Survey Open-File Report* 87-33: 1-116, 1987.

Wheeler, R. L., Earthquakes and cratonward limit of Iapetan faulting in eastern North America. *Geology* 23: 105-108, 1995.

Wheeler, R. L. and Johnston, A. C., Geologic implications of earthquake source parameters in central and eastern North America. *Seismological Research Letters* 63(4): 491-514, 1992.

Wheeler, R. L., Trevor, N. K., Tarr, A. C., and Crone, A. J., Earthquakes in and near Northern United States. *U.S. Geological Survey Geologic Investigations Series* Map I-2737, scale=1:1,500,000, 2001.

Whitman, R. V., Engineering issues of earthquake hazard mitigation in eastern North America, *in* Jacob, K. H. and Turkstra, C. J. (Eds.), *Earthquake hazards and the design of constructed facilities in the eastern United States. New York Academy of Sciences Annals* 558: 11-20, 1989.

Whitman, R. V., Anagnos, T., Kircher, C. A., Lagorio, H. J., Lawson, R. S., and Schneider, P., Development of a National earthquake loss estimation methodology. *Earthquake Spectra* 13(4): 643-661, 1997.

Woodhouse, E. J., Earthquake hazards: a political perspective, *in* Jacob, K. H. and Turkstra, C. J. (Eds.), *Earthquake hazards and the design of constructed facilities in the eastern United States. New York Academy of Sciences Annals* 558: 72-80, 1989.

Yang, J-P. and Aggarwal, Y. P., Seismotectonics of northeastern United States and adjacent Canada. *Journal of Geophysical Research* 86(B6): 4981-4998, 1981.

Youd, T. L. and Idress, I. M., Evaluation of liquefaction resistance of soils: Proceedings of the NCEER Workshop. *National Center for Earthquake Engineering Research*, NCEER-97-0022: 1-40, 1997.

Young, K., *Geology: the paradox of Earth and man.* Houghton Mifflin Company, Boston, Massachusetts, 526pp., 1975.

Zoback, M. D. and Zoback, M. L., *In situ* stress, crustal strain, and seismic hazard assessment in eastern North America, *in* Jacob, K. H. and Turkstra, C. J. (Eds.), *Earthquake hazards and the design of constructed facilities in the eastern United States. New York Academy of Sciences Annals* 558: 54-65, 1989.

Zoback, M. D. and Zoback, M. L., Tectonic stress field of north America and relative plate motions, *in* Slemmons, D. B., Engdahl, E. R., Zoback, M. D., and Blackwell, D. D. (Eds.), *Neotectonics of North America. Geological Society of America, The Geology of North America* Decade Map 1, Chapter 19: 339-366, 1991.

6

Facing Volcanic and Related Hazards in the Neapolitan Area

Giovanni Orsi, Sandro de Vita, Mauro A. Di Vito,
Roberto Isaia, Rosella Nave, and Grant Heiken

INTRODUCTION

An estimated 50 million people living in cities around the world are at risk from volcanic eruptions, and population in volcanic areas is continuing to rise. Mostly for soil fertility and abundance of volcanic rocks that are good building material, cities on or near volcanoes have grown and continue to grow. If, as is the case for the Neapolitan area in Italy, they are in a temperate climate zone, in both commercially advantageous and strategically favorable areas, despite the hazards, humans find good reasons for settlement and development.

With a growing realization of the important link between population density and the natural environment, humanity is faced with the urgent task of the re-establishing and preserving the natural equilibrium. Such a task is even more urgent in areas where nature includes the potential for catastrophic volcanic eruptions. As such eruptions recur in intervals generally exceeding a human life-time, humanity tends to forget them and underestimate the hazards. Ongoing political concerns of fresh water, sanitation, transportation, and industrial growth often exclude planning for volcanic risk, leaving a monumental task for those who recognize volcanoes as the common link to the problems of natural equilibrium.

In this chapter the relationships among volcanism, characteristics of the physical environment, and millenary human inhabitation in the Neapolitan area will be analyzed and discussed. After elucidating the geological history of the area and of each of its three active volcanoes, and their present state, we will run through the history of the development of the city since its foundation in the 7th

Earth Science in the City: A Reader

10.1029/056SP07

century BC with particular reference to quarrying activity and use of volcanic rocks as building materials, and the related hazards. We will then analyze the natural hazards posed by the regional geological setting, the presence of active volcanoes, and the lithological and geomorphological characteristics of the area. Finally we will illustrate the measures taken by the authorities to mitigate the related risks.

GEOLOGY AND GEOMORPHOLOGY OF THE NEAPOLITAN AREA

The Neapolitan area is located within the Campanian Plain, which is bordered by the Southern Apennines (Fig. 1). This mountain chain is the result of the deformation of the African continental margin and is composed of a variety of Mesozoic and Palaeogenic palaeogeographic domains (D'Argenio et al., 1973). The crust is about 10 km thick and composed of a pile of tectonic thrusts made up of Triassic to Pliocene sedimentary rocks overlying a crystalline-metamorphic basement. From Miocene to Pliocene time, variable compressional tectonic phases have deformed both the sedimentary rocks and their basement terrain. Since Quaternary times, mostly extensional tectonic phases have generated the present setting of the Campanian area. The Apennines mountain chain is composed of Mesozoic carbonate and Mio-Pliocene terrigenous sequences, overlain by Quaternary continental deposits generated by both volcanism and sedimentation related to intense erosion during the general building up of the chain. The Campanian Plain, in which lie the active volcanoes, is composed of 2-3,000 m thick sequences of Plio-Quaternary sediments, mostly continental and deltaic and subordinately marine, intercalated with volcanic deposits. It is underlain by a graben formed during activation of NW-SE and NE-SW trending normal faults which, at least during Quaternary times (Brancaccio et al., 1991), have downthrown the western Apennines. The regional stress regime, which has determined the formation of the plain, has also favored generation and rise of the magmas that have fed recent and active volcanism. Geophysical data (Carrara et al., 1973; 1974) and deep wells (Ippolito et al., 1973) have shown the presence of volcanic rocks beneath the sediments filling the plain. These rocks, which are most likely Early-Pliocene to Late-Pleistocene in age, have calc-alkaline compositions, while those exposed and overlying the pile of sediments are alkaline. The active volcanoes of the Neapolitan area, within the Campanian Plain, are the Campi Flegrei (burning fields) and the Somma-Vesuvius. A third active volcanic field is the island of Ischia, located 20 km to the south-west of the city of Naples, at the north-western corner of the gulf of Naples. Between Ischia and the Campi Flegrei is the island of Procida, another volcano whose last eruption occurred about 18 ka bp.

The Neapolitan area is mostly made up of volcanic rocks and subordinately of shallow-sea, coastal and palustrine sediments, as well as rock bodies related to

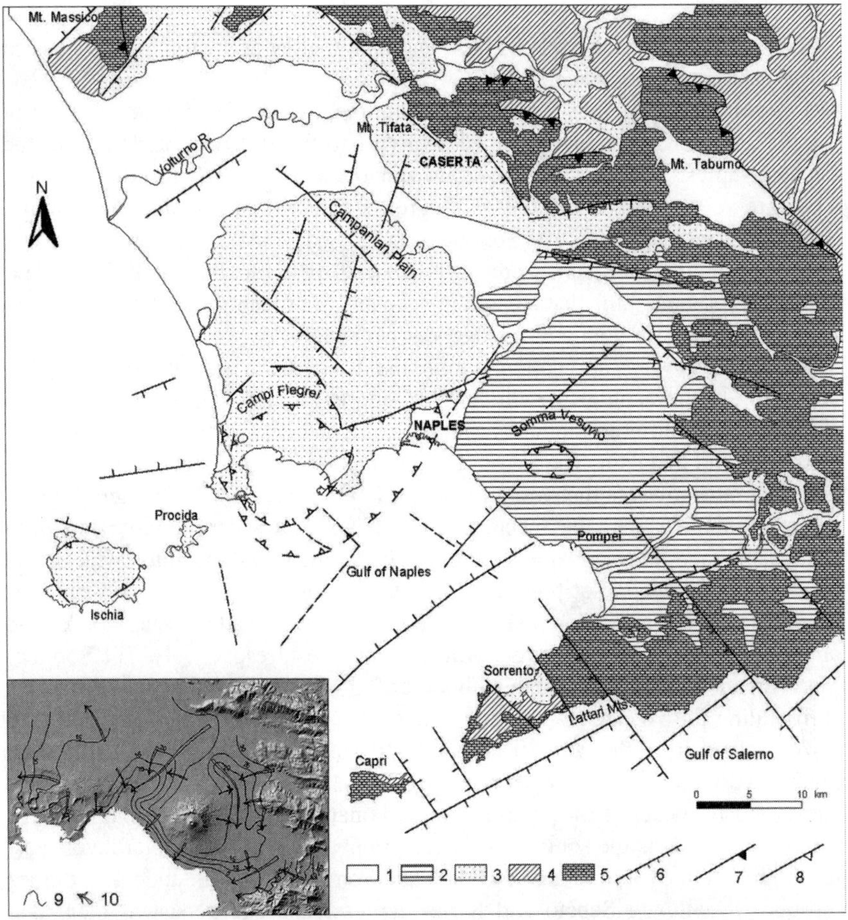

Figure 1. Geological and structural sketch map of the Southern Campanian Plain. In the insert is a hydrogeological map. 1) Quaternary and active terrigenous sediments; 2) Somma-Vesuvius volcanic deposits; 3) Neapolitan-Phlegraean, Procida and Ischia volcanic deposits; 4) Pliocene and Miocene terrigenous sediments; 5) Mesozoic carbonatic units; 6) Faults; 7) Overthrusts; 8) Caldera rims; 9) Isopiezometric curves; 10) Preferential drainage axes.

the geomorphic evolution of slopes (Fig. 1). Deposits related to the long history of human inhabitation in the area are also quite widespread. The large majority of the volcanic rocks in the continental portion of the area have been generated by Campi Flegrei and Somma-Vesuvius volcanic systems, only very few by the Ischia volcano.

The city of Naples lies within the nested Campi Flegrei caldera, with its western periphery inside the still active and restless Neapolitan Yellow Tuff caldera, while the eastern periphery is within the Sebeto valley at the foot of the western slopes of Somma-Vesuvius.

The dominant geomorphic features are the Somma-Vesuvius, the Campi Flegrei and the Campanian Plain, respectively to the east, west and north of the city of Naples. The two volcanoes are separated by the Sebeto alluvial plain. They are very different, the Campi Flegrei is a nested caldera and the Somma-Vesuvius is a strato-cone. The Somma-Vesuvius is a cone 1,281 m high and visible from all the Campanian area, while the Campi Flegrei is more complex, composed of many hills that are either remnants of volcanoes predating the main caldera collapses or younger monogenetic volcanoes.

The ground-water circulation within the Neapolitan area is quite variable according to the geological setting (Corniello et al., 1990; Celico et al., 1998) (Fig. 1). Three different domains can be distinguished: Campi Flegrei, Somma-Vesuvius, and the plain. The Campi Flegrei include the hills that extend from the city of Naples towards the west and the north, into the plain. The eastern limit of this area coincides with the Sebeto depression. Ground-water is confined to loose pyroclastic and non-volcanic sedimentary deposits and, lacking widespread impermeable beds, is not present as isolated aquifers. The ground water flows into the sea, the Campanian plain, the Volla depression, and the Lucrino, Averno and Fusaro lakes towards south, north, east and west, respectively. The Somma-Vesuvius includes a water table with a radial flow into the sea or into the plains surrounding the volcano. A significant part of the water supply of the Vesuvian towns comes from this aquifer. The largest water reservoir of the Campanian plain is within loose sediments underlying the widespread Campanian Ignimbrite and is fed by waters flowing from the carbonate rocks of the Apennines. The water flow towards the south-west is constrained by two high water levels corresponding to the Campi Flegrei caldera and Somma-Vesuvius, and reach the sea through the alluvial Sebeto and Sarno depressions. The persistent magmatic activity induces generation of mineral and thermal waters along tectonic features which facilitate rising gases and heat from depth. Those thermo-mineral waters have been used since Roman times for therapeutic purposes and have also been a source for tourism.

Campi Flegrei Caldera

The Campi Flegrei caldera is the main feature of the Neapolitan-Phlegraean area. It includes a continental and a submerged part (Fig. 2) (Orsi et al., 1996). The shape is very complex, resulting mostly from alternating, sometimes coeval, constructive and destructive volcanic and/or volcano-tectonic events. Variations in time and space of the relationships between sea and earth-surface level have also greatly contributed to the present morphological setting.

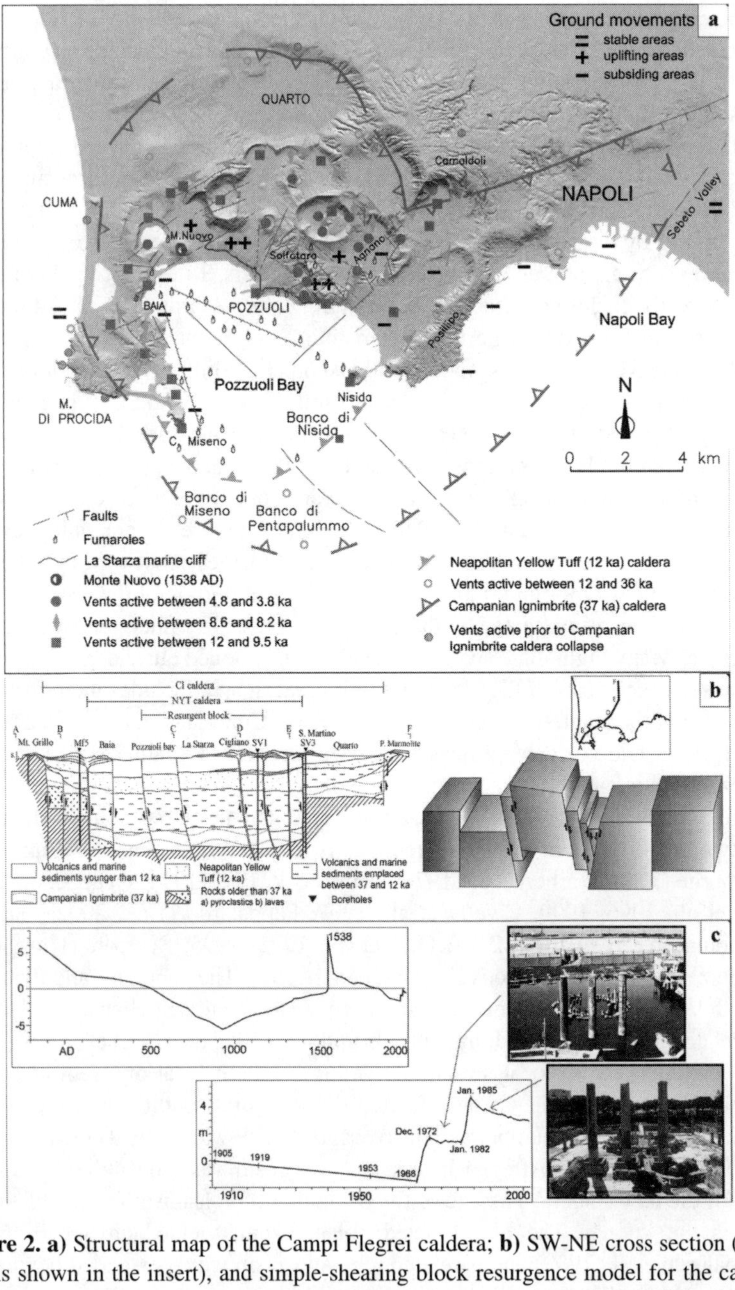

Figure 2. a) Structural map of the Campi Flegrei caldera; **b)** SW-NE cross section (location is shown in the insert), and simple-shearing block resurgence model for the caldera complex; **c)** vertical ground movements at Serapis Roman market in Pozzuoli. In the photographs the market before and after the recent bradyseismic events, showing flooding by the sea and dry land as the area rose above sea level.

The continental portion includes the Campi Flegrei and the city of Naples. The horse-shoe-shaped Campi Flegrei surround Pozzuoli bay and include a central part with a rugged morphology which is the consequence of many overlapping tuff rings and tuff cones, surrounded to the west, north and east by lowlands. To the south the Campi Flegrei connect with the Pozzuoli bay through a narrow, low-lying coast. A steep wave-cut cliff, cut into the La Starza marine terrace, occurs a few hundreds of meters inland. To the west the volcanic field merges into a flat littoral plain broken by the ridges of Monte di Procida and Cuma. The Quarto, Pianura and Soccavo plains are bordered to the north by steep cliffs. To the east the Fuorigrotta plain is partly enclosed by the NE-SW trending steep scarps of the Posillipo hill. The Posillipo hill extends northeastward into a ridge, including the Camaldoli, Vomero, San Martino, Capodimonte and Capodichino hills. The Camaldoli hill is the easternmost and highest (458 m a.s.l.) hill adjacent to the Campi Flegrei; it slopes gently northward into the Campanian Plain. This slope is cut by a dense drainage system. The Megaride islet, and the Pizzofalcone, San Martino, Ponti Rossi, and Poggioreale hills create a southward concave arc bordering westward and northward a large plain. This plain, which includes the Sebeto valley, extends eastward up to the Somma-Vesuvius volcano and merges southward into the Gulf of Naples. The geomorphic setting of the continental portion of the Campi Flegrei caldera has been also affected by the continuous presence of humans over the past millennia. Many high-angle scarps are walls of abandoned quarries.

The submerged part of the Neapolitan-Phlegraean area includes the north-western sector of the Naples bay and the Pozzuoli bay. Continuation underwater of on-land features and volcanic edifices have been recognized on the basis of morphological, seismic, gravimetric and magnetic data.

The Campi Flegrei caldera (Fig. 2a, b) is a resurgent nested structure formed during two major caldera collapses related to large eruptions of the Campanian Ignimbrite (37 ka; Barberi et al., 1978; Deino et al., 1992, 1994; Fisher et al., 1993; Rosi et al., 1996, 1999; Civetta et al., 1997; Pappalardo et al., 2002a) and the Neapolitan Yellow Tuff (12 ka; Orsi et al., 1992, 1995; Scarpati et al., 1993; Wohletz et al., 1995), respectively (Orsi et al., 1996). The geometry and dynamics of both large calderas, as well as of smaller volcano-tectonic collapses such as the one related to the Agnano-Monte Spina eruption (4.1 ka; de Vita et al., 1999), were deeply influenced by local structural setting and both local and regional stress regimes. Each large collapse has effected the structural conditions of the system, including the magma chamber and the overlying shallow crust, and constrained the foci of later volcanism. After each collapse, volcanism was concentrated within the floor of the new caldera. The age of the beginning of volcanism in the Phlegraean area is not known. The oldest dated volcanic unit yielded an age of 60 ka (Pappalardo et al., 1999) and is related to a volcanism which extended beyond the present caldera (Fig. 3).

The Campanian Ignimbrite eruption and caldera collapse was the earliest event to profoundly influence the present geological setting of the area. It erupted

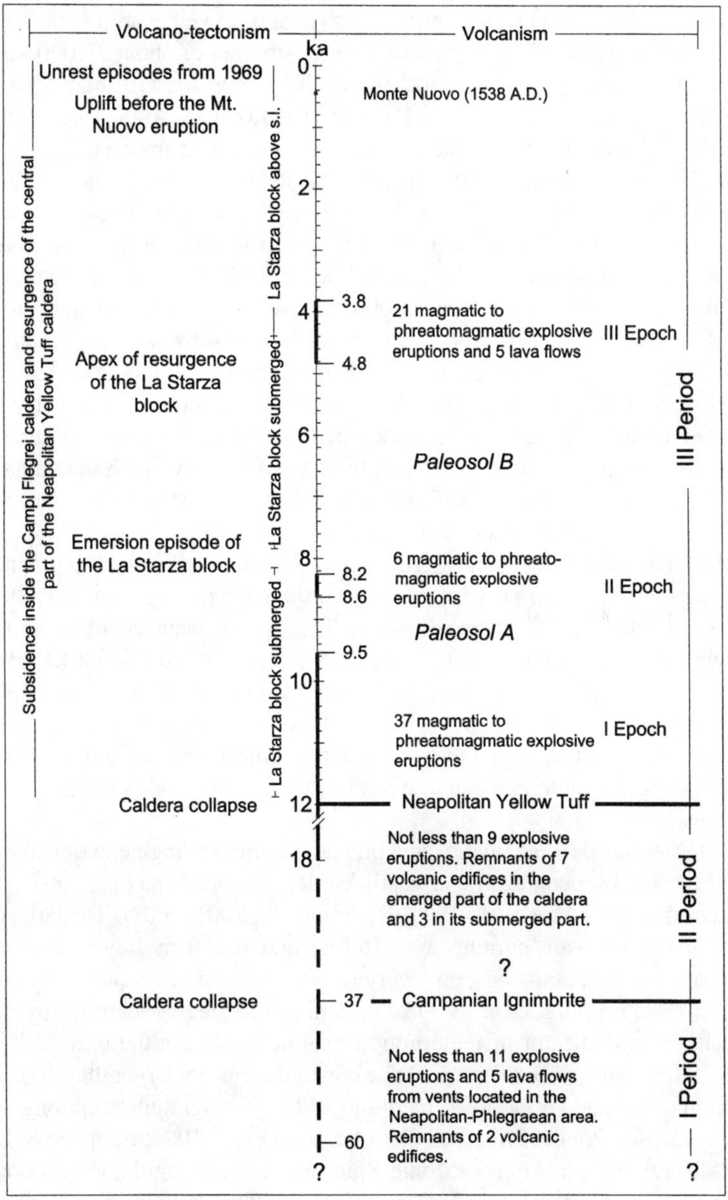

Figure 3. Chronogram of volcanic and deformational history of the Campi Flegrei caldera.

about 150 km^3 of magma (dense rock equivalent [DRE]) and generated a sustained column followed by variable dilute pyroclastic currents, which surmounted ridges with elevations greater than 1,000 m and flowed over the sea. The pyroclastic currents laid down a tuff that covered an area of about 30,000 km^2, significantly modifying the landscape. In proximal areas the tuff includes a densely-welded ignimbrite (known as Piperno [De Lorenzo, 1904]) and lithic-rich breccias. All these lithotypes have been used as building material.

The Neapolitan Yellow Tuff eruption and caldera collapse was the last dramatic event in the history of the caldera. Again the structural setting of the system was significantly modified. The eruption was a phreatoplinian to phreatomagmatic event which erupted about 40 km^3 of magma (DRE). The tuff, which covered an area of about 1,000 km^2, includes a Lower and an Upper Member. The Lower Member was produced mostly by phreatoplinian phases alternating with magmatic explosions. It is generally whitish and mostly composed of cohesive ash beds. The Upper Member, much thicker than the Lower Member, includes a sequence of pyroclastic-current deposits, mostly surges. It is generally zeolitized and acquires a yellowish color from which the name of the unit, the Neapolitan Yellow Tuff, is derived. The Neapolitan Yellow Tuff is commonly used as building material and is the skeleton of the City of Naples.

After the Neapolitan Yellow Tuff eruption, both volcanism and long-term deformation have been very intense within the caldera (Fig. 3) (Isaia, 1998; Di Vito et al., 1999; Orsi et al., 1999a). There have been about 70 eruptions, grouped in three epochs of activity (12.0-9.5, 8.6-8.2 and 4.8-3.8 ka). The last event was in 1538, after about 3,000 years of quiescence, and formed the Mt. Nuovo cone (Di Vito et al., 1987). 65 of the post-Neapolitan Yellow Tuff eruptions were explosive, characterized by alternating phreatomagmatic and magmatic phases, which generated pyroclastic currents and sustained eruption columns.

Contemporaneous magmatic and phreatomagmatic fragmentation dynamics has been widely recognized and has been demonstrated for the Agnano-Monte Spina eruption (de Vita et al., 1999; Dellino et al., 2001, 2003). The large majority of the pyroclastic currents were dilute and turbulent flows generated by phreatomagmatic explosions; only very few were dense flows caused by eruption column collapse. Orsi et al. (2003), assuming the area covered by pyroclastic deposits as an indicator of the eruption magnitude during the past 12 ka, have defined the events whose deposits have covered areas not larger than 20 and 200, and larger than 200 km^2 as low-, medium- and high-magnitude eruptions, respectively. Only the Pomici Principali (10.3 ka; Lirer et al., 1987; Isaia, 1998; Di Vito et al., 1999) and the Agnano-Monte Spina were high-magnitude eruptions and occurred during the 1st and 3rd epochs of activity, respectively. During each epoch, eruptions have followed one another at mean time intervals of a few tens of years. Fallout deposits of the 1st epoch covered the north-eastern sector of the Campi Flegrei caldera and the Camaldoli hill, 15 km from the caldera center.

Only beds of the Agnano-Monte Spina Tephra are as much as 20 cm thick along the western margin of the Apennines, at about 50 km from the vent. Pyroclastic currents traveled within the Neapolitan Yellow Tuff caldera floor and reached the northern slopes of the Camaldoli hill and the Campanian plain, north of Cuma. The western part of the city of Naples and the towns of Pozzuoli, Quarto and Marano were the areas more frequently covered by fallout deposits. The northern and eastern sectors of the Campi Flegrei caldera were more frequently invaded by pyroclastic density currents. The more energetic currents could exit the caldera lowland along deep valleys and low-angle portions of the scarps bordering the caldera, and flow over the northern slopes of the Camaldoli hill. The eruptions of the 2nd epoch were all low-magnitude events and their fallout deposits covered only the caldera and its immediate surroundings. The north-eastern portion of the caldera, including parts of what is now the city of Naples and the towns of Quarto and Marano, were more frequently affected by fallout. The pyroclastic currents deposited most of their load within the caldera lowland; those erupted along the north-eastern portion of the caldera marginal faults, also flowed over the northern and eastern slopes of the Camaldoli hill. The fallout deposits of the 3rd epoch and of the Mt. Nuovo eruption covered the caldera floor and its surroundings (Fig. 4). Only beds of the Agnano-Monte Spina Tephra covered a large area eastward up to the Apennines. The area more frequently affected by fallout extended from what is now the city of Naples to the towns of Qualiano and Bacoli, towards the north and the west, respectively. Pyroclastic currents traveled across the caldera floor and subordinately over the northern slopes of the Camaldoli hill. More frequently they flowed within the central and eastern sectors of the caldera floor, and portions of the city of Naples and of the towns of Pozzuoli and Quarto.

The first eruption of an epoch has never been the largest and there is no relation between length of quiescence and magnitude of the following eruption. The caldera has been affected by structural resurgence through a simple-shearing mechanism (Orsi et al., 1991) that broke its floor in blocks and caused a maximum net uplift of about 90 m at the La Starza marine terrace. The distribution of the vents active through time is a good tracer of the structural conditions of the caldera. A new deformation mechanism was established within the caldera between the 2nd and 3rd epoch of activity, that is not later than 5 ka bp. This mechanism generated a compressive stress regime within the south-western portion of the caldera floor, which corresponds to the Pozzuoli bay, and a tensile stress regime within the north-eastern portion, corresponding to the area between the Agnano and San Vito plains (Fig. 2a, b).

During the past 2 ka, that is since Roman times, the floor of the caldera has been affected by ground movement, documented at the Serapis Roman market in Pozzuoli (Fig. 2c). Since late 1960s, unrest episodes, documented by geophysical and geochemical monitoring systems, have been experienced by the present population. These episodes occurred in 1969-72, 1982-84, 1989, 1994 and 2000

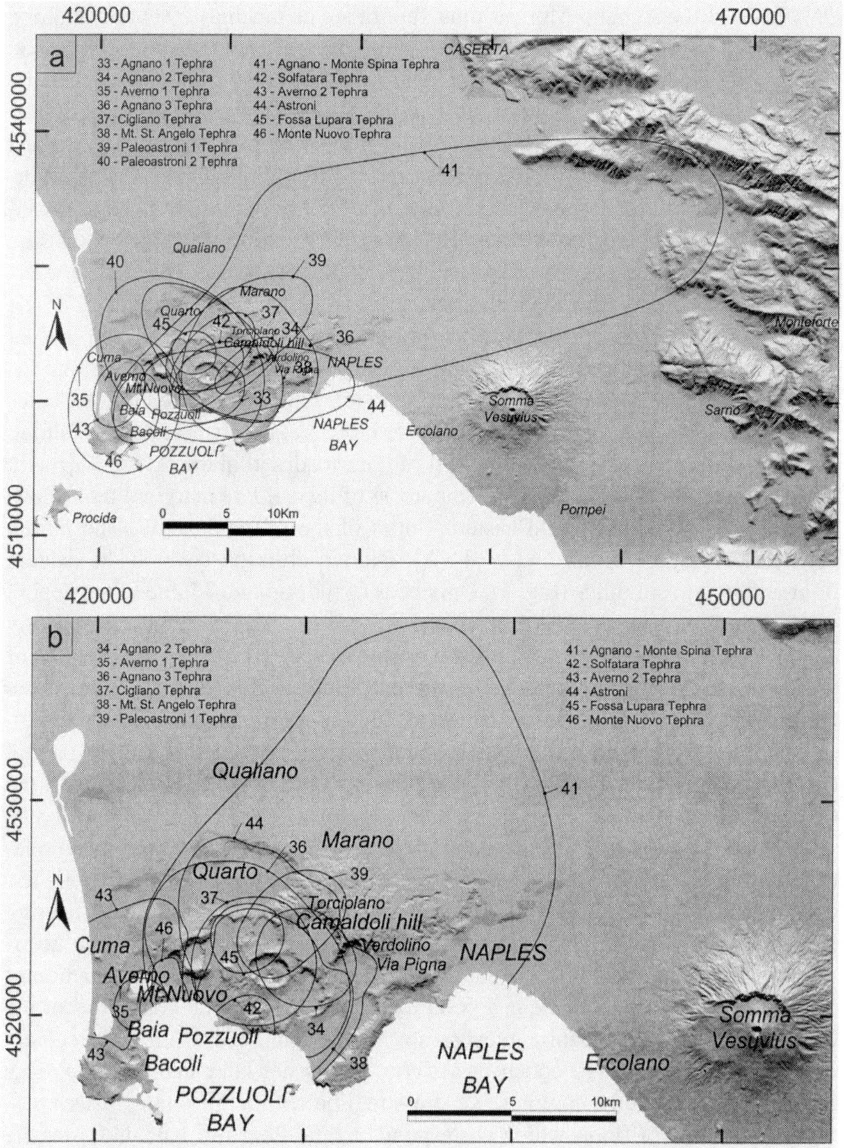

Figure 4. Distribution of the pyroclastic deposits of the 3rd Epoch and Mt. Nuovo (AD 1538). a) 20-cm isopachs of fallout deposits; b) areal distribution of pyroclastic-current deposits.

and have generated uplifts of 170, 180, 7, 1 and 4 cm, respectively. Geometry of this recent deformation is very similar to that of the long-term deformation. This element, together with the earthquake focal mechanisms, suggests that also during the unrest episodes there was a fragile component of the short-term deformation that occurs through a simple-shearing mechanism. Therefore the unrest episodes likely are transient events during the long-term deformation and reflect the stress regime within the caldera, which has not changed over at least the past 5 ka (Orsi et al., 1999a).

The magmatic system of the Campi Flegrei caldera is composed of a shallow, large-volume trachytic magma chamber periodically refilled, since at least 60 ka bp, by batches of magma from a storage zone between 10 and 15 km deep (Orsi et al., 1992; Civetta et al., 1997; D'Antonio et al., 1999; Pappalardo et al., 1999, 2002b). Complex differentiation processes have operated within the shallow chamber. In the past 12 ka three geochemically, and isotopically distinct magmatic components have been erupted: one similar to the Campanian Ignimbrite magma, another similar to the Neapolitan Yellow Tuff magma, and a third is a trachybasalt never erupted before. These data, together with the results of thermal (Wohletz et al., 1999) and magnetic (Orsi et al., 1999a) modeling suggest that a large volume of molten magma beneath the caldera persists even today.

Somma-Vesuvius

The Somma-Vesuvius is composed of an older volcano, Mt. Somma, which was truncated by a summit caldera formed during the AD 79 eruption (Cioni et al., 1999) and a more recent cone, Vesuvius, within the caldera. The growth of the Vesuvius cone has taken place, although with some minor summit collapses, during periods of persistent low-energy open-conduit activity. The last period of such activity occurred between 1631 and 1944 (Andronico et al., 1995; Cioni et al., 1999; Arrighi et al., 2001).

The caldera is elliptical (Fig. 5) with east-west long axis of about 5 km. It is a complex structure resulting from several collapses, each related to a Plinian eruption, the last of which was that of AD 79 (Cioni et al., 1999). The northern caldera margin is a 300 m high scarp, whose summit is at about 1,000 m a.s.l.. The southern portion of the caldera has been filled by lava flows which, since Medieval times, have flowed over the margin and covered almost completely the southern slopes of the Mt. Somma as far as the seacoast. The maximum elevation of this sector is about 700 m a.s.l. Lava flows have filled the caldera and created a wide flat area that connects the Vesuvius cone with the inner slopes of the caldera.

The north-eastern slopes of the volcano are cut by a dense and well-developed radial drainage system in unconsolidated pyroclastic rocks with some local control by NE-SW and NW-SE trending faults. The drainage system of the Vesuvius cone and of the younger slopes of the volcano down to the sea, is also radial but more poorly developed. The slopes vary with increasing elevation from 6 to 40°. Most of the north-eastern slopes are very steep, while the south-western slopes are never more than 25°. The outer flanks of the Vesuvius cone as well as the inner walls of the crater are steep slopes. At the foot of the slopes there are both

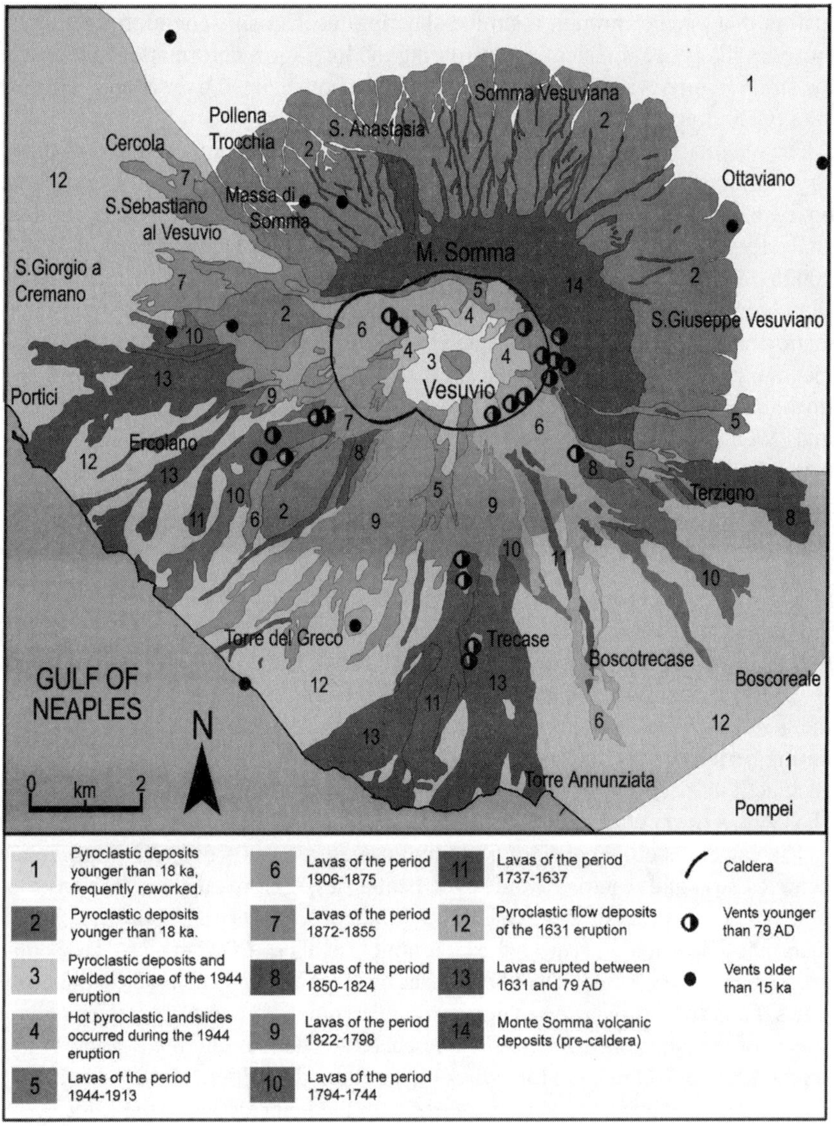

Figure 5. Geological sketch map of Somma-Vesuvius.

pyroclastic fans generated by pyroclastic flows, and wide alluvial fans. Many quarries for extraction of pyroclastic rocks and lavas occur along the medium-low portion of the slopes of the volcano. Some of these quarries have also been used for waste disposal.

Volcanism in the Somma-Vesuvius area was active at least 400 ka bp as testified by lavas intercalated with tuffs and marine sediments, sampled in drill holes at 1,350 m depth (Bernasconi et al., 1981; Santacroce, 1987). The available data do not link the volcanic rocks to a central volcano or to fissure activity. These rocks are overlain by the Phlegraean Campanian Ignimbrite. The growth of Mt. Somma began after emplacement of this widespread tuff deposit (Fig. 6). The activity of Mt. Somma was mostly effusive and subordinately explosive, with low energy events. Eruptions occurred along the central vent of the volcano determining and controlling its growth up to an estimated elevation of 1,600-1,900 m a.s.l. (Cioni et al., 1999). Activity also occurred at lateral vents aligned along faults and fractures as testified by dikes exposed in the caldera wall, and cones occurring along the slopes of the volcano and in the surrounding plains.

The earliest known Plinian eruption (Pomici di Base; 18.3 ka; Arnò et al., 1987; Andronico et al., 1995; Bertagnini et al., 1998; Cioni et al., 1999) determined the beginning of both collapse of the Mt. Somma volcano and formation of the caldera. Since then the history of the volcano as been dominated by three more Plinian eruptions (Mercato, 8 ka; Avellino, 3.8 ka; Pompeii, AD 79) and several sub-Plinian events, and by open-conduit activity characterized by Strombolian explosions and lava flows (Fig. 6).

The Pomici di Base eruption was followed by eruption of lavas that flowed along the eastern slopes of the volcano, and a quiescent period interrupted 15 ka bp by the sub-Plinian eruption of the Pomici Verdoline (Arnò et al., 1987; Andronico et al., 1995). The subsequent long period of quiescence, during which only two low-energy eruptions took place (Andronico et al., 1995), lasted until 8 ka bp, when it was broken by the Plinian Mercato eruption (Johnston Lavis, 1884; Lirer et al., 1973; Delibrias et al., 1979; Rolandi et al., 1993c; Cioni et al., 1999). During the following period of quiescence, interrupted only by two low-energy eruptions, a thick paleosol formed. This paleosol contains many traces of human occupation until the Early Bronze age, and is covered by the deposits of the Plinian Avellino eruption (3.8 ka; Rolandi et al., 1993a; Andronico et al., 1995; Cioni et al., 1995, 1999). This eruption was followed by at least 8 Strombolian to sub-Plinian eruptions, over a relatively short time, and by not less than 7 centuries of quiescence, broken by the Plinian AD 79 eruption (Lirer et al., 1993; Sigurdsson et al., 1985; Cioni et al., 1992, 1999). After the Plinian eruption of AD 79, the volcano has generated only two more sub-Plinian events in AD 472 (Rosi and Santacroce, 1983; Arnò et al., 1987; Lirer et al., 2001) and 1631 (Arnò et al., 1987; Rolandi et al., 1993b; Rosi et al., 1993), and low-energy open-conduit activity between the 1st and 3rd, 5th and 8th, 10th and 11th centuries, and 1631 and 1944 (Andronico et al., 1995; Cioni et al., 1999; Arrighi et al., 2001). This open-conduit activity generated a large amount of lava, which flowed along the south-eastern and south-western slopes of the volcano.

On the basis of the compositional variability of the magmas, Santacroce et al. (2003), divided the history of the volcano in three magmatic periods. The oldest

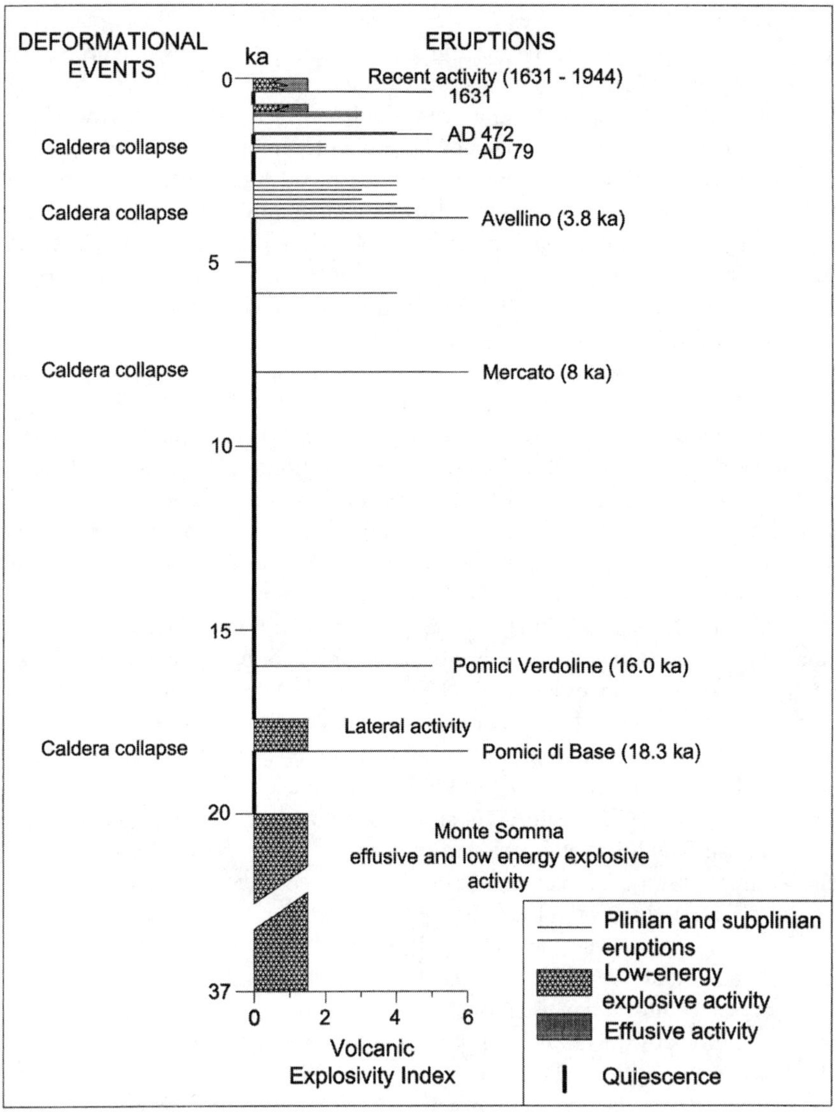

Figure 6. Chronogram of volcanic and deformational history of Somma-Vesuvius.

period, lasted between 19 and 16 ka bp, was characterized by emission of slightly SiO_2 undersaturated lavas and pyroclastic deposits (K-basalt to K-trachyte), intercalated with poorly evolved lavas. During the second period, between 10 ka bp and AD 79, the composition of products ranged from K-phonolitic tephrite to K-phonolite. The youngest period, lasted from AD 79 to 1944, and was characterized by emission of magma ranging in composition from leucititic tephrite to leucititic phonolite.

All Plinian eruptions were characterized by a vent opening, a sustained column and pyroclastic flow and/or surge phases, and they were accompanied by volcano-tectonic collapses. Sustained columns, which reached maximum heights of about 30 km, generated widespread fallout deposits varying in bulk volumes between 1.5 and 4.4 km^3 (Fig. 7). Pyroclastic currents were distributed along the volcano slopes and within the surrounding plains, reaching maximum distances of over 20 km from the vent. The bulk volume of the pyroclastic currents varied between 0.25 and 1 km^3. In proximal areas, thick breccia deposits related to caldera collapse are associated with the Plinian sequences. The quiescence periods preceding the Plinian eruptions have lasted from several centuries to millennia (Fig. 6).

Among the sub-Plinian eruptions of Vesuvius, only the AD 472 and the 1631 events are studied in details (Rosi and Santacroce, 1983; Arnò et al., 1987; Rolandi et al., 1993a; Rosi et al., 1993b; Lirer et al., 2001). They were characterized by alternation of sustained columns and pyroclastic flow and/or surge generation. Sustained columns reached heights of less than 20 km and generated fallout deposits with areal distribution less extended than those of the Plinian eruptions. The pyroclastic currents traveled distances not farther that 10 km from the vent.

Fallout deposits of both Plinian and sub-Plinian eruptions are dispersed to the east of the volcano with dispersal axes direction varying from N50° (Avellino eruption) to N150° (AD 79 eruption) (Fig. 7). The 20 cm isopachs of the fallout beds of the Plinian Pomici di Base and Mercato eruptions include 2,600 and 1,150 km^2, respectively, while those of the sub-Plinian eruptions of AD 472 and 1631 extend over areas of about 1,000 and 400 km^2, respectively. Deposition of Vesuvian fallout deposits over the Apennines chain has generated thick succession of pumice and ash beds, intercalated with paleosols, which overlie Mesozoic carbonate and Mio-Pliocene terrigenous sequences. The pyroclastic successions have been affected by weathering and erosion, especially by surface water, producing floods, hyper-concentrated flood flows and debris flows, which have generated volcaniclastic alluvial fans at the foot of the mountain slopes. The evolution of these fans has been controlled by both climatic conditions and availability of loose pyroclastic deposits overlying carbonate rocks along the slopes. Soon after deposition, the pyroclastic material is transported by water and re-deposited, generating a retrogradation of the alluvial fans. During volcanic quiescence fans prograde by sedimentation of the eroded apical and medial portions. In late Pleistocene, after the Pomici di Base Plinian eruption, during a period of semi-arid climatic conditions and a steppe-type vegetation, erosion of the pyroclastic deposits along the Apennines slopes was very intense (Follieri et al., 1989; Di Vito et al., 1998). Large amounts of loose pyroclastic material, cold climate conditions and scarce arboreal coverage of the Würm last glacial phase, have favored the formation of the large alluvial fans at the foot of the mountain slopes between Sarno and Avella, which were active until deposition of the Phlegraean

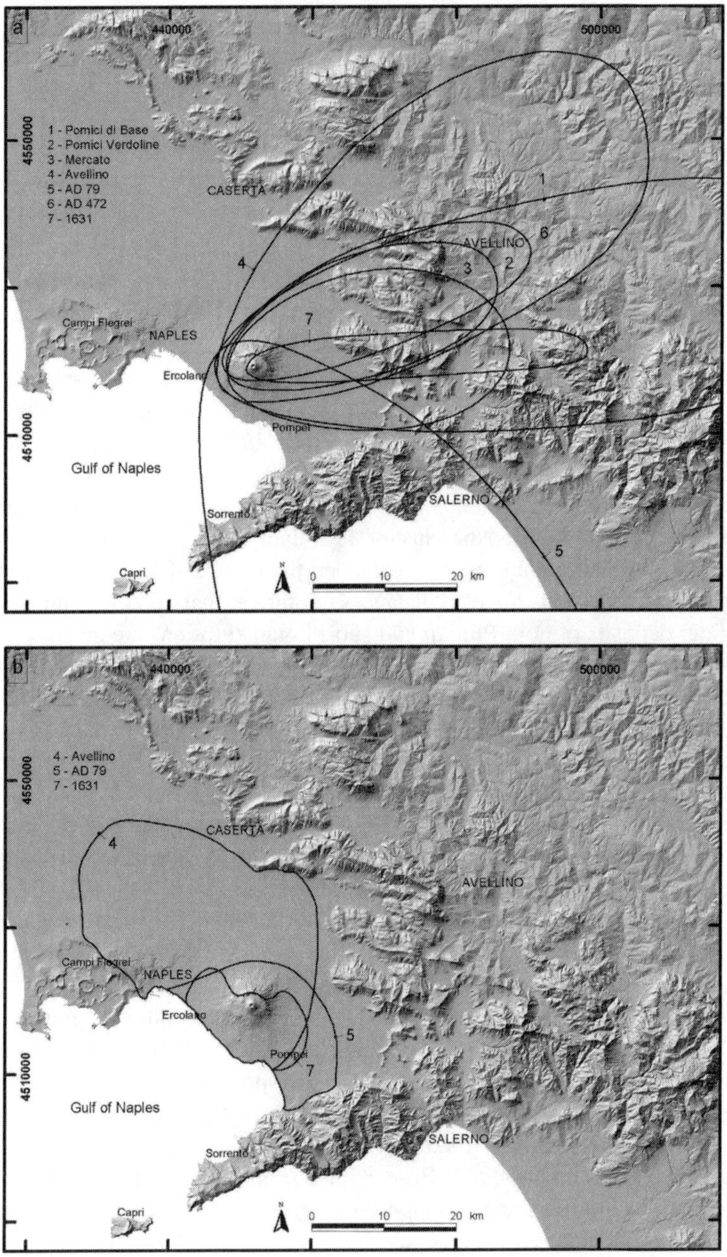

Figure 7. Distribution of selected Plinian and sub-Plinian deposits of Somma-Vesuvius eruptions. a) 20-cm isopachs of fallout deposits; b) areal distribution of pyroclastic-current deposits.

Pomici Principali Tephra (10.3 ka) (Sulpizio et al., 2000; Zanchetta et al., 2002). Smaller Holocene fans, whose activity has been deeply controlled by the deposition of both Phlegraean and Vesuvian fallout beds, particularly those of the Avellino eruption, occur in the same area.

The magmatic system of Somma-Vesuvius is characterized by a deep reservoir extending between 10 and 20 km depth, in which mantle-derived magmas differentiate as evidenced by geophysical (Auger et al., 2001) and petrological (Marianelli et al., 1999; Fulignati et al., 2000) data. From this deep reservoir, magmas rise to a shallow chamber, which is located at 3-5 km depth before Plinian eruptions and at less than 2 km depth before Strombolian activity. In the shallow chamber the new magma batches mix with the magma left by previous eruptions.

Ischia

The island of Ischia is an active volcanic field located in the north-western corner of the Gulf of Naples. It covers an area of about 46.4 km^2 and is dominated by Mt. Epomeo (787 m a.s.l.), located near the island center. A lowland is located in the north-eastern part of the island and delimited by the eastern slope of Mt. Epomeo and by an alignment of peaks toward south-east. The coastline has steep cliffs with interposed promontories on the southern side while elsewhere it slopes gently down to the sea.

Widespread fumaroles and thermal springs, and seismicity (Postpischl, 1985), whose largest event is the catastrophic Casamicciola earthquake of 1883 (Johnston Lavis, 1885), characterize the island since the last eruption, which occurred in 1302 at the end of a period of intense activity, and testify the persistent activity of the system.

Ischia is composed of volcanic rocks, landslide and debris-flow deposits, and subordinate terrigenous sediments. The volcanic rocks belong to the LK-series (Appleton, 1972) and range in composition from trachybasalt to latite, trachyte and phonolite; the most abundant are trachytes and alkalitrachytes. Volcanism at Ischia started more than 150 ka bp (Cassignol and Gillot, 1982) and continued, with centuries to millennia of quiescence, until the last eruption occurred in 1302 (Fig. 8). The oldest exposed rocks belong to a volcanic complex presently partially eroded and covered by later deposits (Fig. 9). The remnant of this complex crops out in the south-eastern part of the island. The products of the subsequent volcanism are small trachytic and phonolitic domes, ranging in age between 150 and 74 ka, which are exposed all around the periphery of the island. After the dome-building activity, a long period of quiescence was interrupted by the eruption of Mt. Epomeo Green Tuff at 55 ka bp. This caldera-forming eruption was followed by simple-shearing block resurgence inside the caldera at least from 33 ka bp (Orsi et al., 1991). Resurgence influenced the later volcanic activity, with

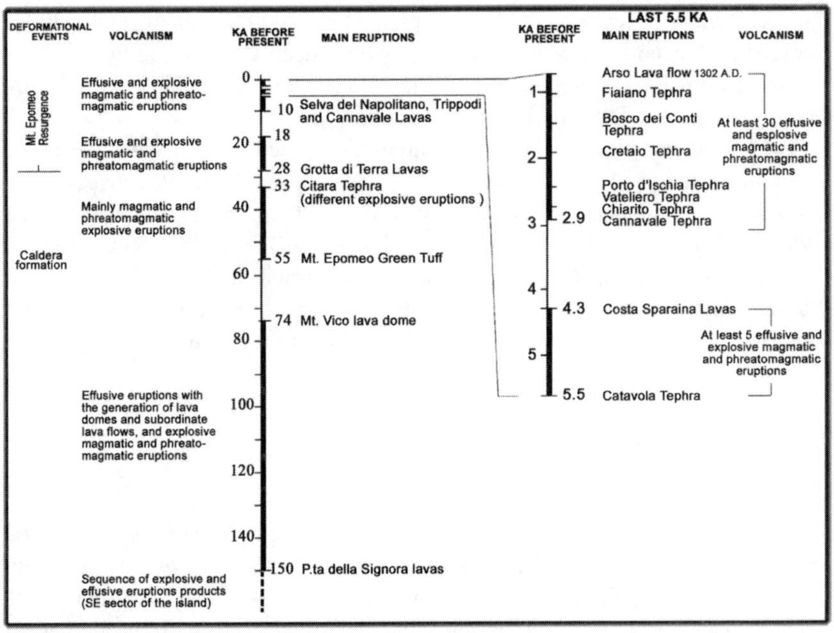

Figure 8. Chronogram of volcanic and deformational history of Ischia.

associated structures guiding magma ascent along the eastern edge of the resurgent block (Fig. 9). Stratigraphic and structural evidence, together with compositional and isotopic variations through time, have allowed subdivision of the geological and volcanological history of the island and its magmatic system in the last 55 ka into three periods of activity (Civetta et al., 1991). Each period was dominated by specific differentiation processes and was characterized by arrival of new, less differentiated magma in the system.

The 1st period of activity began with the eruption of the Mt. Epomeo Green Tuff. This tuff deposit consists mostly of trachytic ignimbrites that partially filled a depression invaded by the sea in what is now the central part of the island. This depression was submerged until the structural resurgence of the Mt. Epomeo block. Volcanism continued with a series of hydromagmatic and magmatic explosive eruptions of trachytic magmas up to 33 ka bp. Most of the pyroclastic rocks from these eruptions are exposed along the southern and western coasts, with vents located along the present south-western and north-western periphery of the island.

The 2nd period of activity started with the eruption of trachybasaltic magma along the south-eastern coast, at about 28 ka bp. Volcanism continued sporadically until 18 ka bp and was characterized by eruptions of trachytic magmas. Hydromagmatic and magmatic explosive eruptions mostly erupted alkali-trachyt-

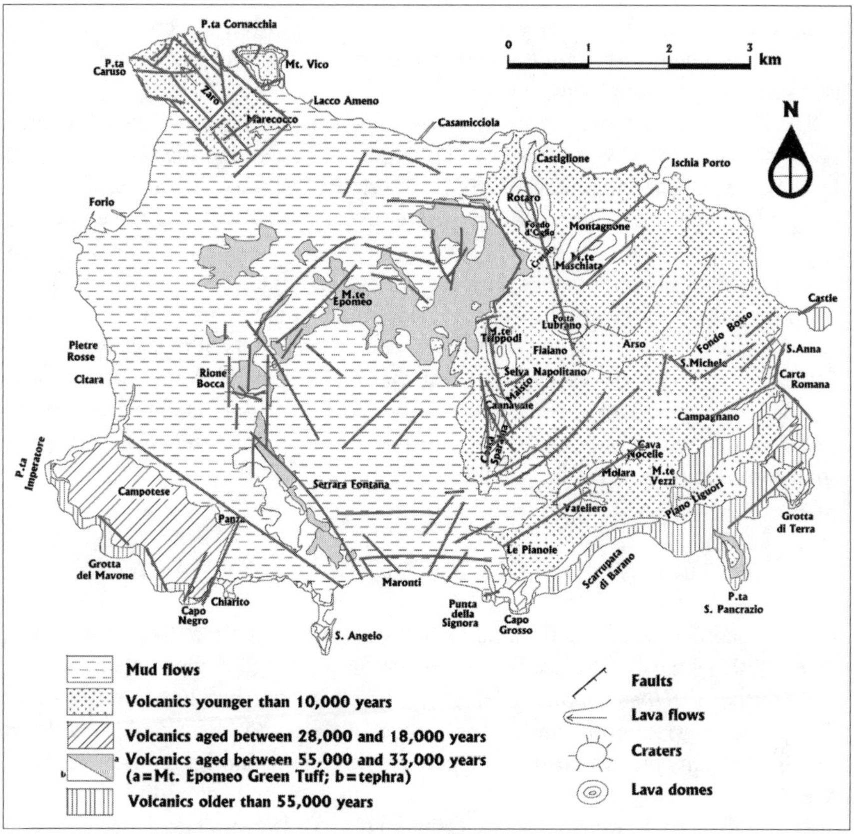

Figure 9. Geological and structural sketch map of Ischia.

ic magmas, while effusive eruptions formed trachytic lava flows. Good exposures of the rocks of this period crop out along the southern coast of the island.

The 3rd period of activity began at about 10 ka bp after a relatively long period of quiescence, and lasted up to very recent times. During this period of activity, volcanism was mainly concentrated in the past 2.9 ka and almost all the volcanic vents are located in the eastern part of the island. Only a few are outside this area, along regional faults, and generated a multi-vent lava field, in the north-western corner of the island, and a pyroclastic sequence exposed in the south-western part of the island. During the past 2.9 ka at least 35 effusive and explosive eruptions took place. Effusive eruptions emplaced lava domes and high-aspect ratio lava flows; explosive eruptions, both magmatic and phreatomagmatic, generated tuff cones, tuff rings and variably dispersed pyroclastic-fall and -flow deposits. The reappraisal of volcanic activity at 2.9 ka bp was accompanied by reactivation of faults and emplacement of landslide and mudflow deposits. These deposits preceded and followed the emplacement of volcanic products, testifying the reactivation of vertical movements likely related to the Mt. Epomeo block resurgence.

The magmatic system has had a variable geochemical behavior during the three periods of activity (Civetta et al., 1991). During the 1st period it acted as a closed system. Before the beginning of the 2nd period of activity, it was refilled by a deeper less-evolved magma, which progressively mixed with the more-evolved resident magma. The last period of activity was preceded by arrival of a new, distinct magma. Inside the chamber complex mingling/mixing processes operated till the last eruption, as testified by isotopic and mineralogical disequilibria.

Geological and petrological studies, as well as the results of modeling of magnetic data (Orsi et al., 1999b), suggest that the magmatic system of Ischia is presently composed of a deep and poorly-evolved magma chamber, interconnected with a number of smaller and most-evolved magma batches, intruded at a shallower depth, that fed the youngest volcanic activity (Fig. 9).

THE URBAN DEVELOPMENT OF THE NEAPOLITAN AREA

The Neapolitan area has been inhabited for a very long time. Recently, Fedele et al. (2002) have suggested that the cultural and biological change from Neanderthal to *Homo Sapiens*, which occurred in southern Italy, and more in general in the Mediterranean Europe, between 40 and 35 ka bp could be related to the environmental variations induced by the Campanian Ignimbrite eruption. Many treatises have been written on the urban development of the Neapolitan area. For this article we mostly cite Capasso (1905), De Seta (1981), Miccio and Potenza (1994), and Napoli (1996). Archaeological excavations have brought to light excellent evidence of the presence of indigenous farmers on low hills close to rivers on the plain north of Naples since Neolithic times (Fig. 10) (Marzocchella, 1998). Eneolithic tombs, excavated at 6 m depth along a gently inclined tuff slope, have been found in the city of Naples (Marzocchella, 1986). Obsidian and chert blades and splinters, together with painted pottery fragments of the "Serra d'Alto" type, dated at the 4th millennium BC, have been found in a palesol on the island of Ischia. On this island, which was along the route from Greece to Tuscany, Sardinia and the island of Elba (Fig. 11), where many different kind of metals were quarried, settlements of metalworkers are known since the 16th century BC. At the end of the 9th and beginning of the 8th centuries BC there was a Rhodian commercial colony on the Megaride islet along the coast of the Mt. Echia in Naples.

The development of an organized society in the Neapolitan area begins with the settlement of Greek colonies in a favorable position along the commercial routes within the Mediterranean. Colonies were established in the first half of the 8th century BC at Ischia, in 730 BC at Cuma, in 680 BC on the Megaride islet

Figure 10. Sledge traces in an Eneolithic human settlement in the plain north of Naples.

and called Parthenope, and in 531 BC at Pozzuoli. At the time of these settlements, the coastline was different from the present and urban areas were all built on promontories overlooking the sea, which had easy landing-places and could be easily defended. The settlement of Parthenope developed into a town that extended across the flat top of the Mt. Echia until 530 BC. In 470 BC the town of Naples (Neapolis - in Greek: new town) was built to the east of Parthenope, on a highland gently sloping toward the sea where the harbor of Naples is now located. The town was delimited by deep gullies on three sides and by a scarp toward the sea. At that time the plains to the east and north were marshes. Parthenope, since then called Palepolis (in Greek: old town), and Neapolis, although not physically linked, were a single town. In 326 BC the town fell under the influence of Rome, keeping its autonomy of government but having the obligation of helping Rome with its fleet until 90 BC when it became a Roman town. After that Neapolis lost its commercial and political importance because the Romans strengthened the harbor of Pozzuoli. The people retained a memory of a still persistent Greek culture, becoming a cultural and entertainment center until the end of the Western Roman Empire (AD 476). The construction of the Greek-Roman town (Fig. 12), and its development and defense required large amount of building material and water resources. Town walls were built in the 5th and 4th centuries BC using blocks of zeolitized Neapolitan Yellow Tuff, likely quarried in the Capodimonte hill. The construction of the earliest aqueduct is not well dated. This structure is named Bolla from the area at the foot of the western slopes of the Somma-Vesuvius from where the water was tapped. Within the

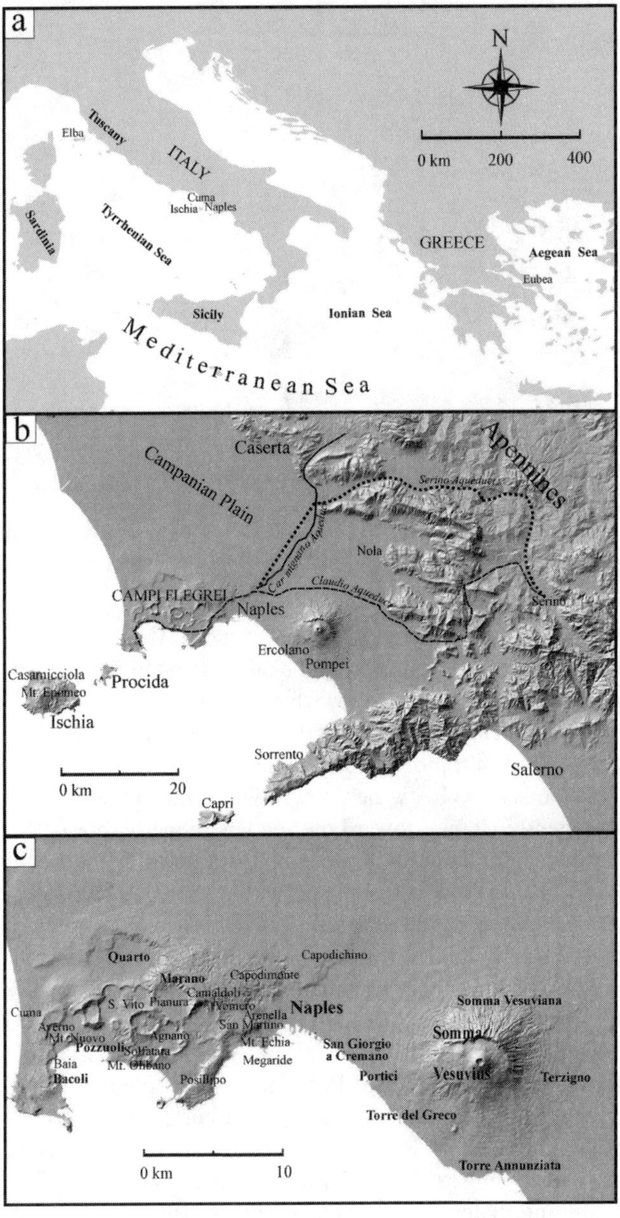

Figure 11. **a)** location of the Greek colonies of the 6th-7th Century BC.; **b)** plan of the main aqueducts, serving the town of Naples from Roman times until the 19th century (modified from Napoli, 1996); **c)** location map of the sites named in the text.

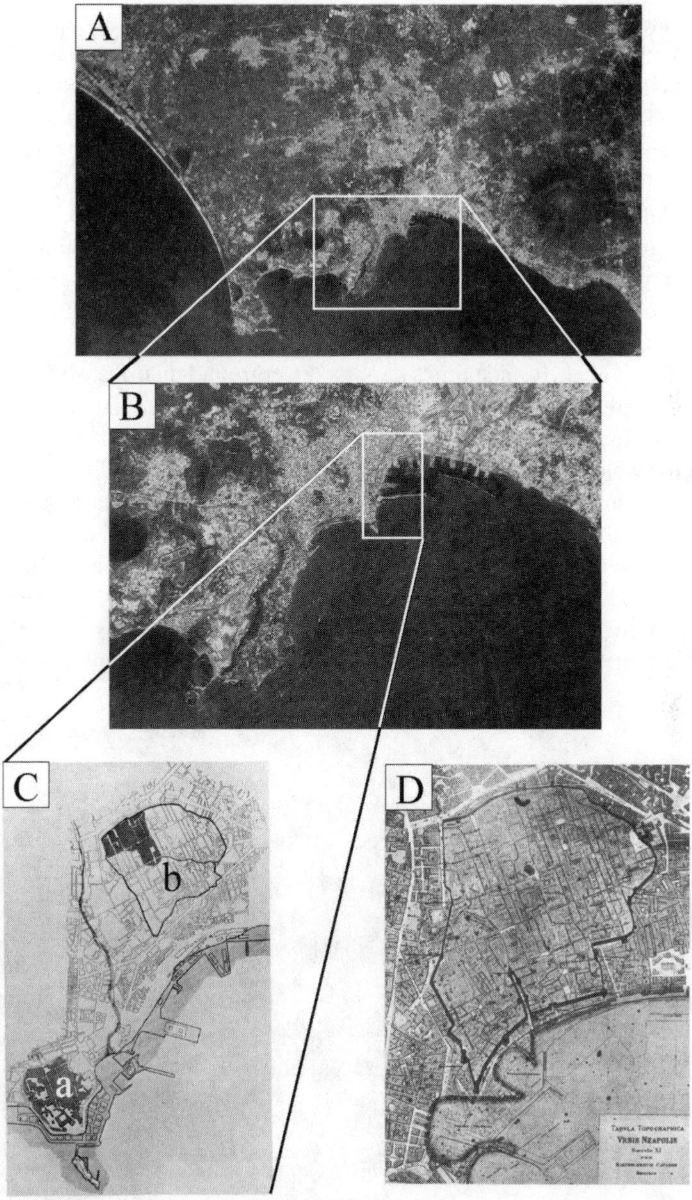

Figure 12. Expansion of the urban area of Naples: **A)** satellite image of the present day extension of the urban area; **B)** zoom on the area colonized by the Greeks in the 7th century BC.; **C)** the two settlements of Paleopolis **(a),** of the 7th Century BC., and Neapolis **(b)** of the 4th Century BC.; **D)** the urban area in the 11th Century AD.

Greek town it is an underground aqueduct composed of a dense network of tunnels cut through the Neapolitan Yellow Tuff and connected to the surface by wells from which water was drawn. In the first half of the 1st century BC, under the Roman emperor Augustus, another aqueduct was built and named the Claudius. This aqueduct, which collected the water from the river Sabato in the area of Serino and transported it to Naples and then to Baia over a distance of 91.6 km (Fig. 11), was in operation until at least the 5th century BC. It reached the town of Naples through Capodichino, Capodimonte, Scudillo, Vomero and Posillipo, from where one branch was directed to Baia and ended into the huge underground reservoir, known as Piscina Mirabilis (Fig. 13). This aqueduct was partly underground and partly constructed on brick arches. Within the town the water was transported through tunnels.

During the Greek-Roman times the first underground quarries were excavated. The excavated materials were mostly Neapolitan Yellow Tuff and loose pyroclastic deposits. The Neapolitan Yellow Tuff was a good material in which to dig tunnels for water storage and aqueducts. The Romans used and traded the loose pyroclastic deposits (pozzulan) for preparation of mortar. The largest cavities of this period are located at the foot of Mt. Echia and its surroundings, through the Posillipo hill with the entrance near to the Virgil's tomb, and through the Capodimonte hill. In this hill there are also catacombs used by both Greeks and Romans as hypogea for burial. In the 6th century there was a demographic decline and an impoverishment of the social framework as consequence of the

Figure 13. The huge reservoir, named *Piscina Mirabilis*, at the end of the Claudio aqueduct.(Photo by F. Rastrelli).

beginning of the barbarian invasions. In AD 440 new town walls were constructed. The center of the town was reduced in the north-eastern part, and the settlements along the coast degraded and those outside the wall were abandoned.

Naples became a Dukedom of the Byzantine Empire in AD 661 and an independent Dukedom in AD 763. Only between the 9th and the 10th centuries population began to rise because of immigration and recovery of commerce and handicraft. In addition to the traditional handicraft such as tanning, yards, joiner's shops, a very important role in the economic development was taken by the linen industry, which developed along the town walls at the termination of the Claudius aqueduct. The urban setting of the town changed into a typical medieval polycentric structure. Likely in this framework the construction of a new branch of the Bolla aqueduct was decided. The recovery of the craftsmanship stimulated the sea trades and many settlements were built up along the town wall near to the sea. In the 9th century the town's wall had to be enlarged. The drainage systems of the Capodimonte and Capodichino hills restricted development of the town eastwards but in that area they fostered a milling industry. Such industry has influenced for centuries the development of the Neapolitan water supply system. At the beginning of the 12th century the population of Naples was around 30,000.

In 1137 the Naples Dukedom became part of the Kingdom of the Two Sicilies with Palermo as capital, under Norman domination. For strategic needs the town expanded inland and towards the plain to the west of Mt. Echia. Sea trade continued to increase and Naples became a major harbor for commerce with France, Spain and Africa.

In 1194 the Kingdom was conquered by the Swabians and, during the reign of Frederic II, the University of Naples was founded (1224).

In 1268 the kingdom was conquered by the Angevins and Naples became the capital. The marshes to the north and east of the town were drained, irrigation channels were built and the town was connected with the countryside through new roads. The water of the aqueduct was purified, collected in large reservoirs, then distributed through a network of underground pipes. Many of the roads were paved and a sewer system was built. After the infrastructure improvements, an extensive development plan was put into effect with urban renewal. The urban setting was renewed with new construction. At the end of the Angevin domination the population was between 40,000 and 50,000.

The structure of the town significantly changed with most construction devoted to the defense of the town under Aragonese rule, which began in 1442. To satisfy military and commercial demands there were extensive works in the harbor. The drainage of the eastern marshes was finished and the area was used for cereal cultivation. An intense earthquake in 1450 severely damaged buildings, mostly in the ancient center of the town. The town grew so rapidly that the walls had to be repeatedly extended. A new residence for the king was built in Poggioreale where a sewer system was constructed and water was transported from the Bolla aqueduct to feed fountains and mills. Intense underground quarrying for building stone took place on the Capodimonte and San Martino hills, for the construction in the older and western part of the city.

In 1504 the kingdom became a province of the Spanish kingdom, governed by a viceroy. During this period there was a population explosion despite the plague epidemics of 1529, 1657 and 1691, which caused a few hundred thousand deaths. Intense development of the religious and private architecture followed immigration of the provincial nobility. In 1534 a new project aimed at paving the roads began, and the town walls were further enlarged following an abrupt population increase. Both sewer and aqueduct systems were reorganized. In 1562 spring waters were connected to the aqueduct because of increased demand. In 1616 the marshy area to the north of the town was reclaimed. Within this framework, in 1629 Cesare Carmignani built an aqueduct, known as the Carmignani aqueduct, which transported water of the Faenza river from Sant'Agata dei Goti to Naples (Fig. 11). For the construction of the Carmignani aqueduct, many tunnels were dug under densely urbanized areas. In 1631, the last sub-Plinian eruption of Vesuvius destroyed the part of the aqueduct between Avella and Casalnuovo, but it was restored in two years. Because of increased population and building construction, large underground and open quarries were activated. Many were excavated directly under the sites on which the buildings were constructed. To decrease the rate of urban development, the use of building material from inside the town limits was prohibited. As a result of this law, many quarters were built using the tuff illegally extracted from the roofs of the aqueduct tunnels. Therefore the aqueduct was transformed into an enormous framework of interconnected underground quarries.

In 1707 the kingdom was influenced by the Austrians for a short period but was still governed by a viceroy. The beginning of the Hapsburg domination was characterized by the contrast between the Empire and the Church of Rome. As the Church was expanding its estate, acquiring buildings and lands or enlarging old edifices and building new churches and monasteries, the viceroy imposed the prohibition on building any kind of edifice. However, following a demographic increase people illegally built new houses, allowing chaotic expansion of the city toward the surrounding hills. In 1718 the construction laws were liberalized and the town continued to expand eastward, along two new roads, opened to link Naples with the Vesuvius area. These roads were paved with blocks of lava from Vesuvius quarries. In 1729 a strong cloudburst damaged a large portion of the town, mainly the quarters located at the foot of the Capodimonte hill.

In 1734 the kingdom of Naples was conquered by the Bourbons and governed by a king residing in town with the title of King of Naples. During the kingdom of Charles III great care was given to public works. During this period, indeed, wonderful monuments were built, such as the Royal Palaces of Capodimonte, Caserta and Portici, with their parks and gardens, the giant Poor People's Home, the Public Granary, the monumental Cemetery of Poggioreale. For these works an enormous amount of Neapolitan Yellow Tuff and Campanian Ignimbrite was needed as building stone, taken from the Naples and Caserta quarries. King Charles III also promoted the restoration of the roads to the Vesuvian area and the seaside way in town, as well as the restoration and the enlargement of the harbor and the Market Square. New roads were built to reach the Royal palace at Caserta. During the second half of the 18th century an extensive program of road restoration was carried out to improve the connection with central Italy and the southernmost regions of the kingdom. However, despite this large restoration effort, the urban expansion continued in a disorganized fashion without a city plan.

Between 1806 and 1815 there was a short French domination linked to the accession and the fall of Napoleon. The attitude of the new government toward science and culture led to the foundation of the Botanic Garden and the beginning of the works for the Astronomic Observatory, which was completed afterward, during the Bourbon restoration.

The kingdom of Naples stayed under the Bourbons from 1816 to the unification of Italy in 1860. Most of the works started under the French domination were concluded but, following the Bourbons tradition, mainly monumental public works were carried out. Important construction works were realized during the kingdom of Ferdinand II. Ferdinand II promoted the development of the iron and steel industry, with the construction of the Napoli-Portici railroad (the first in Italy), scientific research, with the foundation of the Vesuvius Observatory (the first volcanological observatory in the world), and improved the street network of Naples, with the construction of some important roads that still connect the city center with the surrounding hills. However, the big cholera epidemic of 1836 demonstrated the heavy inadequacy of the sewer and water networks, neglected since the 17th century.

In 1860 the kingdom of Italy was unified by the Savoy and until the First World War the history of Naples, from the town-planning point of view, can be correlated with a great slum-clearance enterprise that lasted over half of a century. During this extensive slum-clearance operation the demolition and reconstruction of the old eastern quarters, a large portion of the business center, and some of the Vomero hill settlements was carried out. The shore facing the city center was extended and a coast road was constructed. The two arcades, entitled to the king Umberto I and to the Prince of Naples, and the Stock Exchange building were built, as well as a large number of high class and popular apartments, mainly linked to rent restrictions. Once again the Neapolitan Yellow Tuff was the most widely employed building material.

A long period of neglect characterized the beginning of the 20th century, and culminated with the First World War. Many public works were realized during the following twenty years, including restoration of the sewer network, digging of three tunnels to facilitate connection between the city center and its western periphery, and construction of new panoramic roads along the slopes of the surrounding hills. Most of these works greatly contributed to the knowledge of the underground geology of Naples. Quarrying was still active, although most of the public edifices were built with concrete bricks and faced with non volcanic rocks such as travertine, marble and granite. After the end of the Second World War Naples, strongly damaged by bombing, was reconstructed without any town-plan. The hills surrounding the Greek-roman and medieval town were sites for property speculation. Thousands of cubic meters of tuff were indiscriminately extracted from open and underground quarries within the city. The tuff was used to build both huge houses and embankments needed to mitigate the man-induced slope instability. Moreover deforestation and changes in the drainage network caused by quarrying and intense building construction, together with the obsolescence and inadequacy of the sewer system, contributed to rise up the hydrogeological hazards in the city of Naples, determining the instability of the present slopes.

THE EXTRACTION AND USE OF VOLCANIC ROCKS

The physical and mechanical characteristics of the volcanic rocks of the Neapolitan area and their use for engineering and architectural purposes have been investigated since the 16th century (Palladio, 1570). Among the more recent literature the most comprehensive synthesis is presented by Aveta (1987), who describes lavas and tuffs, distinguishing them on the basis of the variable Neapolitan volcanoes from which they were erupted, and the loose pyroclastic material without reference to the source. The volcanic rocks used as building stone in the Neapolitan area are lavas, variable types of tuff, mostly Campanian Ignimbrite and Neapolitan Yellow Tuff, and loose pyroclastic deposits. This material has been extracted from underground and open quarries.

The Vesuvian lavas have been quarried in the towns of Somma Vesuviana, Terzigno, S. Giorgio a Cremano, Ercolano, Torre del Greco, and Torre Annunziata. They have been used for masonry and ornamental elements, and for road paving. In particular they were used during the urban development promoted by Charles III of Bourbon (1734) either for ornamental elements, as a substitute for piperno (densely welded tuff), or as stone rubble for masonry and as blocks for breakwaters. The Vesuvian lavas have been largely used to pave roads, courtyards, and halls. The upper scoriaceous portions of the lavas, when crushed, were used to construct highly resistant and compact walls because of the strong adhesion between concrete and rough surfaces. The Phlegraean trachytic lavas, quarried at Quarto, Solfatara and Monte Olibano, have rarely been used as building material except by the Romans, who used them to pave the roads outside the town. At the beginning of the 14th century the Angevins used these lavas to pave the roads of the town.

The Campanian Ignimbrite, easy to work and widely exposed within the western portion of the Campanian Apennines and along their slopes towards the

Campanian plain, has been extensively used. In particular it has been used to build balustrades, corbels and cornices. The piperno facies of the Campanian Ignimbrite has been quarried on the southern slopes of the Camaldoli hill and in the Soccavo and Pianura areas. These quarries are presently abandoned. Due to its pleasant color, texture, and workability characteristics, this rock has been largely used for architectural elements such as stairs, cornices, jambs, architraves, brackets, piers, wainscots, and portals (Fig. 14). The Aragonese used it as facing for both town walls and rounded towers on the sides of the town gates as structural and ornamental elements. Because of the stone's poor resistance to wear and tear, the use of this rock has been abandoned in the 20th century. The Neapolitan Yellow Tuff has been used since the first settlement of the Neapolitan area because of its excellent resistance to weathering, the specific weight, abundance and ease of quarrying and working. The tuff, largely used for construction of buildings, has also been used for cornices and roofing of small spaces. It was quarried from underground and open quarries and rough-hewn by hand until the Second World War. Since the end of this war, the tuff has been extracted only from open quarries with cutting machines. Underground quarrying has generat-

Figure 14. a) Colonnade of the S. Marcellino monastery (16th Century) built with *piperno*, a welded facies of the Campanian Ignimbrite; **b)** detail on a column with the typical flattened pumice fragments (*fiamme*) that form the eutaxitic texture of this rock. At the present the monastery hosts the Faculty of Geology.

ed a dense network of tunnels, very often separated by thin walls and supporting posts, and having thin vaults. During the Second World War, many tunnels and old cisterns were adapted as air raid shelters. After the war, the rush to further excavate tuff from underground quarries due to the need for construction stone for new buildings and restoring the damaged ones, deteriorated the already unstable situation. The tunnels excavated for quarrying the tuff and for constructing aqueducts and cisterns, have been later used for variable purposes such as houses, storehouses, garages, and workshops.

The Mt. Epomeo Green Tuff of Ischia has rarely been used because it is very soft and weak. Loose pyroclastic rocks, both Phlegraean and Vesuvian, have characteristics which make them appropriate for use in building industry. The most widely known is the pozzuolan, the fraction of the deposits smaller than 4 mm, which when mixed with lime, plays an active role in production of a very resistant mortar.

VOLCANIC AND RELATED HAZARDS

The Neapolitan area is exposed to a variety of natural hazards. These hazards are related to both geological setting of the territory and millenary human habitation. The dynamics of the Apennines chain generates earthquakes, while that of the Campanian plain generates the conditions for magmatism and volcanism. The volcanic and deformational history and the present state of the Neapolitan volcanoes, indicate that eruptions, probably explosive, may occur. The geomorphic setting of the area has been determined by the volcanic activity. Its present evolution generates erosion and transport with landslides. The use of the territory and volcanic rocks by humans for thousands of years has generated the conditions for further hazards. Fulfillment of basic needs such as development of housing, access to water, building and maintaining fortifications for protection from enemies, required excavation of quarries, wells, and cavities in lavas, tuffs and loose pyroclastic rocks. The results of this activity are sources of hydrogeological hazards such as landslides, collapses, and flooding.

Volcanic Hazards

Campi Flegrei caldera Using geological, volcanological, structural, petrological and geochronological data, Orsi et al. (2003) have attempted to assess the volcanic hazards at the Campi Flegrei caldera in case of renewal of volcanism in short-to-mid terms (from few to tens of years). In 1998 the National Group for Volcanology approved a report, on volcanic hazards assessment at the Campi Flegrei caldera. This report was accepted by the Department for Civil Defense as the scientific reference for drawing up the emergency plan. As shown by the reconstructed volcanic and deformation history, including the recent unrest episodes (Di Vito et al., 1999; Orsi et al., 1999a, c), a new stress regime, which still persists, was established within the caldera between the 2nd and 3rd epoch of activity. Therefore in order to assess volcanic hazards, Orsi et al. (2003) have taken into particular consideration the history of the past 5 ka. They have given answers to the basic questions "when", "where" and "how" the next eruption will occur and have also identified the precursor phenomena which will be felt by the population.

To answer to the question of when, the authors have defined, according to the ongoing deformation dynamics, the structural conditions required for an eruption. A simple-shearing mechanism for the deformation has generated a compressive stress regime in the south-western portion of the Neapolitan Yellow Tuff caldera and a tensile regime in the north-eastern portion, which corresponds to the Agnano-San Vito area. When tensile stress will cause mechanical failure of the rocks, normal faults will form and generate the conditions for magma to rise to surface and erupt. Volcanism could continue until either the total elongation of the portion of the resurgent block under tensile stress is accommodated or all the magma has erupted. Conditions for an eruption to occur could also be initiated along reverse faults at their intersection with other faults of a contemporaneously activated block.

According to this scheme, based on the interactions between magmatism, tectonic and volcanism within the Neapolitan Yellow Tuff caldera, the area with the highest probability of opening of a new vent is the portion of the caldera under a tensile stress regime between the Agnano and San Vito plains. This area, which covers about 12 km^2, has been the site of 16 eruptions over the past 5 ka. It includes the western portion of the city of Naples and part of the town of Pozzuoli. If a future vent were to open at the intersection of two fault systems, an eruption could occur at the north-western corner of the most uplifted portion of the resurgent block, as was the case of the Averno and Mt. Nuovo eruptions (Fig. 15).

The precursor phenomena, which would be felt by the population would include ground deformation, seismicity, and variation in fumarolic activity. Ground deformation would include uplift of the central portion of the caldera with consequent stretching of the area between the Agnano and San Vito plains until fractures and normal faults form. Before activation of these features, seismicity will be very similar to that which occurred during the recent unrest episodes. Mechanical failure of the shallow crust in the Agnano-San Vito area will generate earthquakes characterized by normal fault mechanisms and with hypocentres distributed along fault planes not deeper than 4-5 km. Convective fluid circulation along fractures and fault planes will generate low-frequency swarms which, through time, will increase in number with hypocenters migrating towards the surface. Fracturing and increase in fluid circulation will produce an increase in fumaroles activity in the Agnano-San Vito area. Faulting will

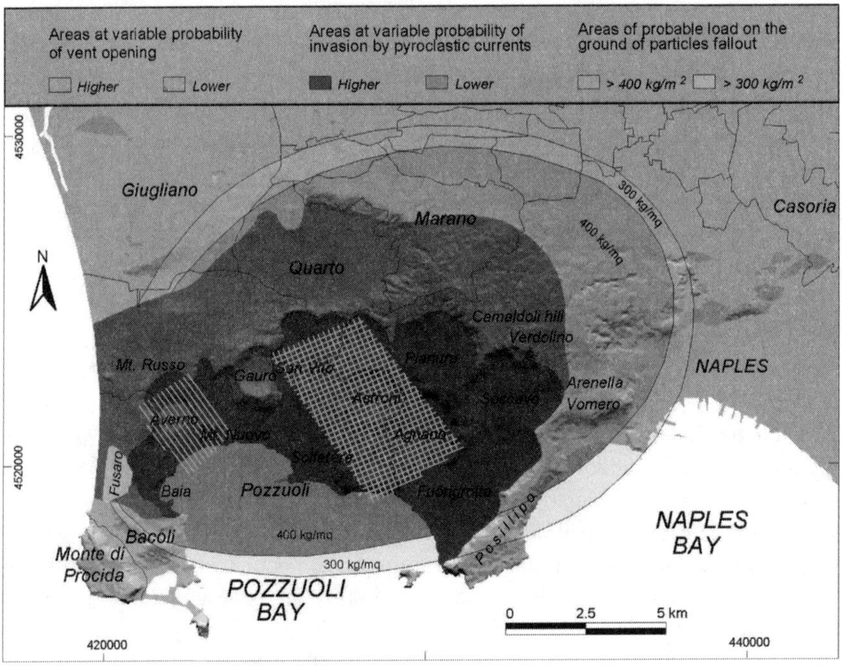

Figure 15. Volcanic hazard map of the Campi Flegrei caldera.

decompress the magma with consequent exsolution of more gases, which will rise to surface, generating an increase in the volume and temperature of emitted fluids and a variation in their composition. In relation to increasing hot fluid circulation at shallow depth, the water table could be overheated and generate phreatic eruptions. Characteristics of the first eruption of each epoch of the past 12 ka such as magnitude, dynamics, composition of the extruded magma and structural position of the vent are not related to the length of the preceding quiescence period. During all three epochs the magnitude of each explosive eruption has never been related to the time sequence of events, and the largest magnitude eruption has never been among the first events of an epoch of activity. Therefore, although a sub-Plinian event, similar in size to the Agnano-Monte Spina, could be assumed as the maximum expected eruption, the authors have assumed as the most probable maximum expected eruption in case of renewal of volcanism in short-to-mid term, an event similar to the medium-magnitude eruptions of the past 5 ka. This event will be characterized by alternation of phreatomagmatic and magmatic explosions. The hazards directly related to such eruption are fallout and pyroclastic density currents. On the basis of the measured thickness and density of the fallout deposits, Orsi et al. (2003) have calculated their ground load. In particular they have constructed frequency maps for ground load in excess of 200, 300, 400 and 500 kg/m^2, the loads under which wood, iron, concrete and reinforced concrete roofs would fail, respectively.

Pyroclastic fallout can be generated during both phreatomagmatic and magmatic explosions. A fallout hazard zoning of the territory in relation to the ground load that could be exerted by falling particles was constructed on the basis of frequency of both deposition and load of particles fallout, direction of dispersal axis and load limits for a variety of roof types (Fig. 15). The portion of the defined area that will be affected by particles fallout will depend on the wind direction at the height reached by the column.

Pyroclastic currents will be generated mostly by phreatomagmatic explosions and will travel at high speed within the Neapolitan Yellow Tuff caldera lowland. Some currents, according to vent location and energy, could exit the lowland in low passes and reach the northern slopes of the Camaldoli hill and the Vomero-Arenella sector of the city of Naples, the towns of Marano and Quarto, and the northern portions of the towns of Pozzuoli and Bacoli (Fig. 15). Along the northern slopes of the Camaldoli hill the currents could be deeply influenced by the complex drainage system.

In the 3rd epoch, as well as in the other epochs, eruptions have followed each other at average time intervals in the order of few tens of years. Therefore it cannot be excluded that the next eruption, especially if the vent will be located in the Agnano-San Vito area, will be followed by other events. The eruption forecast for that area, stimulated by the dynamics of deformation, could also be accompanied by a volcano-tectonic collapse. Deposition of ash and likely heavy rains could induce mud flows along the caldera and the surrounding slopes, including the western margin of the Apennines and the Somma-Vesuvius cone.

Somma-Vesuvius Volcanic hazards in the Vesuvian area have been evaluated by several authors, each using different approaches (Scandone et al., 1993; Lirer and Vitelli, 1998; Lirer at al., 1997; 2001; Santacroce et al., 2003). Scandone et al. (1993) evaluated the volcanic risk of the Vesuvian territory calculating the variable probability of volcanic events with different Volcanic Explosivity Index (VEI, Newhall and Self, 1982) (3, 4 and 5) during a defined time span. The vulnerability value has been obtained exclusively using the number of inhabitants. The area of highest risk is the southern coastal area of Vesuvius. The risk decreases progressively with the increase of distance from the volcano. The result is largely dominated by events with VEI = 3 and 4, whereas the effect on the results of larger magnitude events (VEI=5) is reduced in relation to the low probability of such kind of event.

The hazard and risk of the Vesuvian territory in case of a low explosivity event (VEI = 3), which occurred frequently during the 1631-1944 period of activity, has been analyzed by Lirer and Vitelli (1998). The hazard map has been obtained using the past frequency distribution of lava flows. The risk has been calculated considering the land use crossed with the probability value of occurrence of an event in each considered area. The highest risk values have been obtained for the towns of Torre Annunziata, Portici and Torre del Greco.

A detailed stratigraphic reconstruction of the AD 79 Plinian eruption allowed Lirer at al. (1997) to define the scenarios during the different phases of the eruption. The timing has been derived by comparison between the reconstructed phases and the Pliny's letters. The impact on territory during the different phases has been evaluated using the areal distribution of products. Combining impact and present land use, the authors calculated that after 7 hours from the beginning of the eruption 300 km^2 of the Vesuvian territory would be covered by the eruption products, with roof collapses and partial burial of roads, up to a distance of 30 km southeast from the vent. During the following 11 hours about 500 km^2 of territory should be affected by further accumulation of tephra and occurrence of pyroclastic flows between Portici and Torre del Greco. In this phase only few roads of the northern part of the Vesuvian territory could be used. During the following hours the thickness of the deposits would increase and pyroclastic density currents would enter Torre Annunziata, Pompeii, Boscoreale and Terzigno. The authors calculated that 1,226,000 peoples should be affected by damages induced by an eruption like the AD 79 event.

Recently a hazards evaluation based on the history of the volcano between 8 ka bp and 1906, has been performed by Lirer et al. (2001). On the basis of the distribution of products of all the eruptions in this time span, independently from their magnitude, distribution maps for both pyroclastic-fall and -flow deposits have been constructed. Frequency maps have been obtained by the overlap of distribution of the different pyroclastic flow deposits thicker than 50 cm. Frequency maps were made for fallout deposits exerting a load in excess of 300 kg/m^2. These maps allowed the authors to prepare a hazard map divided in four fields: high, medium and low hazard, and the boundary of the area that could be affected by flowage of pyroclastic currents. A risk map has been obtained by the overlap of the hazard and exposed value (from ISTAT 1994) maps. The highest risk area covers 86 km^2 and includes 984,000 inhabitants, the medium risk area contains an urbanized part of 102 km^2 with 520,000 inhabitants, finally in the low risk area 100 km^2 are urbanized and there are 400,000 inhabitants.

In 1990 the National Group for Volcanology delivered to the Ministry for Civil Defense a document on the scenario of the maximum expected event in case of renewal of volcanism in short-to-mid terms. This scenario, updated by Barberi et al. (1995) and more recently by Santacroce et al. (2003), has been used as the scientific reference by the Department for Civil Defense for preparation of an emergency plan. The maximum expected event is a sub-Plinian eruption similar in size, dynamics and volume of extruded magma to the 1631 eruption. It includes a phreatomagmatic opening phase, which will be followed by a sustained column phase and pyroclastic density currents. The eruption likely will end with phreatomagmatic explosions. The sustained column could reach heights between 12 and 20 km and generate heavy fallout. Pyroclastic currents can be generated by destabilization and collapse of the eruption column. The phreatomagmatic explosions of the waining phase of the eruption could generate ash and mud fallout and mud-hurricanes. Santacroce et al. (2003) described the procedure used to upgrade the hazards zonation of the Vesuvian area. The area, wide sectors of which could be subject to nearly total destruction of buildings by pyroclastic currents and lahars was defined as the Red Hazard Zone (236 km^2) in the 1990 document of the National Group for Volcanology and never modified. It has been delimited combining the areal distribution of the 1631 pyroclastic flows (Rosi et al., 1993) with the results of numerical simulations (Dobran et al., 1993). During the 472 and 1631 eruptions about 40% and 20% of this zone respectively, were ravaged by pyroclastic flows. The area that could be affected by fallout from a sustained column has been defined by combining the areal distribution of the pyroclastic deposits of the major eruptions (magnitude larger than or equal to that of the maximum expected event; 0.2 km^3) occurred during the last 20 ka and the results of numerical simulations. The thickness of the deposits of each eruption has been transformed to ground load, using average bulk density values. On the basis of the past fallout distributions, a frequency map has been prepared for the two load thresholds of 200 and 400 kg/m^2. The wind profiles of the past 15 years have been used to simulate the tephra fallout from columns with heights of between 12 and 22 km. Two particle types have been used: high- and low-vesicularity fragments similar to the 79 and 1631 pumice fragments. The areas with ground loads in excess of 200 and 400 kg/m^2 have been identified. The ratio between the number of simulations of an accumulation at a point which is greater than the fixed threshold and the total number of simulations allowed determination of the probability that the mass will exceed the threshold at that point. The correspondence between the probabilistic distribution maps obtained by numerical simulations and the mapped distribution of the deposits has been evaluated in order to define the fallout hazard zone (1,125 km^2). Sectors of this zone could be affected by heavy (>300 kg/m^2) ash and lapilli fallout, as well as by mud storms, mud fall and flows. In 1631 about 10% of this area was severely damaged. The fallout hazard zone includes an area of 98 km^2 that could be affected also by deposits from large floods. These phenomena could occur, even if with minor intensity, also in case of deposition of thin pyroclastic deposits over the drainage basin of the Nola plain.

On the basis of the knowledge of the behavior of the volcano prior to a large eruption (AD 79, 1631) and of other similar volcanoes, 4 alert levels have been established. They range from "base", which corresponds to present state of the volcano, to "alarm", which is characterized by the appearance and/or evolution of phenomena monitored by the Osservatorio Vesuviano, suggesting a pre-eruption phase.

Ischia Since the last eruption, which occurred in AD 1302 at the end of a period of intense activity, the system has been characterized by widespread fumaroles and thermal springs, and by strong seismicity (Postpischl, 1985), which culminated with the catastrophic Casamicciola earthquake of 1883 (Johnston Lavis, 1885). The slope instability that characterizes a large portion of the island can be enhanced by structural deformation, which could also be considered as an evidence of the activity of the system that could erupt in the future

The island of Ischia has been inhabited by man since Neolithic times and has experienced a complex history of alternating human colonization and volcanic eruptions, which destroyed settlements and drove away the population. Currently the island hosts a population of about 50,000 people which increases significantly during summer. Thriving farms, vineyards and a complex trade-network with the nearby city of Naples, contribute to a high volcanic risk.

Although a detailed evaluation of the volcanic hazards at Ischia has not been yet performed, the results of recent studies aimed at understanding the volcanic and deformation history of the island and the evolution and present state of its magmatic system, permit definition of some constraints to the volcanic hazards assessment.

The area located to the east of the most uplifted block of Mt. Epomeo, has been the site of the volcanic activity for the past 2.9 ka and is characterized by a tensile stress regime induced by the resurgence dynamics, which very likely is still active. As the volcanism was not continuous through time, it has been speculated that resurgence took place during intermittent periods of uplift and tectonic stability. This behavior could be in turn the effect of discontinuous refilling of the magmatic system (Piochi et al., 1999) which has already been demonstrated for the long term (Poli et al., 1987; 1989; Vezzoli, 1988; Civetta et al., 1991). It has been also suggested that, in case of renewal of volcanic activity, the north-eastern part of the area active during the past 2.9 ka is the area with the highest probability of vent opening and invasion by pyroclastic flows, and that all the eastern part of the island could be covered by pyroclastic fallout. The largest recognized explosive eruption, which generated the Cretaio Tephra (1.7 ka bp), has been studied in detail (Orsi et al., 1992) and presently could be assumed as the maximum expected eruption.

Seismic Hazards

The geological-structural and geomorphological setting and the stress regime within the Campanian area define three sectors at variable seismic hazards: the Campanian Apennines, the Tyrrhenian coast, and the volcanic areas. The Campanian Apennines, being one of the most dynamically active areas of Italy, generate many earthquakes. Statistical estimations, based on the catalogue of the historical seismicity, suggest that the maximum expected event is a 6.5-7.0 magnitude shock (Chiarabba and Amato, 1997), which corresponds to the earthquake in 1980. From this seismically-active area the energy radiates to large distances, affecting also the Neapolitan urban area. The Tyrrhenian coast is characterized by very low energy or absence of seismicity. In the volcanic areas seismicity is much lower than in the Apennines because the mechanical characteristics of the volcanic rocks and the high concentration of stress does not allow storage of large amounts of energy. Furthermore the shallow depth of the hypocenters produces effects on a narrow epicentral area which rapidly decline with distance. The Somma-Vesuvius, since the last eruption occurred in 1944, is characterized by a low-magnitude seismicity; the largest shock occurred in October 1999, had a 3.6 magnitude (V° MCS; Mercalli-Cancani-Sieberg). The shocks occur in the caldera area with hypocenters not deeper than 5 km. The recent unrest episodes at the Campi Flegrei caldera have shown that uplifting is accompanied by seismicity, while during subsidence there is no seismicity. Number and magnitude of the events increase with increasing velocity of the ground uplift (Orsi et al., 1999a). The earthquakes have hypocenters not deeper than 4 km. The most intense shock (VII° MCS) occurred in October 1983 and had a magnitude of 4.0. The maximum intensity occurred in the area of Pozzuoli, while within the caldera lowland the intensity was of the VI° MCS.

Seismicity at Ischia is likely related to the Mt. Epomeo resurgence dynamics. Landslides, whose deposits are intercalated with the products of the last period of volcanic activity, likely were triggered by earthquakes generated by volcano-tectonic deformations. Many earthquakes are reported by chronicles in the past centuries, the largest of which occurred in 1288 (IX-X° MCS) and in 1883 (XI° MCS). For this last event a hypocentral depth of 1.0-1.5 km, has been estimated (Luongo and Cubellis, 1998).

The distribution of the effects of the Apennines earthquakes since the 15th century, and the intensity variation in terms of gravitational acceleration, from 0.05 to 0.1, suggest that for the Neapolitan area a VIII° MCS must be considered as the maximum degree of damage.

Buildings were damaged by earthquakes of the Somma-Vesuvius that preceded and accompanied eruptions. The maximum expected magnitude for such events is >5.0 and can generate macroseismic effects of the IX° MCS. The earthquake that occurred in AD 62, few years before the Plinian AD 79 eruption, is the largest known for Vesuvius, had a magnitude of 5.8 and affected the town of Pompeii with an intensity of the IX° MCS.

Earthquakes are expected to occur during the next unrest episode of the Campi Flegrei caldera. The maximum expected magnitude is of 4.5 with effects of the VIII° MCS in the epicentral area. Such an earthquake would generate effects of the VII° MCS within the Campi Flegrei caldera lowland which include the western portion of the city of Naples and the towns of Pozzuoli, Quarto and Bacoli.

Strong earthquakes, with a maximum expected macroseismic effect of the XI° MCS, are also possible at Ischia, mainly in the northern part of the island.

Hydrogeological Hazards

The complex morphology of the Neapolitan area is further increased by a large number of quarries which have significantly modified the original slopes and the drainage pattern. In such a context rock falls, translational slides and earth and debris flows are frequent. Floods occurring in the plains at the end of huge valleys have the ability to transport coarse material. Large amounts of detritus in both superficial and underground drainage channels increases the hazard of floods, because it reduces the flux-capacity of drainage networks. These conditions occur mostly during the flood episodes and lead to overbank flood damage.

Rock fall occurs along the high-angle scarps of the Campi Flegrei caldera, the wave cut cliffs of the Posillipo hill, and the vertical walls of quarries. Such gravitational movements are favored by wind and sea erosion on highly fractured tuffs. Translational slides mobilize the loose pyroclastic deposits mantling high-angle slopes such as caldera walls and flanks of both volcanic edifices and valleys (Fig. 16). These landslides occur during periods of heavy rains, such as those of January 1997 and September 2001. The thickness of loose material mobilized by these landslides is generally less than 1 m. The amount of slide material is generally in the order of tens up to few hundreds of m^3. Earth and debris flows are rare and often result from the evolution of translational slides. Based on the analysis of the characteristics of the landslides that occur in different sectors of the Neapolitan territory, Calcaterra and Guarino (1997) concluded that the most frequent type of landslide is translational slide. Complex landslides (rock falls and translational slides evolving to earth flows) are less frequent, whereas simple rock falls and mudflows are rare.

Figure 16. Detachment scars of translational slides (whitish areas) in the southeastern slope of the Camaldoli hill, northern Campi Flegrei caldera wall.

MEASURES FOR MITIGATION OF THE VOLCANIC AND RELATED HAZARDS

Risk mitigation mostly depends on actively decreasing the vulnerability in relation to the variable hazards to which an area is exposed. In Italy measures aimed at achieving this goal are regulated by national, regional, provincial, and local laws that have resulted in a complex legislative framework. In the past few tens of years, following catastrophic events such as earthquakes or landslides, new regulations and institutions have been decreed. The 1980 earthquake of the southern Apennines was a turning point in the Italian policy of risk mitigation and civil defense. In 1992 the National Service for Civil Defense was established having as main task the prediction and prevention of natural hazards, and organization of the assistance during emergencies. Also the National Committee for the Prevision and Prevention of Great Risks is responsible for identifying research supporting civil defense actions. One of the major goals achieved by this service has been competency subdivision among the various authorities in relation to the severity of the event. For the first time the concept of emergency planning was introduced into Italian environmental policy. In this light, the emergency plans, considered an indispensable measure for risk mitigation, have been and/or are under preparation.

The volcanic emergency plans are drawn up on the basis of the eruption scenario and the alert levels defined for a specific volcano, over a certain period of time. They take into account all the social and economical aspects and the vulnerability of the territory, the population density and the expected human behavior. All these parameters can change through time, so the plans have to be thought as a dynamic tool, to be updated frequently. Each institution involved and the scientific community is responsible for a part of the plan, but a duty common to all is to give constant and accurate information and promote education programs in collaboration with the education system, the media and other partners, in order to put in practice another important risk mitigation measure: the growth of a "culture of the prevention".

Volcanic Hazards

In response to the increasing volcanic risk in the Vesuvius area, on the basis of the eruption scenario of the short-to-mid term maximum expected event, defined in 1990 by the National Group for Volcanology, in 1991 the Italian Minister for Civil Defense appointed a commitee to provide the guidelines for the evaluation of volcanic risk. In 1993, the Minister for Civil Defense appointed a commitee to prepare an emergency plan of the Vesuvian area. The commitee was entrusted to identify the areas exposed to the different hazards, on the basis of the defined maximum expected event, to define the alert levels and corresponding civil defense actions, to develop projects devoted to the information for the population, and to establish links between the scientific community and national and local Civil Defense institutions. The committee presented the results in the "National Emergency Plan for Vesuvius Area" in 1995. In 1996 the deputy Minister for Civil Defense appointed a new Commitee to update the emergency plans of the Vesuvian and Phlegraean areas.

For Vesuvius the Commitee upgraded the hazards zoning map, re-defined the alert levels and improved the actions to be taken. The updated plan, presented in June 2001, proposes a model of civil protection actions related to the different phases of the emergency in correspondence to the alert levels (Fig. 17a). A risk map has been prepared on the basis of the eruption scenario and the hazard zoning (Fig. 17b). In this map the three different hazard zones are delimited by towns administrative boundaries even if only partially included in the hazard zoning. Different measures and civil defense actions are planned for each zone. For the area where pyroclastic flows are expected (Red Risk Zone), there will be the evacuation of about 600,000 people, represented by 18 towns, during the warning alert level. Each of the 18 towns has ties to other communities in another region, where population will be moved in case of emergency. For the area in which exposure to pyroclastic fallout hazards is expected to affect communities represented by 96 towns (Yellow Risk Zone), only part of the population will move to refugee centers inside Campanian Region. For the area exposed to mud flow hazards (Blue Risk Zone) and included in the Yellow Risk Zone, plans are not yet complete.

In the Phlegraean area, a volcanic emergency plan was drawn up in 1984, during the bradyseismic crisis of 1983-84. The Committee appointed in 1996 had the task of updating the plan, on the basis of the eruption scenario, defining the alert levels (Fig. 18b), planning the civil defense actions and completing the seismic vulnerability investigation started in 1984. It presented the results of its work in June 2001, together with the updated Vesuvius plane.

The area exposed to the highest probability of being swept by pyroclastic currents (Fig. 18a) has been defined. This area, called the Red Risk Zone, includes people represented partly by the towns of Naples and Pozzuoli, and by the towns of Bacoli and Monte di Procida.

The risk map, derived from the hazard zoning map, has been constructed taking into account administrative boundaries and road connections among the towns. The limit of the Red Risk Zone has followed administrative boundaries within the towns of Naples and Pozzuoli, because only portions of both large towns lie in the maximum hazard zone. The small town of Bacoli was included in the Red Risk Zone not to fragment the population during an emergency. The town of Monte di Procida, which is outside the zone exposed to the maximum hazard, has been included into the Red Risk Zone because it lies on a promontory and is closely connected with the other towns. Population inside the Red Risk Zone is about 340,000 and will be evacuated to other Italian Regions during the warning alert level.

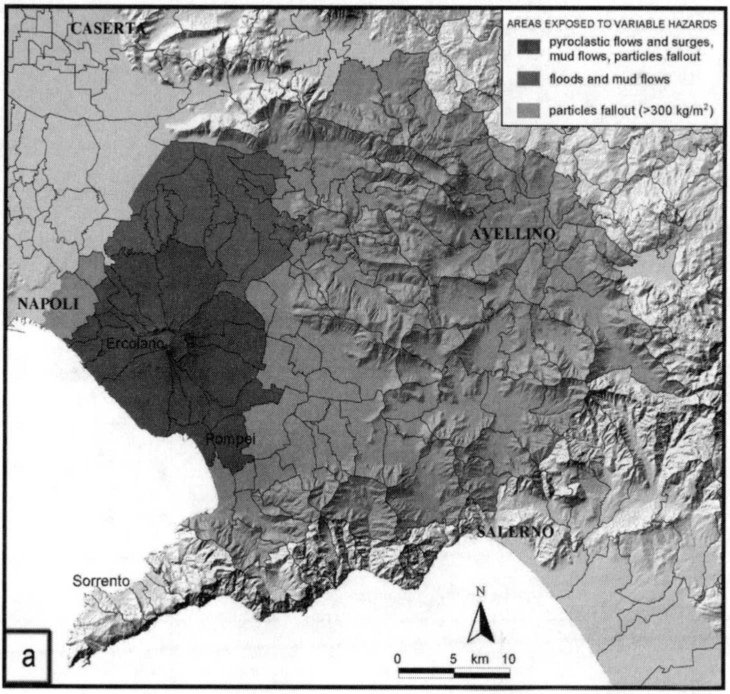

ALERT LEVELS	STATE OF THE VOLCANO	ERUPTION PROBABILITY	TIME TO THE ERUPTION	CIVIL DEFENCE SYSTEM: Scientific Community	CIVIL DEFENCE SYSTEM: Emergency response	PHASES
Base (green)	No significant variation of monitored parameters	Very low	Undefined, not less than several months	Surveillance activity according to schedule	- Ordinary activity	
Attention (yellow)	Significant variation of monitored parameters	Low	Undefined, not less than some months	Technical and scientific alert, and improvement in the monitoring system	- C.C.S.- Napoli activation - Information for people - Autority communications	I phase Attention
Warning (orange)	Further variation of monitored parameters	Medium	Undefined, not less than some weeks	Continuance of the surveillance activity; simulation of the expected eruption phenomena	- Request for emergency staus declaration to the President of Minister Council - Operational Commitee for Civil Protection meeting - DPC on site actiation - CCS activation in the host Italian Regions - Relief teams allocation	II phase Warning
Alarm (red)	Appearance of phenomena and/or evolution of parameters suggesting a pre -eruption dynamic	High	from weeks to days	Surveillance through remote system	- Red zone evacuation - Relief teams leave the red zone going to the yellow zone - Preparation of gates for the use of the roads (traffic regulation) - Police check empty house	III phase Alarm
b	During the eruption			Surveillance through remote system; defining the boundaries of the affected area inside the yellow zone	- Defining boundary of the affected area inside the yellow zone - Yellow zone evacuation - Accomodation in hotels, hostels,.. of Campania	IV phase During the event

Figure 17. a) Risk map of the Somma-Vesuvius; **b**) alert levels.

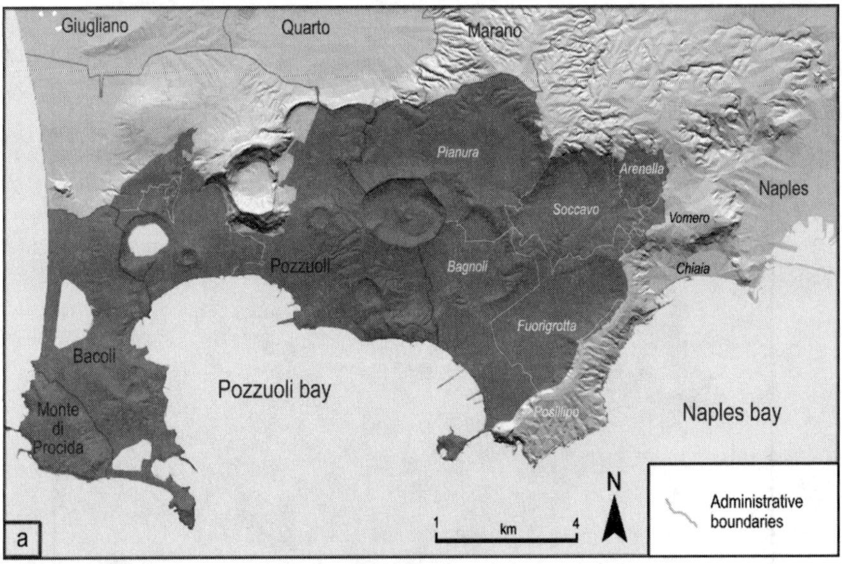

ALERT LEVELS	STATE OF THE VOLCANO	ERUPTION PROBABILITY	TIME TO THE ERUPTION	ACTIONS	COMMUNICATIONS
Base (green)	No significant variation of monitored parameters	Very Low	Undefined, not less than several months	Surveillance activity according to schedule	The Osservatorio Vesuviano publishes six-monthly reports on volcano activity
Attention (yellow)	Significant variation of monitored parameters	Low	Undefined, not less than some months	Technical and scientifical alert, and improvement in the monitoring system	The Osservatorio Vesuviano publishes daily reports and informs the Department of Civil Defense on the volcano state
Warning (orange)	Further variation of monitored parameters	Medium	Undefined, not less than some weeks	Continuance of the surveillance activity; simulation of the expected eruption phenomena	The Osservatorio Vesuviano gives continuously informs the Department of Civil Defense on the volcano state
Alarm (red)	Appearance of phenomena and/or evolution of parameters suggesting a pre-eruption dynamic	High	From days to months	Surveillance through remote systems	The Osservatorio Vesuviano gives continuously informs the Department of Civil Defense on the volcano state

b

Figure 18. a) Risk map of the Campi Flegrei caldera; **b)** alert levels.

Seismic Hazards

Rules and seismic risk mitigation measures have never distinguished between volcanic seismicity. The volcanic seismic hazard has been considered during drawing up of the Vesuvius emergency plan. In order to mitigate the risk related to this hazard a study on seismic vulnerability of the Vesuvian area has been carried out by the National Group for Defense by Earthquakes. As result of this study a seismic vulnerability map of the volcanic Red Risk Zone has been presented.

A study of the seismicity occurring during bradyseismic uplift in order to define mitigation actions and to plan for this particular emergency, that could involve about 10,000 people, is the task of a specific section of the National Emergency Plan for the Phlegraean area, called Abridged Plan for Bradiseismic Emergency.

A regional law of 1983 provides rules on earthquake-resistant buildings and an investigation of features as geology, slope stability, groundwater circulation, that can affect seismic vulnerability of the territory.

Hydrogeological Hazards

The hydrogeological hazards can be defined, more than any other hazard, as diffuse hazards because of the damages caused by large although not catastrophic events, which have a high frequency over the entire territory. Historically, after most of the many damaging events, legislative acts have been issued. An action plan proposed after the catastrophic flooding of the city of Florence in 1966, led to a national law in 1969. Aim of this law is to guarantee soil conservation, water clearance, rational use of the heritage and protection of the related environmental aspects. The law which distinguishes catchments basins of national, interregional and regional importance, introduced the Basin Plan as a tool for planning and putting into effect the actions aimed at mitigating the hydrogeological risk. Such actions include the knowledge of the characteristics of each basin, the installation of surveying and monitoring networks, and the establishment of a unique national informative system. The base for planning of all the actions is the base map which represents the actual situation of the territory. Following the catastrophic landslide events that occurred in the Sarno area in the Campanian Apennines in May 1998, which caused 160 casualties and enormous damages, a new law was issued. This law imposed both the location and definition of the areas exposed to hydrogeological hazards, and the elaboration of plans for mitigating the related risks.

The geological instability of the city of Naples and in particular of its subsoil is in the historical record as edicts enacted already in 1588 and 1615, till the most significant, enacted by the Bourbon government in 1781, which was established to regulate the use of the subsoil. More recently many committees have been appointed, such as the one in 1967, to prepare a comprehensive analysis of the actual problems. The legislative acts have always been issued after hazardous events and are related to management of an emergency. In the past few years, these interventions have been integrated, with actions aimed at planning and mitigating the risk based on investigations leading to a more detailed knowledge of the territory. On the basis of the national regulations, many ordinances have been issued for the Neapolitan area. These ordinances have permitted more efficient actions by the local authorities. After the floods of 1996-97, the mayor of Naples was appointed commissioner for the urgent interventions related to slope stabilization in the city. The commissioner has appointed a technical committee, which is following two approaches: urgent clearance, and hazard prevention and prediction.

The quarrying activity, mostly that carried out within the city of Naples, has been historically a cause of instability. Uncontrolled underground quarrying, which had caused severe damage to the buildings, prompted the authorities in 1926 to issue an ordinance to stop abusive quarrying in the urban area of Naples. Because of the intense excavation after the Second World War, underground quarrying, after thousands of years of activity, was prohibited by law; only open quarrying was permitted. Presently regional laws define the rules for quarrying activity.

CONCLUSIONS

The Neapolitan area is a good example of interaction between three active volcanoes and humanity for thousands of years. The city of Naples and its surrounding towns have been growing through time—despite the volcanic hazards—because of soil fertility, temperate climate, and favorable strategic position within the Mediterranean basin. The presence of active volcanoes is a source of hazards. Their activity accompanied by side effects such as seismicity and diffuse gas emission, and their constructive (new rocks formation) and destructive (volcano tectonic deformation) action have generated conditions for further hazards. Mankind has to find favorable conditions for life, but his presence and his pursuit of livelihood increase the risk he faces. Historically mankind's plans have not taken into consideration that the catastrophic events of a volcano have a longer recurrence time than a human life. In the last decade authorities, solicited by the scientific community, have realized that volcanic hazards have to be assessed and risk mitigation actions have to be planned. Therefore volcanological research and monitoring activity have been supported and emergency plans have been drawn up.

Acknowledgments. V. Augusti, E. Bellucci Sessa and F. Sansivero of the Cartography Laboratory of the Osservatorio Vesuviano are wormly thanked for their effort in preparing the illustrations. G. Valentine and K. Wohletz are acknowledged for a constructive review of the manuscript. The work benefited the support of the Italian National Group for Volcanology.

REFERENCES

Andronico, D., G. Calderoni, R. Cioni, A. Sbrana, R. Sulpizio and R. Santacroce, Geological map of Somma-Vesuvius Volcano, *Per. Mineral,* 64-8, 77_78, 1995.

Appleton, J.D., Petrogenesis of Potassium Rich Lavas from Roccamonfina Volcano, Roman Region, Italy, *J. Petrol.*, 13, 425, 1972.

Arnò V., C. Principe, M. Rosi, R. Santacroce, A. Sbrana, and M. F. Sheridan, Somma-Vesuvius, Eruptive History, *CNR, Quaderni de la Ricerca Scientifica*, XIII, 53_103, 1987.

Arrighi, S., C. Principe, and M. Rosi, Violent Strombolian Eruptions at Vesuvius During post-1631 Activity, *Bull. Volcanol.*, 63, 126_150, 2001.

Auger, E., P. Gasparini, J. Virieux, and A. Zollo, Seismic Evidence of an Extended Magmatic sill Under Mt. Vesuvius, *Science*, 9918, 2001.

Aveta, A., Materiali e Tecniche Tradizionali nel Napoletano. Note per il restauro architettonico. Arte Tipografica, Napoli, 1987.

Barberi, F., C. Principe, M. Rosi, and R. Santacroce, Scenario dell'Evento Eruttivo Massimo Atteso al Vesuvio in caso di Riattivazione a Breve-Medio Termine (Aggiornamento al 20.1.1995), CNR, Gruppo Nazionale per la Vulcanologia, Confidential Report, 19 pp, 1995.

Barberi, F., F. Innocenti, L. Lirer, R. Munno, T. Pescatore, and R. Santacroce, The Campanian Ignimbrite: a Major Prehistoric Eruption in the Neapolitan Area (Italy), *Bull. Volcanol.*, 41 (1), 1_22, 1978.

Bernasconi, A., P. Bruni, L. Gorla, C. Principe, and A. Sbrana, Risultati preliminari dell'Esplorazione Geotermica Profonda Nell'Area Vulcanica del Somma-Vesuvio, *Rend.Soc. Geol. Ital*, 4, 237_240, 1981.

Bertagnini, A., P. Landi, M. Rosi, and A. Vigliargio, The Pomici di Base Plinian Eruption of Somma-Vesuvius, *J. Volcanol. Geotherm. Res.*, 83, 219_239, 1998.

Brancaccio, L., A. Cinque, P. Romano, C. Rosskopf, F. Russo, N. Santangelo, and A. Santo, Geomorphology and Neotectonic Evolution of a Sector of the Tyrrenian Flank of the Southern Apennines, (Region of Naples), Z. Geomorph N. F., 82, 47_58, 1991.

Calcaterra, D., and P.M. Guarino, Fenomeni Franosi Recenti Nell'Area Urbana Napoletana. Il Settore Centro-Orientale. *Geologia delle grandi aree urbane, Atti conv. Bologna,* Nov. 1997. CNR, 1997.

Capasso, B., *Napoli Greco-Romana*, Berisio Ed., Napoli, 1905.

Carrara, E., F. Iacobucci, E. Pinna, and A. Rapolla, Gravity and Magnetic Survey of the Campanian Volcanic Area, S. Italy. *Boll. Geof. Teor. Appl.*, 15 (57), 39_51, 1973.

Carrara, E., F. Iacobucci, E. Pinna, and A. Rapolla, Interpretation of Gravity and Magnetic Anomalies Near Naples, Italy, Using Computer Techniques, *Bull. Volcanol.*, 38 (2), 458_467, 1974.

Cassignol, C., and P. Y. Gillot, Range Effectivness of Unspiked Potassium-Argon Dating, *Numerical dating in stratigraphy*, G. S., Oodin, 160, 1982.

Celico, P., D. Stanzione, L. Esposito, M. R. Ghiara, V. Piscopo, S. Caliro, and P. La Gioia Caratterizzazione Idrogeologica e Idrogeochimica Dell'Area Vesuviana, *Boll. Soc. Geol. It.*, 3_20, 1998.

Chiarabba, C., and A. Amato, Upper-Crust Structure of the Benevento Area (southern Italy) Fault Heterogeneities and Potential for Large Earthquakes, *Geophys. J. Int.*, 130, 229_239, 1997.

Cioni R., P. Marianelli, and A. Sbrana, Dynamics of the A.D. 79 Eruption: Stratigraphic, Sedimentologic and Geochemical Data on the Successions of the Somma-Vesuvius Southern and Easten Sectors. *Acta Vulcanologica*, Marinelli Volume, 2, 109_124, 1992.

Cioni, R., L. Civetta, P. Marianelli, N. Metrich, R. Santacroce, and A. Sbrana, Compositional Layering and Syneruptive Mixing of a Periodically Refilled Shallow Magma Chamber: the AD 79 Plinian Eruption of Vesuvius, *J. Petrology*, 36, 3, 739_776, 1995.

Cioni, R., R. Santacroce, and A. Sbrana, Pyroclastic Deposits as a Guide for Reconstructing the Multi-Stage Evolution of the Somma-Vesuvius Caldera, *Bull. Volcanol.*, 60, 207_222, 1999.

Civetta, L., G. Gallo, and G. Orsi, Sr- and Nd- Isotope and Trace-Element Constraints on the Chemical Evolution of the Magmatic System of Ischia (Italy) in the Last 55 ka, *J. Volcanol. Geotherm. Res.*, 46, 213_230, 1991.

Civetta, L., G. Orsi, L. Pappalardo, R. V. Fisher, G. Heiken, and M. Ort, Geochemical Zoning, Mingling, Eruptive Dynamics and Depositional Processes - The Campanian Ignimbrite, Campi Flegrei Caldera, Italy, *J. Volcanol. Geotherm. Res.*, 75, 183_219, 1997.

Corniello, A., R. De Riso, and D. Ducci, Idrogeologia e Idrogeochimica della Piana Campana. *Mem. Soc. Geol. It.*, 45, 351_360, 1990.

Cubellis, E. and G. Luongo, Sismicità Storica dell'Isola d'Ischia. In: Il terremoto del 28 Luglio 1883 Nell'Isola d'Ischia. *Presidenza del Consiglio dei Ministri, Dipartimento per i Servizi Tecnici Nazionali – Servizio Sismico Nazionale*, 1998.

D'Antonio, M., L. Civetta, G. Orsi, L. Pappalardo, M. Piochi, A. Carandente, S. de Vita, M.A. Di Vito, R. Isaia, and J. Southon, The Present State of the Magmatic System of the Campi Flegrei Caldera Based on the Reconstruction of its Behaviour in the Past 12 ka, *J. Volcanol. Geotherm. Res.*, 91, 247_268, 1999.

D'Argenio, B., T. S. Pescatore, and P. Scandone, Schema Geologico dell'Appennino Meridionale, *Acc. Naz. Lincei, Quad.*, 183, 49_72, 1973.

De Lorenzo, G. L'attività Vulcanica nei Campi Flegrei, *Rend. Acc. Sc. Fis.Mat.*, Napoli, 10, 204_221, 1904.

De Seta, C., Le Città Nella Storia d'Italia: Napoli, Laterza Ed., 1981.

de Vita, S., G. Orsi, L. Civetta, A. Carandente, M. D'Antonio, T. Di Cesare, M. A. Di Vito, R.V. Fisher, R. Isaia, E. Marotta, M. Ort, L. Pappalardo, M. Piochi, and J. Southon, The Agnano-Monte Spina Eruption (4.1 ka) in the Resurgent, Nested Campi Flegrei Caldera (Italy), *J. Volcanol. Geotherm. Res.*, 91, 269_301, 1999.

Deino A. L., G. H. Courtis, J. Southon, F. Terrasi, L. Campajola, and G. Orsi, ^{14}C and $^{40}Ar/^{39}Ar$ Dating of the Campanian Ignimbrite, Phlegraean Fields, Italy, *ICOG* Berkley California, USA, Abstract, p. 77, 1994.

Deino A. L., G. H. Curtis, and M. Rosi, $^{40}Ar/^{39}Ar$ Dating of the Campanian Ignimbrite, Campanian Region, Italy, *29th International Geological Congress, Japan*, Abstracts, 3, 2654, 1992.

Delibrias, G., G. M. Di Paola, M. Rosi, and R. Santacroce, La Storia Eruttiva del Complesso Vulcanico Somma-Vesuvio Ricostruita Dalle Successioni Piroclastiche del Monte Somma, *Rend. Soc. It. Mineral. Petrol.*, 35, 411_438, 1979.

Dellino, P., R. Isaia, L. La Volpe, and G. Orsi, Statistical Analysis of Textural Data from Complex Pyroclastic Sequence: Implication for Fragmentation Processes of the Agnano-Monte Spina Eruption (4.1 ka), Phlegraean Fields, Southern Italy, *Bull Volcanol*, 63, 443_461, 2001.

Dellino, P., R. Isaia, L. La Volpe, and G. Orsi, Interference of Particles Fallout on the Emplacement of Pyroclastic Surge Deposits of the Agnano-Monte Spina Eruption (Phlegraean Fields, Southern Italy), *J. Volcanol. Geotherm. Res.*, submitted, 2003.

Di Vito M. A., R. Sulpizio, G. Zanchetta, and G. Calderoni, The Geology of the South Western Slopes of Somma-Vesuvius, Italy, *Acta Vulcanologica*, 10 (2), 383_393, 1998.

Di Vito, M. A., L. Lirer, G. Mastrolorenzo, and G. Rolandi, The Monte Nuovo Eruption (Campi Flegrei, Italy), *Bull. Volcanol.*, 49, 608_615, 1987.

Di Vito, M.A., R. Isaia, G. Orsi, J. Southon, S. de Vita, M. D'Antonio, L. Pappalardo, and M. Piochi, Volcanic and Deformational History of the Campi Flegrei Caldera in the Past 12 ka, *J. Volcanol. Geotherm. Res.*, 91, 221_246, 1999.

Dobran F., A. Neri, and G. Macedonio, Numerical Simulations of Collapsing Volcanic Columns, *J. Geophys. Res.*, 98, B3, 4231_4259, 1993.

Fedele F., B. Giaccio, R. Isaia, and G. Orsi, Ecosystem Impact of the Campanian Ignimbrite Eruption (~37,000 cal yr B.P.) and its Potential Interference With Biocultural Change in Late Pleistocene Europe: a Progress Report, *Quaternary Res.*, in Press, 2002.

Fisher R. V., G. Orsi, M. Ort, and G. Heiken, Mobility of Large-Volume Pyroclastic Flow – Emplacement of the Campanian Ignimbrite, Italy, *J. Volcanol. Geotherm. Res.*, 56, 205_220, 1993.

Follieri, M., D. Magri,and L. Sadori, Pollen Stratigraphical Synthesis From Valle di Castiglione (Roma). *Quaternary International*, 47/48, 3-20, 1989.

Fulignati, P., P. Marianelli, and A. Sbrana, Glass-Bearing Felsic Noduls From Crystallizing Side-Walls of the 1944 Vesuvius Magma Chamber, *Mineral. Mag.*, 64, 263_278, 2000.

Ippolito, F., F. Ortolani, and M. Russo, Struttura Marginale Tirrenica dell'Appennino Campano: Reinterpretazione di Dati di Antiche Ricerche di Idrocarburi, *Mem. Soc. Geol. It.*, XII, 227_250, 1973.

Isaia, R., Storia Vulcanica ed Evoluzione Morfostrutturale della Caldera dei Campi Flegrei Negli Ultimi 12 ka, Ph.D. thesis, Universita' di Napoli, 1998.

Johnston-Lavis, H. J., The Geology of Mt. Somma and Vesuvius, Being a Study in Volcanology. *Quart. J. Geol. Soc., London*, 40, 35-119, 1884.

Johnston-Lavis, H. J., Monograph of the Earthquakes of Ischia, *Naples and London*, 1885.

Lirer L., R. Munno, P. Petrosino, and A.Vinci, Tephrostratigraphy of the A.D. 79 Pyroclastic Deposits in the Perivolcanic Area of Vesuvius (Italy), *J. Volcanol. Geoth. Res.*, 58, 133_149, 1993.

Lirer, L., and L. Vitelli, Volcanic Risk Assessment and Mapping in the Vesuvian Area Using Gis, *Natural Hazards*, 17, 1_15, 1998.

Lirer, L., G. Mastrolorenzo, and G. Rolandi, Un Evento Pliniano nell'Attività Recente dei Campi Flegrei, *Boll. Soc. Geol. It.*, 106, 461_473, 1987.

Lirer, L., P. Petrosino, I. Alberico, and I. Postiglione, Long-Term Volcanic Hazard Forecasts Based on Somma-Vesuvio Past Eruptive Activity, *Bull. Volcanol.*, 63, 45-60, 2001.

Lirer, L., R. Munno, I. Postiglione, A. Vinci, and L.Vitelli, The A.D. 79 Eruption as a Future Explosive Scenario in the Vesuvian Area: Evaluation of Associated Risk, *Bull. Volcanol.*, 59, 112_124, 1997.

Lirer, L., T. Pescatore, B. Booth, and P. L. Walker, Two Plinian Pumice Fall Deposits From Somma – Vesuvius (Italy), *Geol. Soc. Am. Bull.*, 84, 143_149, 1973.

Marianelli, P., N. Metrich, and A. Sbrana, Shallow and Deep Reservoirs Involved in the Magma Supply of the 1944 Eruption of Vesuvius, *Bull.Volc.*, 61, 48_63, 1999.

Marzocchella, A., L'età Preistorica a Sarno, Le Testimonianze Archeologiche di Foce e San Giovanni. In:, *C., Tremblements de Terre, Erupition Volcaniques et vie des Homes Dans la Campanie Antique,* Edited by Albore Livadie, Centre Jean Bérard, Napoli, 35-53, 1986.

Marzocchella, A., Tutela Archeologica e Preistoria Nella Pianura Campana, *Archeologia e vulcanlogia in Campania*, Arte Tipografica, 1998.

Miccio, B., and U. Potenza, *Gli Acquedotti di Napoli,* A.M.A.N Napoli, 1994.Napoli, M., Napoli Greco-Romana, Colonnese Ed., 1996.

Newhall C., and S. Self , The Volcanic Explosivity Index (VEI): an Estimate of Explosive Magnitude or Historical Volcanism, *J. Geophys. Res.*, 87, 1231_1238, 1982.

Orsi G., L. Civetta, M. D'Antonio, P. Di Girolamo, and M. Piochi, Step-Filling and Development of a Three-Layer Magma Chamber: the Neapolitan Yellow Tuff Case History, *J. Volcanol. Geotherm. Res.*, 67, 291_312, 1995.

Orsi G., M. Piochi, L. Campajola, A. D'Onofrio, L. Gialanella, and F. Terrasi, ^{14}C Geochronological Constraints for the Volcanic History of the Island of Ischia (Italy) Over the Last 5,000 Years, *J. Volcanol. Geotherm. Res.*, 71, 249_257, 1996.

Orsi, G., D. Patella, M. Piochi, and A. Tramacere, Magnetic Modeling of the Phlegraean Volcanic District With Extension to the Ponza Archipelago, Italy, *J. Volcanol. Geotherm. Res.*, 91, 345-360, 1999b.

Orsi, G., G. Gallo, G. Heiken, K. Wohletz, E. and G. Bonani, A comprehensive study of the pumice formation and dispersal: the Cretaio tephra of Ischia (Italy). *J. Volcanol: Geotherm. Res.*, 53: 329-354, 1992.

Orsi, G., G. Gallo, and A. Zanchi, Simple Shearing Block Resurgence in Caldera Depressions. A Model From Pantelleria and Ischia, *J. Volcanol. Geotherm. Res.*, 47, 1_11, 1991.

Orsi, G., L. Civetta, C. Del Gaudio, S.de Vita, M. A. Di Vito, R. Isaia, S. Petrazzuoli, G. Ricciardi, and C. Ricco, Short-Term Ground Deformations and Seismicity in the Nested Campi Flegrei Caldera (Italy): an Example of Active Block Resurgence in a Densely Populated Area, *J. Volcanol. Geotherm. Res.* 91, 415_451, 1999a.

Orsi, G., M. A. Di Vito, and R. Isaia, Volcanic Hazards Assessment at the Restless Campi Flegrei Caldera, *Bull. Volcanol.*, Submitted, 2003.

Orsi, G., M. D'Antonio, S. de Vita, and G. Gallo, The Neapolitan Yellow Tuff, a Large-Magnitude Trachytic Phreatoplinian Eruption: Eruptive Dynamics, Magma Withdrawal and Caldera Collapse, *J. Volcanol. Geotherm. Res.*, 53, 275_287, 1992.

Orsi, G., S. de Vita, and M. A. Di Vito, The Restless, Resurgent Campi Flegrei Nested Caldera (Italy): Constraints on its Evolution and Configuration, *J. Volcanol. Geotherm. Res.,* 74, 179_214, 1996.

Orsi, G., S. Petrazzuoli, and K.Wohletz, The Interplay of Mechanical and Thermo-Fluid Dynamical Systems During an Unrest Episode in Calderas: the Campi Flegrei Caldera (Italy) Case, *J. Volcanol. Geotherm. Res.*, 91, 453_470, 1999c.

Palladio, A., *I Quattro Libri dell'Architettura*, Venezia, 1570.

Pappalardo L., L. Civetta, S. De Vita, M. A. Di Vito, G. Orsi, A. Carandente, and R.V. Fisher, Timing of Magma Extraction During the Campanian Ignimbrite Eruption (Campi Flegrei Caldera), *J. Volcanol. Geotherm. Res.*, in Press, 2002a.

Pappalardo L., M. Piochi, M. D'Antonio, L. Civetta, and R. Petrini, Evidence for Multi-Stage Magmatic Evolution During the Past 60 ka at Campi Flegrei (Italy) Deduced From Sr, Nd and Pb Isotope Data, *J. Petrol.*, in Press, 2002b.

Pappalardo, L., L. Civetta, M. D'Antonio, A. Deino, M. Di Vito, G. Orsi, A. Carandente, S. de Vita, R. Isaia, and M. Piochi, Chemical and Isotopical Evolution of the Phlegraean Magmatic System Before the Campanian Ignimbrite (37 ka) and the Neapolitan Yellow Tuff (12 ka) Eruptions, *J. Volcanol. Geotherm. Res.*, 91, 141_166, 1999.

Piochi, M., L. Civetta, and G. Orsi, Mingling in the Magmatic System of Ischia (Italy) in the Past 5 ka, *Mineralogy and Petrology*, 66, 227_258, 1999.

Poli, S., S. Chiesa, P.Y. Gillot, A. Gregnanin, and F. Guichard, Chemistry Versus Time in the Volcanic Complex of Ischia (Gulf of Naples, Italy): Evidence of Successive Magmatic Cycles, *Contr. Mineral. Petrol.*, 95-3, 322_335, 1987.

Poli, S., S. Chiesa, P.Y. Gillot, F. Guichard, and L. Vezzoli, Time Dimension in the Geochemical Approach and Hazard Estimation of a Volcanic Area: the Isle of Ischia Case (Italy), *J. Volcanol. Geotherm. Res.*, 36, 327_335, 1989.

Postpischl, D. (Ed.), Catalogo dei Terremoti Italiani dall'Anno 1000 al 1980, in *C.N.R.-P.F.G.*, 114-2B, 238 pp, 1985.

Rolandi, G., A. M. Barrella, and A. Borrelli, The 1631 Eruption of Vesuvius, *J. Volcanol. Geotherm. Res.*, 58, 183_201, 1993a.

Rolandi, G., G. Mastrolorenzo, A. N. Barrella, A. Borrelli, The Avellino Plinian Eruption of Somma-Vesuvius (3760 y.B.P.): the Progressive Evolution From Magmatic to Hydromagmatic Style, *J. Volcanol. Geotherm. Res.*, 58, 67_88, 1993b.

Rolandi, G., S. Maraffi, P. Petrosino, and L. Lirer, The Ottaviano Eruption of Somma-Vesuvius (8000 y BP): a Magmatic Alternating Fall and Flow-Forming Eruption, *J. Volcanol. Geotherm Res.*, 58, 43_65, 1993c.

Rosi, M., and R. Santacroce, The A.D. 472 «Pollena» Eruption: Volcanological and Petrological Data for This Poorly Known Plinian-Type Event at Vesuvius, *J. Volcanol. Geotherm. Res.*, 17, 249_271, 1983.

Rosi, M., C. Principe, and R. Vecci, The 1631 Eruption of Vesuvius Reconstructed From the Review of Chronicles and Study of Deposits, *J.Volcanol. Geotherm. Res.*, 58, 151_182, 1993.

Rosi, M., L. Vezzoli, A. Castelmenzano, and G. Grieco, Plinian Pumice Fall Deposit of the Campanian Ignimbrite Eruption (Phlegraean Fields, Italy), *J. Volcanol. Geotherm. Res.*, 91, 179_198, 1999.

Rosi, M., L. Vezzoli, P.Aleotti, and M. De Cenzi, Interaction Between Caldera Collapse and Eruptive Dynamics During the Campanian Ignimbrite Eruption, Phlegraean Fields, Italy, *Bull. Volcanol.*, 57, 541_554, 1996.

Santacroce, R. (Ed.), Somma-Vesuvius, *Quaderni de la Ricerca Scientifica, CNR, 114*, Progetto Finalizzato Geodinamica, Monografie Finali, 8, 1987.

Santacroce, R., R. Cioni, P. Marianelli, Sbrana A., in Press. *Understanding Vesuvius and Preparing for its Next Eruption*. The Cultural Response to the Volcanic Landscape. M.S. Balmuth ed., Tufts University, 2003.

Scandone, R., G. Arganese, and F. Galdi, The Evaluation of Volcanic Risk in the Vesuvian Area, *J. Volcanol. Geotherm. Res.*, 58, 263_271, 1993.

Scarpati, C., P. Cole, and A. Perrotta, The Neapolitan Yellow Tuff - A Large Volume Multiphase Eruption From Campi Flegrei, Southern Italy, *Bull. Volcanol.*, 55, 343_356, 1993.

Sigurdsson, H., S. Carey, W. Cornell, and T. Pescatore, The Eruption of Vesuvius in 79 A.D., *National Geographic Res.*, 1, 332_387, 1985.

Sulpizio, R., M. A. Di Vito, G. Zanchetta, Landscape Response to the Deposition of Airfall Pyroclastic From Large Explosive Eruptions: An Example From the Campanian Area (Southern Italy), *Phys. Chem. Earth (A)*, 25, (9-11), 759_762, 2000.

Vezzoli, L. (Ed.) Island of Ischia. *CNR Quaderni de "La ricerca scientifica"*, 114-10, 122, 1988.

Wohletz, K., G. Orsi, and S. de Vita, Eruptive Mechanisms of the Neapolitan Yellow Tuff Interpreted From Stratigraphic, Chemical and Granulometric Data, *J. Volcanol. Geotherm. Res.*, 67, 263_290, 1995.

Wohletz, K., L. Civetta, and G. Orsi, Thermal Evolution of the Phlegraean Magmatic System. *J. Volcanol. Geotherm. Res.*, 91, 381_414, 1999.

Zanchetta, G., R. Sulpizio, and M. A. Di Vito, Genesis and Evolution of Volcaniclastic Alluvial Fans in the Southern Campania (Italy): Relationships With Volcanic Activity and Climate, *Geol. Soc. of America Bull.*, Submitted, 2003.

7

Tsunami Impact and Mitigation in Inhabited Areas

G. T. Hebenstreit, F. I. González, and J. Preuss

THE TSUNAMI THREAT

Tsunamis are long ocean surface waves generated by the rapid displacement of large volumes of sea water. Most tsunamis result from submarine earthquakes occurring in shallow ocean waters, submarine landslides and, occasionally, exploding volcanoes. More rarely, meteor impacts, such as the one that struck the Yucatan Peninsula 65 million years ago, can also generate tsunamis.

The threat from tsunamis arises when the waves—which are long, low, and fast in the open sea—become shorter, steeper, and slower as they propagate into nearby or very distant coastal areas. If the wave energy is sufficiently high and the offshore topography is conducive to rapid shoaling, tsunamis can reach destructive heights at the shoreline and destroy populated areas.

Tsunami hazards are multi-faceted, and effect damage through a number of physical mechanisms, including:

- *Flooding* due to in-rushing water,
- *Wave-structure impacts* due to pressure effects and wave breaking,
- *Flotation and transport* of heavy objects, turning them into projectiles,
- *Scouring* of subsoils, leading to structural instabilities in dwellings, industrial buildings, and infrastructure,
- *Fire,* and
- *Breaching* of protective barriers sheltering hazardous materials.

Earth Science in the City: A Reader

10.1029/056SP08

Tsunamis can wipe out entire villages and towns in low-lying areas, cause wide-spread destruction even in built-up urban areas, and impair infrastructure enough to limit a locality's ability to cope with the aftermath. The photographs in Figures 1 and 2 provide chilling examples. Notice especially that, at Aonae, debris is scattered over a wide area, blocking access roads and effectively hindering emergency response efforts.

As with most geophysical hazards, destructive tsunamis are irregular and unpredictable events. The last tsunami to cause significant damage in the continental United States arose from the 1964 Prince William Sound earthquake, which struck Alaskan coastal towns and also severely damaged Crescent City, California [Committee on the Alaskan Earthquake, 1972]. A more recent destructive tsunami occurred in Papua-New Guinea in 1998. This event took over 2200 lives and devastated several villages along the PNG coast [Kawata *et al.,* 1999]. In fact, ten destructive tsunamis have claimed more than 4,000 lives since 1990 [González, 1999].

From an historical perspective, destructive tsunamis are most common in the Pacific Ocean basin, surrounded as it is by the Ring of Fire delineated by volcanoes and active plate boundaries that give rise to large earthquakes. However, evidence is growing that shows a long history of such events in the Mediterranean and Caribbean Seas. Little attention has been paid to the possibility of tsunamis on America's east and Gulf coasts, largely because of the presumed low probability of submarine earthquakes of sufficient magnitude. This may change as knowledge of east coast sea floor structure grows.

Also from an historical perspective, tsunamis are most likely to be devastating in small coastal towns and villages where mitigation efforts are often underfunded or, more likely, non-existent. Larger coastal cities in a country like Japan, where the tsunami history is long and well-known, tend to be protected by massive seawalls and other coastal defenses. In contrast, large coastal cities in the United States are, for the most part, unprotected. They have traditionally emerged unscathed from some of the last century's most destructive teletsunamis (that is, tsunamis generated at a great distance). However, we now know from the geologic record that large tsunamis have been generated in the Cascadia Subduction Zone off Washington, Oregon and Northern California, and by the Seattle Fault in Puget Sound. Furthermore, we know that the geophysical processes that generated these tsunamigenic earthquakes have been operating for millennia and continue to this day. Therefore, it is not a question of whether a large, destructive tsunami will occur in these areas ... it is a question of when.

CURRENT TSUNAMI RESEARCH

Tsunamis evolve through three distinct physical processes—*generation* by various geophysical mechanisms, *propagation* through water of variable depth, and *inundation* of land areas. Hebenstreit [1997] provided a short examination of

Figure 1. *(top)* Destruction in Aonae, Okushiri Island, due to wave impacts from the tsunami of 12 July 1993 *(Bottom)* Destruction in Aonae, Okushiri, Island and to fires generated by multiple ignition sources including fishing boat washed ashore and propone tanks, after the tsunami. (Photos from National Geophysical Data Center, Boulder, CO).

Figure 2. A fishing boat was washed onto a fire truck by the 1993 Okushiri tsunami, thereby severely reducing the capacity of the community to fight the devastating fire. (Photo from National Geophysical Data Center, Boulder, CO)

the progress of tsunami research over the last 6 decades. Tsunami study began in earnest in Japan in the 1930's and in the United States and other countries in the late 1940's. Initially, studies focused on the generating mechanisms and on the physics of open-ocean propagation.

Generation research has concentrated in the past on earthquakes, since this is the source mechanism of most tsunamis. However, the importance of subaerial landslides and underwater slumping has become evident in the last decade, and much research is now focused on the dynamics of these generation processes. A great deal remains to be learned about why some earthquakes and landslides cause tsunamis and others do not, and a serious difficulty is the lack of high quality observations for the verification of theory and numerical modeling of tsunami generation. Seismic data inversion and displacement models are the only basis for describing a tsunamigenic earthquake source, but these usually do not provide the fine spatial resolution needed to accurately model a tsunami event. Accurate specification of landslide characteristics are even more problematic, as useful measurements of these events are exceedingly rare.

Propagation is best understood and most accurately modeled. Although not abundant, tsunami records collected by coastal tide gauges and, more recently, by deep-ocean bottom pressure recorders, are available for comparison with model results. Models are generally found to agree well with measurements of the first one or two wave cycles. As a rule, complex scattering and reflection create a complexity in later wave arrivals that models fail to duplicate, although the mean energy levels are frequently reproduced quite well.

Inundation is a highly nonlinear process and only in the last two decades have efforts to understand wave runup processes been able to advance. Especially in the last decade, the research community has successfully organized effective field surveys to acquire accurate measurements of maximum wave runup in the aftermath of a number of tsunami disasters. The subsequent comparisons with numerical models has given confidence that, while not perfect, these models are sufficiently accurate to provide useful products for emergency management purposes.

Tsunami *impact* on coastal structures and populations has only recently begun to receive attention. It is the most complex and least understood process, but advances in this area are essential to an important goal of tsunami research—assessing the destructive impact of a tsunami on a coastal community. To a significant degree the destructiveness of tsunamis is a result of interactions and secondary effects such as debris (cars, boats, buildings etc) and fire.

TSUNAMI HAZARD REDUCTION

Reduction of the threat to coastal populations must In addressing this issue, two fundamentally different tsunami scenarios must be dealt with—*local* tsunamis generated so near a community that the first wave arrives within a few minutes, and *distant* tsunamis (or *tele-tsunamis*) generated so far from a community that the first wave does not arrivc for hours. This distinction is clearly important, although somewhat artificial, since the same tsunami can be experienced by different communities as either local or distant, depending on their distance from the source. Efforts to reduce the tsunami hazard in the United States generally pursue three distinctly identifiable, though closely related, activities—*assessment* of risk, *warning* improvement, and *mitigation* through planning and education. Each component is essential to the overall effort, but its importance and effectiveness depends on whether a tsunami is local or distant. Warning systems can be designed for local and/or distant events, for example, but the technology is clearly more effective in the case of a distant tsunami; similarly, planning and education to ensure a quick, appropriate response by coastal residents is more important in the case of a local tsunami. Brief descriptions of the mixture of technologies needed to effectively mitigate both local and distant tsunami threats in populated areas will be given here. Additional information on different aspects of these efforts, and other relevant references are provided by Bernard *et al.* [1988], and Blackford and Kanamori [1994] and Tatehata [1997].

Community-Specific Hazard Assessment

Hazard assessments take many forms. Several methodologies have been developed in the last few years. For example, Figure 3, from Preuss and Hebenstreit

Figure 3. Critical flood levels for Aberdeen, WA (north side of the Chehalis River) and South Aberdeen (south side of the river) are indicated in gray. Low-lying areas will be subject to flooding during low tide; during high tide the entire urbanized area will be subject to extensive flooding. FEMA 100-year flood boundary is the 10-foot contour. Note that the coastal highway, critical for response as well as search and rescue, is vulnerable to flooding.

[1998], is a base map of Aberdeen, Washington with shaded areas indicating potential tsunami inundation zones based on numerical simulations of a major thrust earthquake occurring offshore in the Cascadia Subduction Zone. This study examined the potential for coupling relatively simple numerical inundation simulation techniques with detailed depictions of locations of hazardous material sites, potential-materials for water borne debris, and lifeline sites such as hospitals, fire stations, and evacuation routes. By bringing these types of information into one presentation, state and local authorities can more readily understand the nature of the hazard they face, its potential impact on communities they govern, and the need for preparedness planning. A similar study by Preuss et al. [1988] produced maps which depict inundation areas in Kodiak, Alaska and the impact of the zones on response capabilities.

Since these studies were completed, a systematic effort has been mounted to develop inundation maps for each U.S. coastal community at risk. In 1997, the U.S. National Tsunami Hazard Mitigation Program (NTHMP) was created as a partnership of the States of Alaska, California, Hawaii, Oregon and Washington with NOAA, USGS and FEMA [Bernard, 1998]. This program established the Center for Tsunami Inundation Mapping Efforts (TIME) to collaborate with and assist numerical modelers in the production of maps such as that shown for Newport, Oregon in Figure 4. Similar efforts have produced numerous such maps in Central and South America [Ortiz, 1996] and Japan.

An even more sophisticated approach is exemplified by Figure 5, from Bernard [1997]. Here, high resolution tsunami runup model results are combined with baseline maps of lifelines as well as estimates of potential zones for liquefaction and landslides induced by ground accelerations due to a local earthquake. Such an integrated view of the risk allows municipal authorities to understand the full spectrum of the potential threat and to lay response plans accordingly.

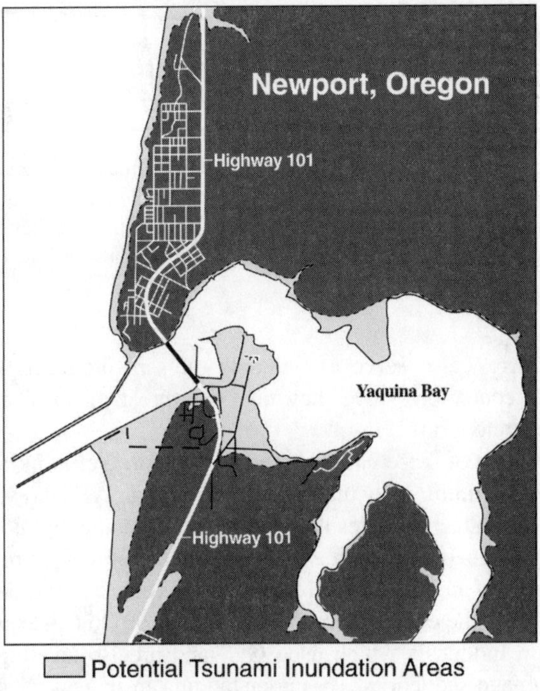

Figure 4. Tsunami inundation map produced for Newport, Oregon under the TIME program.

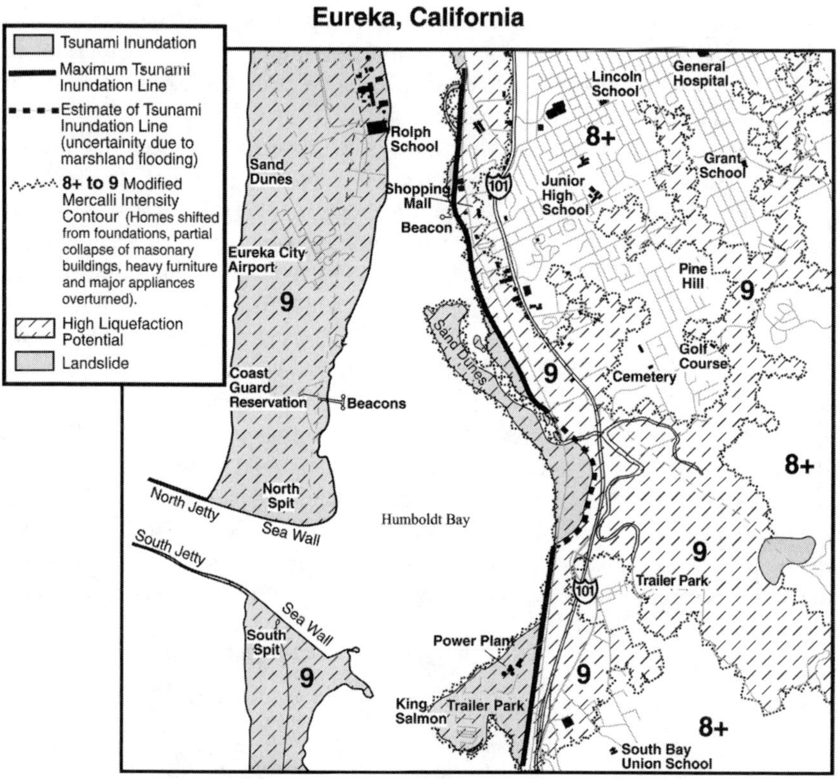

Figure 5. This map identifies areas of tsunami flooding, areas of liquefaction, landslides, and intense ground shaking in Eureka, California. If a local major earthquake near Eureka generates a tsunami, Highway 101 will probably be damaged by liquefied soils to the south, meaning that only northward evacuation would be feasible on the highway. [Bernard, 1997].

These products can also serve as valuable tools in programs for community awareness and education, since they take the threat of an abstract event and depict it in concrete terms.

The effectiveness of these hazard assessment efforts relies in large part on the study and understanding of actual tsunami disasters. As survey, photographic, and communication technologies have improved dramatically in recent years, so has our ability to document such events; international teams are now routinely organized to survey an area struck by significant tsunamis and develop a highly detailed account of the event. Maramai and Tinti [1997] is an example of a post-event survey in Indonesia which used GPS and precise leveling techniques to reconstruct damage sequences. By understanding in increasing detail what *has* happened during tsunamis, scientists and engineers can then develop increasingly informed ideas on what *can* happen in specific localities. These ideas can guide research that focuses more directly on how tsunamis cause damage and lead to refined ideas of how to reduce the damage.

Warning System Improvement

The ultimate goal of efforts to improve a warning system is to provide a swift, reliable, site-specific estimate of tsunami *impact* on structures and population. Since earthquakes are the primary tsunami generating mechanism, estimating the size and location are critical. Thus, the NTHMP has substantially increased the number and quality of seismic network stations and made significant improvements in the speed and reliability of the reports. Current warning systems supplement these seismic data with a sparse network of coastal tide gauges and a historical tsunami database to judge the likelihood of extreme wave generation. If the judgement is made that a destructive tsunami *may* have been generated, then a relatively simple algorithm is employed to forecast site-specific arrival times. But wave height estimates, essential to the development of *impact* forecasts, are not provided.

Existing seismic and coastal tide gauge networks are essential and valuable, but are nonetheless inadequate for reliable, site-specific tsunami forecasting. This is because they do not provide direct measurement of a tsunami in the open ocean, as it propagates from the source to distant coastal communities. To provide this capability, the NTHMP established the Deep-ocean Assessment and Reporting of Tsunamis (DART) Project for the early detection and reporting of tsunami data to warning centers in real-time; this project has established a network of stations directly seaward of active earthquake zones in the North Pacific [Bernard, 1998].

But tsunami energy can be highly directional, so the relatively sparse DART network data must be carefully interpreted, preferably by comparison of numerical simulations with the tsunami wave height measurements. Real-time computations could, in principal, solve the complex set of long wave differential equations which govern tsunami wave height for a specific event, as future computers become more powerful. However, such computations would be of little use to communities close to the source, since the first tsunami wave would strike in minutes. Furthermore, even for distant communities with hours of warning time, such real-time computations would be highly suspect, since accurate specification of the *generation* event typically remains incomplete and unreliable long after the tsunami danger has passed. Currently under development are warning guidance tools which exploit an archived database of pre-computed tsunami model simulations to rapidly provide a scenario that best agrees with real-time tsunami and/or earthquake measurements [Tatehata, 1997; Titov et. al, 1999].

These estimates of offshore wave heights must be extended onto the coast itself through even more complex computations of coastal wave *runup*. Such computations are now routinely used in the production of inundation maps to assess community risk, and R&D efforts are currently underway to exploit this modeling technology for real-time warning guidance. The final step, estimating tsunami *impact* on coastal structures and populations, is the most complex and difficult phase of the tsunami forecasting problem. Currently, our capability in this area tends to be more theoretical and less applied. But the coupling of increased computing power, advances in modeling tsunami dynamics, and improvements in real-time tsunami measurement technology is opening up new avenues for developing improved capabilities for forecasting tsunami impact.

The more distant a community from the source, the more time available for warning guidance to be developed and disseminated, and the effectiveness of a warning system decreases as the source-community distance decreases. Project THRUST (Tsunami Hazard Reduction Using Systems Technology) designed and implemented an effective local warning systems that is still in use today [Bernard *et al.*, 1988]. But in the extreme case of a local tsunami that reaches a community in a few minutes. pre-event planning and educating are of paramount importance. The community must learn the life-saving lesson that "The earthquake is the warning to run inland as soon as possible, preferably to high ground." This educational aspect of mitigation is covered next in more detail.

Planning and Education

Tsunamis are inevitable, and cannot be prevented. However, communities which have come to an understanding of the type of threats a tsunami poses can take steps to minimize those threats by prior planning for effective response. The type of planning required depends on the time scale of the desired response.

The most critical time scale lies in the minutes before a tsunami strikes and the hours after it has ended. This is the time frame during which a community can take steps to leave harm's way and then return to begin rescue and salvage efforts.

The way to approach the minutes before a tsunami strikes is to educate the threatened population on appropriate ways to respond when a warning is given and also what to look for when no warning is possible. The response of the individual citizen should be simple and straightforward: If a warning sounds, head directly for higher ground. If no warning is heard, but an earthquake knocks you over, get up and head directly for higher ground. For a populace to react properly, they must be sensitized well ahead of time to the problem and then reminded regularly of the steps to take. Readily available local maps identifying tsunami hazard zones and evacuation routes is important. Classroom education programs help to maintain awareness and understanding. Roadside signs such as shown in Figure 6 are also proving to be very effective tools in imparting and maintaining response education.

The time period immediately after a tsunami is also key for effective mitigation. If local civil defense authorities have been able to lay plans ahead of time, they can proceed with rescue efforts armed with valuable knowledge about

Figure 6. Roadside tsunami evacuation sign used as part of an education/awareness program.

which portions of their local infrastructure—hazardous materials storage areas, electrical lines, pipelines, fuel tanks, and so on—may be danger spots; what roads, bridges, and railroad lines are likely to be usable or unusable by emergency response equipment; and what lifeline facilities—firehouses, police stations, schools, hospitals, warehouses, etc.—are likely to remain available for mounting relief efforts. Tools such as the integrated hazard maps in Figures 3, 4, and 5 and the inundation map in Figure 7 are fundamental to such response planning. They can form the basis for emergency response drills that allow authorities to develop coordinated tactics. These types of drills are especially important in countries like the United States where multiple agencies may have jurisdiction over various parts of a response effort [Preuss, 1997].

The longer time frame for response planning involves developing policies that guide land use so as to eliminate the tsunami threat altogether. The classic example of this is in Hilo, Hawaii, where much of the downtown area severely damaged by the 1 April 1946 tsunami was replaced by a seaside park. While this option may not be available to many communities, other tools are available to communities which undertake a land use planning process. The first step in the process is to undertake an initial land use inventory. Such a vulnerability assessment will reveal characteristics of uses/occupancies and buildings in the inundation zone which communities can use to define primary risks. Figure 8 is an example of such an inventory. This vulnerability assessment also identifies uses

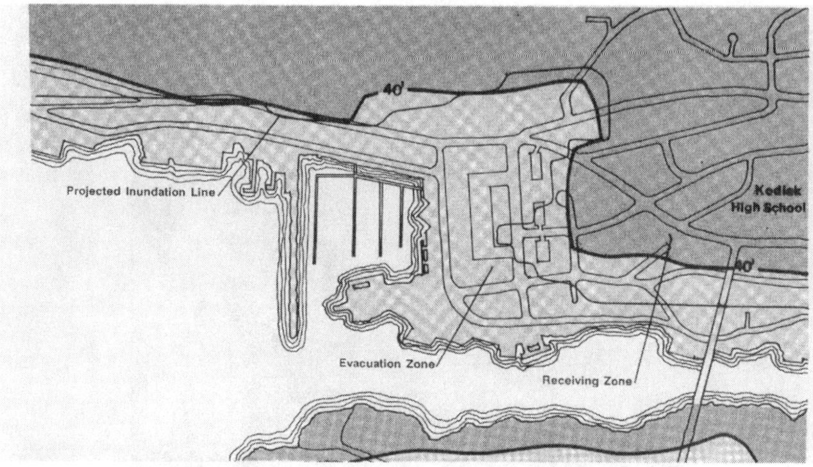

Figure 7. Implications of the tsunami hazards for disaster response in Kodiak, AK. Note that the main coastal evacuation road is subject to flooding and that Kodiak High School could be used as a staging/assembly area [Preuss *et al.*, 1982].

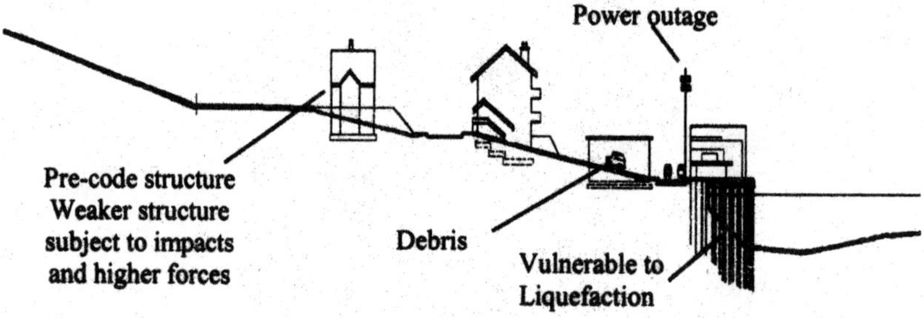

Figure 8. Structures on piers have a high potential for becoming debris, especially if liquefaction occurs. Cars also have a high likelihood of becoming debris and potential ignition sources for fires. Power will be disrupted.

and building conditions which can cause problems (become debris) for nearby structures. All land uses within a projected inundation zone are subject to direct impacts of the earthquake including ground motion, which can be amplified in the saturated soils which frequently characterize coastal areas. Such soils are prone to liquefaction, which can lead to structural failure, making such structures even more vulnerable.

FUTURE TRENDS

The recent advances in scientific understanding of tsunamis and the technologies to observe them, simulate them, and communicate information about them have lead to ever-increasing capabilities to mitigate their effects. The ten significantly destructive tsunamis that have occurred since 1990 have been studied more rigorously and in greater detail than any other previous events. Older conceptions of what tsunamis are and how they behave in the coastal zone are being replaced by more refined paradigms.

In addition, ever more sophisticated laboratory and computational tools are revealing greater levels of detail on wave runup and wave-structure interactions. Better understanding of wave forces on obstacles on the coastline will lead to improved design criteria and tsunami-resistant structures.

We are becoming increasingly aware that tsunami damage results from the moment magnitude and location forces a sound mitigation strategy. Once these effects are better understood in terms of magnitude and location on the structure, appropriate planning and mitigation strategies can be developed.

Understanding of three aspects of the wave impacts are the basis for establishing sound mitigation strategy: size of the moment during the initial impact, magnitude of the force, and subsequently how the moment behaves when the wave turns around to hit the back face of the structure.

Recent research funded by the National Science Foundation have used both numerical and laboratory studies to illustrate representative land use and building conditions located on a gently sloping beach. A three-dimensional numerical simulation initially identified the moment and forces generated by waves striking a square structure which was 6 centimeters square. Subsequently to validate the simulation, Laboratory experiments were conducted using the same assumptions to verify the simulations. The laboratory experiments yielded results almost identical to the numerical simulation (Figure 9). Detailed studies such as these will prove invaluable guides for including tsunami-resistance into future building codes and design practices.

The forward wave creates forces primarily on the front face while the return wave creates forces on the back face. The moment arm of each is about the same. Net forces on the front face are somewhat higher during the forward wave. Forces were measured in Newton (N) which is the measurements for force/acceleration. The conversion factor from Newton's to pounds is 4.448.

Buildings and structures rarely occur in an isolated setting such as the case study. Instead, they are located in communities consisting of many structures, where some face the waterfront, and others are located adjacent to, but behind the first tier of buildings. In many communities, newer buildings have been constructed to recent codes with relatively high levels of resistance to lateral forces, while adjacent sites are occupied by single family and older structures with little built-in resistance. Thus future research will now examine the impacts on the "second tier" of weaker structures.

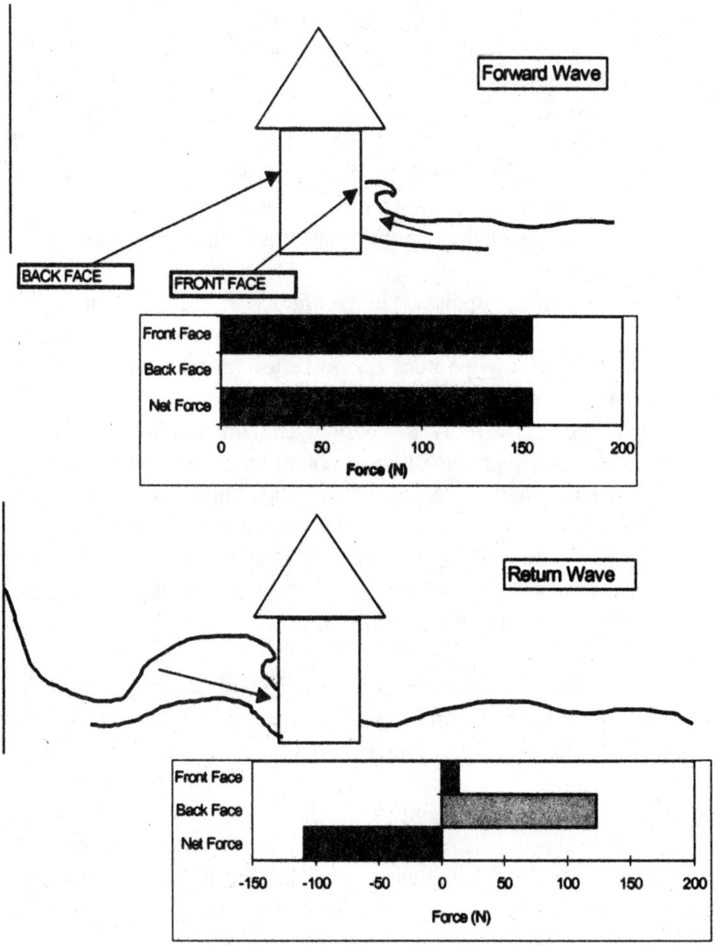

Figure 9. Simulations of the tsunami wave dynamics on a structure indicate the locations and magnitudes of forces. The forward wave creates forces primarily on the front face while the return wave creates forces on the back face. The moment arm of each is about the same. Net forces on the front face are somewhat higher during the forward wave.

Implementation of tsunami hazard mitigation measures will either be through planning policies regulating allowable uses of land or through codes regulating on-site placement on the site of structures or regulating construction specifications. Reduction of damage is thus based on two concepts. One is to minimize exposure through avoidance, e.g.—prohibiting uses in specific areas. The other is to minimize life loss through hardening—i.e. design codes.

More detailed simulations of tsunami flooding will also lead to land use policies that reduce or eliminate potential damages with increasing frequency Comprehensive Plans are being required to address hazardous areas. In some localities the hazard (such as flood risk) is addressed in conjunction with other elements of the plan to minimize exposure, e.g. housing. In other cases, the hazard is addressed with special designations such as a Coastal Zone Management area 200' inland from the HHWL. In most countries, structures, lifelines, and land uses generally must comply with national and local regulations. In some cases, the regulations are quite complex; in other cases important issues can "fall between the cracks." Nonetheless there are some codes, such as national fire codes regulating tanks, which are generally based on a model code.

As tools and techniques are refined, it is inevitable that attention will turn to mitigation efforts in larger urban areas that are threatened by tsunamis. The complexity of the problem grows with the size of the city, because more infrastructures and governing bodies and constituencies enter the picture. Yet the threat will not go away and can only be mitigated by careful planning and preparedness.

REFERENCES

Bernard, E.N., Reducing Tsunami Hazards Along U.S. Coastlines, in *Perspectives on Tsunami Hazard Reduction*, G. Hebenstreit (ed.), Kluwer Academic Publishers, 189-203., 1997.

Bernard, E.N., Program Aims To Reduce Impact Of Tsunamis On Pacific States, *EOS, Trans. AGU*, 79(22), 258 and 262-263, 1998.

Bernard, E.N., R.R. Behn, G. Hebenstreit, F.I. González, P. Krumpe, J.F. Lander, E. Lorca, P.M. McManamon, and H.P. Milburn, On Mitigating Rapid Onset Natural Disasters: Project THRUST (Tsunami Hazard Reduction Utilizing Systems Technology), *EOS, Trans. AGU, 69*(24), 649-661, 1988.

Blackford, M. and H. Kanamori, *Tsunami Warning System Workshop Report (September 14-15, 1994)*, NOAA Tech. Memo., ERL PMEL-103, 80pp.

Committee on the Alaska Earthquake, *The Great Alaska Earthquake of 1964: Oceanography and Coastal Engineering*, National Academy of Sciences, Washington, D.C., 1972.

González, F.I., Tsunami!, *Scientific American, 280 (5)*, 56-65.

Hebenstreit, G. T., An Overview, in *Perspectives on Tsunami Hazard Reduction*, G. Hebenstreit (ed.), Kluwer Academic Publishers, 189-204, 1997.

Kawata, Y., B. Benson, J.C. Borrero, H.L. Davies, W.P de Lange, F. Imamura, H. Lets, J. Nott and C.E. Synolakis, Tsunami In Papua New Guinea Was As Intense As First Thought, *EOS Transactions, AGU, 80* (9), 101-104, 1999.

Maramai, A. and S. Tinti, Coastal Effects And Damage Due To The 3rd June, 1994 Java Tsunami, in *Perspectives on Tsunami Hazard Reduction*, G. Hebenstreit (ed.), Kluwer Academic Publishers, 1-20, 1997.

Ortiz, M., IOC-SHOA-CICESE Course On Numerical Simulation Of Tsunamis: Project TIME, *International Oceanographic Commission Reports of Courses*, #42, 1996.

Preuss, J., P. Raad, and R. Bidoae, Mitigation Strategies Based on Local Tsunami Effects, in *Tsunamis at the End of a Critical Decade*, G. T. Hebenstreit (ed), Kluwer Academic Press, in press, 2001.

Preuss, J., Local responses to the October 4, 1994 tsunami warning: Washington, Oregon, California, in *Perspectives on Tsunami Hazard Reduction*, G. Hebenstreit (ed.), Kluwer Academic Publishers, 35-45, 1997.

Preuss, J. and G.T. Hebenstreit, Integrated Tsunami-Hazard Assessment for a Coastal Community, in Assessing Earthquake Hazards and Reducing Risk in the Pacific Northwest, vol. 2, A.M. Rogers, T.J. Walsh, W.J. Kockelman, and G.R. Priest, eds., US Geological Survey Professional Paper 1560, 517-536, 1998.

Preuss, J., R. Preuss, J. Christensen, and V. Umetsu, *Land Management in Tsunami Hazard Areas*, National Science Foundation, Washington, D.C., 258 pp., 1982.

Preuss, J., R. Preuss, J. Christensen, R. Hodge, S. Farreras, and A. Sanchez, *Planning for Risk: Comprehensive Planning for Tsunami Hazard Areas*, National Science Foundation, Washington, D.C., 246 pp., 1988.

Tatehata, H., The New Tsunami Warning System Of The Japan Meteorological Agency, in *Perspectives on Tsunami Hazard Reduction*, G. Hebenstreit (ed.), Kluwer Academic Press, 175-188, 1997.

Titov, V.V., H.O. Mofjeld, F.I. González, and J.C. Newman, *Offshore forecasting of Alaska-Aleutian Subduction Zone tsunamis in Hawaii*. NOAA Tech. Memo. ERL PMEL-114, 22 pp., 1999.

8

Landslides and Cities: An Unwanted Partnership

Richard J. Pike, David G. Howell, and Russell W. Graymer

Geomorphological processes are natural phenomena that have only become serious hazards because they have increasingly imposed themselves upon a vulnerable, often unsuspecting, and rapidly growing urban community.

R.U. Cooke [1984]

INTRODUCTION

Natural hazards levy an "environmental tax" on society, and the burden is increasing worldwide. While insurance losses ascribed to human actions changed little from 1970 to 1992, compensation for natural disasters rose tenfold [Degg, 1998]. Urban areas (Figure 1) contribute much of this toll [Cooke, 1984; Alexander, 1993; Mileti, 1999], damage in the 1995 Hyogo-ken Nanbu (Kobe) earthquake alone exceeding $100 billion. Overall losses to extreme-weather disasters increased from $3.9 billion per year in the 1950s to $40 billion in the 1990s (U.S. dollars), with 25% of the property damage and 96% of the deaths occurring in non-industrialized countries [Hausmann, 2001]. Insured losses rose from near zero to $9.2 billion annually and twice that if smaller, non-catastrophic, events are included [IPCC, 2001b]. In just the first 11 months of 1998, before Hurricane Mitch, storms, floods, and droughts worldwide inflicted nearly $90 billion in damage, killed 32,000 people, and displaced an additional 300 million.

Seldom recognized as a major hazard [Board on Sustainable Development, 1999, p. 191], landsliding accounts for much of this loss, single disasters claiming as many as 20,000 lives [Sidle *et al.*, 1985; Alexander, 1989; Schuster and Highland, 2001]. Guzzetti [2000] found that over the past 50 years annual fatalities from landslides in Italy equal those from earthquakes and far exceed flood deaths. Damaging landslides are ubiquitous [Brabb and Harrod, 1989].

Earth Science in the City: A Reader

10.1029/056SP09

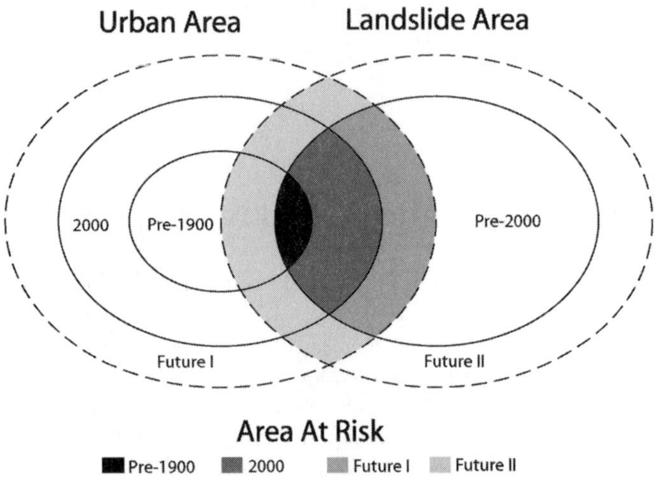

Figure 1. Why urban landslides? Conceptual model of evolving slope-instability threat to the world's cities. Urban space (left) and hazard space (right) intersect to define area at risk (four shaded bands), which was small (black ellipse) pre-1900 (smallest oval, left) but grew (dark-gray band) with 20th-century urbanization (large solid oval, left). Hazard space has remained unchanged; present defined as 2000 (solid oval, right). Future scenario I: more and larger cities expand urban space (dashed oval, left), adding area at risk (medium-gray band). Future scenario II: land degradation and effects of global warming expand hazard space (dashed oval, right), adding more area at risk (light-gray band). Areas not to scale.

Especially at risk are cities in less-developed parts of the tropics [Ahmad *et al.*, 1999; Smyth and Royle, 2000], earthquake-prone regions everywhere [Japan Landslide Society, 1988; Wasowski *et al.*, 2000], and Mediterranean climates—where rainfall, ground saturation, and minimum evaporation peak in the same season [Oliver, 1993; Luzi and Pergalani 1999; Flentje *et al.*, 2000]. No environment is spared [Eyles *et al.*, 1978; Duncan *et al.*, 1980; Siebe *et al.*, 1996]. Landslides occur in all 50 of the United States and its island territories, annual fatalities averaging 25 to 50 and economic losses reaching $1.5–$2 billion [Committee on Ground Failure Hazards, 1985; Schuster, 1996]. Despite popular association of slope problems with cities in the U.S. West—Los Angeles, San Francisco, Seattle [McPhee, 1989; Bell, 1999]—landslides are a problem in such perceived "safe" communities in the East as Cincinnati and Pittsburgh [Pomeroy, 1982; Bernknopf *et al.*, 1988].

Civil engineers and geologists have developed methods to reduce the losses from landsliding [Rogers, 1992]. From the earliest attempts at mitigation, preceding Collin's [1846] studies of canal embankments, engineering works concentrated on keeping transportation corridors open to commerce [Sharpe, 1938;

Terzaghi, 1950; Eckel, 1958]. By the 1960s, however, landslide damage to newly constructed suburbs in the U.S. began to expose the vulnerability of residential areas [Radbruch and Weiler, 1963; Leighton, 1966; Nichols and Campbell, 1971]. A decade later, chapters on landslides and other hazards were appearing in earth-science textbooks and symposia on engineering geology [Leggett, 1973, p. 392–487; Leighton, 1976; Utgard *et al.*, 1978, p. 29–138; Leveson, 1980]. Volumes devoted wholly to landsliding followed as the wide extent of the problem became appreciated [Veder, 1981; Varnes, 1984; Committee on Ground Failure Hazards, 1985; Sidle *et al.*, 1985; Crozier, 1986; Brabb and Harrod, 1989; Flageollet, 1989]. Now recognized as a major threat, landslides are stimulating research worldwide [Dikau *et al.*, 1996; Schrott and Pasuto, 1999; Wasowski *et al.*, 2000; Wieczorek and Naeser, 2000]. But despite the wealth of accumulated understanding [Turner and Schuster, 1996], cities remain at risk.

This chapter reviews the complexities of urban landsliding and examines its persistence. We then explore the prospects for landslide-resistant cities, where slope hazards might be prevented from becoming landslide disasters. Although we can not raise every issue germane to landsliding, appended references provide entry into the literature as well as detail for the topics we do address. We draw heavily on our experience and that of U.S. Geological Survey (USGS) and other professional colleagues in California's San Francisco Bay area. Not only is this urban region subject to both rainfall- and earthquake-generated landslides, but it illustrates two global trends that profoundly influence landsliding: a growing population and its concentration in expanding cities (Figure 2), many of them situated in susceptible environments.

LANDSLIDES, AN URBAN PROBLEM

Arising from the intersection of geomorphic processes with those of human settlement (Figure 1), the problem of landsliding in cities is not one of geology and meteorology alone [Heim, 1932; Schuster and Highland, 2001]. The hazard does not reduce to a nice set of physical systems or engineering specifications, but rather is an untidy phenomenon also shaped by geography, economics, politics, cultural traditions, and psychology. After an introduction to population growth and the nature of urbanization, this section reviews slope instability and its contributing factors, describes landslide disasters triggered by different processes, and discusses some aspects of economic loss.

Urban landsliding originates in poor land-use practices, commonly the result of a growing population confined by local geography [Leggett, 1973, p. 4–13]. The natural setting of many cities limits the supply of level terrain that is geotechnically stable and otherwise suited to human occupation [Eisbacher and Clague, 1981; Cooke, 1984; Clarke *et al.*, 1997]. Expansion into surrounding hillsides and their bordering lowlands, terrain more likely to fail or be overrun by landslides, destabilizes slopes through grading, road-building, and residential construction.

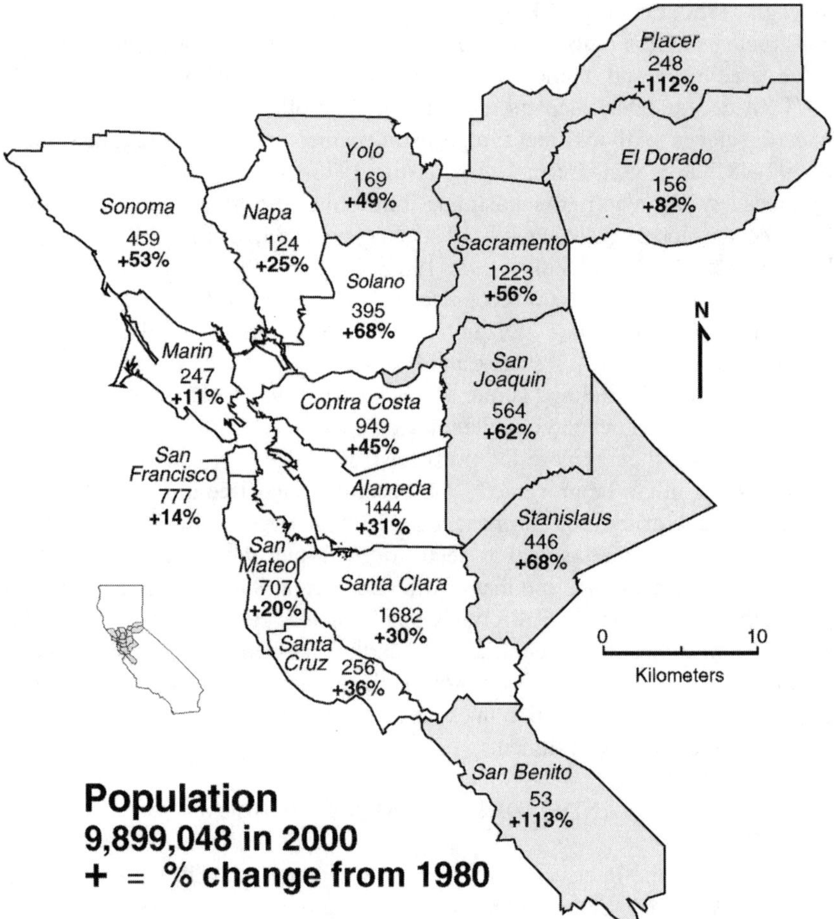

Figure 2. The demographic half of the urban landslide problem exemplified: the new central California megalopolis. Relation of urban sprawl to 20-year population trends in the greater San Francisco Bay region. While the 10 inner counties (lighter shading) grew by 31% to 7,039,362 inhabitants over the last two decades, the seven satellite counties (darker shading) increased by 65%, to 2,859,686. Year 2000 county populations in 000s of residents. Source, U.S. Census Bureau.

Where zoning is inadequate, owing to political disinterest or reluctance to regulate land use, unstable hillsides, alluvial fans, and other marginal sites are developed as a matter of convenience—or because economic pressures to build override long-term concerns [Burby, 1998; Ghilardi *et al.*, 2000; Southern California Studies Center, 2001]. At risk are cities of every size, communications linkages, transportation infrastructure, conduits for water, fuel, and energy, as well as the natural environment that ultimately sustains these urban elements.

Population Growth

The human population has risen exponentially since onset of the Enlightenment and late in 1999 reached 6 billion [United Nations Population Division, 1999; Bongaarts and Bulatao, 2000]. Technological and cultural changes—in medicine, sanitation, literacy, and education—following the industrialization of much of Western society upset the equilibrium between fertility and mortality rates that previously had limited the world's numbers [Thomas, 1956, p. 970; Dyson, 2002]. The resulting increase in population "... seems to have affected everything, yet is seldom held responsible for anything" [Smail *et al.*, 1997]. Clearly, it has brought about landsliding, soil erosion, accelerated sedimentation, and other harmful effects detailed in Thomas [1956, p. 641], McCall [1998], Turner *et al.* [1990], and Schuster and Highland [2001]. Although the rate of increase recently fell, absolute growth has continued: 80 million people were added to the planet each year from 1995 through 2000 [United Nations Population Division, 2001].

Disparity in the distribution of wealth has accompanied this growth [Board on Sustainable Development, 1999]. Aggravated by a dysfunctional social order in many oligarchic states, economic polarization widened both within and between countries. In 1970, income of the richest 20% of the world's population averaged 32 times that of the poorest 20%. By 1995, it was 78 times greater. In 1965, average per-capita income of the seven most developed (the G7) countries was 20 times that of the world's poorest seven; by 1995, it was 39 times as much. The inequity has led to an increase in both the number and proportion of people living in extreme poverty, many of them urban residents [Hausmann, 2001].

Urbanization

Cities house half the Earth's inhabitants and consume 75% of its resources [Fornos, 2000] while occupying just 2% of its land surface. The term "city" defies definition [Thomas, 1956, p. 382], and size alone is insufficient. Population thresholds that officially designate an area as urban vary by country, from >200 inhabitants (Sweden), >1000 (Australia), and >2,500 (U.S.), to >5000 (India), >10,000 (Italy, Turkey), and >30,000 (Japan). Archaeological criteria, which specify a relatively large agglomeration of people and activities not based upon a subsistence economy, yield an arrangement that resembles a city rather than a rural village. Legal criteria normally result in a smaller settlement. Increasingly, the area that functions as a modern city no longer is a compact cluster of people and activities, but a vast, interrelated "daily urban system" (Figure 2) that extends beyond the initial legal entity, the old central city [Batty and Longley, 1994; Southern California Studies Center, 2001]. Where the rural highways, railways, and utility lifelines that connect and support cities are vulnerable to landsliding, we extend "city" to include this resource and transportation network.

Cities form, grow, and change by a complex process that has many determinants—physical, economic, and cultural [LeGates and Stout, 1999; Fainstein and Campbell, 2001]. Human concentration in permanent settlements began with a shift from hunting-gathering and nomadic herding to agriculture and a market economy about 8,000 years ago [Powelson, 1989]. This consolidation has continued save for a pause during the Medieval period. The shift now underway is from subsistence farming and local marketing to urban industrialization, and in highly developed countries to a post-industrial urban and suburban economy. The attraction of large modern cities is opportunity—for steady employment, wages higher than can be earned in the countryside, education, health care, and a range of social services unavailable elsewhere.

Three accelerating trends mark the transition from rural to urban life: in the number and size of cities and concentration of this growth in developing countries. The percentage of Earth's population that is urban rose dramatically from 37% in 1970 to 45% just 25 years later. (This number stabilizes at 80% to 90% in the most industrialized countries [Board on Sustainable Development, 1999, p. 63]). Whereas in 1800 only London and Beijing claimed a million inhabitants, the number of large (1–10 million) cities had risen to 81 by 1950 and to 270 by 1990. Over the same 40 years, still larger *megacities* (>10 million) increased to 21 from two (New York and London). "Megalopolis" no longer designates just the northeastern United States [Gottmann, 1961] but any multi-centered urban area of over 10 million having much low-density settlement and complex networks of economic specialization, for example, Tokyo-Nagoya-Osaka, and recently the greater San Francisco Bay region (Figure 2). Urbanized during the Industrial Revolution, all ten of the world's largest cities in 1900 were North American or European, but by 2000 seven were in developing countries. In 1800, the 100 largest cities totaled 20 million inhabitants; by 1990 the 100 largest urban areas contained 500 million people, nearly half of them in 19 megacities—15 of which are in the developing world [Fornos, 2000].

According to United Nations criteria, over half the urban inhabitants of Asia, Africa, and Latin America live in poverty. Many rural-to-urban migrants are unqualified for employment in cities, but accept menial work at low wages or worse [Bongaarts, 2002]. At least 120 million of the 2.8 billion people in the global workforce are unemployed and another 700 million earn too little to meet basic needs. Concentration of the poor in nonindustrial cities, especially along the seacoasts and rivers that provide ready access to trade, can lead to high death tolls from natural disasters [Mitchell, 1999]. The slums and shantytowns that now house 25%–30% of the world's urban population commonly are in marginal terrain, where their low-income inhabitants suffer disproportionately from flooding and landslides. Loss of homes, possessions, and often livelihood to an earthquake or storm leads to further impoverishment.

Compounding this privation is a low rate of urban home and workplace ownership, up to 70% in Peru, for example [McLaughlin and Palmer, 1996]. Property registration in the cities of developing nations is complex, inefficient, and prohibitively expensive for the poor. A dozen different agencies can be involved in the generation of each land title. Many marginal neighborhoods are owned publicly and the government may wish to pass ownership to its occupants, but existing institutions may be unable to complete the transfer. Denied the use of real estate as collateral for raising capital to improve property or invest in a business, many city dwellers can not afford to leave marginal neighborhoods that are prone to landslides and floods.

Landsliding and its Causes

Landslides constitute one class of ground failure, the disruption or dislodgment of topography, usually involving both vertical and horizontal displacement (Figure 3). Specifically, the term "landslide" denotes "the movement of a mass of rock, debris or earth down a slope" [Cruden and Varnes, 1996, p. 36]. Included are landslides in coastal bluffs that have been eroded or undercut by wave action [Moore *et al.*, 1995; Pike *et al.*, 1998] and submarine landslides, which affect cities when they cut seafloor cables and pipelines or when the tsunamis generated by their movement damage coastal communities [Ahmad *et al.*, 1999]. Failures that are not normally regarded as landslides, although they may be associated with them, include snow avalanche, ground settlement, cracking due to soil expansion or earthquake shaking, as well as seismically induced rupture of the surface and ridge-crest splitting [Turner and Schuster, 1996].

The block diagram in Figure 4 shows an idealized landslide and general terms that describe the gross anatomy of many failures [Varnes, 1978]. The mass of *displaced material* that has moved downslope is distinguished from the unfailed slope, or *original ground surface*. The displaced mass, which may be in a deformed or undeformed state, occupies two distinct zones. In the *zone of depletion*, defined by the *surface of rupture* along which the mass moved, the displaced material lies below the original surface. Where no displaced material remains on the surface of rupture (the *main scarp*) or where flowage rather than rupture has occurred this surface is better termed the *source area*. In the *zone of accumulation*, or area of deposition, the *foot* of the displaced material overlies undeformed ground, the *surface of separation*, along which no failure has occurred. Specific terms further distinguish individual styles of failure [Cruden and Varnes, 1996; Dikau *et al.*, 1996].

Landslides result from a local, either sudden or continuing, imbalance among the factors that maintain stability in the landscape. Over time hillsides tend toward a metastable condition, or dynamic equilibrium, among tectonic setting, rock type and structure, surface gradient and curvature, vegetation type, climatic regime, and soil properties—especially texture, particle size, and moisture content. Slopes become unstable and fail when one of two fundamental changes, a rise in shear stress or a reduction in material strength, upsets the equilibrium [Terzaghi, 1950; Záruber and Mencl, 1982].

Figure 3. Complex slump-earth flow, a common type of landslide (see Figures 4 and 5). The 4 March 1995 La Conchita landslide, Ventura County, southern California [Sarna-Wojcicki, 1996]. The 200,000 m^3 failure within an existing landslide in marine sediments destroyed or damaged nine houses beneath the coastal bluff. A 1999 civil lawsuit by the affected residents against owners of the bluff-top property failed to win compensation. Photograph by R.L. Schuster, USGS.

The initiating changes may be natural or anthropogenic [Selby, 1993; Wieczorek, 1996]. An imbalance in shear stress can result in several ways: removal of support by excavating the toe of a landslide deposit to build a road, addition of overlying material by constructing a pad for residential housing, imposition of transitory stresses from an earthquake or passing motor vehicles, and crustal uplift or tilting due to tectonism. Some earth materials are inherently of low strength: sand, clays, decomposed rock, and soil or rock saturated by water or weakened by faults, joints, bedding planes, or other discontinuities. Severe

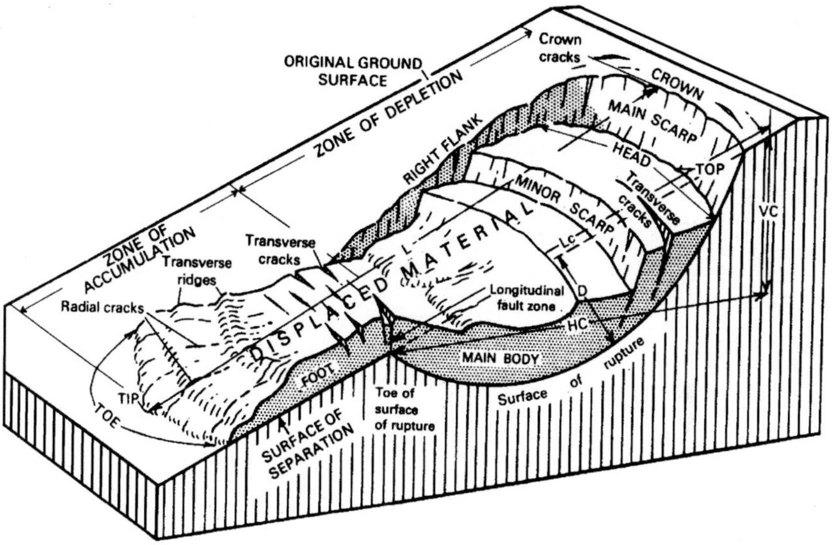

Figure 4. Anatomy of a landslide. Diagram of a slump-earth flow showing features typical of a complex failure and their nomenclature. Compare with Figure 3. Modified from Varnes [1978].

weathering, developed under a change in climatic regime or from release of confining pressure by construction and excavation, can further reduce strength. The most significant factor, however, is saturation of the substrate. Arising through many conditions—high rainfall, snowmelt, a perched water table, interruption of natural drainage or insufficient storm drainage, impoundment of reservoirs, leaking septic systems and swimming pools, soil erosion, removal or change in vegetative cover—ground saturation can lower soil and rock strength by increasing pore-water pressure relative to the surrounding area.

The Variety of Slope Failure

The landslides caused by these changes mobilize in various ways. Although Collin [1846] early distinguished between shallow and deep failure, among the first classifications of landslides into distinct types were those of Heim [1932] and Sharpe [1938]. There are now several such categorizations [Selby, 1993; Dikau *et al.*, 1996; Hungr *et al.*, 2001]. That devised by David Varnes [Eckel, 1958, p. 20-47] and later refined [Varnes, 1978; Cruden and Varnes, 1996] is widely accepted, especially in North America (Figure 5). Varnes' bivariate scheme combines five types of movement with three types of material, for example, *rock fall*, *earth flow* (also *earthflow*), *debris slide*. The nomenclature can be elaborated further according to rapidity of movement and water content.

TYPE OF MOVEMENT			TYPE OF MATERIAL		
			BEDROCK	ENGINEERING SOILS	
				Predominantly coarse	Predominantly fine
FALLS			Rock fall	Debris fall	Earth fall
TOPPLES			Rock topple	Debris topple	Earth topple
SLIDES	ROTATIONAL	FEW UNITS	Rock slump	Debris slump	Earth slump
	TRANSLATIONAL	MANY UNITS	Rock block slide Rock slide	Debris block slide Debris slide	Earth block slide Earth slide
LATERAL SPREADS			Rock spread	Debris spread	Earth spread
FLOWS			Rock flow (deep creep)	Debris flow (soil creep)	Earth flow
COMPLEX			Combination of two or more principal types of movement		

Figure 5. The variety of landsliding. Abbreviated classification of slope movements by mode of motion and type of material. Redrawn from Varnes [1978].

Landslides move as *falls*, *topples*, *spreads*, *slides*, and *flows* [Varnes, 1978]. Falls are masses of soil or rock that dislodge from steep slopes and then free-fall, bounce, or roll downslope. Topples move by the forward pivoting of a displaced mass around an axis located below its center of mass. Lateral spreads, which mobilize from the liquefaction of unconsolidated sediments or fill in an earthquake, move by horizontal extension accommodated by shear or tensile fractures. Slides displace masses of material along one or more discrete planes; movement may be either rotational or translational. In rotational sliding the plane is curved and the mass rotates backwards around a common axis parallel to the slope. In translational sliding the failure surface is near-planar or gently undulating, and the mass moves roughly parallel to the ground surface. Flows are masses that mobilize as a deforming, viscous unit without a discrete failure plane [Hungr *et al.*, 2001]. More than one type of displacement may characterize a failure, in which case the movement is classified informally as *complex* (Figures 3, 4).

The affected materials are *rock*, *soil*, or a combination of the two [Cruden and Varnes, 1996]. Rock refers to hard or firm bedrock that was intact and in place before slope movement. Soil, either residual or transported material, is used in the engineering sense to mean loose, unconsolidated particles or poorly cemented rock or inorganic aggregates. Soil is distinguished further on the basis of texture as *debris* (20%–80% of fragments >2 mm) or *earth* (≥80% of the fragments ≤ 2 mm).

Other landslide classifications overlap with that of Varnes [1978]. On rocky slopes, for example, Selby [1993] distinguished three types of movement and ten specific mechanisms: fall (rockfall, slab failure, topple, and avalanche), sliding (slump, planar slide, wedge failure, and block slide), and creep (cambering and curving of strata). Claiming that the nomenclature of flow phenomena [Hungr *et al.*, 2001] is inconsistent in distinguishing flow from slide, Selby [1993] further advocated broader terms based on rheology. The classifications of Varnes, Selby, and that described in Dikau *et al.* [1996, p. 2–9], are suited to failure recognition and mapping. Although numerical expression of slope movement and materials might yield a more physically based ordering, such a scheme would require too many geotechnical measurements to be widely applicable.

Among the most easily recognized failures are earth flows and slides, which commonly leave large deposits of a few hectares to several square kilometers (Figures 3–5). These landslides involve surficial mantle and bedrock down to depths of a meter to over 50 m, distort the ground when they move, and remain as recognizable masses that can persist for thousands of years [Nilsen *et al.*, 1979; Sidle *et al.*, 1985]. Although usually they move slowly and thus seldom threaten life directly (for an exception, see Cotton and Cochrane [1982]), earth flows and slides can pose a hazard to property. When they move, in response to one or more of the destabilizing changes described above, they can offset roads, destroy foundations, and break underground gas pipes and water mains as well as override property downslope (Figure 6).

A contrasting type of landslide, the debris flow (Plate 1), is more closely associated with human fatalities [Costa and Wieczorek, 1987; Watanabe *et al.*, 1996; Wieczorek and Naeser, 2000]. Although debris flows move under physical conditions that differ from those initiating deeper failure, both types may coexist

Figure 6. A small but expensive landslide. Disrupted surface of old earth flow near town of Aromas, northern San Benito Co., California (Figure 2), reactivated on 23 April 1998 by El Niño rains. The landslide cut natural gas supplies to 60,000 Santa Cruz Co. residents, costing the gas utility $10 million to service pilot lights and repair the two severed pipelines. Photo by W.R. Cotton.

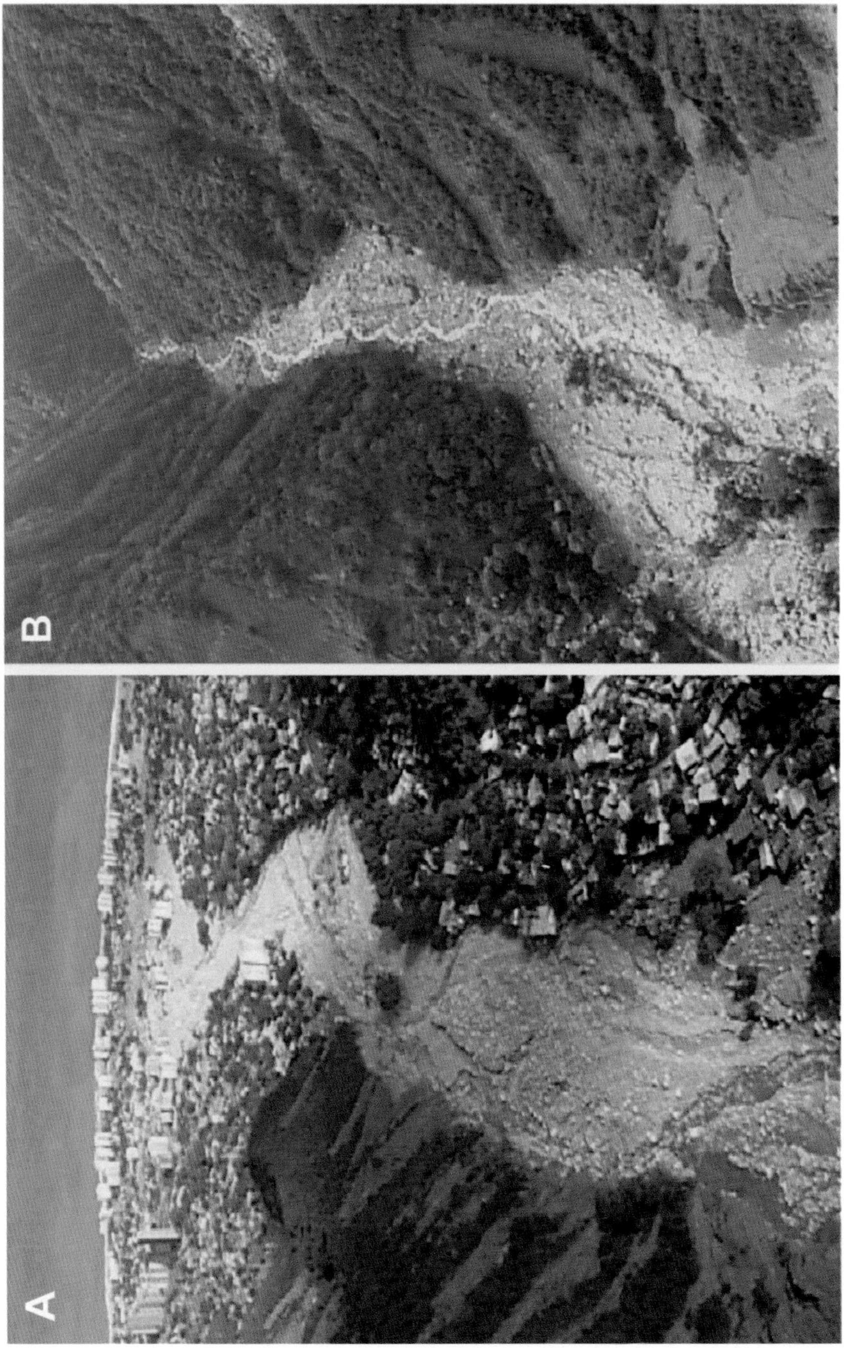
A
B

in the same environment. Debris flows are shallow mixtures of water and soil that mobilize suddenly on steep hillsides during brief but heavy rains, run rapidly downslope and form thin, usually ephemeral, deposits that do not markedly distort the ground surface [Campbell, 1975; Ellen, 1988]. Most debris flows are small. Although vegetation quickly and deceptively obscures their recognizable features, over time debris flows may recur in the same location (Plate 1). Among their lethal variants are *lahars*, volcaniclastic slurries that may include melted snow and ice. The remoteness of many volcanoes from cities creates a false sense of security, for lahars that originate in eruptions on steep slopes can travel great distances at high speed [Scott *et al.*, 1995; Siebe *et al.*, 1996; Watanabe *et al.*, 1996; Sisson *et al.*, 2001].

Some Landslide Disasters

The human and economic toll of landslides on cities has been chronicled for hundreds of years. In Italy alone, 840 events since 1279 have killed over 10,000 people [Guzzetti, 2000]. Although individual landslides can be spectacularly destructive, they overshadow the myriad smaller and less dramatic failures that add up to greater cumulative damage [Schuster and Fleming, 1986; Schuster, 1996]. The 12 landslide episodes reviewed here illustrate the variety of failure as well as the losses incurred and to be anticipated as urbanization continues. For details and additional accounts, see reports cited in Schuster and Highland [2001] and other references.

Precipitation. High rainfall is the most frequent immediate cause of landsliding. Rio de Janeiro, for example, sustains repeated damage from storm-generated instability of its overdeveloped hillsides [Jones, 1973; Smyth and Royle, 2000]. Many of the granite slopes surrounding the Brazilian city have been logged for fuel and lumber and cleared for agriculture, leaving the deep residual soil exposed to potential failure. The steep hillsides house a million of Rio's six million inhabitants in *favelas* or shanty towns, and roads built to the shacks have further steepened the slopes. The 2000 lives lost in debris and mud flows during the stormy seasons of 1966 and 1967 failed to prompt measures to reduce the risk. In a subsequent episode during February 1988, 12 cm of rain fell on Rio de Janeiro in 4 hours. The resulting floods and debris slides killed 200 people and

---

Plate 1. Inappropriate land-use and the inevitable outcome. **A**. Situated on ancient debris-fan deposits, the northern Venezuelan city of Caraballeda was devastated in a two-day December 1999 storm. Nearly 2 million tons of sediment spread throughout the city (background) from debris flows (scars on hillsides, foreground) and stream-channel deposits (foreground). **B**. View, upstream from same location as **A**, of steep valley walls scarred by debris flows; the large catchment area has 2000 m of relief. Photos by M.C. Larsen, USGS.

left 20,000 without housing—which cost $100 million to rebuild [Alexander, 1989]. Another 70 people died in rainstorm-driven mud flows over the Christmas holiday in 2001.

Although sliding and earth flow were well understood processes around California's San Francisco Bay [Radbruch and Weiler, 1963; Taylor and Brabb, 1972; Taylor *et al.*, 1975; Nilsen *et al.*, 1979], the risk from debris flow was all but unrecognized until 1982. The catastrophic rainstorm of 3–5 January dropped up to half the Bay region's mean annual precipitation in 32 hours [Ellen and Wieczorek, 1988]. Over 18,000 shallow landslides mobilized into debris flows that moved rapidly down hillslopes and drainages with little warning, damaging at least 100 houses and killing 14 residents [Cotton and Cochrane, 1982; Ellen, 1988]. Estimated damage exceeded $66 million (1982 dollars) from landslides alone [La Vopa-Creasy, 1988]. Research stimulated by the storm led to rainfall thresholds for debris flow [Cannon and Ellen, 1985], a real-time warning system [Keefer *et al.*, 1987; Wilson *et al.*, 1993], statistical models of debris-flow probability [Mark, 1992; Mark and Ellen, 1995], mitigation measures [Brabb, 1984; Baldwin *et al.*, 1987], and a plan for addressing the landslide hazard nationally [USGS, 1982; Committee on Ground Failure Hazards, 1985].

Unrelenting precipitation from Hurricane Mitch, the fourth most powerful Atlantic storm on record, devastated much of Central America in October 1998. The week-long catastrophe took over 11,000 lives and left hundreds of thousands of people without shelter. Landslides in Honduras stripped overgrazed and overcropped hillsides, filling the coastal lowland with sediment and debris, although slopes in native forest were little affected [Molnia and Hallam, 1999]. In Nicaragua, a large rock and mud avalanche broke out high on the deeply dissected Casita volcano. The steep failed slopes in altered and fractured rock transformed into a debris flow that destroyed towns and settlements several kilometers downslope, killed 1600 people while displacing hundreds more, and disrupted the Pan American Highway at many bridges [Molnia and Hallam, 1999, p. 44–69]. Exposed land mines unearthed by landslides created a new and unanticipated type of urban hazard.

In one of Latin America's worst disasters, twice the mean annual precipitation in two days spawned thousands of debris flows and flash floods along 300 km of coastal Venezuela in December 1999 [Larsen *et al.*, 2001]. Nearly 32,000 people died, 140,000 were left homeless, and another 200,000 lost employment. In some areas mud and debris contaminated the drinking water of up to 70% of the population. Along the Caribbean coast, debris flows (many on otherwise undisturbed, forested, slopes) denuded hillsides and hundreds of other landslides destroyed or damaged sections of the coastal highway. Entire villages vanished, and several cities were devastated. The coastal resort of Caraballeda paid dearly for its attractive location on an active debris fan at the foot of a mountain range [Ghilardi *et al.*, 2000; Larsen *et al.*, 2001]. An estimated 1.8 million tons of fresh sediment that included truck-size boulders buried parts of this city of 22,000 beyond its one-story rooftops (Plate 1). The ruined nearby town of Carmen de Uria, in a steep narrow canyon susceptible to flash flooding and landslides, had 4800 of its 6000 inhabitants swept out to sea and is unlikely to be rebuilt.

Earthquake. Seismic shaking is the second leading cause of failed slopes in many cities [Wasowski *et al.*, 2000]. The following four examples show how the geology and the moisture content of disturbed materials can influence landsliding. Remarkably little damage, for example, resulted from landslides in two other U.S. temblors, the M6.9 Loma Prieta, California, earthquake of 17 October 17 1989, which struck in the dry autumn season [Keefer, 1998], and the M6.8 Seattle, Washington, earthquake of 28 February 2001, which occurred during a dry winter.

On 31 May 1970, a M8.0 earthquake detached a 1.0 km × 1.5 km block of rock and ice from the steep western flank of Nevado Huascaran, the highest peak in the Peruvian Andes. Initial movement of the resulting debris avalanche, the volume of which is estimated to have been as great as 10 million m^3, was nearly vertical. Perhaps airborne for several seconds, the avalanche reached an average velocity of about 60 m/s before burying 20,000 people in Yungay, a town 3 km below the summit and 15 km distant. The landslide left little debris but evidently incorporated enough water to move as a flow 100 m thick and to deposit single boulders of 700 to 14,000 tons. The 50,000 added fatalities from seismic shaking rank the Huascaran event as South America's worst known natural disaster [Watanabe *et al.*, 1996, p. 91–95].

The M6.7 Northridge earthquake of 17 January 1994 scored a direct hit on a modern American city, bringing down landslides that destroyed homes, blocked roads, and damaged oil-field facilities and a dam [Harp and Jibson, 1995]. Eleven hundred failures across greater Los Angeles, up to 90 km from the epicenter, were concentrated in a mountainous 1,000-km^2 area to the northwest. Because the steep hillsides were thinly settled, landsliding caused little of the $20 to $40 billion in total damage (among the highest in U.S. history) and few of the 57 deaths. (Conditions for the earthquake were dry, albeit during the normally wet winter). Most landslides were small, shallow falls and slides in weak young sediments, but hundreds were deeper rotational slumps and block slides. Clouds of airborne dust raised by landslides in the Santa Susana Mountains had unexpected consequence for the town of Simi Valley: an outbreak of valley fever (coccidioidomycosis) contracted by inhaling dust that contained fungal spores carrying the disease. From landslides inventoried by Harp and Jibson [1995], Jibson *et al.* [1998] developed a technique of susceptibility mapping that creates multiple scenarios for slope failure in simulated earthquakes.

Exactly one year after Northridge, a M6.9 earthquake beneath the populous city of Kobe in southern Japan killed 5500 people, wrought over $100 billion in damage, and created 20 million tons of waste debris. Because, again, the Hyogoken-Nanbu earthquake occurred in the dry season, lateral accelerations of $\leq$0.8g brought down just 675 landslides, mainly rock slides and falls and rock-debris avalanches on the susceptible slopes of Rokko mountain and terraced fill behind the city. The largest failure, a debris flow-slide in Plio-Pleistocene lake and marine deposits of

Plate 2. A preventable tragedy. The earthquake-triggered landslide of 13 January 2001 in the Las Colinas neighborhood of Santa Tecla, El Salvador. Densely packed houses were built adjacent to the potentially unstable slopes of El Balsamo ridge. Photo by E.L. Harp, USGS.

granitic sand and clay, crushed 11 houses at the base of the mountain, killing 34 people [Sassa *et al.*, 1996]. More ruinous were the lateral spreads (some with displacements on the order of 3 m), in recent but inappropriate granitic gravel fill, which crippled facilities along a 20-km strip of Japan's chief commercial port. Although few lives were lost to landsliding in Northridge and Kobe, a subsequent earthquake proved truly lethal.

On 13 January 2001, a M7.6 earthquake 100 km off the Pacific coast, followed by a M6.6 event and 4000 aftershocks, brought down hundreds of landslides in weak volcanic deposits across El Salvador. The largest failure was a 100 m × 500 m sector of a steep ridge of andesitic cinders and tephra above Santa Tecla, a suburb of San Salvador. The resulting slump-earthflow rapidly moved 700 to 800 m into the middle-class neighborhood of Las Colinas, destroying 400 closely spaced dwellings and burying 500 people beneath 7–8 m of debris (Plate 2). Seismic shaking, exceeding 0.7 g, was amplified by the topographic effect of the ridge. The immediate cause of the failure evidently was liquefaction of a buried ash deposit in a perched water table overlying a weathered soil [E.L. Harp, 2001, unpublished USGS data]. (Although more visible, the removal of native vegetation from the hillsides and ridge to accommodate construction of upscale homes—over protests by residents of the now-buried neighborhood—may have contributed little to the failure.) Poor engineering performance of slopes on volcanic deposits also caused major landsliding in the 1976 Guatemala earthquake [Harp *et al.*, 1981].

Volcanic Eruption. Landsliding and volcanism can combine to cataclysmic effect, as in the 1980 rock slide-debris avalanche at Washington State's Mount St. Helens [Glicken, 1996]. In the two months before the eruption of 18 May, magma moved high into the edifice, bulging out the north flank by 150 m. Within seconds of a M5.1 earthquake, a massive landslide displaced the bulge and the summit and inner core of the mountain, triggering explosions by the sudden release and expansion of gas in the magma and the flashing into steam of superheated ground water. As the landslide moved down the volcano at nearly 80 m/s, the explosions grew in violence, directing the entrained gas, steam, and rock northward in a lateral blast. Accelerating to 150 m/s in 60 sec, the blast overtook the landslide, which divided in three. One part of the slide entered a lake, raising waves that reached 250 m up surrounding ridges; the second surged over a 400-m-high ridge 8 km from the volcano. Diverted by this ridge, the third and largest part of the landslide moved 22 km down a river valley, within 10 minutes filling it with rock debris to a depth of 45 meters. Mud flows scoured other flanks of the mountain, depositing material in valleys up to 40 km away.

This large debris flow, in which only 57 people died, is a credible scenario for a future eruption at nearby Mt. Rainier [Scott *et al.*, 1995; Sisson *et al.*, 2001], Popocatépetl in central Mexico [Siebe *et al.*, 1996], and other calc-alkalic stratovolcanoes that lie closer to population centers of the Pacific Rim than does Mount St. Helens. Five years later, activity resumed in the dormant Colombian volcano

Nevado del Ruíz, west of Bogota. On the afternoon of 13 November 1985, gas and cinders from the summit crater alerted Armero, a city built on earlier lahar deposits 74 km down-valley from the edifice. Although an emergency committee urged evacuation of the city, no action ensued. In a major eruption of the volcano about 9:00 P.M., water released from melting ice and snow near the summit combined with weak volcanic deposits to send waves of muddy debris down into the surrounding valleys. At 11:25 P.M. the first of several pulses reached Armero; by 1:00 A.M. the city was under 2 to 5 m of lahar deposits and 23,000 of its 29,000 inhabitants were dead [Watanabe *et al.*, 1996, p. 57, 108–110].

Dams and Reservoirs. Blocking drainage or entering previously impounded water can add to the damage from landslides induced by the three processes described above. In April 1983, ground-water buildup from high rainfall the previous September and the melting winter snowpack (150% to 400% of normal) reactivated an old landslide in north-central Utah [Anderson *et al.*, 1984]. The remobilized mass moved across the nearest canyon, raising a natural dam 60 m high that backed up the Spanish Fork river, inundating the small town of Thistle and blocking a major railroad and highway. The damage, which led to the state's first presidential disaster declaration and required $200 million (1983 dollars) to repair, marks the Thistle Landslide as the costliest single U.S. slope disaster to date. The failure might have been averted by lowering the water table through a modest $550,000 system of surface and subsurface drains [Slosson *et al.*, 1992].

The greatest loss of life to a dam-related landslide occurred 9 October 1963, in the Dolomite Region of the Italian Alps after heavy rains had lubricated fractured limestones and clays above the newly created Vaiont reservoir [Alexander, 1993, p. 362]. The steep valley wall collapsed, releasing 250 million m^3 of debris that filled much of the reservoir. The world's highest (265 m) thin-arch dam did not fail, but impounded water displaced by the landslide generated a 90-m-high wave that overtopped the dam and spilled into the valley below, drowning 2600 villagers. Even though geologists and engineers had located the dam in a geotechnically unfavorable site and the valley wall was known to be unstable, closer monitoring of the still-filling reservoir would have prevented the tragedy. The warning signs had been ample. A 700,000 m^3 landslide had ruptured the unstable slope in 1960, and continual creep, measured the same year at 4 cm/day, had increased to 25 cm/day just before the 1963 failure.

The Nature of Economic Loss

Compounding the human toll of a damaging landslide are its economic effects (Figures 7, 8). Despite efforts to strengthen city infrastructure and better prepare for disasters, the societal costs of earthquakes and extreme-weather events are rising [IPCC, 2001b]. The upward trend in urban losses to landsliding over the last 50 years reflects population growth, encroachment of cities on vulnerable

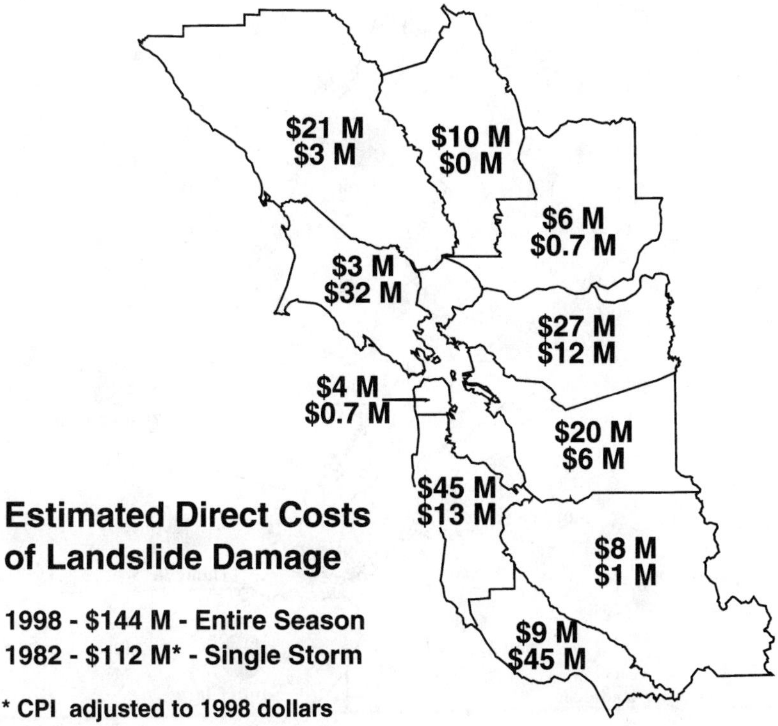

Figure 7. Contrasts in landslide loss: one storm and one season. Map of the 10 inner counties of the San Francisco Bay region, comparing damage from failures over the entire 1997–98 El Niño storm season (upper number) with landslide damage incurred in the storm of 3–5 January 1982 (lower number, 1998 dollars). For county names and public versus private costs in 1998 see Figure 8. Compiled by authors.

terrain, and the geographic concentration of increasing wealth [Schuster and Highland, 2001]. The various costs are categorized as direct or indirect and public or private.

Fleming and Taylor [1980] and Schuster and Fleming [1986] define direct cost as that of "replacement, repair, or maintenance due to damage to installations or property within the boundaries of the responsible landslide." Direct costs include materials and labor to rebuild, repair, or replace roads, homes, buildings, and sewer, water and electrical lines and other components of the built environment, as well as the expense of removing landslide debris. All other costs are indirect. Commonly they far exceed direct costs but can be difficult to quantify. Among indirect costs are interrupted utilities and transportation (Figure 6), lengthened travel time, lost wages and profits, moving expenses, decline in property values

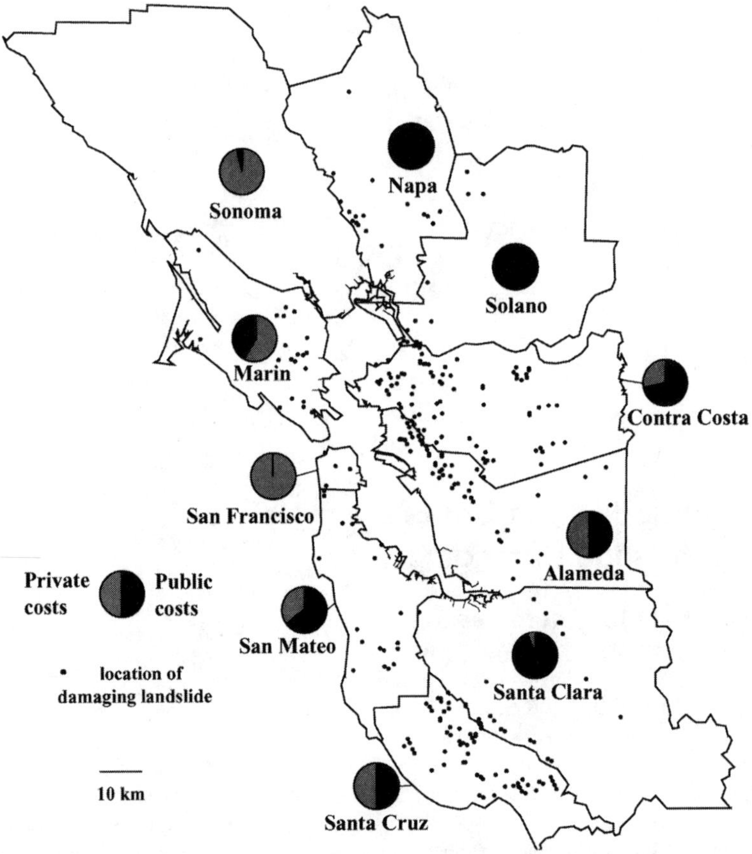

Figure 8. Who pays for landslide damage: public versus private losses. Map of the 10 inner San Francisco Bay region counties showing locations damaged by landslides in the 1997–98 El Niño storms. Proportions of public and private costs varied greatly by county. For total costs see Figure 7. The city of Oakland is in western Alameda County. From Godt and Savage [1999].

and tax revenues, litigation expenses and court settlements, emergency services, the environmental effects of stream sedimentation (Plate 1), engineering work to prevent damage from future landslides, and various social losses.

The public cost of landsliding, largely to maintain roads [Eckel, 1958; Fleming and Taylor, 1980], can be difficult to assess. Because the amount spent to repair or reopen a specific road is known to the local department that does the work, costs are easier to establish for damage by isolated slides than for stretches of highway affected by multiple failures. Charges for recurring landslides along highways may be folded into maintenance budgets and escape post-disaster recording. Agencies also may know only the total outlays following a storm or earthquake, and landslide damage may be combined with that caused by flooding and fallen trees. Other public expenditures include response to emergency and relief by fire, police, and medical personnel, as well as building inspections, hazard assessments, and neighborhood evacuations. Public costs are borne by local, regional, and national governments, but ultimately by taxpayers (Figure 8).

Private costs are to individuals and businesses that incur loss or damage to homes, commercial buildings, and other property (Figures 3, 8). Condemnation of a piece of property and the added requirement of demolition compound the toll. In some cases, the value of land affected by a landslide is reduced or lost (Figure 6). Costs involving real estate, construction, and demolition can be determined if the data are available, although litigation and fear of property condemnation may make this information difficult to obtain in practice. Because insurance for residential real estate rarely covers landsliding [Olshansky and Rogers, 1987], few private losses are redressed [Steiner, 2002]. The Federal Emergency Management Agency (FEMA) did assist owners of homes lost to landsliding in the rural California towns of La Honda and Rio Nido during the El Niño storms of 1998 [Pike *et al.*, 1998], but the < $140,000 payment per dwelling was well below pre-storm market value. Court-awarded compensation for the more expensive homes destroyed or damaged by a 1993 landslide in a Los Angeles suburb [Barrows *et al.*, 1993] was even less, < $36,000 per household [Rodrigue, 2001].

PERSISTENCE OF URBAN LANDSLIDING

The losses of life and property to landslides have become a fixture of the urban condition. Rather than declining with the worldwide spread of modernity, the threat worsened throughout the 20th century (Figure 1). The underlying causes remain as much economic, cultural, and political as they are geologic, meteorologic, or tectonic. In this section we note the failure to arrest the two growth trends identified previously (Figure 9), the most important consequence of which is that clearing land of its native vegetation to provide the necessary living space continually exposes more terrain to landsliding [Burby, 1998]. We then examine public awareness of the hazard and the communication of technical information, discuss the roles of engineered mitigation and landslide mapping, review some recent legal trends, and close with a summary of the problem.

Population Growth Continues

Earth's population has not stabilized (Figure 9). By 2025 it is expected to rise by almost two billion, the number of people added 1975–2000 [Board on Sustainable Development, 1999; United Nations Population Division, 1999, 2001]. The increase mimics a demographic transition first observed in Europe over the last 200 years, as births and deaths decreased to their current low rates, with deaths declining first. Europe's fertility rate fell to the replacement level (zero growth) of about 2.1 offspring per woman, yet its population grew rapidly for some time thereafter. The cur-

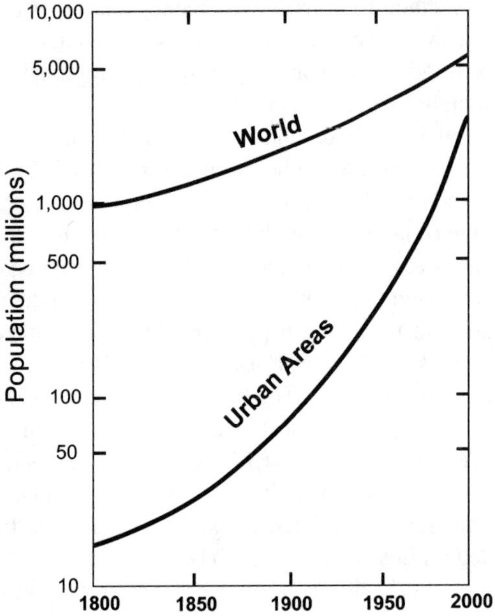

Figure 9. Enlarging the extent of potentially unstable urban terrain that is modeled schematically in Figure 1. Semi-logarithmic relation shows that cities are growing more rapidly than the global population. From United Nations data.

rent, global, transition is more pronounced. Birth and death rates in non-industrialized countries dropped faster than expected, births per woman falling from six at the early 1960's peak to three today. However, although by 1995 the *rate* of growth declined from this peak of 2.04% per year to 1.33%, *absolute* growth remained high and will continue to do so for several decades because the declining rate applies to a still-increasing population. At the close of the twenty-first century the world's inhabitants are projected to number 8.5–9.5 billion, about 97% of the growth occurring in less-developed countries [Bongaarts and Bulatao, 2000; Lutz *et al.*, 2001].

Three factors continue to raise the absolute numbers: an unmet demand for family-planning services, a persistent desire for large families in some developing regions, and the demographic momentum of nearly three billion people expected to reach reproductive age over the next 25 years [Fornos, 2000]. Actual growth will depend on prospective parents' choices of family size and their ability to make these choices. In some non-industrialized regions, government policies to encourage smaller families conflict with lingering cultural norms. Societal views in some industrialized nations, moreover, have limited the resources needed to slow population growth in the developing world [Bongaarts, 2002]. At the important 1994 United Nations Conference on Population and Development in Cairo [Finkle, 2002], participating nations requested funding of $17 billion per year by

2000, rising to $22 billion by 2015. By 2000, developing countries seeking to restrain fertility were committing about $7.5 billion annually, but such wealthier sources of funding as the World Bank only $2 billion. In failing to achieve anticipated reductions in the rate of population growth, the developed world's $7.5 billion shortfall may have resulted in an annual toll of 100–200 million unintended pregnancies, largely in the world's sprawling cities [Fornos, 2000].

Urbanization is Accelerating

The rate of urban growth exceeds that of the planet's human inhabitants (Figure 9). As many people now live and work in cities as occupy the countryside [United Nations Population Division, 2001], and the urban share of Earth's population will be 60% within one generation. This increase, together with overall growth, projects at least 50 million new urban dwellers annually every year throughout the eventual transition to global sustainability late in this century. By 2015, million cities are expected to number about 500 and megacities 30. Less-developed countries alone will host at least 20 megacities (up from none in 1950), which will house 12% of their urban population, between 350 and 400 million people. By 2020, Dhaka, Karachi, and Jakarta will have displaced New York and Los Angeles from the world's 10 largest cities (Figure 10). The greatest increase over 2001–2020, however, is expected in cities with populations ranging from 250,000 to one million. The combined growth of these smaller cities, now about 28 million people each year, will exceed 30 million annually in another 15 years.

Many of these added tens of millions of city dwellers are poor. As the population grows beyond the capacity of urban economies to provide employment, thereby depressing wages and adding to impoverishment, the oversupply of residents drives up the cost of food, housing, and other necessities. Further contributing to poverty in developing countries, low ownership of private property remains widespread despite beginning efforts to formalize the registration of city residences and businesses [McLaughlin and Palmer, 1996]. Urban privation is spreading and deepening. Six of every 10 children in the nonindustrialized world are projected to live in cities by 2025, and over half of them will be poor [Fornos, 2000], forcing added millions into marginal neighborhoods that are susceptible to landslides and floods. Uncertainties in the food supply becloud any optimism [Bongaarts, 2002]. Even if agriculture keeps pace with population, its unequal distribution will steepen the downward spiral of poverty, substandard habitation, and deteriorating use of the land.

Long-term trends of a different kind—the eight warmest years of the past century all postdate 1990—further threaten these marginalized residents [IPCC, 2001b]. Large cities are concentrating energy in urban heat islands [Thomas, 1956, p. 596–601; Zoback, 2001], thermal sinks that contribute to the alteration of Earth's heat balance now underway from increases in CO, CO_2, and oxides of nitrogen [IPCC, 2001a]. Motor vehicles are the primary sources of such "greenhouse gases," but burning fossil fuels to generate electricity for air con-

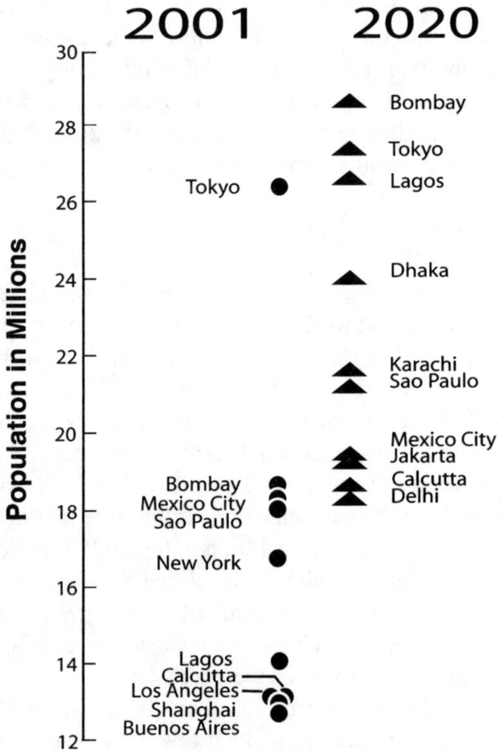

Figure 10. Threat to the urban inhabitants most vulnerable to natural disasters. Graph shows likely increase in size and number of cities in less-developed countries. World's 10 largest cities, in 2001 (circles) and projected for 2020 (triangles), in millions of inhabitants. By 2020, the median population of the largest cities will have risen 40% and the industrialized world will be represented only by Tokyo. From United Nations data.

ditioning and other energy-intensive technologies adds to these emissions. Atmospheric warming potentially can raise sea level, modify vegetation patterns, and intensify the El Niño-Southern Oscillation and severe storms conducive to landsliding [Schrott and Pasuto, 1999]. Although expansion of the cities themselves is perceived as a greater threat to the environment than landslides [Mitchell, 1999], slope instability will become more evident a problem as these two urban pathologies increase together. Thus far, however, landsliding remains almost invisible.

Perception is Poor, Awareness Low

Landslides are not taken seriously, by either the urban public or most professionals. But then, neither are natural hazards overall. In eight recent assessments of 30 environmental hazards worldwide, four types of geologic and weather-spawned disasters—flooding, drought, severe storm, earthquake—as a group ranked last in importance [Board on Sustainable Development, 1999, p. 188–191]. Not only did the evaluating experts consider land-water-air pollution, toxic contamination, and resource depletion more of a danger to global sustainability than naturally occurring events, but none of the assessments explicitly identified landsliding as a natural disaster. Contributing to this poor visibility is the concealment of landslide fatality and damage statistics within those of the initiating storms and earthquakes. Although in nonindustrialized societies landslides may be endured as an unavoidable force of nature [Jones, 1973], such fatalism is not unique to rural areas or the developing world.

In Great Britain, for example, knowledge of landslide character, distribution, and significance remains poor throughout the population. This unawareness persists despite the 1966 Aberfan mine-tip tragedy [Alexander, 1993, p. 266], the many failures in the urbanized Rhondda Valleys of Wales [Jennings and Siddle, 1998], and the 200-year-old slope problems of Ventnor and other towns on Britain's south coast [Moore *et al.*, 1995]. Because much of the country seems too flat to incur major sliding and most steeply sloping uplands are sparsely settled, the risks appear low. These appearances are deceptive. Not only are new damaging landslides reported each year, but thousands of old (ancient) landslides can move again if disturbed [Jones and Lee, 1994]. Old slides are more common in the British landscape than previously thought and remain a danger to dwellings and other structures.

Historically in the U.S., landslides have been an incidental problem to be addressed locally [USGS, 1982; Committee on Ground Failure Hazards, 1985]. This fragmented response to slope failure contrasts with the concerted attention accorded earthquakes, hurricanes, and floods—typically regarded as regional and national responsibilities [Board on Sustainable Development, 1999, p. 191]. Only where the landslide hazard is extreme or persistent, do the resulting losses gain more than local attention. Public agencies in the U.S. occasionally launch a project to assess the hazard, but sustained programs at the national level are rare and minimally funded [Highland, 1997]. Nor is there a recognized forum for establishing a program to alter the uncoordinated response to landsliding [Spiker and Gori, 2000].

Indifference in the U.S. extends to the local level, even in areas known to have a landslide problem. In the urbanized San Francisco Bay region, federal, state, and local geologists and engineers have been assembling data on the nature of the hazard since the early 1970's [Brabb *et al.*, 1972; Nilsen, 1973; Taylor *et al.*, 1975], but the effect of the information varies by county and municipality [Mader and Crowder, 1971]. Over the past 25 years several counties and a number of cities ceased to staff a full-time geologist, and landslide problems increasingly have been left to the private sector. In 1995, the USGS abandoned the region's debris-flow warning system [Keefer *et al.*, 1987], and for a time landslide mapping by federal and state agencies all but ceased.

Not only does urban landsliding attract little civic interest but its consequences are viewed differently by the affected parties, which maneuver to avoid indemnity and shift the responsibility for damage [Rodrigue, 2001]. A related difficulty is establishing the level of risk that individuals and society are willing to bear before committing resources to reduce the losses [Susskind and Field, 1996, p. 112]. Even then, remedial action follows only when mitigation costs are justified in the context of perceived, as distinct from actual, risk. Howell *et al.* [1999b] observed that neither government ordinances nor public and individual attitudes are consistent in recognizing the relation of reduced landsliding to prudent land-use, research, mapping, and redress for damage.

Brabb [1984] and Alexander [1989], among others, found that information on landslide risk was not reaching the people who most need it. Municipal planners and engineers often lack the data needed to foresee the consequences of hillside grading. They may have insufficient knowledge of landsliding to avoid the hazards or to remediate them prior to development. Consulting specialists can evaluate slope conditions locally [Leighton, 1971; Scullin, 1994], but the regional distribution of landslides is poorly publicized. Accordingly, government policy-makers and insurance underwriters have little basis for estimating the risks across a metropolitan area, and neither private citizens nor officials responsible for emergency management are warned sufficiently in advance to prepare for damaging landslides [Degg, 1998]. Some of the low awareness stems from difficulties in the transfer of technical information.

Impeded Communication of Risk

Landsliding is not widely understood. Besides the public's general lack of experience with scientific questions that can have only indefinite answers, specific obstacles hinder the transfer of knowledge from geologists and engineers to potential recipients. One barrier is language required to describe the variety and subtlety of slope processes. Undue insistence by specialists on the precise usage of technical nomenclature can irritate rather than clarify [Susskind and Field, 1996, p. 116]. The minutiae of landslide taxonomy may bore or confuse news media and other nonspecialists. From experience around the San Francisco Bay area [Pike *et al.*, 1998; Howell *et al.*, 1999b], we found that getting a lay audience to grasp even the most basic observation on the region's landslide hazard, the contrast between debris flow and deep-seated failure, was a major accomplishment.

Nor do all professionals agree on landslide terminology. Although nomenclature should be used consistently when talking with the media and the public or serving as an expert witness, such harmony is not always realized. "Mudslide," for example, a term omitted from most classifications of slope movement (Figure 5) and disliked by some U.S. geologists, not only is common parlance among journalists and the general populace everywhere but is standard geologic nomenclature in the U.K. and other countries [IPCC, 2001b, p. 15; Dikau *et al.*, 1996, p.

103]. In the U.S. the confusion is institutionalized. "Mudslide" was added to the U.S. National Flood Insurance Program (NFIP) legislation in 1982 and since defined in the courts, so that insurance now covers losses to a landslide that has a high water content. Most "mudslides" are simply earth flows [Figure 5; Varnes, 1978], and an acceptable alternative, mudflow, has been in use for many years [Watanabe *et al.*, 1996; Dikau *et al.*, 1996, p. 181].

A colloquialism that we found helpful in educating the public and the media about contrasting styles of failure is "bedrock landslide." Although absent from standard classifications, and perhaps more palatable to some specialists if rephrased as "landslides in bedrock," the term succinctly conveys to lay audiences that large, deep-seated slides and flows can involve "safe," intact, strata; that is, "bedrock can move." Many such failures have not shifted in hundreds of years, but can remobilize under the right conditions (Figures 3, 6). Although most landslides in bedrock remain unnoticed while dormant, much development has occurred on them in ignorance of the potential risk [D.J. Rodgers, letter to Howell, Dec. 9, 1997]. To some extent this risk can be reduced through technology.

Engineering Slopes is Not Enough

Slope problems in the cities are best avoided through care in permitting construction sites, but even actively sliding hillsides can be stabilized. A variety of corrective measures are available to modify the forces that drive or resist landsliding [Eckel, 1958; Veder, 1981; Záruber and Mencl, 1982; Slosson *et al.*, 1992; Popescu, 2001]. Although understanding of the effective stress and progressive failure that leads to slope instability has changed little since the theory established by Terzaghi [1950], methods of repair have matured from the initial attempts to arrest landsliding along the canal and railroad embankments of England and France [Collin, 1846]. As earthwork equipment improved, removal and recompaction of entire slide masses (Figure 11) supplemented the early techniques of toe buttressing and dewatering. Geotextile and membrane products recently have broadened the options for repair, especially under conditions of restricted access [Rogers, 1992].

Although effective in such affluent cities as Hong Kong (with over 50,000 cuts and fills, the world's greatest concentration of engineered slopes) and many in the U.S. (Figure 11), under most circumstances "hard engineering" is too costly to apply more than locally. In developing-world cities it remains an impossibility. Terracing, deep drainage wells and channels, pumps, excavation and regrading, geotextile nets, flexible barriers, rock bolts, gunnite, debris stilling-basins and weirs, osmotic and cathodic electrical systems that reduce soil moisture all are expensive. Alexander [1993, p. 256–257] claims that maintenance of these systems can account for much of the total expenditure. Retaining walls alone cost Rio de Janeiro $50 million each year. To control landslides and limit damage, Japan spends over $4 billion annually, mostly on myriad *sabo* dams that intercept debris flows on steep hillsides, [Japan Landslide Society, 1988; Watanabe *et al.*, 1996].

Figure 11. An urban lifeline imperiled and relieved. Stabilization work on a small landslide in San Mateo Co., California (Figure 2), showing endangered dwelling and failed concrete wall emplaced in earlier, unsuccessful, attempt at mitigation. The 96-inch water main buried at the toe of this ancient landslide (behind the camera), which reactivated during the 1997–98 El Niño storms, supplies 2.5 million city dwellers. Repairs, including excavation of the entire slide mass to below the slip surface, cost $3.5 million. Photo by D.W. Ramsey, USGS.

Lower-cost measures that are being explored, particularly in the cities of non-industrialized countries, include such innovations as retaining walls built of recycled materials [Shore, 1999]. Entirely non-structural, "soft engineering," techniques warrant wider application [Degg, 1998; Popescu, 2001]. Although well-designed surface drainage is effective and less costly than most engineering approaches, large tracts of terrain that are subject to shallow failure can be stabilized even morc cconomically by vegetation. Once established, well chosen plantings improve soil cohesion and lower moisture content. Not all replacement land-cover is effective; shallow-rooted trees can promote mass movement by blowing down in high winds (post-hurricane landsliding); success also depends upon depth of weathering, hillslope drainage, and movement of active landslides [Sidle *et al.*, 1985]. Before any attempt at broad-scale repair of slopes, however, provenance of the hazard must be established.

Extent of the Threat is Poorly Known

A spatial phenomenon, landsliding is most usefully characterized by maps. Landslide maps are essential for making sound decisions in managing urban land [Varnes, 1984; Leroi, 1996]. They also help prepare communities for emergencies through training exercises and by identifying areas likely to fail during an earthquake or severe rainstorm or after a season or successive seasons of high precipitation. Besides portraying the seismic shaking or critical rainfall that initiate landsliding [Wilson and Jayko, 1997; Peterson *et al.*, 2000; Miles and Keefer, 2000], maps characterize five aspects of the threat:

- Landslide-*inventory* maps show evidence of prior failure. Prepared by interpreting minor features on the Earth's surface, inventories of existing landslides are the first step in identifying the location of future failure [Nilsen, 1973; Wieczorek, 1984; Amanti *et al.*, 1996]. They may incorporate historical records of the timing of past movement [Harp and Jibson, 1995; Coe *et al.*, 2000].
- A landslide *hazard* is an unstable condition arising from the presence or likely future occurrence of slope failure. Two types of landslide-hazard maps are distinguished:
    - Landslide *susceptibility* is likelihood of failure at a site based on its local properties, including rock or soil strength, vegetation, terrain steepness and exposure, and evidence of prior failure [Brabb *et al.*, 1972; Pike *et al.*, 1994; Parise and Jibson, 2000].
    - Landslide *potential* couples site susceptibility with landslide *opportunity*, probability of a triggering event [Coe *et al.*, 2000]. Because landslide inventories incorporate these events, although a specific cause may not be known, many susceptibility maps also are maps of potential.
- Landslide *risk* is the expected loss due to landsliding. Risk combines the landslide hazard with the estimated number of lives lost, persons injured, damage to property, or disrupted economic activity should a landslide occur [Bernknopf *et al.*, 1988].
- Landslide *zones*, declared officially to have a high probability of landsliding, are areas within which specific actions are legally mandated before development [Mader and Crowder, 1971; Wilson *et al.*, 2000]. The criteria may be landslide potential or susceptibility, but some zones are based only on slope gradient or a landslide inventory. Zone maps are planning tools intended for non-technical people.

Landslide maps are invaluable, but their collective coverage is meager. The maps require professional expertise and are costly to prepare even for towns and small cities. Several quantities may need to be measured in the field [Miller, 1995; Fernández *et al.,* 1996; Massari and Atkinson, 1999]. The detailed data required for a fine-scale map of a small area [Wieczorek, 1984] are prohibitively expensive for a regional survey. Nor has the automation afforded by geographic information systems (GIS) reduced the time and expense required to identify landslides by field observation and airphoto interpretation. Systematic mapping to assess landsliding across a large city entails a major commitment of resources that rarely is forthcoming [Brabb and Harrod, 1989]. We return to this point in a later section.

Landslide Maps Have Limitations

Even where available, landslide maps do not reveal the full extent of the hazard [Nilsen, 1973]. Slope instability, like earthquake and other natural processes, is difficult to predict [Haneberg, 2000; Oreskes, 2000]. Maps do not indicate when a slope will fail, and their spatial forecasts are oversimplified because not all controls on the process can be included. To cover large areas, landslide maps must omit factors that require local measurement of terrain, geology, soil, and ground-water conditions; still other factors remain unknown. Moreover, most areas mapped as prone to failure include some sites that are not dangerous. Conversely, and more germane to public safety, areas estimated as being of low susceptibility or potential are not without hazard. Sites mapped as unlikely to host large slides and earth flows, for example, may be subject to debris flows—frequent in the area but not included in the data on which the map was based. Finally, many landslides can not be anticipated. Innumerable failures result from chance blockage of surface drainage or other consequences of hillside development, as well as from the random operation of natural processes.

Maps show the relative importance of landsliding and thus overall stability, but are only a guide to future movement. Geologic and climatic changes over the last few hundred millennia have altered conditions under which old slides formed [Reneau *et al.*, 1986]. Deforestation, grading, road-building, and other human activities can so disturb the equilibrium of a landscape that previously unfailed terrain may become susceptible. Because many public codes require detailed investigations to judge the stability of a site, hazard maps are no substitute for a report by a licensed engineering geologist or soils engineer [Nilsen, 1973; California Division of Mines and Geology (CDMG), 1997]. A full assessment of the risk may require other maps—showing the likelihood of flooding, earthquake ground-motion, and soil liquefaction [Haydon *et al.*, 2000; Peterson *et al.*, 2000].

Finally, the maps may require expert guidance in applying them to public safety and land-use planning [Pike *et al.*, 1998; Howell *et al.*, 1999b]. Landslide maps can appear abstract to nonspecialists or be otherwise difficult to understand and thus not communicate well to the public officials who must use them [Brabb, 1997]. Maps derived statistically from a computer model can so lack "transparency" that their input data are difficult to correlate with existing landslides and other field observations [Dhakal *et al.*, 2000]. Landslide maps, especially inventories, also can have unforeseen—commonly litigious—consequences through their very portrayal or omission of a landslide.

The Legal Conundrum—Who Pays?

To recoup economic losses to a landslide, victims have little recourse but to sue [Olshansky and Rogers, 1987]. The lack of landslide-insurance protection for private property usually leads to claims and legal proceedings against builders, owners of neighboring property, local jurisdictions, and insurance carriers [Shuirman and Slosson, 1992; Leighton, 1992]. Only New Zealand has a national policy of compensation [Swanston and Schuster, 1989]. Scullin [1994] notes that the adversarial nature of landslide cases avoids the core issues while provoking attempts to shift liability to other parties. The usual area of dispute is prior knowledge of an unstable condition, or the cause or even the occurrence of a landslide. Excepting the term "mudslide" in the U.S., no definition of a landslide suffices for legal purposes [Griffiths, 1999]. Despite all the technical data that have been amassed on landslide damage, case law has yet to satisfactorily clarify the legalities surrounding the origin and existence of a failed slope [Olshansky, 1990].

Lawsuits that stem from ill-defined responsibility for landsliding can persist for years and yield little satisfaction. In 1993, for example, a condition of instability known to local officials prior to failure generated lengthy court battles and murky politics over damage from a single storm-triggered landslide in Anaheim, California [Rodrigue, 2001; Steiner, 2002]. The circumstances described by Barrows *et al.* [1993] surrounding this and other Los Angeles-area landslides caused by the same storm reveal a chronic failure to protect the public. One symptom is the uneven quality of reviews by local jurisdictions that evaluate the technical reports required before development. Another is the poor familiarity of local officials with the California law authorizing Geologic Hazard Abatement Districts to mitigate landsliding [Kockelman, 1986]. Litigation over loss-compensation also can generate political or economic pressure to relax the conservative standards of engineering practice that ensure safe development [Scullin, 1994].

The Problem in Microcosm

Summarizing this section, the results of a 1999 workshop in the West Indies [Howell *et al.*, 1999a] illustrate how landsliding is perceived by decision-makers and the public. Kingston, Jamaica, faces slope problems much like those in the San Francisco Bay region [Ahmad *et al.*, 1999]. Meeting to share knowledge of the hazard, experts and officials representing a variety of functions in Kingston responded to a long questionnaire on local land-use concerns. The detailed answers compiled by Howell *et al.* [1999a], although unique to Jamaica, have close counterparts everywhere and thus echo widespread views on urban landsliding:

- Landslides compete poorly for the attention of professionals charged with many responsibilities: other hazards, environmental management, water quality, land use, deforestation, zoning and development, road construction, waste disposal, ground-water recharge, wildlife management, agricultural practices, and industrial location.
- The economic toll of landsliding is so poorly documented that the hazard may not be taken seriously. Estimates of annual damage in Jamaica ranged over a factor of six.
- Losses may be misperceived. Although both landslide damage and population are stable trends in Jamaica, popular resentment against wealthy Jamaicans who are building upscale homes in the hills elicited various causes for a nonexistent increase in damage: developing vulnerable or marginal terrain, squatters, deforestation, El Niño, hillside terracing, and road construction.
- No one agency either records landslide activity or provides information on reducing its effects, and the two functions are not linked. Among parties perceived as responsible are Unit for Disaster Studies (local university), Mines and Geology (national government), Disaster Preparedness and Emergency Management Office, Kingston Public Works, Agriculture Ministry, public utilities, town planners, Natural Resources Conservation Authority, parish councils, Meteorological Service, Soil Conservation Office, the Red Cross, and the Province Geologist.
- The kinds of information available to reduce urban landsliding are understood but vary in quality, coverage, and access: public-awareness programs; remedial engineering; technical reports from Mines and Geology, Office of Disaster Planning, and the university; data on meteorology, hydrology, and seismic shaking; and maps of landslides (especially near reservoirs), "preliminary hazards," and land capability.
- Officials responsible for reducing the hazard may not win budgetary support. Research, airphoto surveys, meteorological and seismic data, warning systems, improved engineering standards, education, and maps—of geology; geotechnical properties; land use; elevation; landslide frequency, hazard, and risk; and slope at regional, local, and site scales—compete poorly for funds.
- All participants supported national landslide-insurance, but many felt that low-income Jamaicans could not afford it and some suggested that the central government should fund it.

From their guided interrogation of municipal staffers and decision-makers in Jamaica, Howell *et al.* [1999a] distilled a few general truths. Landslides involve many public functions and agencies. Yet despite broad interest and an array of existing information that could help reduce landslide losses—in Jamaica, in the San Francisco Bay region, and other areas—land-use decisions continue to be made without the benefit of scientific data. Perhaps four factors contribute to this neglect:

- No single entity is recognized or designated as the lead authority on the landslide hazard. Responsibility for collecting and dispensing information is diffused.
- Available information on landslides originates across multiple agencies and usually is packaged at cartographic scales too coarse to be useful for site evaluation.
- Many people resist government regulation of private land-use.
- Large landslides are too infrequent to sustain public or government interest. Slope instability is less visible a societal problem than crime, health, education, and road traffic.

TOWARD LANDSLIDE-RESISTANT CITIES

Although landslides pose a deepening threat to the world's cities, the loss of life and property can be contained. In this section we consider a spectrum of remedial measures, including some recent developments in mapping the hazard. These factors are a mix of the demographic, technological, economic, perceptual, and political. Few of them are new. Perhaps most urgent are a heightened awareness of the landslide problem and the collective will to address the many interwoven questions.

Stabilize Population Growth

Reducing urban landslides and other natural hazards by reining in the world population is a complex undertaking that lies beyond the scope of this chapter, but some closing remarks are warranted. Earth's inhabitants are multiplying at an annual rate of about 1.2%, some 77 million each year [United Nations Population Division, 2001]. The most optimistic models predict a total population peaking at 9 billion before starting to decline after about 2070, as well as a 60% likelihood of 10 billion people before 2100 and only a 15% chance that the 2100 population will be smaller than it is now [Lutz *et al.*, 2001]. Until the rate of growth falls to zero, the rise in raw numbers will continue to sustain urban sprawl and the attendant effects described earlier [Bongaarts and Bulatao, 2000].

Moderating this gloomy prospect is a growing consensus among demographers that more women in developing regions are deciding independently to have fewer children [Finkle, 2002]. In India and the other moderate-fertility nations that together account for over 75% of the world's population, one- and two-child families will soon be the norm [Dyson, 2002]. Announced at the March 2002 United Nations expert-group meeting in New York [Caldwell, 2002], this optimistic forecast may reflect some of the prior recommendations for limiting the world's numbers: family planning, rising prosperity, better education of women, and inclusion of more women in the workforce [Smail *et al.*, 1997]. Fertility has declined even in some poor nations with limited literacy. In Tunisia, where nearly 40% of the women can not read, the fertility rate is now at the replacement level. Some experts, however, fear that the new projections may discourage donor nations from maintaining the attitudes, policies, and patterns of expenditure that have sustained the fertility decline [Caldwell, 2002].

Manage Urbanization More Effectively

Because the world's numbers are certain to rise by at least 50%, any current decline in fertility will have little immediate effect on urbanization and the landslide hazard. Under favorable circumstances, however, the proliferation and growth of cities can be slowed, those that must be built can be properly planned, and land use within existing cities can be improved [Mitchell, 1999].

The pace of urban growth will slacken if cities attract fewer rural migrants. This reversal can be encouraged by expanding economic opportunity in the countryside, through political stability, modern infrastructure, industries suited to rural societies, and job incentives for local inhabitants. Achieving success in developing countries may require land reform, more private ownership of property, and reducing the disparate distribution of wealth. Whereas an equitable system of land tenure is essential to a healthy, dynamic rural society [Prosterman and Riedinger, 1987; Powelson, 1989], in many countries a few owners hold vast expanses of land. If that minority also controls access to credit and marketing, rural life remains less attractive than that in cities. Powelson and Stock [1990] note further that land reform has the best chance of regenerating rural economies if it is tailored to individual societies and avoids augmenting centralized state power at the expense of its intended beneficiaries. Besides leading to fewer new cities, these changes encourage better land-use practices and reduce soil erosion and landsliding in rural settlements.

Landsliding can be forestalled in cities yet to be built through initial care in siting and designing neighborhoods [Burby, 1998; Ghilardi *et al.*, 2000]. In the ideal scenario, planning identifies undesirable areas and zones them for non-urban use, while legislation requires the new communities to include landslide hazards in the planning process and to apply the resulting laws uniformly. Scullin [1994] and Brabb [1997] emphasize the importance of proper procedures and their enforcement from the outset. Well-crafted regulations can reduce the disturbance of landslide-prone terrain while ensuring that engineering and geological expertise guide development in areas that are permitted. Hazard maps and technical information further assist planners and other decision-makers in building landslide-resistance into the new cities [Varnes, 1984].

Reducing conditions that foster landsliding in existing cities requires the determination to alter established practices. The necessary changes are well known [Schuster, 1995]. They include zoning and grading ordinances, watershed management, added capacity for storm drainage, public agencies prepared for landslide damage, and using data on landslide losses to quantify the benefits of mitigation. Successful examples abound. In the U.S., Leighton [1966] documented reduced losses to landsliding in Los Angeles following enactment of grading codes. Moore *et al.* [1995] described the nationwide changes in England and Wales, where formal Planning Policy Guidance issued by the Department of the Environment now advises local authorities, landowners, and builders on the role of planning regulations as a tool for landslide control.

Wider ownership of property in developing-world cities promotes better land use in poor neighborhoods. The Urban Property Rights Project is a joint venture between the World Bank and the Instituto Libertad y Democracia in Peru to formalize titles to urban land. The program offers residents greater security by encouraging investment in property, real-estate sales, and collateral for borrowing—which creates wealth by raising land values [McLaughlin and Palmer, 1996]. This radical approach to institutional reform addresses legal, administrative, and technical problems. New agencies sustain the reform by bypassing agencies that are corrupt or incompetent; a low-cost process "mass formalizes" ownership through automated, neighborhood-based, titling. Although this program has been controversial because of the legal changes and its quasi-common-law approach to establishing property rights, a growing number of countries are seeking similar reforms.

More drastic measures must be imposed in other cities. Uprooting established neighborhoods and relocating them to reduce landslide losses has become a last resort, particularly where the urban poor have been excluded from locations that are geotechnically stable. Velasquez [1998] describes how abandoning unstable land to open space can shift development from areas of high risk. The municipality of Manizales, Colombia, controlled expansion of its built-up area and prevented construction in terrain prone to landslides by creating a series of eco-parks where grading is severely restricted. Owners of buildings in areas of high risk were offered land exchanges that enabled them to move to safer ground. Future recourse to such severe measures could be reduced by better anticipating the growth of cities.

Forecast Areas of Urban Growth

Identifying slope problems before they develop enables planners to reduce landslide damage through zoning and regulation. Gaining this advantage requires knowing where urbanization is likely to encroach upon potentially unstable terrain (Figure 1). Landslide-hazard maps can indicate the susceptible areas, but the locations of future of neighborhoods are less certain. Forecasts of city extent as a function of time would reduce this uncertainty and suggest which of the predicted built-up areas to set aside as open space.

Recent developments in urban geography provide a basis for time-dependent spatial modeling. Cities once were thought to grow and take shape according to the central-place theories of Walter Christaller and August Lösch, based on minimum market size and maximum distance-to-market [Bunge, 1966, p. 129]. A

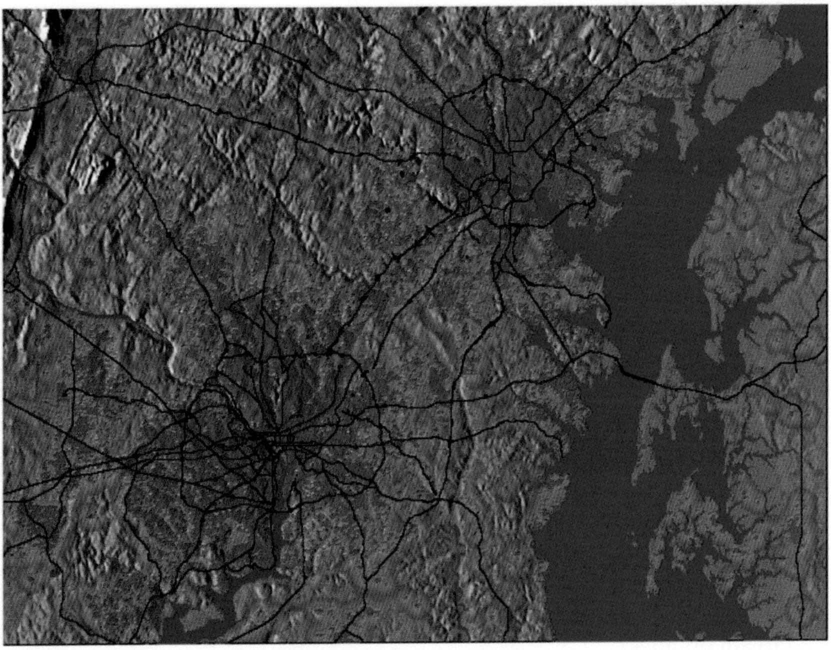

Plate 3. Looking ahead to learn where not to build. Map showing the Baltimore-Washington, D.C., region in 2050 forecast by a computer model of city growth [Clarke *et al.,* 1997]. Red-orange, built-up area in 1992. Probability of urbanization in 2050 decreases in five 10% increments from 90–100% (yellow-orange) to 50–60% (darkest green). Some locations not now urban contain slopes unsuitable for development. Map is about 135 km across. Courtesy of K.C. Clarke, University of California at Santa Barbara.

new model is challenging this analogy with classical gravitation. Batty and Longley [1994] describe the self-organizing process wherein locally unrelated decisions yield broader patterns that define city size and shape. Constrained by local-decision rules, urban areas modeled by diffusion-limited aggregation can grow from a single seed at the geographical origin of development into fractal clusters that resemble cities with dominant central business districts. Makse *et al.* [1995] found the resulting forms consistent with power laws that describe population mass-radius, area, and density, as well as the rank-size relations that govern city dimensions and spacing.

Experiments in computer mapping based on these principles are predicting the extent of cities. From an understanding of prior geography and demography, Clarke *et al.* [1997] modeled the spatial growth of the San Francisco Bay region by a time-dependent cellular automaton. Model inputs were several outlines of the built-up area at intervals dating to 1850, interpreted from archival maps, in addition to terrain slope, water boundaries, and transportation routes. Running the algorithm for the period 1850 to 2000 accurately recapitulated the region's spatial development [Clarke *et al.*, 1997; Clarke and Gaydos, 1998]. Modeling beyond the present predicted urbanization spreading into valleys in outlying counties and filling in between San Francisco and Sacramento (Figure 2). The similar forecasts for the Baltimore-Washington D.C. area shown in Plate 3 could be made for any city that has the necessary historical records and maps. Simulations described by Parish and Müller [2001] add detail within the predicted built-up area. However, neither urban models nor other technical data that bear upon the landslide problem will influence planning decisions unless they are effectively communicated.

Improve Risk Communication

Vital to reducing slope failure, information transfer to non-technical audiences [Kockelman, 1986] demands skills that were not part of the training received by most landslide professionals. Through its experience in communicating uncertainty to a broad range of audiences, the U.S. National Weather Service (NWS) has learned that five perceptual thresholds must be crossed for technical information to have the desired effect: first, ensuring that the message—for example, urban landsliding is a costly hazard that must be reduced—is HEARD, next making certain that it is UNDERSTOOD, then convincing recipients to BELIEVE the information, largely by packaging it in ways that so PERSONALIZE the message that individuals finally will ACT on it.

Successful execution of this sequence entails sharing landslide expertise and engaging its recipients, through information designed to raise civic awareness and educate the wider community, as well as training for the participating scientists, engineers, and government officials [Spiker and Gori, 2000]. Brabb [1997] identified credibility and persistence as essential in influencing the decision-making process. The most effective experts will accommodate their lay audiences by making information accessible in different ways and through different channels. To spread its message, the landslide community must build trust with potential recipients, learn to work comfortably with the media, and avoid memorable blunders that can destroy credibility.

Interaction with nonspecialists should increase understanding, not confusion, particularly when explaining probability and other vague concepts [Zoback, 2001]. As Haneberg [2000] points out, the public is familiar with the clear-cur answers typically forthcoming from technology and engineering, not the iterative findings of the scientific process. Important information about uncertainty and relative risk must be conveyed in terms that can be grasped by a wide listenership. Successful discourse need not "dumb down" science for the untutored or unprofessionally pander to the sensational, but bringing the insights of a technical field to outsiders often calls for plain words rather than the specialty's detailed vocabulary [Susskind and Field, 1996, p. 116]. In tailoring explanations of slope instability to general audiences, perhaps three factors emerge as the key to listener reception:

- Credibility: does the presenter really know the information, or is he or she a spokesman relating it second-hand? Especially at the site of a fatal or damaging landslide, those present want—and deserve—an expert.
- Clarity: is the information making sense? The focus should be on a minimum number of important facts and how they interrelate.
- Utility: why the information is useful in the immediate circumstances and why it may be important to the listener. Where does it lead; what is the next step?

These factors matter most when the media are present, for the resulting television, radio, or newspaper coverage can reach—and influence—a vast audience.

Broadcast the Message

Because neither local discussions nor technical publications prepared by experts are sufficient to create a public appreciation of landsliding, the hazard needs broader exposure. Popular science-writer John McPhee [1989] published one of the most accessible accounts of slope failure in the U.S. Besides relating how debris flows result from heavy rainfall and antecedent local conditions, he describes the debris basins and check dams constructed in southern California canyons in attempts to reduce the hazard. Another well-researched description of landsliding by freelance writer Brenda Bell [1999] focuses on a 1997 family tragedy in the Pacific Northwest. Both articles, however, appeared in low-circulation literary magazines. Better use could be made of such media as Sunday newspaper supplements, which are aimed at a wider audience. The full-color 24-page booklet on earthquake preparedness distributed by San Francisco Bay region newspapers after the 1989 Loma Prieta earthquake [Ward, 1990] reached over 2.5 million households. Although the most direct medium, network television, shows landsliding only in brief coverage of graphic episodes, the Internet now offers an alternative path to high visibility.

The World Wide Web is well suited to the dissemination of hazard information, particularly data in map form. In just six months, the USGS Web site that hosts digital-map databases created in anticipation of 1997–98 El Niño rainstorms attracted 185,000 visits [San Francisco Bay Landslide Mapping Team, 1997]. The potential influence of private Web sites is illustrated by the Los Angeles-area controversy alluded to earlier. Thirty-two luxury homes destroyed in January 1993 by a rainfall-triggered landslide in the Anaheim Hills neighborhood [Barrows *et al.*, 1993] had been constructed with full knowledge of prior ground movement. Rodrigue [2001] discusses implications of the extraordinary resource of landslide information and government maps in the 2000-page Web site built by one of the displaced homeowners [Steiner, 2002]. Revealing a weakness in the requirement for real-estate disclosure that led to faulty hazard-perception by Anaheim Hills residents, the site exposes dialogue involving geologic risk-assessment and the politicized processes of risk management and local decision-making. The site further alerts prospective home buyers to potential risks that might pass unheeded, especially if the hazards-disclosure statement for real-estate transactions requires no landslide information. The Steiner Web site not only is well informed but also entertaining and popular by virtue of its sarcastic and one-sided tone. Although Internet activism of this sort can complicate the tasks of local officials responsible for risk management, Rodrigue [2001] suggests that it may raise public awareness and increase the pressure for pre-landslide intervention.

Mitigate Pre-, Not Post-Disaster

Measures to reduce the losses from landsliding began locally in the U.S. Unprepared for a cyclical shift to rainier winters in 1951–52, the city of Los Angeles sustained $7.5 million in landslide and related damage to its new postwar suburbs. Leighton [1976] described how the grading ordinances that resulted have greatly reduced subsequent losses. After $112 million in damage from storm-driven landsliding in the San Francisco Bay area in 1982 (Figures 7, 8), the California legislature passed the Landslide Hazard Identification Act. This addition to the Public Resources Code established a program, including a mapping and advisory effort, to assist local and state agencies in land-use and permitting decisions for areas subject to landsliding [Barrows, 1993]. Prepared by the CDMG (now the California Geological Survey), maps of urban and urbanizing areas identify landslide hazards at a scale suited to local planning [Wills and Majmundar, 2000; Wilson *et al.*, 2000].

Nationwide, U.S. losses to natural disasters have risen alarmingly (Figure 12), landslide damage from 1997–98 storms in the San Francisco area alone exceeding $140 million in direct costs (Figures 7, 8). Since 1989, the federal government has dispensed over $20 billion in aid. Federal assistance has become so expensive that mitigation, rather than response, is being adopted to break the damage-and-loss cycle [Mileti, 1999; Butler, 2000; Platt *et al.,* 1999]. With

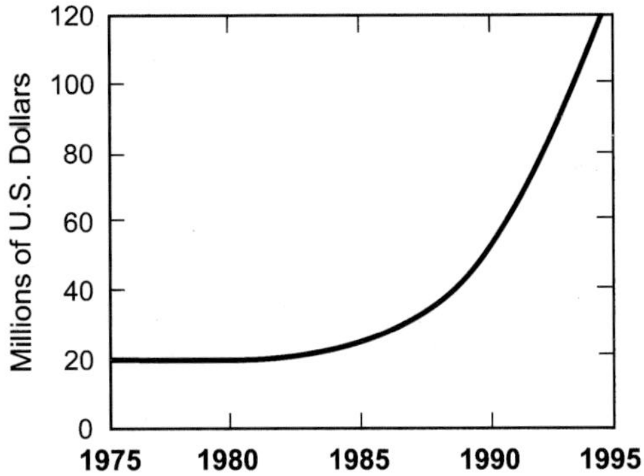

Figure 12. The rising natural-hazards tax on one industrialized society. Graph, which shows cost of recent U.S. natural disasters, in constant 1994 dollars lost per million inhabitants, suggests that prevention of landslides and other natural calamities would be less expensive than post-disaster relief. Redrawn after Mileti [1999].

Congressional support FEMA established Project Impact, a joint undertaking with the private sector to develop disaster-resistant communities through emergency preparedness, response, and recovery, and by applying scientific expertise and information [FEMA, 2000]. When the Alameda County city of Oakland (Figure 8, Plate 4) was selected for Project Impact, the USGS entered into an agreement with the California Office of Emergency Services (OES), the CDMG, and FEMA to assess the area's earthquake and landslide hazards [Miles and Keefer, 2000; Wilson *et al.*, 2000; Pike *et al.*, 2001].

Because cities can not be made landslide-proof, only resistant, losses are inevitable. Preventive measures do not reduce the need for help in coping with the aftermath of a fatal or damaging landslide [Alexander, 1993, p. 406–494; Platt *et al.,* 1999]. In addition to the customary emergency-response and humanitarian outreach, post-landslide assistance includes technical aid and information. Expert help and landslide data need to be available to governmental agencies at all levels, as well as to the general public [Spiker and Gori, 2000]. Recent experience has shown that disparate groups with a common interest in the public safety of a city can work together to collect, interpret, and disseminate landslide-hazard information.

Coordinate Local Response

An El Niño forecast for the northern California winter of 1997–98 led three government agencies to cooperate in preparing the 10-county San Francisco Bay region (Figure 8) for local emergencies [Pike *et al.*, 1998; Howell *et al.*, 1999b]. The initial plan comprised three elements: USGS maps showing localities susceptible to landslides and the precipitation amounts likely to trigger debris flows,

NWS information on rainfall for specific storms, and a communications network activated by the State of California's Region II OES (Figure 13).

To publicize areas of potential landsliding, the USGS reassigned staff to prepare a folio of digital maps [San Francisco Bay Landslide Mapping Team, 1997]. Its 33 individual sheets, completed before the onset of severe storms in February 1998, show hydrography, relief, roads, distribution of large slides and earth flows, rainfall thresholds, and likely source-areas of debris flows. Each county was mapped at 1:125,000, the accuracy of many of the data and the scale suited to emergency planning. A cross-referenced index provided access to larger-scale USGS and CDMG maps of local landslides. Released as a database on a USGS Web page, the folio also was obtainable from the agency on digital tape [San Francisco Bay Landslide Mapping Team, 1997]. Complementing the folio, a post-storm reconnaissance evaluated damage and estimated its direct costs [Godt and Savage, 1999]. Because historically, all large storms in the San Francisco Bay area cause landsliding [Taylor and Brabb, 1972; Ellen and Wieczorek, 1988], the 1997 folio remains useful in preparing for future winters as well as for general land-use planning.

During the El Niño emergency, USGS geologists delivered paper copies of the maps to all ten counties and explained the information. Local officials used the maps in exercises to train emergency-service crews, and some USGS data were incorporated into county and municipal planning for disaster relief. In the severe storms of early 1998, the USGS and OES jointly monitored NWS data (Figure 13). Guided by ongoing appraisals of weather conditions, the OES distributed

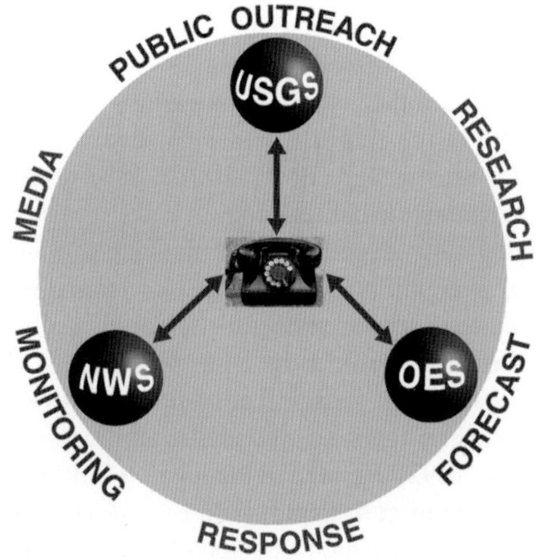

Figure 13. Interagency cooperation toward fostering a landslide-resistant city. The USGS, NWS, and OES in the San Francisco Bay region maintained an ongoing telephone linkage during the 1997–98 El Niño winter. The six elements that frame the diagram comprise any successful program to reduce landslide risk.

landslide advisories through their county network; the USGS received comments on landslide activity from OES operations across the San Francisco Bay area; and the USGS, NWS, and OES coordinated by telephone during major storms or when storms were predicted. Combining NWS precipitation data with USGS rainfall-threshold and landslide maps gave OES response units a sense of the likelihood of landsliding, particularly debris flow, as well as indicating localities where the activity might be expected. Other USGS staff assisted FEMA at its Disaster Field Office in the state capitol. Because coordinated efforts can reduce but not eliminate landslide damage, the losses that do occur need to be redressed.

Insure Landslide Losses

Insurance against landslide damage, a common goal that could unite interests in reducing the hazard [Olshansky and Rogers, 1987; Olshansky, 1990; Howell *et al.*, 1999a, b], is not an immediate prospect. Reasonable premiums and a realistic schedule of risk-apportionment could create a financial resource to encourage wise land-use investment while denying public reimbursement for reckless development in areas known to be susceptible to landsliding—a lesson learned previously in the U.S. by the National Flood Insurance Program. However, California's experience with earthquake insurance after the costly 1994 Northridge temblor—high premiums, large deductibles, and minimal coverage make the program unattractive to many homeowners—indicates some of the difficulties that face the introduction of landslide insurance.

The insurance and reinsurance industries themselves are increasingly vulnerable: the ratio of weather-related losses to premiums collected from property- and casualty-insurance worldwide tripled between 1985 and 1999. Expanding cities are one factor that has raised the actuarial uncertainty in assessing risk. Another, still unknown, variable is the future increase in severe weather-related events perceived to reflect climate change [IPCC, 2001b]. The result could be pressure to raise insurance premiums or declare certain risks uninsurable. The growing uncertainty also could slow expansion of financial services into developing countries, reduce the availability of insurance for spreading risk, and require added government-funded compensation for natural disasters.

In the absence of insurance protection, legal proceedings will continue to assign responsibility for landslide losses. The courts, however, are shifting from rigid doctrines of liability to a more case-specific application of *reasonable care* [Olshansky and Rogers, 1992]. Related to the legal definition of *negligence*, failure to exercise the degree of care appropriate to a situation, reasonable care is defined as "That degree of care which a person of ordinary prudence would exercise in the same or similar circumstances. Failure to exercise care is ordinary negligence." Both negligence and reasonable care are relative terms. Whether a litigant has behaved reasonably is up to a jury, according to the facts of the case and courtroom testimony—often by contradictory experts. Despite vagueness of the law—for the courts offer little guidance as to what constitutes reasonable care—the recent trend away from absolute rules makes good geologic sense.

To take advantage of the move toward case-by-case review, the landslide community needs to adopt uniform guidelines on slope stability [Olshansky and Rogers, 1987]. Current legal climate offers the opportunity for geologists and engineers to educate lawyers, property owners, realtors, and themselves: What warning signs portend instability? What constitutes reasonable care for owners of hillside property? What are appropriate practices in landscaping and maintenance? What is a competent site investigation? Accepted standards would carry weight in expert testimony and help to define for juries the state of knowledge within the profession. Until landslide insurance and procedural standards are reality, urban developers would do well to heed Scullin's [1994] precautions to avoid litigation: hire competent builders, institute quality assurance and control, execute sound geotechnical contracts, obtain insurance protection, and define areas of responsibility for all participants.

Investigate Landslide Behavior

Besides clarifying legal proceedings, knowledge of landslide likelihood for different environments would provide a basis for establishing landslide insurance [Bernknopf *et al.*, 1988]. This knowledge can come only from research into the nature of failure processes [Wieczorek, 1996]. Here we describe four examples of recent work to advance the predictive understanding of landslides: experiments in debris-flow physics, instrumentation of active slides, evaluation of earthquake-induced failure, and digital terrain-modeling.

To study all stages of the debris-flow process, from initiation through deposition, the USGS and the U.S. Forest Service built an instrumented flume that simulates shallow landsliding under controlled conditions [Denlinger and Iverson, 2001]. This unique facility tests the various controls that trap, deflect, or channel debris flows on steep slopes and determine the form and extent of deposition on the gentler ground below (Figure 14). The 96-m-long concrete chute slopes at 31°, typical of terrain hosting natural failures [Ellen, 1988]. A gate at the head of the structure suddenly releases saturated sediment of a specified composition, or a sloping mass of dry sediment is watered until it fails; the resulting slurry descends the flume and forms a thin deposit at its base. Glass windows enable flows to be observed and photographed as they move, data-collection ports in the floor record the forces exerted on sliding and colliding particles, and other instruments measure acceleration. Iverson and Denlinger [2001] describe how these experiments improved theoretical knowledge of debris flows, particularly the difficult-to-model distribution of runout.

Instrumenting naturally occurring landslides enables failure processes to be studied in the field while providing warnings to local communities. During the 1997–98 El Niño, debris flows originating in an active landslide above the small

Figure 14. Basic research to expand knowledge of landsliding. The instrumented flume was built by the USGS and the U.S. Forest Service to conduct controlled experiments (one of which is shown in progress) on debris flow initiation, movement, and run-out. Photo by R.M. Iverson, USGS.

California town of Rio Nido destroyed or damaged a dozen houses in the canyon below [Pike *et al.*, 1998]. The residual slide block (10 times the volume of the original debris flows) remained unstable, prompting the evacuation of 140 additional dwellings. To monitor rainfall and ground water and detect changes in landslide movement that might presage massive failure, Reid and LaHusen [1998] installed an automatic data-collection system on the perched block. Within five days of authorization, data on the Rio Nido slide were accessible to Sonoma Co. geologists on their Intranet. Sampled every second and transmitted routinely every 10 minutes, the information is sent immediately in the event of ground vibrations associated with landslide movement. Similar systems monitor landslides above U.S. Highway 50 in El Dorado County (Figure 2), after debris flows spawned by rains in January 1997 blocked the road and caused $27 million in damage [Reid and LaHusen, 1998].

Landslides generated by earthquakes, which strike cities with little warning, are less easily monitored [Keefer, 1998; Wasowski *et al.*, 2000, p. *v*]. The coupling of the two processes complicates attempts to estimate the geographic extent of their effects. Because emergency-management decisions based on hazard zonation could have widespread social and economic consequences in a large earthquake, the methods used to map the seismic performance of a hillslope are critical. Miles and Keefer [2000] found that different executions of the Newmark-displacement model [Jibson *et al.*, 1998] yield maps showing differing degrees of hillside stability. These results reveal, among other things, the potential for making incorrect decisions based on a landslide map printed from a single computer model. Miles and Keefer [2000] suggest that the traditional paper map may no longer be the best medium for hazard zonation. Computer simulations of multiple models, which simultaneously evaluate many earthquake scenarios under varied conditions, provide a more flexible alternative. Among the spatial data essential to these dynamic models are slope gradient and other measures of the ground surface.

Quantitative analyses of hillside form contribute to an understanding of how topography controls landsliding. The traditional approach treats landslides and their enclosing drainage basins as discrete landforms. Hylland and Lowe [1997], Jennings and Siddle [1998], and Luzi and Pergalani [1999], for example, correlated the dimensions, azimuth, shape, and volume of individual landslides with substrate properties, local hydrometeorology, and other characteristics to isolate causal factors. Mapping technology and mass-produced terrain data have made possible a second, regional, approach to surface modeling. GIS-based procedures compute gradient, slope curvature, and other measures over continuous terrain and combine them with non-topographic data to identify areas at risk [Brunori *et al.*, 1996; Cross, 1998; Rowbotham and Dudycha, 1998]. Although measurable by field survey [Ellen, 1988, p. 85] or on contour maps [Pike, 1988; Ellen and Wieczorek, 1988, p. 118, 140], surface form increasingly is described from gridded arrays of terrain heights, or digital elevation models (DEMs) [Pike, 1988]. New technologies, such as laser range-finding (LIDAR) and the February 2000 global Shuttle Radar Topography Mission, are so improving DEM accuracy and availability that digital terrain-modeling eventually will be able to help quantify the landslide hazard over extensive areas.

Identify Hazardous Terrain

Research on landslide process and form is developing a spatial understanding of failure by synthesizing hazard, potential, and zone maps from information stored in digital-map databases [Brabb, 1984]. Pike *et al.* [1994], for example, estimated the relative susceptibility to lateral-spread failure over a large tract of coastal California; their statistical model combined age and sand content of geologic units with ground slope and distance to the nearest stream. Computer-compiled maps of California cities now show zones subject to earthquake-initiated landslides, where engineering reports are mandated before development can begin [Wilson *et al.*, 2000]. Freed from many of its qualitative limitations by GIS technology, the mapping of landslide hazard and risk has spread worldwide [Aniya, 1985; Brunori *et al.*, 1996; Fernández *et al.*, 1996; Soeters and van Westen, 1996; Cross, 1998; Carrara *et al.*, 1999; Dhakal *et al.*, 2000]. Here we examine three aspects of slope-instability mapping.

Inventory Existing Failures. One clue to the location of future landslides is the distribution of past movement (Plate 4B) [Nilsen *et al.*, 1979; Wieczorek, 1984; Jennings and Siddle, 1998]. The State of California's DMG responded to the 1983 legislative mandate [Barrows, 1993] by publishing sets of 1:24,000-scale landslide maps for urban areas. In addition to an interpretive map divided into areas of varying hazard, each set includes an inventory map that shows old landslides—including both the scar and its deposit. CDMG classifies every landslide by level of activity (from active or historic, through dormant, to relict), materials involved, and type of movement [Wills and Majmundar, 2000]. Geologists deduce these properties from landslide morphology [Keaton and DeGraff, 1996], the distinctness of which further ranks each interpretation as definite, probable, or questionable [Cruden and Varnes, 1996]. Reconnaissance inventories lack such fine detail, but have the advantage of covering much larger areas [Nilsen, 1973]. All inventory maps are valuable. Recast as a digital-map database [Amanti *et al.*, 1996; Roberts *et al.*, 1999], for example, an inventory can be put on the Web with an overlay of labeled streets, so that city residents can locate their homes in relation to known landslides [Howell *et al.*, 1999b].

Predict Location of Deep-Seated Failure. An inventory reveals the extent of past movement and thus the probable locus of future activity within old landslides, but not the likelihood of failure for the much larger area between them. Inventories, especially of the larger, deeper slides identified by airphoto interpretation [Nilsen, 1973], can be combined with other data to prepare derivative maps showing landslide potential or susceptibility [Brabb *et al.*, 1972; Wilson *et al.*, 2000; Pike *et al.*, 2001]. To interpret relative potential from airphotos and

---

Plate 4. Creating an urban landslide-hazard map. Part of the city of Oakland, California; maps are about 2 km across. **A**. Geology, showing actively creeping NNW-striking Hayward Fault Zone (at eastern margin of unit KJfm) and the area's complexity; units identified in Table 1. **B**. Inventory of old landslide deposits (orange polygons) and locations of post-1970 landslides (red dots) on uplands east of the fault and on gentler terrain to the west; shaded relief from 10-m DEM. **C**. Landslides and deposits (from **B**) and 1995 land use at 100-m resolution. Yellow, residential; green, forest; tan, scrub; blue, major highway; pink, school; orange, commercial; brown, public institution; white, vacant and mixed use; road net in gray. **D**. Values of relative susceptibility at 30-m resolution computed as continuous variable and mapped from low to high as gray, 0.00; purple, 0.01–0.04; blue, 0.05–0.09; green, 0.10–0.19; yellow, 0.20–0.29; light-orange, 0.30–0.39; orange, 0.40–0.54; red, ≥0.55. After Pike *et al.* [2001].

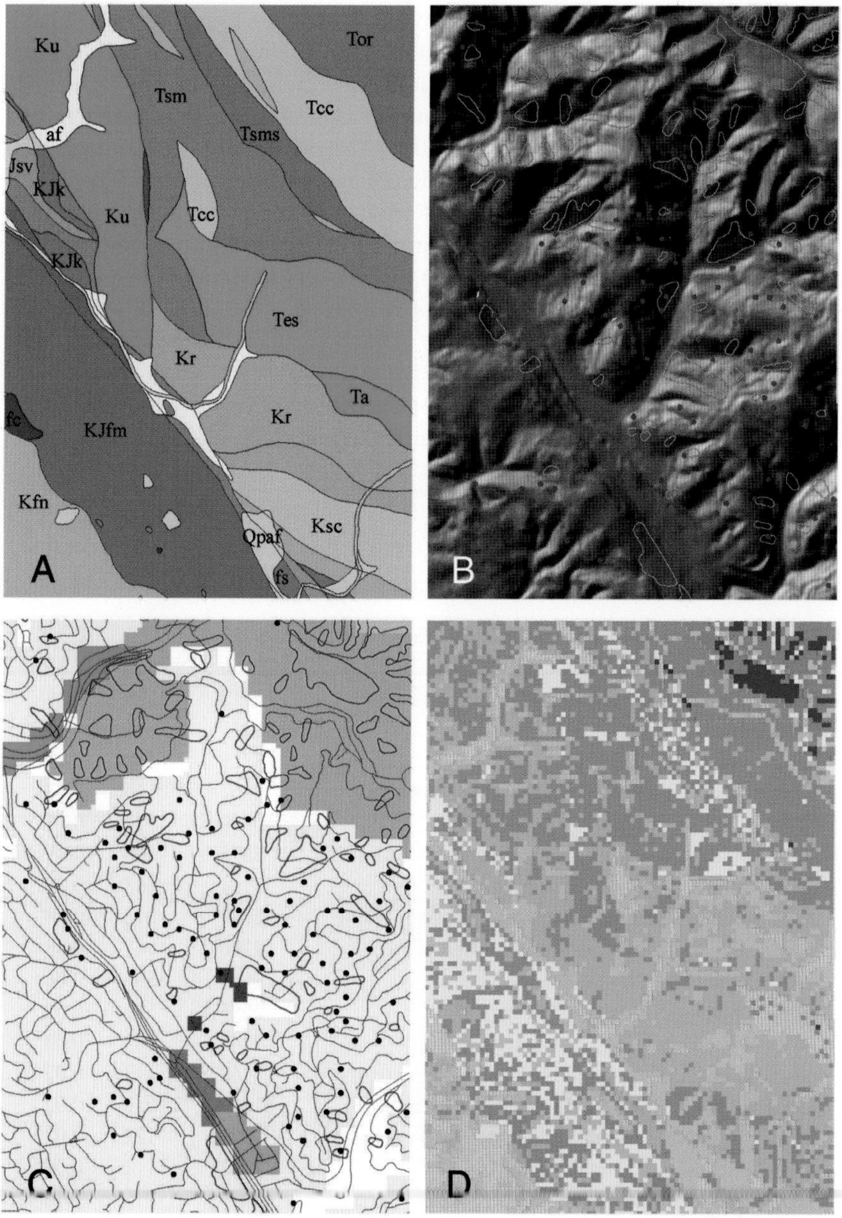
Ku
Tor
Tsm
Tcc
af
Tsms
Jsv
KJk
Ku
Tcc
KJk
Tes
Kr
Ta
fc
KJfm
Kr
Kfn
Qpaf
Ksc
fs
A
B
C
D

field observations, CDMG considers bedrock type and lithologic properties related to slope failure, structural framework, and the behavior of slopes. Actively moving material is assumed weakest and thus most likely to fail, whereas apparently stable alluvium and low-relief bedrock are assumed from their lack of landslide features to have the lowest potential. Stability between extremes is evaluated subjectively from field data. A four-fold scale indicating relative capacity of slopes to resist failure divides the map into areas of very low, low, moderate, and high potential [Wills and Majmundar, 2000]. Other types of landslide maps estimate hazard in greater detail and by different approaches.

Pike *et al.* [2001] devised an index of relative susceptibility to landsliding for the Project Impact city of Oakland, an area of diverse land-use, terrain, and geology. The statistical model is based on the frequent observation that rock type and then surface slope are the two site characteristics that most closely control large failures by sliding and earth flow [Brabb *et al.*, 1972; Wieczorek, 1984;

TABLE 1. Mean Spatial Frequency of Old Landslide Deposits in Selected Geologic Units in the City of Oakland, California

Map symbol	Geologic Map Unit [Graymer, 2000]	Number of 30m Grid Cells		
		In map unit	On deposits	Deposits / map unit
Tus	unnamed sedimentary & volcanic rocks	99233	35956	0.36
Tor	Orinda Formation	35166	9682	0.28
KJfm	Franciscan Complex mélange - undivided	12212	2559	0.21
Th	Hambre Sandstone	26341	3454	0.13
Tsm	unnamed glauconitic mudstone	3389	438	0.13
Tsms	Tsm, but mudstone-siltstone & sandstone	362	46	0.13
Tcc	Claremont Shale	10590	1177	0.11
Ksc	Shephard Creek Formation	5675	508	0.09
KJk	Knoxville Formation	8164	663	0.08
Jsv	keratophyre, qtz keratophyre above ophiolite	15627	1212	0.08
Ku	Great Valley sequence - undifferentiated	12706	965	0.08
Kr	Redwood Canyon Formation	27503	1697	0.06
Tes	Escobar Sandstone (Eocene)	2513	141	0.06
fs	Franciscan Complex sandstone	3441	109	0.03
Ta	unnamed glauconitic sandstone	163	3	0.02
Qpaf	alluvial fan & fluvial deposits (Pleistocene)	61867	1010	0.02
Kfn	Franciscan Complex, Novato Quarry terrane	7879	122	0.02
fc	Franciscan Complex chert	323	1	0.00
af	artificial fill (Historic)	65934	15	0.00

For explanation see Plate 4A, Figure 15, and text

Massari and Atkinson, 1999]. GIS-based calculations (Figure 15, Plate 4) combine 120 geologic units [Graymer, 2000], ground slope from a 30-m DEM [Graham and Pike, 1998], and 6700 old landslide deposits exclusive of debris flows [Nilsen, 1973]. The resulting index of susceptibility, displayed as an eight-color map, is computed as a continuous variable over the 870 km^2 area at a comparatively fine 30-m resolution (Plate 4D). The model improves upon landslide inventories and other maps of susceptibility by distinguishing the degree of hazard both between and within landslide deposits.

After Brabb *et al.* [1972], Pike *et al.* [2001] quantified susceptibility as the spatial frequency of old landslide deposits (Table 1), adjusted locally by steepness of the topography. Susceptibility of terrain between the old deposits is read from a histogram for each geologic unit, as the percentage of 30-m cells in each one-degree slope interval that coincide with the deposits (Figure 15). Susceptibility within landslide deposits is this percentage raised by a multiplier derived from the spatial frequency of recent failures (Plate 4B, C). Although applicable anywhere the three basic ingredients—geology, slope, and existing landslides (Plate 4A–C)—exist in digital-map form, the model can be refined from more detailed inventories and by adding measures to better predict recent failures in developed terrain (Plate 4C). Further predictive power may reside in seismic shaking, distance to nearest road (a measure of human changes to the landscape), and terrain elevation, aspect, relief, and curvature [Pike, 1988; Miles and Keefer, 2000].

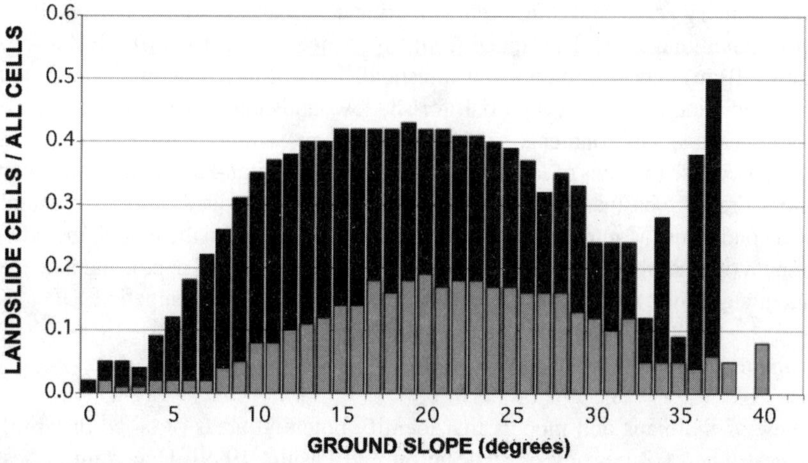

Figure 15. Landslide susceptibility of two geologic units in the Oakland, California, area contrasted graphically by spatial frequency of prior failure. Number of 30-m grid cells on old landslide deposits in the unit divided by all cells in the unit, as a function of ground slope in 1° intervals. Susceptibility varies markedly with slope (for mean spatial frequencies see Table 1). Hambre Sandstone (gray, map unit Th) is less susceptible than the Tertiary sedimentary and volcanic rocks (black, unit Tus; both units crop out NE of area in Plate 4A). After Pike *et al.* [2001].

Predict location of debris flow. The likelihood of shallow landsliding is modeled differently from deeper failure [Costa and Wieczorek, 1987]. Because debris flows commonly mobilize within steep hillside concavities, terrain geometry can indicate potential sites of future activity [Pike, 1988; Mark and Ellen, 1995; Campbell *et al.*, 1998]. Several DEM-based techniques have been proposed.

The physically-based SHALSTAB model of Dietrich *et al.* [1993] and Montgomery *et al.* [1998, 2001] routes surface water through convergent and divergent topography identified by measures of slope gradient and curvature. This approach estimates soil moisture and strength. It does not define the hazard itself, but creates a map of the relative steady-state precipitation needed to raise soil pore-pressures to the level where debris flow is likely. Locations with the lowest *critical rainfall*, or amount of precipitation required for instability, are assumed to be the most likely to fail. Low-, medium-, and high-hazard categories are assigned empirically from the frequency of landslide scars observed in each range of critical rainfall on the map. The high-hazard category includes most areas revealed by field or airphoto survey to have hosted shallow landslides historically, plus other areas that resemble the unstable locations. Medium-hazard areas are those remaining that are likely to contribute shallow landslides. Low-hazard areas, which have not failed and are unlikely to fail, are considered unconditionally stable.

To avoid designating too large an area as high hazard, the model requires fine-scale, preferably 2-m-resolution, DEMs derived from airborne LIDAR [Montgomery *et al.*, 2001], as well as a critical-rainfall threshold [Wilson, 1997] below a nominal range determined from experience to date. Coarser DEMs (resolution ≥10 m) may warrant a lower threshold. Tests of the model indicate reasonable prediction of areas at high risk for shallow landsliding over a wide range of lithologies and environments, including an urban setting [Montgomery *et al.*, 2001]. Because the rates of landsliding associated with high-hazard categories vary among drainage basins, specific risk-assessments require added calibration against local conditions [Montgomery *et al.*, 1998]. Enhancements to the model, including strength of plant roots [Schmidt *et al.*, 2001], prediction of soil depth, and a debris-runout algorithm, require further measurements from field investigation.

Estimate Temporal Probabilities

Few of the maps and models that identify potential areas of slope instability suggest when failure may occur [Schrott and Pasuto, 1999]. Uncertain at best, predictions of timing for earth-science phenomena risk raising public expectations that cannot be met [Oreskes, 1999]. Landslides resemble earthquakes in that their occurrence is inevitable and the overall location of the activity is known, but specific times and places are not known. Faced with this uncertainty, landslide scientists have little recourse except probability [Haneberg, 2000; Zoback, 2001].

Campbell *et al.* [1998], for example, modeled time-to-failure for debris flow over the course of a single rainstorm in a hilly part of the Oakland Project Impact area. Based on regression analysis and a statistical survivor-function, their spatial model incorporates data on slope, shear resistance, thickness of colluvium, and a time-dependent variable representing the cumulative effect of rainfall duration and rate. The input observations are from 11 debris-flow sites distributed across the San Francisco Bay region [Ellen and Wieczorek, 1988]. Each site has a known time of failure during the 3–5 January 1982 storm and a continuous record of precipitation from a nearby rain gauge. Highly storm-dependent and based on a small sample, this probabilistic model of shallow landsliding has yet to be tested by a major storm in the Oakland area. Conceivably the Campbell *et al.* [1998] approach could be adapted to forecast deeper failures, which respond to seasonal and multi-year fluctuations in groundwater levels.

Predicting slope instability begins with a record of mapped landslides. Accurate times of movement are included in few inventories not made soon after a failure [Reneau *et al.*, 1986; Godt and Savage, 1999] and are elusive for older (prehistoric) slides. Most records of recent landsliding, even in cities (where failures are likely to have been disruptive and thus reported), are too short and fragmentary to identify dominant patterns of rainfall or seismicity. Documentation is improving [Lang *et al.*, 1999; Coe *et al.*, 2000; Guzzetti, 2000], and recently Glade [1996], Schrott and Pasuto [1999], and Krapiec and Margielewski [2000] were able to correlate past episodes of landsliding with specific storms and floods. When eventually climate records establish time-frequency probabilities for major storms or high-rainfall seasons, and geophysical research better constrains the recurrence intervals of earthquakes, rainstorm and seismic records can be compared to model which mechanism may have caused the more landslides.

Provide Real-Time Warning

In the absence of temporal predictions, automated data-collection systems can alert the urban public to potential landsliding. A network described by Keefer *et al.* [1987] and Wilson *et al.* [1993], operated in the San Francisco Bay region from 1986 to 1995, provided public warnings when storm rainfall approached or reached levels that trigger debris flows. Precipitation thresholds likely to cause landsliding (Figure 16) were determined from the correlations observed between landslide occurrence and the intensity and duration of rainfall [Cannon and Ellen, 1985]. The warning system collected data on soil moisture from USGS instruments, installed for another project, and rainfall data from a regional array of radio-telemetered gauges sponsored by the NWS [Keefer *et al.*, 1987]. The arrangement required only a modest initial outlay for equipment; operations and maintenance were staffed by USGS personnel assigned to current projects. NWS operations also were handled by existing personnel. The media, government officials, and the public came to rely on the warnings, which resulted in specific

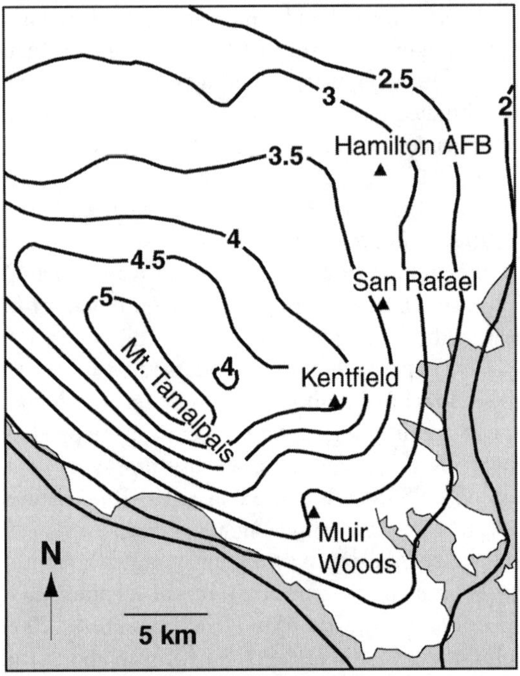

Figure 16. A tool for landslide prediction and hazard warning. Map showing rainfall thresholds calculated for southern Marin County, San Francisco Bay region, California (Figure 2). Different amounts of rain, here ranging from 50 mm (2 in) to 125 mm (5 in), in the same six-hour period are required to upset the slope-stability equilibrium in different localities—thereby triggering debris flows. Rainfall contours at 12.5 mm (0.5 in) intervals. Triangles are rain-gauge locations. After Wilson and Jayko [1997].

actions, such as evacuating neighborhoods deemed at risk. The system was terminated in 1995 after budget cuts and staff reductions in both agencies.

Technological advances in GIS [Lazzari and Salvaneschi, 1999], meteorology [Wilson, 1997], and communications [Reid and LaHusen, 1998] have led to second-generation warning systems. In the U.S., precipitation estimates from NEXRAD Doppler radar can be telemetered immediately by NWS to network centers for analysis, while cooperating agencies issue warnings of potential landslide activity. Likely channels for communication are NOAA weather radio, pagers, telephone broadcast systems, commercial radio, television, and the Internet. These new technologies have yet to be deployed on the scale of the former array in the San Francisco Bay region. Besides the monitoring of Highway 50 in eastern California [Reid and LaHusen, 1998], current installations in the U.S. include a system to warn of volcanic mudflows (lahars) in southwest Washington. Scott *et al.* [1995] and Sisson *et al.* [2001] have described the multiple lahar deposits, some as young as 600 years, that flowed from active Mt. Rainier and underlie the densely populated valley of the Puyallup River. Upon detection of a lahar by an acoustic monitor, the new system is designed to automatically notify local county officials and prompt immediate, preplanned, evacuation and other emergency responses. Warning systems are only the most recent element of strategies to reduce landslide risk.

Adopt a Comprehensive Plan

Cities can not reduce landslide losses solely through science and technology, nor will remedial measures be effective if carried out piecemeal. The various urban interests threatened by landsliding need to join in a common purpose [Spiker and Gori, 2000]. Scientists and engineers must learn to work with their constituents in government, the private sector, and among the general public, and all participants need to understand their different, often conflicting, roles and responsibilities. Success requires the political will to enact legislation and commit the necessary resources. Also needed is a distributed, less centralized, approach to organization that builds on grass-roots support among city dwellers and encourages a sense of personal and civic accountability. An effective program includes technical, planning, and policy elements ranging from basic research to site remediation [Swanston and Schuster, 1989; Schuster, 1995]. Few of them are new [USGS, 1982; Committee on Ground Failure Hazards, 1985; Jones, 1992].

Several of these elements are present in an effort to arrest mudflows on the steep hillsides of Rio de Janeiro. A team of university researchers from Brazil and Canada developed a way to recycle discarded automobile tires into retaining walls at less than one-third the cost of conventional walls [Shore, 1999]. Although local inhabitants initially were skeptical, the scrap-tire walls withstood torrential rain better than some of their concrete counterparts, whereupon *favela* residents began to upgrade their dwellings and even redeveloped a local square (Praca Projeto Pneus; Tire Project Square). Civil engineers on the project reported that social scientists from a local university played a key role in this success. The academics' lengthy association with the slum community, their knowledge of local organizations and perceptions, and efforts to ensure that hillside residents understood and accepted the new technology helped the engineers to access the construction site and protected the tire walls from vandalism.

A program to reduce landslide loss in the world's cities probably is best carried out by individual nations, or as Hausmann [2001] observes in the case of Central America, by groups of neighboring small countries that share a common hazard but have limited productive capacity. Coordination of a global program of loss reduction by the United Nations or other international organization is unlikely and perhaps unworkable. The optimal approach to mitigation will vary by country and region, according to economic and societal circumstances, but as generalized for the U.S. [Spiker and Gori, 2000] and other industrialized regions, might include:

- a national forum linking science, engineering, the business community, policy- and decision-makers, the media, and the public to raise and resolve key issues surrounding the landslide threat;
- a comprehensive, widely endorsed, plan that addresses all or most of those issues and involves all parties from the outset;
- a nationwide program to execute the plan, as a flexible venture among the private sector and governments at every level; and
- evaluation of results, readjusting priorities and resources to attain the desired objectives and extend the program where appropriate.

Scientists and engineers have a dual role in such a plan. Besides expanding the knowledge of slope instability and its distribution, landslide professionals contribute to loss reduction by narrowing the gap between technical knowledge and policymaking needs. Howell *et al.* [1999a] suggest that this second responsibility comprises three overarching tasks:

- demonstrate the savings in both public and private resources that follow prudent land-use planning and mitigation;
- communicate the message all the way through the decision-making process, at all levels; and
- provide the technical information, landslide maps or computer scenarios, and hazard warnings requested by local officials.

CONCLUDING OBSERVATIONS

The unwanted partnership between landsliding and the world's cities will continue to levy its environmental tax on property and human life, most heavily in less-developed regions [Hausmann, 2001, p. 52]. Three uncertainties cloud a more optimistic prospect. Stabilization and eventual reduction of the growing world population that accounts for at least half the problem will not occur for several decades and cannot be guaranteed [Smail *et al.*, 1997]. Nor can we accurately foresee the effects of an increasing variability of the climate, which could include more severe and numerous storms and other extreme-weather conditions. Finally, much of the progress toward landslide-resistant cities remains at the whim of human nature and the inherent unpredictability of the political process. Despite these uncertainties, the danger can be eased.

A sustained commitment worldwide will be required to slow the growth in the number and size of cities and the commensurate rise in landslide damage. Needed throughout this century are technical assistance, political oversight, land reform, continued investment in population management, and changes in governance that promote social stability. Operationally, a good beginning can be made through simple avoidance: improving land-use by refusal to develop hazardous terrain could reduce urban landsliding as effectively and perhaps at a lower cost than any other means. Such a clear break with past practice will entail some sacrifice, and any benefit-cost judgments must include the value of land set aside for non-urban use.

Although scientists and engineers can not be expected to address demographic components of the landslide problem, the efforts of technical professionals are essential to safer cities. Living with landslides has become a reality in many communities, and urban hillsides can continue to be graded for development where property values sustain the cost. Equally important is the identification of past movement and prediction of future landslides. As the relation of failure processes to geology, soil moisture, terrain, and other landscape attributes is better understood and techniques of hazard-mapping mature, the numerical ranking of slope stability will become feasible over large areas. Although prediction of seismically generated failure requires more accurate forecasting of earthquakes, some warning of storm-triggered landsliding already is possible through real-time monitoring and satellite-borne measurements of atmospheric conditions.

Reducing the landslide risk through technological advances can succeed only within a receptive political, economic, and social environment, which is contingent upon a number of conditions. Preventing landslide hazards from becoming landslide disasters may be a realistic expectation for the world's cities, providing:

- natural hazards attract the persistent attention now accorded such civic concerns as crowding, poverty, crime, and competing environmental problems;
- the true cost to society of landsliding itself, as distinct from the more attention-commanding events that initiate it, becomes widely realized;
- landslide experts communicate hazard and risk with enough success to elicit concerted action from government, the private sector, and the public;
- cities prepare for landslides as deliberately as many communities now anticipate earthquakes, hurricanes, volcanic eruptions, and floods;
- the knowledge derived from methods that identify landslide-prone terrain is translated into enforceable regulation of urban land-use;
- landslide insurance can reduce the burden of private liability by distributing the financial risk; and
- a growing collective sense of individual responsibility eventually replaces the top-down regulation that at first may be needed to reduce landsliding.

Acknowledgments. We dedicate this chapter to Earl Brabb, Bob Schuster, and the memory of David Varnes, each in his own way a master of our craft. The work was funded by the Landslide Hazards Reduction Program and the National Cooperative Geologic Mapping Program of the U.S. Geological Survey. We thank Ray Wilson and John Williams for helpful reviews, and Steven Sobieszczyk and Zenon Valin for technical support. Gerry Wieczorek presented our "landslides and cities ..." poster at the 2000 Spring AGU meeting.

REFERENCES

Ahmad, Rafi, J.V. DeGraff, and J.P. McCalpin, Landslide Loss Reduction: A Guide for the Kingston Metropolitan Area, Jamaica, Unit for Disaster Studies, *Publication 6*, University of the West Indies, Mona, Kingston, 1999: http://www.oas.org/en/cdmp/document/kma/udspub6.htm.

Alexander, D.E., Urban landslides, *Progress in Physical Geography, 13,* 157-191, 1989.

Alexander, D.E., *Natural Disasters*, 632 pp., Kluwer, Dordrecht, 1993.

Amanti, Marco, Nicola Casagli, Filippo Catani, Maurizio D'Orefice, and Guido Motteran, *Guida al Censimento die Fenomeni Franosi ed Alla Loro Archiviazione*, Servizio Geologico, Istituto Poligrafico a Zecca Stato, Roma, 109 pp., 1996.

Anderson, L.R., J.R. Keaton, and W.G. Wells II, *The Utah Landslides, Debris Flows, and Floods of May and June 1983*, 96 pp., National Academy Press, Washington, D.C., 1984.

Aniya, Masamu, Landslide-susceptibility mapping in the Amahata River basin, Japan, *Annals of the Association of American Geographers, 75,* 102-114, 1985.

Baldwin, J.E. II, H.F. Donley, and T.R. Howard, On debris flow / avalanche mitigation and control, San Francisco Bay area, California, *Reviews in Engineering Geology, 7,* 223-236, 1987.

Barrows, A.G., California's Landslide Hazard Identification Program, *California Geology, 46,* 132-140, 1993.

Barrows, A.G., S.S. Tan, and P.J. Irvine, Damaging landslides related to the intense rainstorms of January-February 1993, southern California, *California Geology, 46,* 123-131, 1993.

Batty, Michael, and Paul Longley, *Fractal Cities: A Geometry of Form and Function*, 394 pp., Academic Press, San Diego, CA, 1994.

Bell, Brenda, The liquid earth, *The Atlantic Monthly, 283, January*, 58-72, 1999.

Bernknopf, R.L., R.H. Campbell, D.S. Brookshire, and C.D. Shapiro, A probabilistic approach to landslide hazard mapping in Cincinnati, Ohio, with applications for economic evaluation, *Bulletin of the Association of Engineering Geologists, 25,* 39-56, 1988.

Board on Sustainable Development, *Our Common Journey: A Transition Toward Sustainability*, 363 pp., National Research Council, Policy Division, National Academy Press, Washington, D.C., 1999.

Bongaarts, John, Population: ignoring its impact, *Scientific American, 286, no. 1*, 67-69, 2002.

Bongaarts, John, and R.A. Bulatao (Eds.), *Beyond Six Billion: Forecasting the World's Population*, 348 pp., Commission on Behavioral and Social Sciences and Education, National Academy Press, Washington, D.C., 2000.

Brabb, E.E., Innovative approaches to landslide hazard and risk mapping, in *International Landslide Symposium Proceedings, Toronto, 1*, pp. 307-323, 1984; reprinted in *Proceedings, IVth International Conference and Field Workshop in Landslides, Tokyo, Japan,* pp. 17-22, 1985.

Brabb, E.E., Hazard maps are not enough, in *Proceedings, Second Pan-American Symposium on Landslides, Rio de Janeiro, November 10-14, 1997*, 3, 277-282, 1997.

Brabb, E.E., and B.L. Harrod (Eds.), Landslides: Extent and economic significance, *Proceedings, Symposium on Landslides, 28th International Geological Congress, 17 July, 1989, Washington, D.C.,* Balkema, Rotterdam, 385 pp., 1989.

Brabb, E.E., E.H. Pampeyan, and M.G. Bonilla, Landslide susceptibility in San Mateo County, California, U.S. Geological Survey, *Miscellaneous Field Studies Map MF-360*, scale 1:62,500, 1972.

Brunori, F., Nicola Casagli, S. Fiaschi, C.A. Garzonio, and S. Moretti, Landslide hazard mapping in Tuscany, Italy: an example of automatic evaluation, in *Geomorphic Hazards*, edited by Olav Slaymaker, pp. 56-67, Wiley, Chichester and New York, 1996.

Bunge, William, *Theoretical Geography*, 2nd ed., 285 pp., Gleerup, Lund, Sweden, 1966.

Burby, R.J. (Ed.), *Cooperating With Nature: Confronting Natural Hazards With Land-use Planning for Sustainable Communities*, 368 pp., Joseph Henry Press (National Academy Press), Washington, D.C., 1998.

Butler, David, *Disasters by Design, a Reassessment of Natural Hazards in the United States: a Bibliography*, 2000: http://www.colorado.edu/hazards/ assessbib.html#history

Caldwell, J.C., The contemporary population challenge, *Expert Group Meeting on Completing the Fertility Transition, March 11-14,* UN/POP/CFT/2002/BP/1, 12 pp., Population Division, United Nations Secretariat, New York, 2002: http://www.un.org/esa/population/publications/completingfertility/Caldwellpaper.PDF

California Division of Mines and Geology (CDMG), Guidelines for Evaluating Seismic Hazards in California, Department of Conservation, Sacramento, *Special Publication 117,* 74 pp., 1997.

Campbell, R.H., Soil slips, debris flows, and rainstorms in the Santa Monica Mountains, Southern California, U.S. Geological Survey, *Professional Paper 851*, 51 pp., 1975.

Campbell, R.H., R.L. Bernknopf, and D.R. Soller, Mapping time-dependent changes in soil slip-debris flow probability, U.S. Geological Survey, *Miscellaneous Investigations Map I-2586,* 16 pp., 1998.

Cannon, S.H., and S.D. Ellen, Rainfall conditions for abundant debris avalanches, San Francisco Bay region, California, *California Geology, 38*, 267-272, 1985.

Carrara, Alberto, Fausto Guzzetti, Mauro Cardinali, and Paola Reichenbach, Use of GIS technology in the prediction and monitoring of landslide hazard, *Natural Hazards, 20,* 117-135, 1999.

Clarke, K.C., and L.J. Gaydos, Loose coupling a cellular automaton model and GIS: long-term growth prediction for San Francisco and Washington/Baltimore, *International Journal of Geographical Information Science, 12,* 699-714, 1998.

Clarke, K.C., Stacy Hoppen, and L.J. Gaydos, A self-modifying cellular automaton model of historical urbanization in the San Francisco Bay area, *Environment and Planning B: Planning and Design, 24*, 247-261, 1997.

Coe, J.A., J.A. Michael, R.A. Crovelli, and W.Z. Savage, Preliminary map showing landslide densities, mean recurrence intervals, and exceedence probabilities as determined from historic records, U.S. Geological Survey, *Open-File Report 00-303*, 25 pp., scale 1:25,000, 2000: http://greenwood.cr.usgs.gov/pub/open-file-reports/ofr-00-0303/

Collin, Alexandre, *Recherche Expérimentales sur les Glissements Spontanés des Terrains Argileux, Accompagnées de Considérations sur Quelques Principes de la Mécanique Terrestre*, 168 pp., 21 atlas plates, Carilian-Goeury et Dalmont, Paris, 1846. (Translated by W.R. Schriever, with a memoir on Alexandre Collin by A.W. Skempton, as *Landslides in Clays*, 160 pp., University of Toronto Press, 1956).

Committee on Ground Failure Hazards, *Reducing Losses From Landsliding in the United States*, Commission on Engineering and Technical Systems, National Research Council, National Academy Press, Washington, D.C., 41 pp., 1985.

Cooke, R.U., *Geomorphological Hazards in Los Angeles, a Study of Slope and Sediment Problems in a Metropolitan County*, 206 pp., Allen & Unwin, London, 1984.

Costa, J.E., and G.F. Wieczorek (Eds.), Debris flows / avalanches: process, recognition, and mitigation, *Reviews in Engineering Geology, 7*, 239 pp., 1987.

Cotton, W.R., and D.A. Cochrane, Love Creek landslide disaster January 5, 1982, Santa Cruz County, *California Geology, 35,* 153-157, 1982.

Cross, Martin, Landslide susceptibility mapping using the matrix assessment approach: a Derbyshire case study, in Geohazards in Engineering Geology, edited by J.G. Maund and Malcolm Eddleston, pp. 247-261, *Engineering Geology Special Publication 15,* Geological Society, London, 1998.

Crozier, M.J. (Ed.), *Landslides: Causes, Consequences, and Environment*, 252 pp., Croom Helm, London, 1986.

Cruden, D.M., and D.J. Varnes, Landslide types and processes, in *Landslides: Investigation and Mitigation, Special Report 247*, edited by R.L. Schuster and A.K. Turner, pp. 36-75, Transportation Research Board, National Academy of Sciences, Washington, D.C., 1996.

Degg, M.R., Natural hazards in the urban environment: the need for a more sustainable approach to mitigation, in Geohazards in Engineering Geology, edited by J.G. Maund and M. Eddleston, pp. 329-337, *Engineering Geology Special Publication 15*, Geological Society, London, 1998.

Denlinger, R.P., and R.M. Iverson, Flow of variably fluidized granular masses across three-dimensional terrain I. Numerical predictions and experimental tests, *Journal of Geophysical Research, 106B,* 553-566, 2001.

Dhakal, A.S., Takaaki Amada, and Masamu Aniya, Landslide hazard mapping and its evaluation using GIS: an investigation of sampling schemes for a grid-cell based quantitative method, *Photogrammetric Engineering and Remote Sensing, 66,* 981-989, 2000.

Dietrich, W.E., C.J. Wilson, D.R. Montgomery, and James McKean, Analysis of erosion thresholds, channel networks, and landscape morphology using a digital elevation model, *Journal of Geology, 101,* 259-278, 1993: http://socrates.berkeley .edu/~geomorph/shalstab/

Dikau, Richard, Denys Brunsden, Lothar Schrott, and M.L. Ibsen (Eds.), *Landslide Recognition: Identification, Movement and Causes*, 251 pp., Wiley, Chichester, 1996.

Duncan, J.M., G. Lefebvre, and P. Lade, *The Landslide at Tuve, Near Göteborg, Sweden, on November 30, 1977*, 25 pp., National Academy Press, Washington, D.C., 1980.

Dyson, Tim, On the future of human fertility in India, *Expert Group Meeting on Completing the Fertility Transition, March 11-14,* UN/POP/CFT/2002/CP/6, 20 pp., Population Division, United Nations Secretariat, New York, 2002: http://www.un.org/esa/population/publications/completingfertility/Dysonpaper.PDF.

Eckel, E.B. (Ed.), *Landslides and Engineering Practice*, *Special Report 29*, 232 pp., Highway Research Board, National Research Council, Washington, D.C., 1958.

Eisbacher, G.H., and J.J. Clague, Urban landslides in the vicinity of Vancouver, British Columbia, with special reference to the December 1979 rainstorm, *Canadian Geotechnical Journal, 18*, 205-216, 1981.

Ellen, S.D., Description and mechanics of soil slip/debris-flows in the storm, in Landslides, floods, and marine effects of the storm of January 3-5, 1982, in the San Francisco Bay Region, California, edited by S.D. Ellen and G.F. Wieczorek, pp. 63-112, U.S. Geological Survey, *Professional Paper 1434*, 1988.

Ellen, S.D., and G.F. Wieczorek (Eds.), Landslides, floods, and marine effects of the storm of

January 3-5, 1982, in the San Francisco Bay region, California, U.S. Geological Survey, *Professional Paper 1434*, 310 pp., 1988.

Eyles, R.J., M.J. Crozier, and R.H. Wheeler, Landslips in Wellington City, *New Zealand Geographer, 34*, 58-74, 1978.

Fainstein, Susan, and Scott Campbell (Eds.), *Readings in Urban Theory*, 672 pp., Blackwell, Cambridge, 2001.

Federal Emergency Management Agency (FEMA), *Project Impact: Building a Disaster-Resistant Community*, 2000: http://www.fema.gov/impact/impact00.htm.

Fernández, Tomás, Clemente Irigaray, and José Chacón, GIS analysis and mapping of landslides determinant factors in the Contraviesa area (Granada, Southern Spain), in *Landslides, Proceedings of the 8th International Conference and Field Trip, 8th, Granada, September 27-28*, edited by José Chacón, Clemente Irigaray, and Tomás Fernández, pp. 141-151, Balkema, Rotterdam, 1996.

Finkle, J.L., Impact of the 1994 International Conference on Population and Development, *Expert Group Meeting on Completing the Fertility Transition, March 11-14*, UN/POP/CFT/2002/BP/2, 8 pp., Population Division, United Nations Secretariat, New York, 2002:
http://www.un.org/esa/population/publications/complet-ingfertility/Finklepaper.PDF.

Flageollet, J.C., *Les Mouvements de Terrain et Leur Prévention*, 224 pp., Masson, Paris, 1989.

Fleming, R.W., and F.A. Taylor, *Estimating the Costs of Landslide Damage in the United States*, U.S. Geological Survey, *Circular 832*, 21 pp., 1980.

Flentje, P.N., R.N. Chowdhury, and P. Tobin, Management of landslides triggered by a major storm event in Wollongong, Australia, in *Debris-flow Hazards Mitigation: Mechanics, Prediction, and Assessment*, edited by G.F. Wieczorek and N.D. Naeser, pp. 479-487, Balkema, Rotterdam, 2000.

Fornos, Werner, *World Population Overview 2000: Population and the Urban Future*, 18 pp., The Population Institute, Washington, D.C., 2000.

Ghilardi, P., L. Natale, and F. Savi, Debris-flow propagation on urbanized alluvial fans, in *Debris-flow Hazards Mitigation: Mechanics, Prediction, and Assessment*, edited by G.F. Wieczorek and N.D. Naeser, pp. 471-477, Balkema, Rotterdam, 2000.

Glade, Thomas, Establishing the frequency and magnitude of landslide-triggering rainstorm events in New Zealand, *Environmental Geology, 35*, 160-174, 1996.

Glicken, Harry, *Rockslide-Debris Avalanche of May 18, 1980, Mount St. Helens Volcano, Washington,* U.S. Geological Survey, *Open-File Report 96-677*, 98 pp., 1996: http://vulcan.wr.usgs.gov/Projects/Glicken/framework.html.

Godt, J.W., and W.Z. Savage, El Niño 1997-98: direct costs of damaging landslides in the San Francisco Bay region, in *Landslides*, edited by J.S. Griffiths, M.R. Stokes and R.G. Thomas, pp. 47-55, Balkema, Rotterdam, 1999.

Gottmann, Jean, *Megalopolis, the Urbanized Northeastern Seaboard of the United States*, 810 pp., The MIT Press, Cambridge, 1961.

Graham, S.E., and R.J. Pike, Slope maps of the San Francisco Bay region, California: a digital database, U.S. Geological Survey, *Open-file Report 98-766*, 17 pp., 1998: http://wrgis.wr.usgs.gov/open-file/of98-766/index.html.

Graymer, R.W., Geologic map and map database of the Oakland metropolitan area, Alameda, Contra Costa, and San Francisco Counties, California, U.S. Geological Survey, *Miscellaneous Field Studies Map MF-2342*, scale 1:50,000, 2000:

http://geopubs.wr.usgs.gov/map-mf/mf2342/.

Griffiths, J.S., Proving the occurrence and cause of a landslide in a legal context, *Bulletin of Engineering Geology and the Environment, 58,* 75-85, 1999.

Guzzetti, Fausto, Landslide fatalities and the evaluation of landslide risk in Italy, *Engineering Geology, 58,* 89-107, 2000.

Haneberg, W.C., Deterministic and probabilistic approaches to geologic hazard assessment, *Environmental and Engineering Geoscience, 6,* 209-226, 2000.

Harp, E.L., and R.W. Jibson, Inventory of Landslides triggered by the 1994 Northridge, California, earthquake, U.S. Geological Survey, *Open-File Report 95-213,* 17 pp., map scales 1:100,000 and 1:50,000, 1995: http://geology.cr.usgs .gov/pub/open-file-reports/ofr-95-0213/TABLE.HTML.

Harp, E.L., R.C. Wilson, and G.F. Wieczorek, Landslides from the February 4, 1976, Guatemala earthquake, U.S. Geological Survey, *Professional Paper 1204-A,* 35 pp., map scale 1:50,000, 1981.

Hausmann, Ricardo, Prisoners of geography, *Foreign Policy, Jan.-Feb.,* 44-53, 2001.

Haydon, W.D., Elise Mattison, and K.B. Clahan, Liquefaction zones, in Seismic Hazard Evaluation of the Cities of Oakland and Piedmont, Alameda County, CA, California Division of Mines and Geology, *Open-File Report 99-11, 1,* 3-23, scale 1:24,000, 2000: ftp://ftp.consrv.ca.gov/pub/dmg/shezp/evalrpt/OFR99-11.pdf.

Heim, Albert, *Bergsturz und Menschenleben,* 218 pp., Fretz and Wasmuth, Zürich, 1932; translated by Nigel Skermer, 1989, as *Landslides and Human Lives,* BiTech Publishers, Vancouver, 195 pp., 1989.

Highland, L.M., Landslide hazard and risk: current and future directions for the United States Geological Survey's landslide program, in *Landslide Risk Assessment,* edited by D.M. Cruden and Robin Fell, pp. 207-213, Balkema, Rotterdam, 1997.

Howell, D.G., E.E. Brabb, and Rafi Ahmad, Interest in landslide hazard information: parallels between Kingston, Jamaica and the San Francisco Bay region, in *Landslides,* edited by J.S. Griffiths, M.R. Stokes and R.G. Thomas, pp. 73-79, Balkema, Rotterdam, 1999a.

Howell, D.G., EE. Brabb, and D.W. Ramsey, How useful is landslide hazard information? Lessons learned in the San Francisco Bay region, *International Geology Review, 41,* 368-381, 1999b.

Hungr, Oldrich, S.G. Evans, M.J. Bovis, and J.N. Hutchinson, A review of the classification of landslides of the flow type, *Environmental and Engineering Geoscience, 7,* 221-238, 2001.

Hylland, M.D., and Mike Lowe, Regional landslide-hazard evaluation using landslide slopes, western Wasatch County, Utah, *Environmental Engineering and Geoscience, 3,* 31-43, 1997.

IPCC, *Climate Change 2001: The Scientific Basis,* Intergovernmental Panel on Climate Change, Working Group I, Summary for Policy Makers, Geneva, 20 pp., 2001a.

IPCC, *Climate Change 2001: Impacts, Adaptation, and Vulnerability,* Intergovernmental Panel on Climate Change, Working Group II, Summary for Policy Makers, Geneva, 22 pp., 2001b.

Iverson, R.M., and R.P. Denlinger, Flow of variably fluidized granular masses across three-dimensional terrain I. Coulomb mixture theory, *Journal of Geophysical Research, 106B,* 537-552, 2001.

Japan Landslide Society, *Landslides in Japan,* 54 pp., National Conference of Landslide Control, Tokyo, 1988: http://www.tuat.ac.jp/~sabo/lj/index.htm.

Jennings, P.J., and H.J. Siddle, Use of landslide inventory data to define the spatial location of landslide sites, South Wales, U.K., in Geohazards in Engineering Geology, edited by J.G. Maund and Malcolm Eddleston, pp. 199-211, *Engineering Geology, Special Publication 15*, Geological Society, London, 1998.

Jibson, R.W., E.L. Harp, and J.A. Michael, A method for producing digital probabilistic seismic landslide hazard maps, *Engineering Geology, 58,* 271-290, 1998.

Jones, D.K.C., Landslide hazard assessment in the context of development, in Geohazards: Natural and Man-Made, edited by G.J.H. McCall, D.J.C. Laming, and S.C. Scott, Association of Geoscientists for International Development (AGID) *Special Publication Series 15*, pp. 117-142, Chapman and Hall, London, 1992.

Jones, D.K.C., and E.M. Lee, *Landsliding in Great Britain*, 381 pp., Department of the Environment, H.M. Stationery Office, London, 1994.

Jones, F.O., Landslides of Rio de Janeiro and the Serra das Araras Escarpent, Brazil, U.S. Geological Survey, *Professional Paper 697*, 42 pp., 1973.

Keaton, J.R., and J.V. DeGraff, Surface observation and geologic mapping, in Landslides, Investigation and Mitigation, edited by A.K. Turner and R.L. Schuster, pp. 178-230, Transportation Research Board, National research Council, *Special Report 247*, Washington, D.C., 1996.

Keefer, D.K. (Ed.), The Loma Prieta, California, Earthquake of October 17, 1989: Landslides, U.S. Geological Survey, *Professional Paper 1551-C*, 128 pp., 1998.

Keefer, D.K., R.C. Wilson, R.K. Mark, E.E. Brabb, W.M. Brown III, S.D. Ellen, E.L. Harp, G.F. Wieczorek, C.S. Alger, and R.S. Zatkin, Real-time landslide warning during heavy rainfall, San Francisco Bay region, California, 12-21 February, 1986, *Science, 238,* 921-925, 1987.

Kockelman, W.J., Some techniques for reducing landslide hazards, *Bulletin of the Association of Engineering Geologists, 23,* 29-52, 1986.

Krapiec, Marek, and Wlodzimierz Margielewski, Dendrogeomorphological analysis of mass movements in the Polish flysch Carpathians, *Geologia, 26,* 141-171, 2000.

La Vopa-Creasy, C., Landslide damage: a costly outcome of the storm, in Landslides, Floods, and Marine Effects of the Storm of January 3-5, 1982, in the San Francisco Bay Region, California, edited by S.D. Ellen and G.F. Wieczorek, pp. 195-203, U.S. Geological Survey, *Professional Paper 1434*, 1988.

Lang, Andreas, José Moya, Jordi Corominas, Lothar Schrott, and Richard Dikau, Classic and new dating methods for assessing the temporal occurrence of mass movements, *Geomorphology, 30,* 33-52, 1999.

Larsen, M.C., G.F. Wieczorek, L.S. Eaton, B.A. Morgan, and Heriberto Torres-Sierra, Venezuelan debris flow and flash flood disaster of 1999 studied, *Eos, Transactions, American Geophysical Union, 82,* 572-573, 2001.

Lazzari, Marco, and Paolo Salvaneschi, Embedding a geographic information system in a decision support system for landslide hazard monitoring, *Natural Hazards, 20,* 185-195, 1999.

LeGates, Richard, and Frederic Stout (Eds.), *The City Reader*, 592 pp., Routledge, New York, 1999.

Leggett, R.F., *Cities and Geology*, 624 pp., McGraw-Hill, New York, 1973.

Leighton, F.B., Landslides and hillside development, in *Engineering Geology in Southern California*, edited by Richard Lung, and R.V. Proctor, pp. 149-206, Association of Engineering Geologists, Special Publication, Glendale, CA, 1966.

Leighton, F.B., The role of consulting geologists in urban geology, in *Environmental Planning and Geology*, edited by D.R. Nichols and C.C. Campbell, pp. 82-89, Proceedings of the Symposium on Engineering Geology in the Urban Environment, Assoc. of Engineering Geologists, October 1969, San Francisco, U.S. Dept. of Housing and Urban Development and U.S. Dept. of Interior, USGPO, Washington, D.C. 1971.

Leighton, F.B., Urban landslides: targets for land-use planning in California, in Urban Geomorphology, edited by D.R. Coates, pp. 37-60, Geological Society of America, *Special Paper 174*, Boulder, CO, 1976.

Leighton, F.B., *Mitigation of Geotechnical Litigation in California*, 274 pp., Munson Book Associates, Huntington Beach, 1992.

Leroi, Eric, Landslide hazard-risk maps at different scales: objectives, tools and developments, in *Landslides*, edited by R. Senneset, pp. 35-51, Balkema, Rotterdam, 1996.

Leveson, David, *Geology and the Urban Environment*, 386 pp., Oxford University Press, New York, 1980.

Lutz, Wolfgang, Warren Sanderson, and Sergei Scherbov, The end of world population growth, *Nature, 412,* 543-545, 2001.

Luzi, Lucia, and Floriana Pergalani, Slope instability in static and dynamic conditions for urban planning: the 'Oltre Po Pavese' case history (Regione Lombardia - Italy), *Natural Hazards, 20,* 57-82, 1999.

Mader, G.C., and D.F. Crowder, An experiment in using geology for city planning: the experience of the small comunity of Portola Valley, California, in *Environmental Planning and Geology*, edited by D.R. Nichols and C.C. Campbell, pp. 176-189, Proceedings of the Symposium on Engineering Geology in the Urban Environment, Assoc. of Engineering Geologists, October 1969, San Francisco, U.S. Dept. of Housing and Urban Development and U.S. Dept. of Interior, USGPO, Washington, D.C., 1971.

Makse, H.A., Shlomo Havlin, and H.E. Stanley, Modelling urban growth patterns, *Nature, 377,* 608-612, 1995.

Mark, R.K., Map of debris-flow probability, San Mateo County, California, U.S. Geological Survey, *Miscellaneous Investigations Series, Map I-1257-M,* scale 1:62,500, 1992.

Mark, R.K., and S.D. Ellen, Statistical and simulation models for mapping debris-flow hazard, in *Geographical Information Systems in Assessing Natural Hazards*, edited by Alberto Carrara and Fausto Guzzetti, pp. 93-106, Kluwer, Dordrecht, 1995.

Massari, Remo, and P.M. Atkinson, Modeling susceptibility to landsliding: an approach based on individual landslide type, *Transactions, Japanese Geomorphological Union, 20,* 151-168, 1999.

McCall, G.J.H., Geohazards and the urban environment, in Geohazards in Engineering Geology, edited by J.G. Maund and M. Eddleston, pp. 309-318, *Engineering Geology Special Publication 15*, Geological Society, London, 1998.

McLaughlin, John, and David Palmer, Land registration and development, *ITC Journal, 96-1*, 10-18, 1996.

McPhee, John, Los Angeles against the mountains, Chapter 3 in *The Control of Nature*, pp. 183-272, Noonday Press, Farrar, Straus and Giroux, New York, 1989; originally published in *The New Yorker*, 26 Sept. and 3 Oct. 1988.

Miles, S.B., and D.K Keefer, Evaluation of seismic slope-performance models using a regional case study, *Environmental and Engineering Geoscience, 11,* 25-39, 2000.

Mileti, Dennis, *Disasters by Design: a Reassessment of Natural Hazards in the United States*, 376 pp., Joseph Henry Press, Washington, D.C., 1999.

Miller, D.J., Coupling GIS with physical models to assess deep-seated landslide hazards, *Environmental and Engineering Geoscience, 1,* 263-276, 1995.

Mitchell, J.K. (Ed.), *Crucibles of Hazard: Mega-Cities and Disasters in Transition*, 549 pp., United Nations Press, New York and Tokyo, 1999.

Molnia, B.F., and C.A. Hallam, Open Skies aerial photography of selected areas in Central America affected by Hurricane Mitch, U.S. Geological Survey, *Circular 1181*, 82 pp. and CD-ROM, 1999.

Montgomery, D.R., Kathleen Sullivan, and H.M. Greenberg, Regional test of a model for shallow landsliding, *Hydrological Processes*, *12*, 943-955, 1998.

Montgomery, D.R., H.M. Greenberg, W.T. Laprade, and W.D. Nashem, Sliding in Seattle: test of a model of shallow landsliding potential in an urban environment, in *Land Use and Watersheds: Human Influence on Hydrology and Geomorphology in Urban and Forest Areas*, edited by M.S. Wigmosta and S.J. Burges, pp. 59-73, American Geophysical Union, Washington, D.C., 2001.

Moore, R., E.M. Lee, and A.R. Clark, *The Undercliff of the Isle of Wight: A Review of Ground Behaviour*, 68 pp., South Wight Borough Council, UK, Cross Publishing, Ventnor, 1995.

Nichols, D.R., and C.C. Campbell (Eds.), *Environmental Planning and Geology*, 204 pp., Proceedings of the Symposium on Engineering Geology in the Urban Environment, Assoc. of Engineering Geologists, October 1969, San Francisco, U.S. Dept. of Housing and Urban Development and U.S. Dept. of Interior, Washington, D.C. USGPO, 1971.

Nilsen, T.H., Preliminary photointerpretation map of landslide and other surficial deposits of the Concord 15-minute quadrangle and the Oakland West, Richmond, and part of the San Quentin 7.5-minute quadrangles, Contra Costa and Alameda Counties, California, U.S. Geological Survey, *Miscellaneous Field Studies map MF-493*, scale 1:62,500, 1973.

Nilsen, T.H., R.H. Wright, T.C. Vlasic, and W.E. Spangle, Relative slope stability and land-use planning in the San Francisco Bay region, California, U.S. Geological Survey, *Professional Paper 944,* 96 pp., map scale 1:125,000, 1979.

Oliver, Stuart, 20th-century urban landslides in the Basilicata region of Italy, *Environmental Management, 17,* 433-444, 1993.

Olshansky, R.B., *Landslide Hazard in the United States: Case Studies in Planning and Policy Development*, 176 pp., Garland Publishing, New York, 1990.

Olshansky, R.B., and J.D. Rogers, Unstable Ground: Landslide Policy in the United States, *Ecology Law Quarterly, 13,* 939-1006, 1987.

Olshansky, R.B., and J.D. Rogers, The concept of "reasonable care" on unstable hillsides, in Landslides / Landslide Mitigation, edited by J.E. Slosson, A.G. Keene and J.A. Johnson, *Reviews in Engineering Geology, 9*, 23-27, 1992.

Oreskes, Naomi, Why predict? Historical perspectives on prediction in the earth sciences, in *Prediction—science, decision making, and the future of nature*, edited by Daniel Sarewitz, R.A. Pielke, Jr., and Radford Byerly, Jr., pp. 23-40, Island Press, Washington, D.C., 2000.

Parise, Mario, and R.W. Jibson, A seismic landslide susceptibility rating of geologic units based on analysis of characteristics of landslides triggered by the 17 January, 1994, Northridge, California earthquake, *Engineering Geology, 58,* 251-270, 2000.

Parish, Y.I.H., and Pascal Müller, Procedural modeling of cities, *Computer Graphics, 35,* 301-308, 2001: http://www1.acm.org/pubs/articles/proceedings/graph/ 383259/p301-parish/p301-parish.pdf.

Peterson, M.D, C.H. Cramer, G.A. Faneros, C.R. Real, and M.S. Reichle, Potential ground shaking, in Seismic hazard evaluation of the cities of Oakland and Piedmont, Alameda

County, CA, California Division of Mines and Geology, *Open-File Report 99-11, 3,* 55-64, 2000: ftp://ftp.consrv.ca.gov/pub/dmg/shezp/evalrpt/ OFR99-11.pdf.

Pike, R.J., The geometric signature: quantifying landslide-terrain types from digital elevation models, *Mathematical Geology, 20*, 491-511, 1988.

Pike, R.J., R.L. Bernknopf, J.C. Tinsley III, and R.K. Mark, Hazard of earthquake-induced lateral-spread ground failure on the central California coast modeled from earth-science map data in a geographic information system, U.S. Geological Survey, *Open-File Report 94-662*, 46 pp., scale 1:62,500, 1994.

Pike, R.J., S.H. Cannon, S.D. Ellen, S.E. Graham, R.W. Graymer, M.A. Hampton, J.W. Hillhouse, D.G Howell, A.S. Jayko, R.L. LaHusen, K.R. Lajoie, D.W. Ramsey, M.E. Reid, B.M. Richmond, W.Z. Savage, C.M. Wentworth, and R.C. Wilson, Slope failure and shoreline retreat during northern California's latest El Niño, *GSA Today, 8,* 1-6, 1998.

Pike, R.J., R.W. Graymer, Sebastian Roberts, N.B. Kalman, and Steven Sobieszczyk, Map and map database of susceptibility to slope failure by sliding and earth flow in the Oakland area, California, U.S. Geological Survey, *Miscellaneous Field Studies Map 2385*, scale 1:50,000, 2001: http://geopubs.wr.usgs.gov/map-mf/mf2385/.

Platt, R.H. (Ed.), with K.B. O'Donnell and David Scherf, contributors, *Disasters and Democracy: The Politics of Extreme Natural Events*, 335 pp., Island Press, Washington, D.C., 1999.

Pomeroy, J.S., Landslides in the Greater Pittsburgh region, Pennsylvania, U.S. Geological Survey, *Professional Paper 1229*, 48 pp., 1982.

Popescu, M.E., A suggested method for reporting landslide remedial measures, *Bulletin of Engineering Geology and the Environment, 60*, 69-74, 2001.

Powelson, J.P, *The Story of Land: A World History of Land Tenure and Agrarian Reform*, 347 pp., Lincoln Institute of Land Policy, Cambridge, MA, 1989.

Powelson, J.P, and R.D. Stock, *The Peasant Betrayed: Agriculture and Land Reform in the Third World*, 402 pp., Cato Institute, Washington, D.C., 1990.

Prosterman, R.L., and J.M. Riedinger, *Land Reform and Democratic Development*, 313 pp., Johns Hopkins University Press, Washington, D.C., 1987.

Radbruch, D.H., and L.M. Weiler, Preliminary report on landslides in a part of the Orinda Formation, Contra Costa County, California, U.S. Geological Survey, *Open-File Report 689*, 35 pp., map scale 1:24,000, 1963.

Reid, M.E., and R.G LaHusen, Real-time monitoring of active landslides along Highway 50, El Dorado County, *California Geology, 51,* 17-20, 1998.

Reneau, S.L., W.E. Dietrich, R.I. Dorn, C.R. Berger, and Meyer Rubin, Geomorphic and paleoclimate implications of latest Pleistocene radiocarbon dates from colluvium-mantled hollows, California, *Geology, 14,* 655-658, 1986.

Roberts, Sebastian, M.A. Roberts, E.M. Brennan, and R.J. Pike, Landslides in Alameda County, California, a digital database extracted from Preliminary photointerpretation maps of surficial deposits by T.H. Nilsen in U.S. Geological Survey, Open-File Report 75-277, U.S. Geological Survey, *Open-File Report 99-504*, 20 pp., 1999: http://wrgis.wr.usgs.gov/open-file/of99-504/.

Rodrigue, C.M., Impact of internet media in risk debates: the controversies over the Cassini-Huygens mission and the Anaheim Hills, California, landslide, *The Australian Journal of Emergency Management, 16*, 53-61, 2001: http://www.ema.gov.au/5virtuallibrary/pdfs/vol16no1/rodrigue.pdf.

Rogers, J.D., Recent Developments in Landslide Mitigation, in Landslides / Landslide

Mitigation, edited by J.E. Slosson, A.G. Keene and J.A. Johnson, *Reviews in Engineering Geology, 9*, 95-118, 1992.

Rowbotham, D.N., and Douglas Dudycha, GIS modelling of slope stability in Phewa Tal watershed, Nepal, *Geomorphology, 26,* 151-170, 1998.

San Francisco Bay Landslide Mapping Team, San Francisco Bay Region, California, Landslide Folio, U.S. Geological Survey, *Open-File Report 97-745 A-F*, variously paged, 1997: http://wrgis.wr.usgs.gov/open-file/of97-745/.

Sarna-Wojcicki, A.M., Landslides: an example in southern California, in Geologic processes at the land surface, edited by H.G. Wilshire, K.A. Howard, C.M. Wentworth, and Helen Gibbons, pp. 2-4, U.S. Geological Survey, *Bulletin 2149*, 1996.

Sassa, Kyoji, Hiroshi Fukuoa, Gabrielle Scarascia-Mugnozza, and Stephen Evans, Earthquake-induced landslides: distribution, motion, and mechanism, *Soils and Foundations, Special issue on geotechnical aspects of the January 17, 1995, Hyogoken-Nanbu Earthquake*, 53-64, 1996.

Schmidt, K.M., J.J. Roering, J.D. Stock, W.E. Dietrich, D.R. Montgomery, and T. Schaub, The variability of root cohesion as an influence on shallow landslide susceptibility in the Oregon Coast Range, *Canadian Geotechnical Journal*, 38, 995-1024, 2001.

Schrott, Lothar, and Alessandro Pasuto (Eds.), Temporal stability and activity of landslides in Europe with respect to climatic change (TESLEC), *Geomorphology, 30,* 211 pp., 1999.

Schuster, R.L., Reducing landslide risk in urban areas: experience in the United States, in *Urban Disaster Mitigation, the Role of Engineering and Technology*, edited by F.Y. Cheng and M.S. Sheu, pp. 217-230, Elsevier, Oxford and New York, 1995.

Schuster, R.L., Socioeconomic significance of landslides, in Landslides, Investigation and Mitigation, edited by A.K. Turner and R.L. Schuster, pp. 12-35, Transportation Research Board, National research Council, *Special Report 247*, Washington, D.C., 1996.

Schuster, R.L., and R.W. Fleming, Economic losses and fatalities due to landslides, *Bulletin of the Association of Engineering Geologists, 23,* 11-28, 1986.

Schuster, R.L., and L.M. Highland, Socioeconomic and environmental impacts of landslides in the Western Hemisphere, U.S. Geological Survey, *Open-File Report 01-276*, 47 pp., 2001: http://geology.cr.usgs.gov/pub/open-file-reports/ofr-01-0276/

Scott, K.M., J.W. Vallance, and P.T. Pringle, Sedimentology, behavior, and hazards of debris flows at Mount Rainier, Washington, U.S. Geological Survey, *Professional Paper 1547*, 56 pp., 1995.

Scullin, C.M., Ethical concerns in geotechnical practice in our hillside urban environment, *Bulletin of the Association of Engineering Geologists, 31,* 404-408, 1994.

Selby, M.J., *Hillslope Materials and Processes*, 451 pp., Oxford University Press, Oxford and New York, 1993.

Sharpe, C.F.S., *Landslides and Related Phenomena: A Study of Mass Movements of Soil and Rock*, 137 pp., Columbia University Press, New York, 1938.

Shore, K.J., Stopping landslides in Rio: recycling scrap tires into retaining walls, International Development Research Centre, Ottawa, *Project 94-1005 Report,* http://www.idrc.ca/reports/read_article_english.cfm?article_num=385, April 30, 1999.

Shuirman, Gerard, and J.E. Slosson, Malibu landslide: massive litigation, in *Forensic Engineering: Environmental Case histories for Civil Engineers and Geologists*, pp. 18-86, Academic Press, New York, 1992.

Sidle, R.C., A.J. Pearce, and C.L. O'Loughlin, Hillslope stability and land use, American

Geophysical Union, *Water Resources Monograph Series, 11*, 140 pp., Washington, D.C., 1985.

Siebe, Claus, Michael Abrams, J.L. Macías, and Johannes Obenholzner, Repeated volcanic disasters in prehispanic time at Popocatépetl, central Mexico: past key to future?, *Geology, 24*, 399-402, 1996.

Sisson, T.W., J.W. Vallance, and P.T. Pringle, Progress made in understanding Mount Rainier's hazards, *Eos, Transactions, American Geophysical Union, 82, 113,* 118-120, 2001.

Slosson, J.E., D.D. Yoakum, and Gerard Shuirman, Thistle landslide: was mitigation possible?, in Landslides / Landslide Mitigation, edited by J.E. Slosson, A.G. Keene, and J.A. Johnson, *Reviews in Engineering Geology, 9*, 83-93, 1992.

Smail, J.K., V.D. Abernethy, Tim Dyson, Lindsay Grant, Betsy Hartmann, Carl Haub, R.D. Lamm, T.F. Flannery, Wolfgang Lutz, Norman Myers, Jack Parsons, David Pimentel, Marcia Pimentel, M.S. Swaminathan, R.F. Tang, Bruce Wallace, C.F. Westoff, and David Willey, The case for dramatically reducing human numbers, *Politics and the Life Sciences, 16,* 183-236, 1997: http://www.gn.apc.org/eco/pubs/ smail.html.

Smyth, C.G., and S.A. Royle, Urban landslide hazards: incidence and causative factors in Niteroi, Rio de Janeiro State, Brazil, *Applied Geography, 20,* 95-118, 2000.

Soeters, Robert, and C.J. van Westen, Slope instability recognition, analysis, and zonation, in Landslides, investigation and mitigation, edited by A.K. Turner and R.L. Schuster, Transportation. Research Board, NRC, *Special Report 247*, pp. 129-177, Washington D.C., 1996.

Southern California Studies Center, Sprawl hits the wall: confronting the realities of metropolitan Los Angeles, *Atlas of Southern California, 4*, 56 pp., University of Southern California, Los Angeles, CA, 2001: http://www.brook.edu/es/urban/la/ blackandwhite.htm.

Spiker, E.C., and P.L Gori, National landslide hazards mitigation strategy: a framework for loss reduction, U.S. Geological Survey, *Open-File Report 00-450*, 49 pp., 2000: http://geology.cr.usgs.gov/pub/open-file-reports/ofr-00-0450/.

Steiner, G.M., *The Anaheim Hills (Jan. 18, 1993) Landslide Update*, 2002: http://anaheim-landslide.com.

Susskind, Lawrence, and Patrick Field, *Dealing With an Angry Public*, 276 pp., The Free Press, New York, 1996.

Swanston, D.N., and R.L. Schuster, Long-term landslide hazard mitigation programs: structure and experience from other countries, *Bulletin of the Association of Engineering Geologists, 26*, 109-133, 1989.

Taylor, F.A., and E.E. Brabb, Maps showing distribution and cost by counties of structurally damaging landslides in the San Francisco Bay region, California, winter of 1968-69, U.S. Geological Survey, *Miscellaneous Field Studies Map MF-327*, scales 1:500,000 and 1:1,000,000, 1972.

Taylor, F.A., T.H. Nilsen, and R.M. Dean, Distribution and cost of landslides that have damaged man-made structures during the rainy season of 1972-1973 in the San Francisco Bay region, California, U.S. Geological Survey, *Miscellaneous Field Studies Map MF-679*, scales 1:500,000 and 1:1,000,000, 1975.

Terzaghi, Karl, Mechanism of landslides, in *Application of Geology to Engineering Practice*, edited by Sidney Paige, pp. 83-123, Geological Society of America, 1950.

Thomas, W.L., Jr. (Ed.), *Man's Role in Changing the Face of the Earth*, 1193 pp., University of Chicago Press, Chicago, 1956.

Turner, A.K., and R.L. Schuster, (Eds.), Landslides, Investigation and Mitigation, Transportation Research Board, National Research Council, *Special Report 247*, 673 pp., Washington, D.C., 1996.

Turner, B.L. II, W.C. Clark, R.W. Kates, J.F. Richards, J.T. Mathews, and W.B. Meyer (Eds.), *The Earth as Transformed by Human Action: Global and Regional Changes in the Biosphere Over the Past 300 years*, 713 pp., Cambridge University Press, Cambridge, UK, 1990.

U.S. Geological Survey (USGS), Goals and tasks of the landslide part of a ground-failure hazards reduction program, U.S. Geological Survey, *Circular 880*, 49 pp., 1982.

United Nations Population Division, *The World at Six Billion*, Department of Economic and Social Affairs, United Nations Secretariat, 63 pp., New York, 1999: http://www.un.org/esa/population/publications/sixbillion/sixbillion.htm.

United Nations Population Division, *World Population Prospects—The 2000 Revision, I, Comprehensive Tables*, Department of Economic and Social Affairs, United Nations Secretariat, New York, CD-ROM, 2001: http://www.un.org/esa/population/ publications/wpp2000/wpp2000h.pdf.

Utgard, R.O., G.D. McKenzie, and Duncan Foley, *Geology in the Urban Environment*, 355 pp., Burgess Publishing Co., Minneapolis, 1978.

Varnes, D.J., Slope movement types and processes, in Landslides: Analysis and Control, edited by R.L. Schuster and R.J. Krizek, pp. 11-33, U.S. National Academy of Sciences, National Research Council, Transportation Research Board, *Special Report 176*, 1978.

Varnes, D.J., Landslide hazard zonation: a review of principles and practice, UNESCO, *Natural Hazards, 3,* 63 pp., Paris, 1984.

Veder, Christian, *Landslides and Their Stabilization*, 247 pp., Springer-Verlag, New York, 1981.

Velasquez, L.S., Agenda 21: a form of joint environmental management in Manizales, Colombia, *Environment and Urbanization, 10,* 9-36, 1998: http://www.iied.org/ pdf/urb_la4a_velasquez.pdf.

Ward, P.L. (Ed.), *The next big earthquake in the Bay Area may come sooner than you think: are you prepared?,* U.S. Geological Survey, in cooperation with other public service agencies, 24 pp., 1990: http://geopubs.wr.usgs.gov/fact-sheet/fs094-96/fs094-96.pdf.

Wasowski, Janusz, D.K. Keefer, and R.W. Jibson, (Eds.), Landslide hazards in seismically active regions, *Engineering Geology, 58*, 398 pp., 2000.

Watanabe, Masayuki, R.P. Brenner, and S. Malla, *Mudflows: Experience and Lessons Learned From the Management of Major Disasters*, 139 pp., United Nations, Department of Humanitarian Affairs, New York and Geneva, 1996.

Wieczorek, G.F., Preparing a detailed landslide-inventory map for hazard evaluation and reduction, *Bulletin of the Association of Engineering Geologists, 21,* 337-342, 1984.

Wieczorek, G.F., Landslide triggering mechanisms, in Landslides, Investigation and Mitigation, edited by A.K. Turner and R.L. Schuster, pp. 76-90, Transportation Research Board, National Research Council, *Special Report 247*, 1996.

Wieczorek, G.F., and N.D. Naeser (Eds.), *Debris-Flow Hazards Mitigation: Mechanics, Prediction, and Assessment*, 608 pp., Balkema, Rotterdam, 2000.

Wills, C.J., and H.H. Majmundar, Landslide hazard Map of southwest Napa County, CA, California Division of Mines and Geology, *Open-File Report 99-06*, scale 1:24,000, 2000.

Wilson, R.I., T.P. McCrink, J.R. McMillan, and W.D. Haydon, 2000, Earthquake-induced landslide zones, in Seismic hazard evaluation of the cities of Oakland and Piedmont,

Alameda County, CA, California Division of Mines and Geology, *Open-File Report 99-11, 2*, pp. 35-54, scale 1:24,000, 2000: ftp://ftp.consrv.ca.gov/pub/ dmg/shezp/eval-rpt/OFR99-11.pdf

Wilson, R.C., Normalizing rainfall-debris-flow thresholds along the U.S. Pacific coast for long-term variations in precipitation climate, in *Proceedings, First International Conference on Debris-Flow Hazards Mitigation*, edited by C.L. Chen, pp. 32-43, Hydraulics Division, American Society of Civil Engineers, 1997.

Wilson, R.C., and A.S. Jayko, Preliminary maps showing rainfall thresholds for debris-flow activity, San Francisco Bay region, California, 20 pp., U.S. Geological Survey, *Open-File Report 97-745 F*, 1997: http://wrgis.wr.usgs.gov/open-file/ of97-745/of97-745f.html.

Wilson, R.C., R.K. Mark, and G. Barbato, Operation of a real-time warning system for debris flows in the San Francisco Bay area, California, in *Hydraulic Engineering '93, Proceedings of the Conference, July 25-30, 1993, San Francisco, CA*, edited by H.W. Shen, S.T. Su and F. Wen, *2*, pp. 1908-1913, Hydraulics Division, American Society of Civil Engineers, 1993.

Záruber, Quido, and Vojtech Mencl, *Landslides and Their Control*, 2nd ed., 324 pp., Elsevier / Academia, New York and Prague, 1982.

Zoback, M.L., Grand challenges in earth and environmental sciences: science, stewardship, and service for the twenty-first century, *GSA Today, 12,* 41-47, 2001.

SECTION III

URBAN HYDROLOGY

All too frequently cities in poor health do not have the infrastructure to deliver and distribute potable water. Residents are forced to use shallow, polluted wells, springs, cisterns, or streams loaded with waste. Even modern cities tap into local aquifers, especially in the suburbs, which are drastically affected by construction, pavement, and runoff contaminated by industrial chemicals to lawn fertilizer.

From Roman aqueducts to the California Water Project, city growth is directly linked to the delivery of water from distant sources. Yet even these distant sources are threatened by contaminated runoff from farms and septic tanks at vacation homes.

To mediate pollution by runoff, integrated information from the geosciences, atmospheric sciences, and engineering should be required by every city with concern for their residents. Geoscientists should work toward an understanding and protection of the waning hydrological resources required by urban agglomerations.

9

Effects of Urbanization on Groundwater Systems

J. M. Sharp, Jr., J. N. Krothe, J. D. Mather, B. Garcia-Fresca, and C. A. Stewart

INTRODUCTION

In 1900, only 10% of the Earth's population lived in cities (United Nations, 1991), but today more than 50% dwell in urban areas. In developing countries, much urban growth occurs in illegal and unplanned squatter settlements. Urban growth rates are due to both immigration from the rural areas and high birthrates. Many developing countries have population doubling times on the order of 25 years, but even in the United States, with a relatively low rate of overall population growth, urban populations (e.g., the Austin-San Antonio region in Texas) are expected to double in the same time frame. Undoubtedly, urban areas will accommodate most of the projected increases in population for both developed and developing countries. Rates of areal expansion (urban sprawl) are also impressive. In the New York City metropolitan region, there was a 61% increase in urbanized area between 1964 and 1989, but the population growth was much less. Bangkok has been expanding at a rate over 3,200 hectares per year (Lowe, 1992). Even smaller urban and suburban areas are "exploding." North of Austin, Texas, in 1997, ranch land was being subdivided for residential and commercial development at a rate of 3.3 hectares per day.

The effects of urbanization on shallow hydrogeologic systems are similar to the effects of karstification. Zones of greatly enhanced permeability are created with highly variable properties and connectivities. Recent studies on the hydrogeological effects of urbanization demonstrate that there are significant alterations to rates of recharge and permeability distributions. These alterations may affect groundwater pollution and its remediation, water resource availability, urban stream flows, and other geotechnical applications.

Earth Science in the City: A Reader

10.1029/056SP10

Below we discuss generally how to meet the demands for water in the urbanizing world; the reasons that groundwater is an underutilized resource; the hydrogeological effects of urbanization; the effects of utility systems with both field data and numerical models; and possible technological and policy developments to utilize groundwater resources more efficiently.

Meeting the Demands for Water in an Urbanizing Earth

People in urban and urbanizing areas require water; groundwater supply issues stem from these demands. Specific solutions are complex because of competing political and economic interests, scientific uncertainties, and inadequate funding for both resource development and scientific study. In principle, however, all solutions must follow one or more of these options:

1. Increase water supplies;
2. Reduce water demands; or
3. Use available water resources more efficiently.

Increased supplies can come from new surface-water or groundwater sources, harvesting rainwater, or utilization of presently unpotable or other underutilized waters (Sharp, 1997). This last option includes both desalination and the use of unpotable water for bathing, washing, irrigation, sewage disposal, and industrial use. Dual-distribution systems are useful in this regard, but they are costly to install and/or retrofit.

Decreased demands for water can be achieved by limiting population growth (and concomitant water use), increased water prices, rationing, or public appeals to reduce water consumption. Limiting urban population growth is an ideal solution, but it has not yet been achieved. Increased water prices carry a political price so that rationing becomes the means of allocation in drought times. In the United States, there have been cases in which citizens voluntarily reduced water usage during drought, but the water prices had to be increased proportionately because operating costs for the water supply system are mostly for personnel and construction financing.

Finally, we need to find ways to operate more efficiently with existing water resources by conjunctive use, use of treated wastewaters, artificial recharge, and water conservation measures. The more efficient utilization of groundwater is an option that will have to be considered seriously and on a case-by-case basis; this will be useful in those urban areas where groundwater is presently an under-utilized water resource.

Why Groundwater is Underutilized in Some Urban Areas

London is an excellent example (Lucas and Robinson, 1995) of an urban area that previously relied heavily on groundwater, but largely switched to surface water because of economies of scale, resource adequacy, and public health concerns. Many cities were previously serviced by multiple shallow wells; pollution in these shallow aquifers and the resulting health threats led to the establishment of centralized water-supply and treatment systems. Chlorination of water supplies in American cities became a requirement at the turn of the century (Havlick, 1974), and it was more economical to limit the number of water treatment plants. Furthermore, the costs of surface-water development (in the United States) were and still are commonly subsidized by the state and federal governments; well-field costs are typically borne by local communities. However, there are also perception problems. When the public learns that water levels in supply wells have dropped, there is concern that the wells will "dry up." Water stored in a reservoir somehow seems more reassuring than a well in the ground.

Groundwater, however, offers a number of unique advantages. It may be widely available, less vulnerable to climatic variability, of superior quality, and cheaper to develop and to distribute than surface water. Groundwater exists everywhere beneath the land surface, whereas major surface-water bodies are rare in arid and semi-arid zones. Where local surface waters are not sufficient to meet urban demands, construction of reservoirs and long pipelines may be required. Urban areas in southern California obtain much of their water by transfer from the Colorado River and from northern California. In other areas, surface waters are badly polluted. In much of the world, the locations of the major aquifers are known. For instance, in the U. S., these were inventoried and mapped by Meinzer (1923). In Heath's (1988) more recent map, major aquifers are essentially the same as those identified by Meinzer. However, even in areas where major aquifers do not exist, wells can be used to meet individual domestic needs.

In urbanizing areas, detailed hydrogeological maps are required to reduce uncertainty and to allow more efficient utilization of groundwater. This is especially critical when various water-supply options are considered. Detailed hydrogeological maps are also important for urban planning issues. Protected recharge zones in Germany, Hong Kong, and Sweden demonstrate how hydrogeologic knowledge can be utilized propitiously in urban areas to protect and replenish groundwater resources. These zones can also serve as parklands and wildlife refuges. In many areas, however, there are insufficient data to delineate protection zones adequately. In areas of limited recharge, urban areas may have to locate well fields many kilometers distant from the city, much like surface-water reservoirs.

POTENTIAL HYDROGEOLOGICAL EFFECTS OF URBANIZATION

The hydrogeological effects of urbanization have important ramifications for municipal water management policy. Possible effects include: increased rates of groundwater extraction, increased surface water runoff, increased or decreased groundwater recharge, a highly altered permeability structure, altered vegetation, urban irrigation, and increased potential for water-quality degradation. There are similarities between urban areas and karstic system settings. Karstic systems have surface streams that flow after heavy rainfalls analogous to paved urban drainage ways; internal drainage systems (dolines) are analogous to storm drainage systems on streets and parking lots; groundwater flow directions in both karstic and urban systems may not correspond to topographic gradients; and permeability structures in karst are dominated by caves, fractures and conduits that are analogous to utility trenches, tunnels, and other conduits (Sharp et al., 2001). The construction of utility systems creates a network of interconnected conduits that are infilled with material that commonly have highly contrasting permeabilities compared to original subsurface materials. In both urban and karstic settings, such features evolve rapidly in the geologic sense. Urbanization commonly buries stream channels and springs, but these features still exist in the subsurface and can create unforeseen hydrogeological complications. This is especially critical in older cities (Barton, 1992).

Commonly acknowledged effects of urbanization on groundwater are: 1) aquifer depletion from increased groundwater overexploitation; 2) subsidence; 3) salt-water intrusion; 4) deterioration of special environments; and 5) water quality deterioration (Sharp, 1997).

Aquifer Depletion

In general, aquifers run little chance of total exhaustion. Exceptions could occur in arid or semi-arid regions where an aquifer receives minimal recharge and is essentially being mined. However, these cases are rare, and conjunctive use and long-range planning can allow additional resources to be brought on-line as "mining" depletes others. Although permanent depletion is not generally a threat, growing demands can exceed the safe or permissive yield of an aquifer. Of concern are situations in which water levels drop so far that pumping becomes too expensive, water yields are severely diminished, or concomitant negative effects occur. For instance, artesian aquifers in Waco and Dallas, Texas, formerly provided free flowing wells, but their hydraulic heads have fallen many tens of meters. These cities have switched to surface water resources for their potable supplies. Shallow wells in unconfined aquifers may no longer prove reliable where water levels drop very low, as has commonly happens during droughts. In El Paso, the Hueco Bolson aquifer can no longer meet the projected demands of the growing city; and El Paso must now look elsewhere for additional water supplies.

Subsidence

Overdraft has caused severe subsidence in many coastal cities, including Houston (Figure 1), Jakarta, Shanghai, Venice, and Calcutta. In Houston, a decline since subsidence rates in the 1980s is the result of limiting the pumpage of groundwater (Holzschuh, 1991). Houston, like some other coastal areas, switched to surface waters and relocated well fields farther inland. However, dif-

Figure 1. House in the Brownwood subdivision near Houston that was abandoned because of subsidence, which was caused primarily by groundwater extraction, and the ensuing inundation.

ferential subsidence in inland cities (e.g., Mexico City, Mexico, and Albuquerque, New Mexico) may create problems with drainage systems and lead to localized flooding. Differential subsidence can be managed by regulating the locations and rates of pumping to minimize subsidence effects. The negative effects of coastal subsidence and inundation can also be minimized or eliminated by land-use planning. One solution is very simple – limit construction and development in low-lying areas. This also makes sense when we consider the potential effects of both coastal storms and long-term sea-level fluctuations. Nevertheless, in many countries the development of low-lying coastal areas continues because of demands for land by a growing population, the desirability of being located on the coast, or the fact that these areas may possess very fertile deltaic soils.

Salt-Water Intrusion

Overexploitation can cause the intrusion of poor-quality water. This is especially important for cities on oceanic islands or in close proximity to the coastline where salt water underlies or is otherwise adjacent to the fresh waters. In Britain, extensive industrial development along many river estuaries has caused intrusion of poor-quality brackish water. This has occurred along the Thames and Humber estuaries where the Cretaceous Chalk aquifer has been affected. Comprehensive management schemes have been developed which establish a balance between recharge and abstraction rates (e.g., University of Birmingham 1987). Once intrusion has occurred, it takes a much longer period of time of reduced pumping for the aquifer to recover. However, salt-water intrusion can also occur inland. For instance, south of Kansas City, Missouri, overdraft of the Ordovician carbonate aquifers has induced downwards intrusion of saline water from overlying Pennsylvanian clastic rocks. In El Paso, Texas, and adjacent Juarez, Mexico, overdrafts have reversed the hydraulic gradient with the Rio Grande, and poor- quality, brackish river water is infiltrating into the El Paso's major aquifer, the alluvium of the Hueco Bolson. In addition, the city is pumping water from the Hueco Bolson alluvial aquifer at rates in excess of recharge. Recent and projected pumping rates and water levels declines are even greater in Juarez, Mexico, the city abutting El Paso along the Rio Grande (Hibbs et al., 1997).

Responses to salt-water intrusion vary. The ideal response to salt-water intrusion is desalination, which is yet very expensive. However, brackish groundwater, depending upon the technology employed, may be easier to desalinate than seawater. Strategies may utilize a combination of water importation, shifting pumping strategies, limiting groundwater extraction, creating hydraulic barriers through injection wells or infiltration galleries, or desalination of brackish or saline water. In Brighton on the south coast of England, tourism led to a major increase in groundwater abstraction and increased salinities in several Chalk sources. An aquifer management policy was introduced that makes maximum use of boreholes located along the coastal margin during the winter months when aquifer recharge and outflow to the sea occur. As spring and summer progress, abstraction from these coastal boreholes is reduced and more is pumped from inland sources where groundwater levels have been allowed to build up over the winter (Headworth and Fox, 1986; Owen, et al., 1991). In El Paso, the city has experimented with artificial recharge of reclaimed water from their sewage treatment plants and is examining the feasibility of importing groundwater from sources up to 200 kilometers distant. Surface-water solutions appear unlikely; groundwater will have to be utilized; and the long-term prospects for water supply in the El Paso-Juarez metroplex remain uncertain at this time.

Protection of Special Environments

The protection of wetland environments that are fed by groundwater discharge (including those of threatened or endangered species) is becoming an important issue (Sharp and Banner, 2000). A prime example is the Edwards aquifer of Texas. This aquifer supplies over two million people, including the city of San Antonio. Natural aquifer discharge is to a number of large springs, but, with increasing pumpage, spring flows have decreased. Federal courts have ruled that minimum discharges of >8.5 cubic meters/second (300 cubic feet per second) must be maintained at two of the major spring systems to ensure the survival of several species of flora and fauna that only exist in waters emanating from the springs (McKinney & Sharp, 1995). Data from 1934 to the present show wide variations in recharge to the aquifer and increasing discharge by wells, largely for San Antonio (Figure 2). These data are plotted as 5-year, linearly weighted running averages to minimize recharge variability and show the general trends.

It is manifest from Figure 2 that when conditions similar to those of the drought of record (1947-1956) occur again, the court-prescribed springflows and current levels of pumping cannot both be maintained. Because San Antonio is projected to double in population in the next 25 to 50 years, drought effects will be exacerbated. There are few unused surface-water resources available, and there is a need to maintain fresh-water inflows to bay and estuarine environments of the Gulf of Mexico. Therefore, San Antonio's options are limited: long-distance surface-water importation, reduction of agricultural irrigation that also tap the aquifer, growth controls, water conservation, reuse of treated sewage, or augmentation of springflow. The last option maintains court-prescribed minimal springflows by one or more of several options (Uliana & Sharp, 1996): injection wells, enhanced natural recharge along losing streams that largely recharge the Edwards aquifer, exfiltration galleries near the springs, or direct addition of water to the spring lakes. Because all options for San Antonio are either expensive or leave a large volume of high quality water unused in the aquifer, no consensus course of action has yet been achieved.

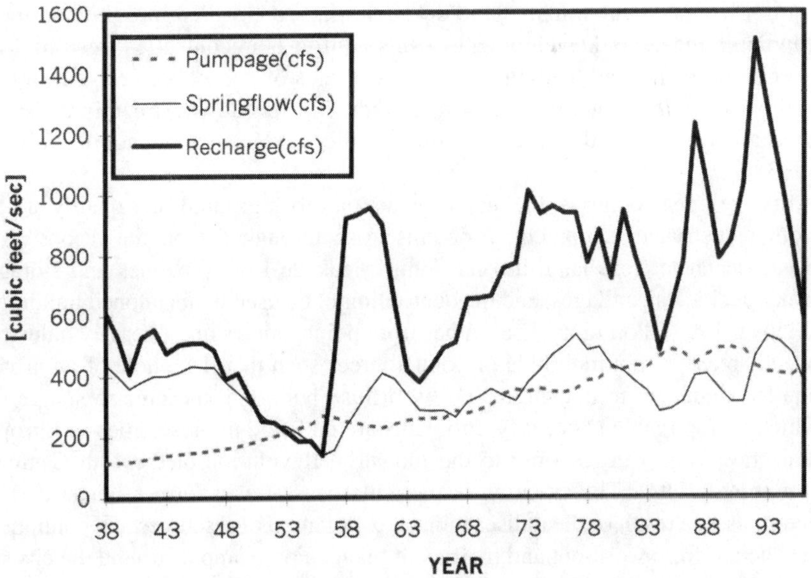

Figure 2. Five-year linearly-weighted, running averages for recharge and discharge by wells and springs for the San Antonio segment of the Edwards aquifer, Texas (modified from Technical Advisory Panel, 1990, with data from Slattery and Thomas, 2000).

Water Quality

Water-quality considerations are an important issue for urban groundwater supply. Shallow aquifers and surface waters in urban settings are subject to pollution by runoff from paved surfaces, leaky storage tanks, surface spills, illegal dumping of hazardous waste, leaky sewage lines, and lack of sanitation facilities. With the spread of urbanized areas, contamination of shallow aquifers is a major threat. In the urban areas of many developing nations, the provision of sewerage systems lags behind population growth and the provision of mains for water supply. In many cases only small areas in the centers of cities are sewered. In the unsewered areas, more than 90% of domestic water supply may be disposed of via pit latrines, cesspits or septic tanks representing a significant source of poor quality groundwater recharge (Mather et al., 1996). This limits the use of shallow wells as a source of drinking water.

Declining water levels can also cause water-quality deterioration from encroachment of poor-quality water or by changing hydrochemical conditions. As shallow aquifers become contaminated they are often abandoned in favor of deeper aquifers that are relatively protected from pollution. In some urban environments this has induced the downward movement of water from the upper layers. A typical example is the Yemeni capital, Sana'a. Here traditional wells abstracted water from an upper alluvial aquifer that is nearly 300m thick. With increasing urbanization from the 1970s onwards, this aquifer became contaminated and deeper wells were drilled into the confined Tawilah Sandstone. Development of this aquifer is uncontrolled, resulting in over exploitation and leading to drawdown rates of nearly 5m per year in the main well-field area. This, in turn, is inducing the downward migration of contaminants from the alluvial aquifer into the Sandstone aquifer (Fara, 1993).

In developed countries, the major impact on urban groundwater quality arises from industrial development. Problems arise through the on-site disposal of waste, leachate from landfills and dumps, leakage from pipelines and storage tanks, accidental spillages, and the demolition of disused or abandoned buildings (Mather, 1993). Pollution arises either from point sources or, in heavily industrialized areas, from a multitude of point sources such that the whole of an urban aquifer is subject to contamination by diffuse pollution, the precise source of which is impossible to identify. Problems are endemic in those cities in Europe that grew largely in response to the Industrial Revolution of the 19th Century (Lumsden, 1994). The situation is particularly severe in former Eastern Bloc countries as a result of the indiscriminate exploitation of resources and continued production from worn-out and inefficient plants. For example, around the city of Ploesti in Romania, up to 5m of oil floats on the water table as a result of leakage from pipelines and storage tanks belonging to a network of refineries (Mather et al., 1996).

Modern water-treatment plants are sophisticated and compact and, coupled with the fact that groundwater may require less sophisticated treatment, ease the need for highly centralized treatment and distribution systems. Increase of impervious cover and storm sewers can reduce recharge and lower water tables. On the other hand, urban areas can suffer from rising water tables if they switch from groundwater to surface-water resources or if storm drainage disposal is not properly engineered. Urban development must be controlled or managed to protect future water resources and environmentally sensitive areas. This requires detailed hydrogeological maps, data, and analyses before decisions are made.

A key point is to understand how urbanization has affected or can affect hydrogeologic processes. For instance, utility lines, tunnels, and utility trenches backfilled with gravel or riprap create the equivalent of a karst conduit. Contaminated groundwater, if intercepted by utility conduits, may flow rapidly to areas that do not appear to be down gradient. This makes prediction and control of contamination plumes difficult. Sump pumps, leaky water and sewer mains, and storm-sewage disposal can create significant local hydraulic gradients. Paved surfaces limit and alter the natural distribution of areal recharge from precipitation. Consequently, prediction of contaminant transport and contaminant plume geometry may be difficult in urban areas.

Leakage From Mains and Sewerage Systems

Leakage from water mains and sewerage systems, as well as urban irrigation, can increase recharge despite increases in "impervious" cover and storm drainage. Leakage rates from water distribution systems ranges from about 8% to over 60% (Foster, 1996; Sharp et al., 2000). Table 1 is a compendium of estimates of water main leakage in a variety of urban areas. We infer equivalent rates of infiltration will occur from leaky sewage lines; sewer lines in the U.S. are designed with a planned 10 % leakage rate, but few quantitative data are available. During years of low rainfall and drought conditions, water and wastewater leakage becomes more significant and may dominate recharge to urban aquifers. Lorenzo-Rigney and Sharp (1999) estimated that in low rainfall years up to 12% of water flowing to Barton Springs, an important recreational asset in Austin, Texas, recharges from leaky utility systems. Sharp et al. (2000) estimated up to 30% of the annual recharge to the Edward aquifer in San Antonio region could arise from the same. Finally, urban settings (including plantings of non-native vegetation) themselves can alter the local climate with unintended hydrological effects.

An excellent example of leakage problems is recorded by George (1992) for the city of Riyadh, Saudi Arabia. City water utilities were undeveloped prior to 1950; water was abstracted from hand-dug wells. In the 1950s, boreholes were drilled in both shallow and deep aquifers beneath the city. These wells were abandoned in the 1960s, as the city grew, because of contamination and large drawdowns. With increasing urbanization in the 1970s water was imported into

Table 1. Estimates of water main leakage in urban areas. Data are from Knipe et al. (1993), Morris et al. (1994), Foster (1996), Mather et al. (1996), Geomatrix (1997), Lerner (1997), Austin-American Statesman (1998), Vodokanal (unpub. data), Escolero et al. (2000), Basu and Main (2001), Gleick (2001), Sharp and Krothe, 2002, and Krothe et al., 2002)

Urban Area	Water Main Loss [%]	Recharge Increase [%]	Recharge [mm/yr]
Hong Kong	8		
San Antonio, TX	8.4		
Austin, TX	12		
Birmingham, UK		-4	132
Round Rock, TX	26		
St. Petersburg, Russia	~30		
General rates, UK	~30		
Third world urban areas	30-50		
Calcutta India	36		
San Marcos, TX	37		
Los Angeles, CA	6 - 8	~260-600*	150-300
Santa Cruz, Bolivia		71	290
Hat Yai, Thailand		118	370
Merida, Mexico	~50	233	600
Lima, Peru	>60	very high	740

*The Geomatrix (1997) study estimated recharge from "delivered water" relative to recharge from precipitation under current conditions.

the city from external well fields followed by desalinated water. Very large leakage rates resulted in groundwater rising to the surface where it formed springs and contributed to surface flow in storm-water sewage systems. This in turn caused structural damage and endangered public health.

EFFECTS OF UTILITY TRENCHES

The study utilizes both field permeability tests and numerical models to assess the effects of utility systems on urban hydrogeology.

Field Permeability Tests

The construction of utility systems changes the permeability both within and overlying the trenches in which utility conduits and pipes have been placed. The basic construction of a trench is given below (Figure 3). The excavation of the trench may or may not include the creation of a lateral gradient in the trench bottom due to the nature of the utilities to be contained. For example, sewer and storm water systems commonly rely on gravity drainage. Thus, the entrenchment usually has a specified grade or slope. The bottom of the trench is generally filled with well-sorted, clean sand or gravel to protect the pipes from the rock walls and settlement. Above this sand pack and pipe (or conduit), backfill material is placed. This varies greatly from location to location and on the type of utility being installed. In some cases, the excavated soil and rock bits from the trench are simply returned. In other cases, sand, gravel, or both may be mixed into the excavated material in order to give it a desired consistency. Topsoil, concrete pavements, or other surficial coverings may be added.

At some of our field sites, the trenches were entirely filled entirely with sand. We conducted permeability tests on the materials both inside (backfill) and outside (undisturbed) the trenches. Locations in the Austin, Texas, vicinity were chosen both for their distinctive geologic setting and the availability of site access. Two of the major hydrogeological systems in the Austin metropolitan area were evaluated – settings underlain by the Cretaceous Glen Rose and Edwards aquifer limestones (Sharp and Banner, 1997) and settings underlain by Quaternary terrace and alluvial deposits associated with the Colorado River and its tributaries.

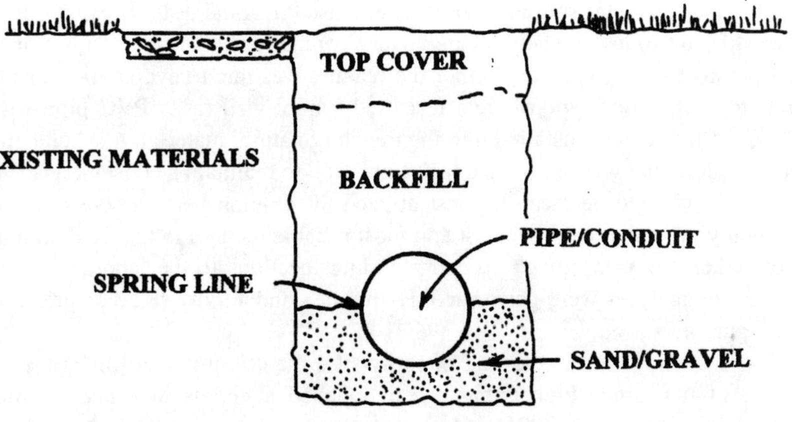

Figure 3. Elements of a utility trench. Distinct permeability zones exist: the existing natural geologic or anthropic materials; the sand or gravel pack; the utility pipe or conduits, which may or not be breached; the trench backfill; and topsoil or surficial cover.

Sites located within the Glen Rose limestone consist of the following basic characteristics. The soils tend to be very thin, ranging from 0" to 6" where present. Many of the soils are clay rich, likely due to landscaping. In places where landscaping had not been done, the soils are more of a silt loam with rock fragments. Beneath the soil zone, the sections consist of weathered limestone. This is the basic epikarstic type terrain. The weathering becomes less prominent the farther down we dug until solid bedrock is encountered. This epikarstic zone ranged from a few inches to over several feet. It is very difficult to gauge the thickness accurately without major excavations on the sites, and generally no permission was given for such an undertaking. Of the Glen Rose sites visited, only one was under construction, which allowed us to view the types of fill being used to take grab samples. The other sites had all been fairly well established with the buildings and parking areas having been there for more than 5 years. The trenches in these areas appeared to be filled with a clean sand/gravel pack to either the spring line on the pipe or half way up the trench. The remainder was a mixture of sand and crushed material taken from the trench during digging.

The Colorado River alluvium was chosen in order to get a wider range of conditions for the study. The alluvium consists of inter-bedded sands, silts, and clays. Some areas also have a large amount of gravel mixed into the sediment. The soils above the alluvium ranged in thickness from near zero in areas with little developed soil and over 15 cm (6 inches) in areas of good soil development. The soil tends to be very organic rich clay. Beneath the soil zone, the alluvium consists of layers of varying thickness, alternating between sands (with some gravel) and silts. Some exposures of material in the area also show this same sequence and there are areas that are heavily mined for the sands and gravel. These sites are mostly along the roadways and in some trailer park communities.

Several field tests were used to determine the hydraulic conductivities. A single 1-meter diameter ring infiltrometer was used in some tests. However, it was very difficult to use the ring at sites with very clay- rich soils. At times, it was not possible to distinguish whether the reading was due to evaporation or infiltration. The second type of field test used a mini-well (a 2" PVC pipe with a .010" slotted screen) installed into the trench or natural material. A special bucket with plumbing was mounted on the mini-well to maintain a constant head. The third type of field permeability test utilized the Guelph Permeameter. This test generally proved to be the fastest and most reliable method in the field. Samples were taken from the trench and surrounding locations to the laboratory where grain-size analyses were performed. Hydraulic conductivity (K) data are shown on Figure 4.

The data show that in the vast majority sites the utility trench fills are several orders of magnitude higher than the untrenched materials, whether natural or anthropic (Underwood, 2001). Most of the increases are greater than 2 orders of magnitude. In the alluvial materials, the general increase in permeability was larger because soil development on fine-grained overbank flood materials created low

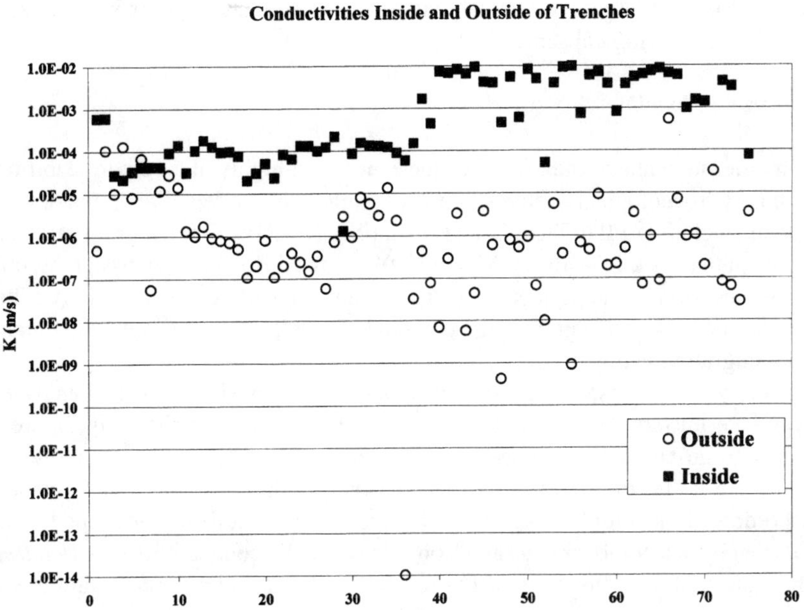

Figure 4. Hydraulic conductivity (K) data for pairs of measurements (outside and inside/above utility trenches) in the Austin, Texas, urbanized area: 1—38 are sites in the Quaternary alluvial terrace deposits associated with the Colorado River; and 39-77 are sites that overlie Glen Rose and Edwards aquifer limestones.

permeability surficial zones. At these sites, the backfills and surficial materials were generally well-sorted alluvial sands and fine gravels. In the limestone soils of the Glen Rose Limestones and Edwards aquifer limestones (the Georgetown, Person, and Kainer Formations, Sharp, 1990), there were occasional fine-grained clay rich backfills used so that in 4 sites the trench material was less permeable than the natural materials. One had a very low permeability clay rich material outside the utility trench so the trench materials were 10 orders of magnitude more permeable. The inhomogeneity of the trench materials (backfill and top cover) is great, although in our data the untrenched materials appear to be even less homogeneous. The backfill is generally the most accessible local materials. The sand/gravel fill that is generally placed in the trench bottom up to the spring line of the pipes and conduits would be expected to have hydraulic conductivities that range between 10^{-5} to 10^{-2} m/sec. Urbanization in these two settings has increased the permeability of the surficial hydrogeological system. The magnitude and importance of this effect will depend upon the number, size, continuity, and orientation of the trenches. Because of the linear nature of the trenches, the system is also probably highly anisotropic. Furthermore, these changes have occurred in each area over a few years or decades. This is an extremely rapid increase in a geological sense.

Numerical Modeling

In order to evaluate the effects of these new and highly altered permeabilities on the hydrogeological systems, we used numerical models. We tested several models, including FRAC3DVS, and Visual MODFLOW. Our trial runs led us to use the latter for this analysis. MODFLOW allowed flexibility in assigning utility system designs and permeability patterns and the particle tracking subroutines permit an easy visualization of the potential effects of utility systems of contaminant migration pathways.

There are, of course, many scenarios that can be envisioned in such analyses. Typical results are depicted in Figure 6 for the simple setting shown in Figure 5. This is a one-kilometer by one-kilometer grid, with the black areas denoting the path of an interconnected utility trench system. In this case, the trench material is 4 orders of magnitude more permeable than the undisturbed, untrenched materials that have a homogeneous and isotropic hydraulic conductivity of 10^{-6} m/sec. The pattern of the connected utility trenches is shown in Figure 6. A simple single layer with a saturated thickness of 1 meter is assumed. The flow system is

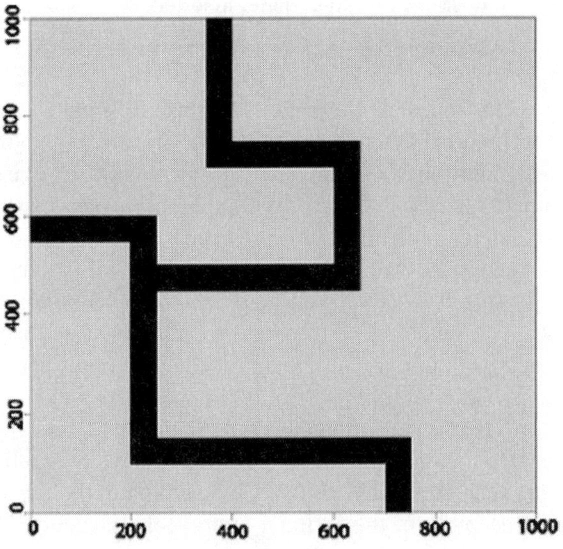

Figure 5. A hypothetical utility system is superposed in a simple, naturally one-dimensional flow system (top to bottom). A permeability contrast of 4 orders of magnitude exists between the utility trenches and the natural geological materials.

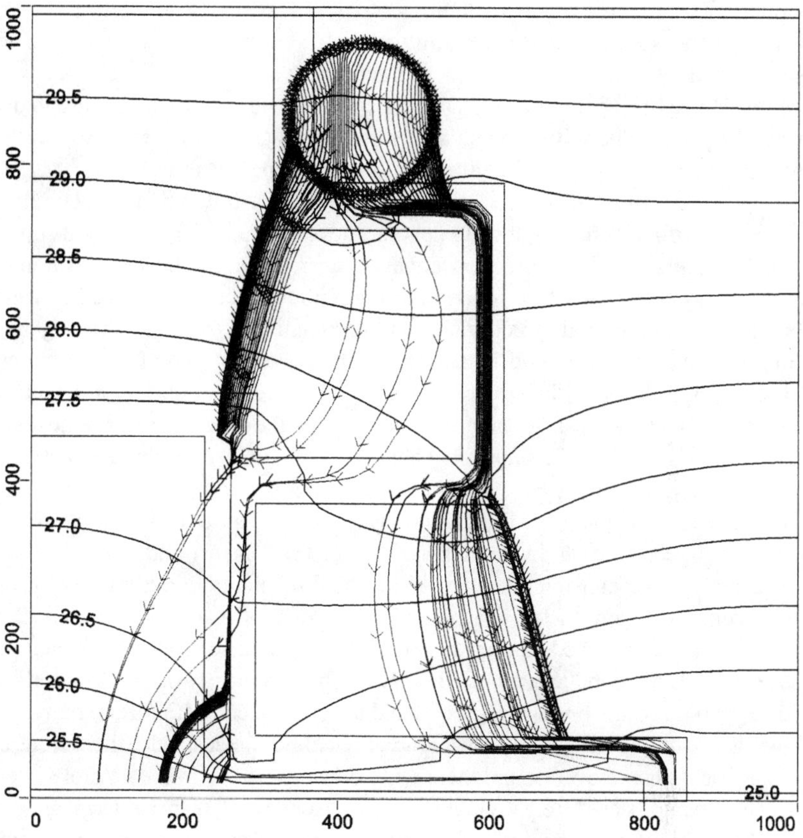

Figure 6. Pathlines originating from a single ring-shaped source. The utility trenches accelerate solute/contaminant transport and create multiple plumes that migrate differently than might be predicted from standard analyses. The trench system can be come a locus for point sources of contamination.

one-dimensional from top to bottom with no-flow lateral boundaries. Constant head boundaries of 30 and 25 m are imposed at the upper and lower boundaries, respectively. This yields a hydraulic gradient of 5×10^{-3} that would be uniform under undisturbed conditions. The flow system is in a steady state.

The perturbation in flow systems for a conservative (nonsorbing, nondecaying) solute or contaminant introduced at the ring site is shown in Figure 6. The transport pathlines are depicted. Several interesting patterns are evident. First, although the general transport direction coincides with the hydraulic gradient, transport directions in some areas diverge over 85 degrees. This is also not uncommon in fractured or karstic hydrogeological systems. Second, the transport rates once the solute reaches the utility trench system are much more rapid. Again, this is similar to observations in fractured, karstic settings. Third, several plumes have been generated from a single source and that are migrating in differing directions. The utility trench system becomes a potential point source for development of new plumes in either shallow or deeper systems.

In a previous study (Halihan et al., 2000), we demonstrated that a fractured or equivalently urbanized medium can generate several plumes from a point source if the hydraulic gradient direction changes over time. In the case examined above, the system is in steady state; the hydraulic gradients are fixed in time, although they are altered directionally and in magnitude from pre-existing conditions by the effects of the utility trenches. Finally, we ignored in the above scenario the possibility that the pipe or conduit itself could be ruptured or breached. This could create even a greater permeability in the trench system and even greater unpredictability in transport because of possible interchanges between the trench materials, the pipes, and the pre-existing materials.

Figures 7 and 8 depict flow lines a two-layer saturated system. The utility trench system of Figure 5 exists in the upper (5-meter thick layer) and beneath it is a homogeneous 20-meter thick layer with identical boundary conditions as for the above scenario. Again, the trench material is 4 orders of magnitude more permeable than the undisturbed, untrenched materials that have a homogeneous and isotropic hydraulic conductivity of 10^{-6} m/sec. The complexity of the flow paths is evident on both the plan view (Figure 7) and the cross section (Figure 8) that show all the flow paths. Note that the hydraulic gradient has been steady; there were no site-specific recharge events; and there was no heterogeneity in the natural system as would be expected (e.g., Galloway and Hobday, 1997; Koltermann and Gorelick, 1996). Consequently, predicting contaminant transport directions could be problematic in urban settings similar to those represented in the model.

In other model scenarios, the utility trenches functioned as system drains and collected solutes or contaminants from up gradient sources. Consideration of multiple levels of such heterogeneities as might exist in older cities, unsaturated-saturated zone interactions, the effects of local recharge sources (lawn irrigation or leaky water or sewage mains), and variations in gradient direction create yet more complications. The models make it clear that the utility trenches can dramatically affect both rates and directions of solute or contaminant transport. Our ability to make predictive models useful in remediation may also be greatly diminished in areas of shallow groundwater and high-permeability utility conduits or other urban hydrogeologic systems.

WATER MANAGEMENT TOOLS

Underutilized groundwater resources will be needed to meet the demands of increasing global urbanization. New technologies will offer new opportunities. A

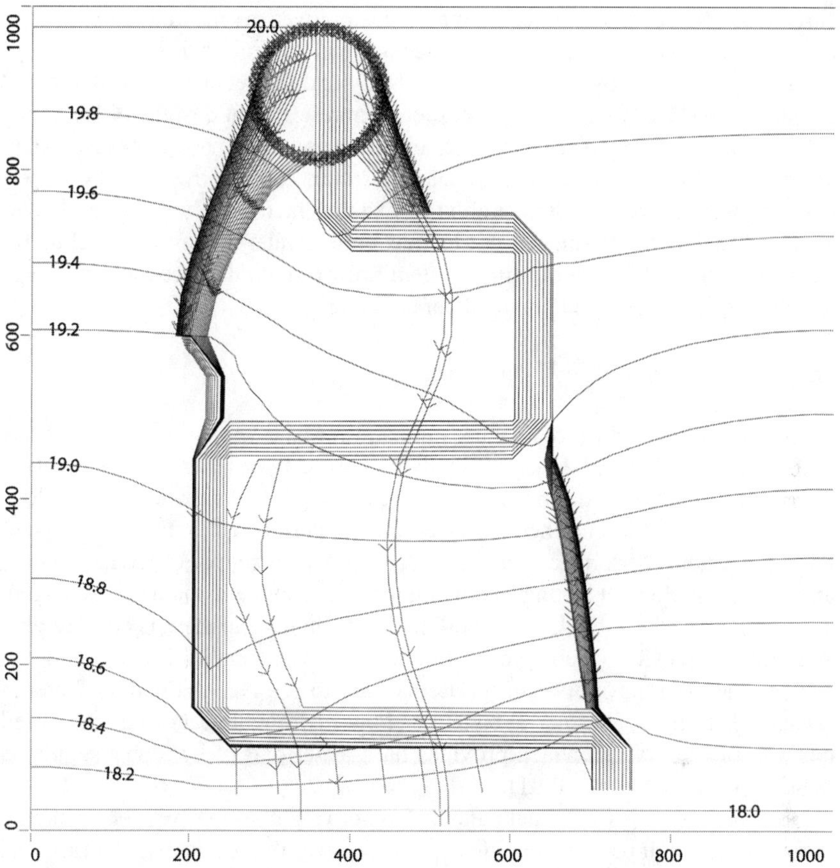

Figure 7. Pathlines originating from a single ring-shaped source. The system is similar to that shown in Figure 6 except that there is a homogeneous lower layer. Note that deeper flow paths may bypass the concentrated flow in the utility trench system.

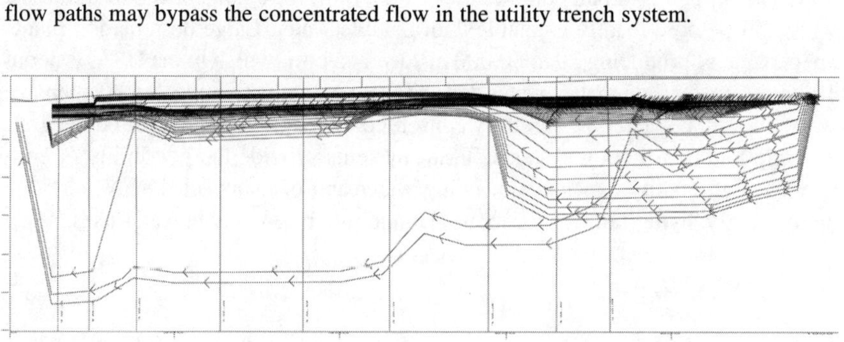

Figure 8. Cross section showing flow paths. Figures 6 through 8 demonstrate that utility systems can make the prediction of solute (contaminant) transport exceedingly difficult.

variety of management strategies must be employed to maximize our use of urban groundwater systems and to protect critical natural environments. These strategies will basically increase the flexibility and the redundancy of the water resources used by cities or they will create more efficient designs of the urban infrastructure. In the first category, we include conjunctive use of groundwaters and surface waters, water banking, aquifer storage and recovery (ASR), inter-basin transfers, and flexible operation of existing facilities. In the second category, we include increasing or optimizing groundwater recharge, recycling/re-use of waters, and design of utility system infrastructure to account for leakage and transport of water and potential contaminants.

Increasing Flexibility and Redundancy

Aquifer storage and recovery uses the aquifer as a temporary storage reservoir. Suitable quality water is artificially recharged by spreading basins or, typically, injection wells during periods of excess precipitation or stream flow, stored in the aquifer, and then extracted during periods of drought. The essential requirements for artificial recharge are a plentiful supply of recharge water and a hydrogeological setting to contain that water once it has been recharged. The London Basin in the UK has both these attributes and there is considerable potential for ASR to supplement London's water supplies. An operational scheme has been developed in the Lee Valley to the NE of London. Water is recharged at times of surplus to be pumped out in drought years when the stored quantity of river water, which provides the greater part of London's supply, is inadequate (Owen et al., 1991).

Desalination or partial desalination of otherwise unusable waters, including groundwaters, will become more feasible because of the relatively great financial resources of the urban areas. Such waters can be mixed with other waters to reach a suitable quality. Desalination is a very energy-intensive process, but it is economic in supplying urban communities in the oil-rich regions of the Middle East where oil is more readily available than potable water. Large desalination plants are capable of producing 20 to 30,000 m^3/day at a cost, which in the USA, is about 10 times greater than that of water derived from any other source. Such expensive water should, however, be carefully conserved; this option might best be integrated with dual-distribution systems. These systems provide the possibility of conserving high quality expensive drinking water and of using brackish waters that are currently "lying fallow" for industrial and other nonpotable water uses.

More Efficient Design

In El Paso, tertiary-treated sewage is injected into the alluvial aquifer and pumped out down the flow path. In temperate areas, warm water from industrial plants can be injected in the summer and then pumped in the winter months and its heat extracted. Karstic aquifers can serve as natural canals or pipelines for transmission of water from areas of recharge to the cities. Where feasible, this will save construction and associated costs. Finally, we speculate that *a priori* assessment of the effects of utility systems and detailed hydrogeologic mapping could lead to efficient urban designs that encapsulate the inevitable leakage and alterations of the shallow hydrogeologic system. The designs could consider interception of contaminants or shallow aquifer flow systems, directing flow to critical areas, and consideration of recharge from manmade alterations to the urban environment as an important water resource.

CONCLUSIONS

Our data document orders of magnitude increases in permeability along utility trenches in carbonate rock and alluvial settings. These equate to an exceedingly rapid rate of porosity and permeability development. Permeabilities are predicted to be even higher if the utility pipes/conduits leak, but this is not tested in our data and models. It may be difficult to determine directions for groundwater flow and contaminant transport in such inhomogeneous and anisotropic systems. Numerical models of flow representing high-permeability trench systems show complex or multiple plumes may develop even without changes in the direction of the regional hydraulic gradient. The problem may be exacerbated in older cities that may have had many generations of manmade alteration, including installation of utility systems, to the hydrogeological environment.

Opportunities for the use of groundwater in urban areas are being created by evolving technologies, including aquifer storage and recovery, groundwater heat pumps, dual distribution systems, artificial recharge, and use of aquifers to reclaim treated sewage effluent. Many large cities in the Unites States underutilize groundwater so these waters provide a significant future urban water resource. Careful design of trenches and tunnels could manage recharge from utility system leakage and shallow aquifer flow systems. By understanding the network of channels that urbanization creates, we may be better able to assess and remediate groundwater contamination from near surface sources. Groundwater offers unique advantages for supplying water to the urbanizing Earth. Careful aquifer development and conjunctive use can provide low-cost, redundant systems to address growing demands. Deep aquifers are relatively insulated from urban pollution and from climatic extremes. They can also provide insurance against droughts and catastrophic contamination events. In addition, the costs of groundwater alternatives are often less than for new supplies from surface-water resources.

Hydrogeological data are required to prevent water-quality deterioration and the critical overdraft of the aquifers with potential deleterious secondary effects. These secondary effects include subsidence, salt-water intrusion, degradation of groundwater quality, and loss of critical natural habitats, as well as aquifer depletion. New and continually evolving technologies create opportunities for the use of groundwater in urban areas. These now include aquifer storage and recovery, groundwater heat pumps, integration of groundwater supply into dual distribution systems, artificial recharge, and using aquifers to reclaim treated sewage waters. Detailed hydrogeological mapping and analyses must be accomplished before key decisions are made so that we can effectively use all of our water resources. Too often we are forced to make decisions in the absence of key data and the eventual consequences can be costly. We must begin to evaluate our hydrogeological resources on a scale more detailed than has been commonly done previously.

Acknowledgments. We acknowledge The National Science Foundation (Grant #EAR-9902899) and an AT&T Industrial Ecology Fellowship for support of this research. The Owen-Coates Fund and the Chevron Centennial Professorship of the Geology Foundation, The University of Texas, supported for manuscript preparation. We deeply appreciate the landowners and companies who let us sample the permeability of utility trenches on their properties.

REFERENCES

Austin-American Statesman, 16 July 1998, http://archives. statesman.com/, 1998.

Barton, N. J., *The Lost Rivers of London*, (Republished 1982) Historical Publications Ltd, New Barnet, Herts., UK, 1962.

Basu, S. R., and H. A. C. Main., Calcutta's Water Supply: Demand, Governance and Environmental Change, *Applied Geography*, *21*, 23_44, 2001.

Escolero, O., L. E. Marin, B. Steinich, and J. Pacheco, Delimitation of a Hydrogeological Reserve for a City Within a Karstic Aquifer: the Merida, Yucatan, Example, *Landscape and Urban Planning*, *51*, 53_62, 2000.

Fara, M., Sewage Water Disposal and Aquifer Pollution in Arid Lands: A Case Study: Sana'a Waste Water Discharge, in *Hydrogeology of Hard Rocks* (Part 2), Memoires of the 24th Congress, International Association of Hydrogeologists (eds., D. Banks and S. Banks), Oslo, Norway, 813_823, 1993.

Foster, S., Groundwater Quality Concerns in Rapidly Developing Countries, in *Hydrology and Hydrogeology of Urbanizing Areas* (eds., J. H. Guswa, et al.): American Inst. Hydrology, St. Paul, MN, MIU12_MIU26, 1996.

Galloway, W. E., and D. K. Hobday, *Terrigenous Clastic Depositional Systems* (2d ed.), Springer-Verlag, Berlin, 1996.

Geomatrix, Inc., *Conceptual Hydrogeologic Model, Charnock Wellfield Regional Assessment*, unpublished report, Los Angeles, CA, 1997.

George, D .J. 1992. Rising Groundwater: A Problem of Development in Some Urban Areas of the Middle East, in *Geohazards - Natural and Man-made* (eds., G. J. H. McCall, D. J. C. Laming, and S. C. Scott), 171_182, Chapman and Hall, London, 1992.

Gleick, P. H., 2001, Making Every Drop Count: *Scientific American*, February, p. 43.

Halihan, T., C. N. Hansen,, and J. M. Sharp, J. M., Jr., LUST Characterization in Urban Fractured Karstic Aquifers, *Joint Annual Meeting, Texas Sections of Association of*

Engineering Geologists and American Institute of Professional Geologists, Austin, Texas, Program and Abstracts, 12, 2000.

Havlick, S. W., *The Urban Organism*, New York: Macmillan, New York, 1974.

Heath, R. C., Hydrogeologic Setting of Regions, in *Geology of North America, Hydrogeology* (eds., W. Back, J. S. Rosenshein, and P. R. Seaber), Geological Society of America, Boulder, CO, *O-2*, 1988.

Headworth, H. G., and G. B. Fox, The South Downs Chalk Aquifer: Its Development and Management, *Jour. Institution. Water Engineers and Scientists*, *40*, 345_361, 1986.

Hibbs, B. J., R. N. Boghici, M. E. Hayes, J. B. Ashworth, A. T. Hanson, Z. A. Samani, J. F. Kennedy, B. J. and Creel, B. J., *Transboundary Aquifers of the El Paso/Ciudad Juarez/Las Cruces Region*, Report to U. S. Environmental Protection Agency, 1997.

Holzschuh, J. C., Land Subsidence in Houston, Texas, USA, in *Fourth International Symposium on Land Subsidence*: Houston, 1991.

Knipe, C. V., J. W. Lloyd. D. N. Lerner, and R.Greswell, Rising Groundwater Levels in Birmingham and the Engineering Implications, *CIRIA Special Pub.*, *92*, 1993.

Koltermann, C. E., and S. M. Gorelick, Heterogeneity in Sedimentary Deposits: A Review of Structure-imitating, Process-imitating, and Descriptive Approaches, *Water Resources Research*, *32*, 2617_2658, 1996.

Krothe, J. N., B. Garcia-Fresca, B., and J. M. Sharp, Jr., Effects of Urbanization on Groundwater Systems, in *Groundwater and Human Development*, Proceedings of the 32nd Congress, International Association of Hydrogeologists, Mar del La Plata, Argentina, in press, 2002.

Lerner, D. N., Too Much or too Little: Recharge in Urban Areas, in *Groundwater in the Urban Environment: Problems, Processes and Management*, (ed. J. Chilton): Proceedings, 27th Congress, Int. Assoc. Hydrogeologists, Rotterdam, A. A. Balkema, *1*, 67_78, 1997.

Lorenzo-Rigney, B., and J. M. Sharp, J. M., Jr., Urban Recharge in the Edwards Aquifer, *Geol. Soc. America, Abs. with Programs* (South-Central Sec.), *31*, A-12, 1999.

Lowe, M. D., Shaping Cities: in *State of the World 1992*, W. W. Norton, New York, 1992.

Lumsden, G.I., *Geology and the Environment of Western Europe*, Clarendon Press, Oxford, 1994.

Lucas, H. C., and V. K. Robinson, Modelling the Rising Groundwater Levels in the Chalk Aquifer of the London Basin, *Quart. Jour. Eng. Geol.*, *28*, S51_S62, 1995.

Mather, J. D., The Impact of Contaminated Land on Groundwater, A United Kingdom Appraisal, *Land Contamination and Reclamation*, *1*, 187_195, 1993.

Mather, J. D., I. M. Spence, A. R. Lawrence, and M. J. Brown, M. J., Man-made Hazards, in *Urban Geoscience* (eds., G. J. H. McCall, E. F. J. de Mulder, and B. R. Marker), A.A. Balkema, Rotterdam, 127-161, 1996.

McKinney, D. C., and J. M. Sharp, Jr., *Springflow Augmentation of Comal and San Marcos Springs, Texas: Phase I - Feasibility Study*, The University of Texas Center for Research in Water Resources Report 247, Austin, TX, 1995.

Meinzer, 0. E., Groundwater Regions of the United States, *U. S. Geological Survey Water-Supply Paper 2242*, 1923.

Morris, B. L., A. R. Lawrence, and M. E. Stuart, *The Impact of Urbanisation on Groundwater Quality (Project Summary Report)*, Technical Report WC/94/56, British Geological Survey, Nottingham, U. K., 1994.

Owen, M., H. G. Headworth, and M. Morgan-Jones, Groundwater in Basin Management, in *Applied Groundwater Hydrology - a British Perspective* (eds., R. A. Downing and W. B. Wilkinson), pp. 16_34, Clarendon Press, Oxford, 1991.

Sharp, J. M., Jr., Stratigraphic, Geomorphic and Structural Controls of the Edwards Aquifer, Texas, U. S. A., in *Selected Papers on Hydrogeology* (eds., E. S. Simpson and J. M. Sharp, Jr.), *1*, 67_82, International Association of Hydrogeologists, Heise, Hannover, 1990.

Sharp, J. M., Jr., 1997, Ground-water Supply Issues in Urban and Urbanizing Areas, in *Groundwater in the Urban Environment: Problems, Processes and Management*, Proceedings of the 27th Congress, International Association of Hydrogeologists (ed. J. Chilton), *1*, 67_74, Nottingham, A. A. Balkema, Rotterdam, 1997.

Sharp, J. M., Jr., and J. L. Banner, 1997, The Edwards Aquifer - a Resource in Conflict, *GSA Today*, *7*, 1_9, 1997.

Sharp, J. M., Jr., and J. L. Banner, The Edwards Aquifer, Water for Thirsty Texans: in *The Earth Around Us: Maintaining a Livable Planet* (ed., J. S. Schneiderman), W. H. Freeman, pp. 154_165, 2000.

Sharp, J. M., Jr., C. N. Hansen, and J. N. Krothe, Effects of Urbanization on Hydrogeological Systems: The Physical Effects of Utility Trenches, in *New Approaches Characterizing Groundwater Flow*, Proceedings of the 31st Congress, International Association of Hydrogeologists (eds., K.-P. Seiler, and S. Wohnlich), supplement vol., Munich, Germany, 2001.

Sharp, J. M., Jr., C. N. Hansen, J. D. Mather, and C. A. Stewart, Effects of Urbanization on Groundwater Resources [abs.], *EOS*, *81*, no. 19, S9, 2000.

Sharp, J. M., Jr., and J. N., Krothe, 2002, Anthropogenic Effects on Water Budgets in Urban Areas, in *Balancing the Groundwater Budget*, International Association of Hydrogeologists, Darwin, Australia, Abstracts, *1*, 69, 2002.

Slattery, R. N., and D. E. Thomas, *Recharge to and Discharge From the Edwards Aquifer in the San Antonio Area, Texas, 1999*, U. S Geological Survey [web-only publication], http://tx.usgs.gov/reports/dist/dist-2000-01, 2000.

Technical Advisory Panel, *Technical Factors in Edwards Aquifer Use and Management*, Final Report to the Special Joint Texas Senate and House of Representatives Committee on the Edwards Aquifer, 1990.

Uliana, M. M., and J. M. Sharp, Jr., Springflow Augmentation Possibilities at Comal and San Marcos Springs, Edwards Aquifer, *Transactions, Gulf Coast Assoc. Geol. Societies*, *46*, 423_432, 1996.

Underwood, J. R., Jr., Anthropic Rocks as a Fourth Basic Class, *Environmental and Engineering Geoscience*, *7*, 104_110, 2001.

United Nations, *World Urbanization Prospects, 1990*. New York, 1991.

University of Birmingham, *North and South Chalk Modelling Study*: Final Report, Anglian Water, Huntingdon, UK, 1987.

10

Integrated Environmental Modeling of the Urban Ecosystem

Timothy N. McPherson, Steven J. Burian, Michael J. Brown,
Gerald E. Streit, and H.J. Turin

INTRODUCTION

The health of human populations and ecological systems is increasingly tied to the quality of the urban environment. The United Nations Population Division projects the world population will undergo a major transition by 2005 from a rural population to a primarily urban population. This trend has already occurred in the United States (US) where 80% of the population lives in cities [US Census Bureau, 2001]. Urban environments are complex aggregations of interrelated social, economic, physical, and biological systems. The structure and nature of these interrelationships can have significant consequences for human health and local and regional environmental quality [World Resources Institute *et al.*, 1996]. The increasing importance of these integrated systems can be noted by the recent decision of the National Science Foundation to fund two Urban Long Term Ecological Research projects in Baltimore, MD and Phoenix, AZ [LTER Network, 2001].

The definition of an urban environment is constantly evolving. The US Census Bureau defines an urbanized area as at least one central place and adjacent densely settled surrounding territory that together have a minimum population of 50,000 people [US Census Bureau, 1999]. In the Census Bureau definition, the densely settled area typically consists of a continuous residential development and a general overall population density of at least 1,000 people per square mile. Accompanying the increased population density is a large amount of physical

Earth Science in the City: A Reader

10.1029/056SP11

infrastructure such as buildings, road networks, sewer lines and storm drains. This physical infrastructure coupled with the increased waste load generated by the agglomeration of a large number of people in a relatively small area can significantly degrade environmental quality. The development and increasing growth of urban populations can increase the pressure on local resources by increasing the local demand for goods and services and by transforming the associated ecosystems to ones that are more conducive to human settlement. Often, these alterations in land use adversely affect the surrounding ecological systems. Sauvajot *et al.* [1998] noted a trend between urbanized areas in Southern California and species distributions in the Santa Monica Mountains in Los Angeles, CA. They found a significant alteration in species composition near roadways with lower densities of mammals endemic to chapparal vegetation and higher densities of disturbance tolerant species. Land use alteration also has been shown to degrade water resources in watersheds of the Pacific Northwest by increasing channel instability and altering channel morphology [Bledsoe and Watson, 2001; Finkenbine *et al.*, 2000]. These ecological systems provide essential services to human societies and their continued degradation is of concern [Costanza *et al.*, 1996].

Cities have been shown to impact the local weather by perturbing wind, temperature, moisture, turbulence, and surface energy budget fields. Numerous investigations have shown that buildings and urban landuse significantly modify microscale and mesoscale flow fields [Bornstein, 1968; Hosker, 1987]. In the late 1960's and early 1970's, a number of studies on urban air quality and circulation found the urban climate system was multi-dimensional and complex with numerous feedback mechanisms between components. In addition, it was apparent that drag and turbulence created by the roughness of buildings were large enough to reduce the strength of the mesoscale wind and enhance boundary-layer-scale mixing. [Ackerman, 1974; Angel *et al.*, 1971; Atwater, 1972; Bornstein, 1975; Clarke, 1969; Ludwig, 1970; McElroy, 1973; Myrup, 1969; Oke and East, 1971]. Smith *et al.* [1998] found that buildings can act as loci for the development of thermal eddies that are capable of altering tracer dispersion in a simulated environment. In a follow-up study, Smith *et al.* [2001] further evaluated the complex effects of a single building on atmospheric flow. In that study, they found thermal heating of the building and ground surface coupled with the vortex circulation associated with the eddies alongside of the building produced a significant convergence of air within the cavity zone of the building, which in turn promoted the lofting of the air mass immediately downstream of the building.

Urban areas have also been shown to have different temperature profiles than rural areas due to multiple factors such as decreased longwave radiation loss due to reduced sky factor (i.e., the building walls trap, or intercept, infrared radiation that would otherwise escape up into the sky), increased downward longwave radiation from the warmer air above the city, increased shortwave absorption, decreased evapotranspiration due to less vegetation and moisture availability, increased anthropogenic heat input, increased heat storage by canopy elements and reduced heat transport due to smaller wind speeds and turbulent mixing in the urban canopy [Oke, 1987]. The differences in wind flows brought about by roughness and altered heat exchange in urban areas affect the transport of contaminants. Urban environments are also significant sources of air contaminants due to the wide range of industrial, commercial, and transportation activities occurring within them. The altered transport of airborne contaminants could have important ramifications for urban citizens.

Urban development has also been shown to alter the hydrology of watersheds turning them into net exporters of contaminants. Changes in land use in urban environments often increase the imperviousness of land surfaces. When those hydrologic alterations are coupled with the greater utilization of hazardous materials in industrial, commercial, and residential land uses increased pollutant loading can occur during wet weather conditions. The 1983 NURP study [Athayde, 1983] found that increased imperviousness coupled with land-use changes made nearly all cities studied exporters of nutrients, toxics, or conventional pollutants, such as biochemical oxygen demand and total suspended solids. The effect of urban systems on water quality and ecosystem health has been noted especially in coastal systems in the US [Hicks, 1997; Hinga, 1991].

An added complication to management of urban environments is the cross-media nature of many pollutants. Franz *et al.* [1998] noted that PCB ambient air concentrations were greater in the Chicago area than in the surrounding regions and that those concentrations were a source of contamination to Lake Michigan via dry deposition. Hornbuckle *et al.* [1992] found PCB concentrations over Lake Michigan were greater closer to Green Bay than the concentrations in more rural regions of the Lake. Those elevated concentrations were found to be the result of volatilization from the more contaminated underlying water column in the Green Bay area.

Another anthropogenic cross-media impact to water quality that has been targeted in recent years is atmospheric deposition of nutrients [Greenfelt and Hultberg, 1986; Hicks 1997; Hicks, 1998]. Studies have shown a significant proportion of contaminant loads to many coastal water bodies originates from atmospheric deposition [Hicks 1998; Paerl *et al.*, 2000]. Pitt [1987] stated that atmospheric deposition, deposition from activities on paved surfaces (auto traffic, material storage, etc.), and the erosion of material from upland connected areas are the major sources of pollutants in urban runoff. A detailed understanding of the urban ecosystem is needed to account for its relationship to the total waste load and the local and regional transport of that pollutant load. The integrated nature of urban infrastructures and ecological systems and the capacity for contaminants to cross media following discharge requires integrated assessment tools for effective management.

Stormwater runoff, solid waste, hazardous waste, wastewater, receiving-water quality, and air quality typically fall under the purview of different agencies and departments at the federal, state, and local levels. Therefore, the tools derived to assist in the management of these issues have predominantly focused on a single medium. Detailed deterministic and stochastic models have been developed to address each problem, but few of these models have considered the connection between different environmental media, i.e., air, surface water, groundwater, sediment, soil, and urban infrastructure. Due to the cross-media nature of many contaminants, there has been growing interest in integrated airshed and watershed modeling [Hicks, 1998].

In this paper, we describe a tool for the integrated assessment of air and water quality in an urban environment. First, we discuss issues in linked environmental modeling. Secondly, we present a demonstration of a linked environmental modeling framework in Los Angeles, CA. The linked environmental modeling framework may be used to study the complex interactions between air, water, and land surfaces. In this demonstration, the framework has been applied to study atmospheric deposition of nitrogen compounds and urban stormwater runoff.

INTEGRATED ENVIRONMENTAL MODELING BACKGROUND

Numerous multimedia models have been developed to address the connectivity between environmental compartments, such as surface water, air, or sediments [MacKay and Paterson, 1981; MacKay and Paterson, 1991]. These models are based on the fugacity (or escaping pressure) of a chemical in an environmental compartment. Early versions of this approach were based primarily on chemical thermodynamics using steady-state assumptions for mass transport. Recent work has included greater specification of mass transport phenomenon [Harner and MacKay, 1995; Paraiba *et al.*, 1999; Suzuki *et al.*, 2000], which can be significant in an urban environment. Fugacity and other partitioning models are best suited to toxic contaminants such as persistent organic pollutants.

In recent years, advances in computing capabilities have increased the interest in integrating preexisting detailed deterministic models into a framework for simulating real-time dynamics in urban and agricultural environments. In the US, significant advances have been made in the integration of airshed, watershed, and water body models as part of studies on the Chesapeake Bay. Models of the Chesapeake Bay airshed, watershed, and tidal waters have been created and linked to model daily atmospheric deposition loading and the impacts on Bay water quality and resources (e.g., underwater grasses, benthic communities, pelagic fish habitat) [USEPA, 1997]. In particular, the Regional Acid Deposition Model (RADM) has been used to delineate the airshed contributing nitrate to the Chesapeake Bay watershed and water surface [Dennis, 1997]. Further research has estimated the direct and indirect loadings of nitrogen compounds to the Bay using monitoring data and linked airshed-watershed-water body modeling [Linker, *et al.*, 1993; Linker and Thomann, 1996; Wang, *et al.*, 1997].

A significant quantity of integrated analysis of the indirect atmospheric pollutant loading via riverine exports from watersheds was conducted as part of an assessment of nitrogen loads to US estuaries [Alexander *et al.*, 2000]. This work utilized the Spatially Referenced Regression on Watershed Attributes (SPARROW) model system. SPARROW uses a statistical approach to watershed loading. The calibrated model indicated atmospheric deposition contributed 4 to 330 kg/km^2-yr to watershed nitrogen exports. This represented 4% to 35% of the total nitrogen load from the studied watersheds. Researchers in the Lake Michigan Mass Balance Study (LMMBS) have also been conducting a significant amount of integrated modeling and assessment [USEPA, 2000]. The LMMBS focuses on PCBs, trans-nonachlor, atrazine, and total mercury due to the historical water quality problems related to those contaminants. The LMMBS includes extensive monitoring and the development of a mass balance model that links transport and transformation sub-models to study the changes in concentrations in the air, water, soil, and biota that would result from changes in watershed or atmospheric pollutant loading. The Rouge River Wet Weather Demonstration Project (RRWWDP) is another integrated airshed, watershed, and water body modeling and management project. This project has primarily focused on the integration of the US EPA Storm Water Management Model (SWMM) and the US EPA Water Quality Analysis Simulation Program (WASP5). This research has produced linkages between the SWMM and WASP5 that may be used to study a wide range of issues. In the RRWWDP, the SWMM to WASP5 linkage has been used to evaluate the effects of combined-sewer overflow policy and best management practice implementation on the water quality in the Rouge River [RRWWDR, 1993]. The relative importance of atmospheric deposition has also been assessed by incorporating monitoring data into the analyses [Amargit, 1994].

Coupling legacy environmental models requires dealing with many issues. These issues include but are not limited to the following: 1) differences in the spatial scale of each model, 2) differences in the processes simulated and 3) differences in the model structure. Brandmeyer and Karimi [2000] discuss approaches to integrating legacy environmental models in detail. Those researchers suggest a protocol for developing integrated modeling and discuss significant issues in resolving the spatial differences and temporal scale of the different models. One of the more important issues is model interoperability. When linking legacy codes, the models must have similar spatial and temporal scales to allow modeling. Often this interoperability is achieved by transferring data following manual or automated input/output modifications.

In linking meteorological, atmospheric chemistry, watershed, and receiving water models, there are a number of important input/output issues to be addressed. Atmospheric chemistry models often have significant domain sizes due to their dependency on local and regional meteorology. Removal processes in those models are typically included to estimate mass loss and maintain mass balance. Often this parameterization accounts only for the total mass lost and not the multitude of processes that produce the mass load to the underlying surface. The process that dominates the mass loading to a particular surface will determine the future fate and transport of the contaminant following deposition. Furthermore, in atmospheric chemistry models, deposition is output as a flux (mass/area) at each time step and water quality models require a load time series (mass/time). Also watershed loading models utilize concentration in precipitation and mass on the watershed surface to determine pollutant discharge to the downstream receiving water. Each of these issues in linking air chemistry models to watershed and receiving water models requires preprocessing which can be simple to complex depending on the watershed and issue at hand.

The primary processes connecting atmospheric compartments to underlying media are wet and dry deposition, resuspension, and volatilization. The transfer of atmospheric deposition through the terrestrial biosphere is presently an area of great uncertainty [Valigura, 1996]. Dry deposition involves the turbulent and gravitational transfer of pollutants from the air to the underlying surface during dry weather [Hicks, 1998]. Wet deposition refers to the droplet processes that scavenge material from the atmosphere during precipitation events and deposits them on the surface. Removal processes on the surface (e.g., street sweeping, resuspension and relocation, plant uptake, nuisance flows) can reduce the amount of deposited material available for washoff during the next rainfall-runoff event. Linkage of air chemistry codes, watershed models, and receiving water models requires the solution of the differences in the spatial scale of each model, the processes simulated and the structure of each model.

INTEGRATED ENVIRONMENTAL MODELING FRAMEWORK

In this demonstration, we describe the initial development of a modeling system that addresses the connectivity between the air, watershed, and surface water while still addressing the large scale effects of the urban environment on the transport and transformation of a given constituent. The modeling framework is at present a one-way data transfer, as described by Brandmeyer and Karimi [2000], where modelers interface with each model and manually transfer data. Several interface support codes have been developed to facilitate data transfer. The linked urban airshed-watershed modeling framework is comprised of five primary models: RAMS, HOTMAC, CIT, SWMM, and WASP5. These models, described below, simulate dry-weather meteorology, wet-weather meteorology, air chemistry, urban stormwater runoff, and receiving-water quality (Figure 1).

The Regional Atmospheric Modeling System (RAMS) [Pielke *et al.*, 1992] and the Higher-Order Turbulence Model for Atmospheric Circulation (HOTMAC) [Yamada and Kao, 1986] are both 3-d prognostic mesoscale meteorological models. Employing finite difference schemes, they solve the geophysical fluid dynamics conservation equations for mass, momentum, heat, and moisture, as well as surface energy budget equations. In this linkage, RAMS simulates wet

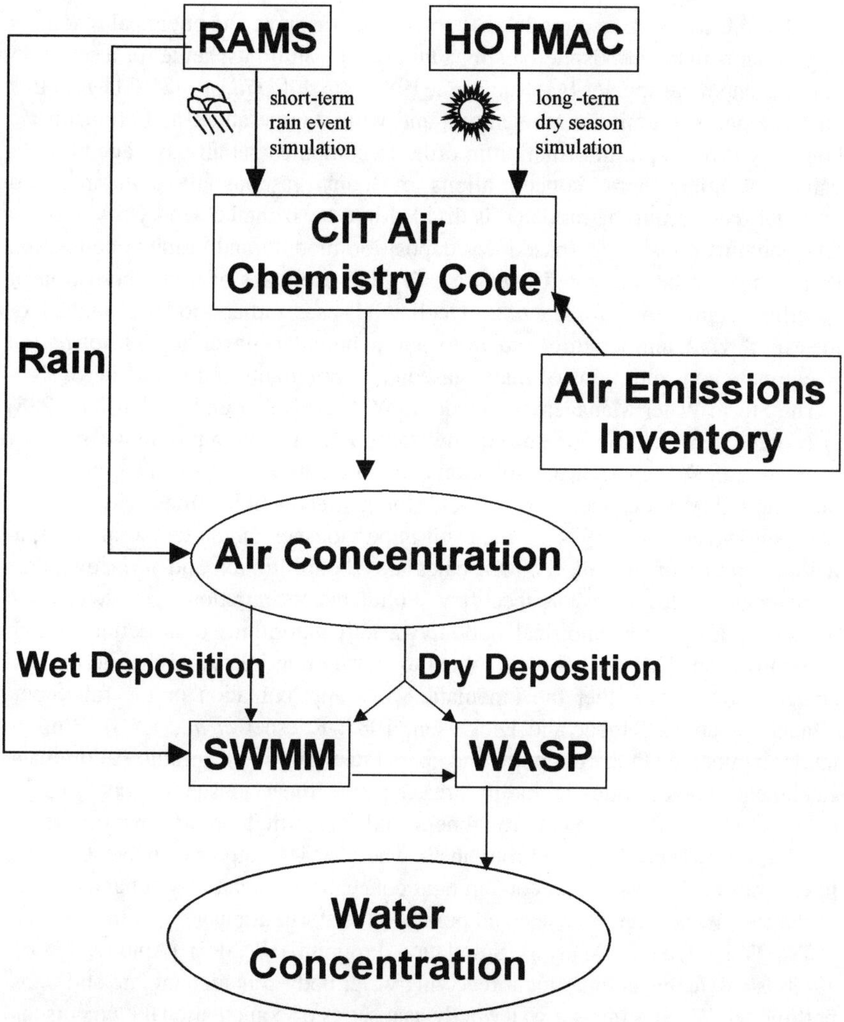

Figure 1. The Urban Air to Water Modeling Framework.

weather conditions using a nested grid approach in order to cover the synoptic scale weather over the Pacific Ocean and Western US and to resolve the region of interest. For the dry weather simulations, HOTMAC was run in hydrostatic mode using a nested grid scheme with an urban canopy parameterization to account for the effect of sub-grid urban effects. The outer-most grid covered the southern third of California and the Pacific Ocean, while the intermediate grid matched the CIT air chemistry domain.

CIT is a Eulerian-based model that solves the transport and chemical reactions of pollutants in the atmosphere using a numerical solution scheme for a set of 35 reacting chemical species [McRae *et al.*, 1982; Russell *et al.*, 1988]. CIT requires land use data, an emissions inventory, and wind, temperature, and atmospheric boundary layer depth information in order to compute spatially averaged hourly values of atmospheric concentrations of many gaseous air pollutants and aerosols. The area to be modeled is divided into horizontal 5 km by 5 km grids. CIT contains a resistance-based dry deposition module and predicts deposition fluxes that can be integrated over time to yield estimates of total deposition of specified compounds for each grid cell. CIT uses either model-produced or measured wind, temperature, and atmospheric boundary-layer depth information to compute concentrations of many gaseous air pollutants and nitrate aerosols.

The Storm Water Management Model (SWMM) [Huber and Dickinson, 1988] is a comprehensive deterministic stormwater runoff simulation program capable of simulating the transport of precipitation and pollutants washed off the ground surface, through pipe/channel networks and storage/treatment facilities, and finally to receiving waters [Nix, 1994]. Given soil imperviousness and a temporal and spatial distribution of rainfall, SWMM calculates the infiltration and surface storage of water and routes the rest as sheet flow. Pollutant concentrations are calculated in the sheet flow using empirical buildup/washoff algorithms or a defined rating curve function. The sheet flow is routed to storm drain inlets and then to the discharge point using either the kinematic wave approximation or the full Saint-Venant Equations [Huber and Dickinson, 1988; Roesner *et al.*, 1988]. Primary model inputs are (1) the temporal and spatial distribution of rainfall, (2) drainage catchment characteristics including area, percent imperviousness, slope, depression storage, and drainage path roughness, and (3) storm drain information including drain geometry, slope, and roughness. The SWMM outputs of interest include time series of flow rates and contaminant concentrations and other characteristics of the stormwater runoff at selected points in the storm drainage system.

The Water Quality Analysis Simulation Program (WASP5) [Ambrose *et al.*, 1983] is a dynamic compartment receiving water body contaminant fate and transport model. WASP5 treats a water body as a series of computational elements and solves equations based on the conservation of mass to determine the fate and transport of chemical constituents using a finite difference solution scheme. Environmental properties and chemical concentrations are considered spatially constant within segments. WASP5 contains EUTRO5, a kinetics module that predicts the effects of nutrients and organic matter on dissolved oxygen and phytoplankton dynamics by simulating the transport and transformation reactions of up to eight state variables within four interacting systems: phytoplankton kinetics, the phosphorus cycle, the nitrogen cycle and the dissolved oxygen balance. WASP5 also has a hydrodynamics module (DYNHYD5) for simulating systems with complex flow regimes. DYNHYD5 is developed around the principles of the conservation of mass and momentum. WASP5 inputs include hydrogeometry, inflows, tidal heights, wind speed, advective and dispersive transport, boundary concentrations, point and diffuse source waste loads, kinetic parameters, constants and time functions, and initial concentrations.

The framework can be implemented with any combination of the models listed in Figure 1. The implementation would likely depend on the goal of the study. We linked the five models in the air-water urban transport modeling framework by developing procedures for transforming output from one model into input for another model. As noted above, the linkage between the models is not automatic. RAMS, SWMM and WASP5 are highly interoperable. The linkages from RAMS to SWMM and SWMM to WASP5 are relatively straightforward. The calculated accumulated rainfall from RAMS can be inserted directly into the SWMM input file, and the flow rates and pollutant concentrations calculated by SWMM at specified discharge locations can be inserted into the WASP5 input file with some unit conversions. The complicated linkage occurs between the CIT and SWMM models and the CIT and WASP5 models. Each of these linkages involves accurately assessing dry and wet deposition. In this case, the CIT dry deposition fields are modified to account for different processes affecting the fate of the deposited material. While many processes exist which can affect the material balance between a watershed and the overlying air column, e.g., nuisance flows, volatilization, re-suspension and relocation, plant uptake, and street sweeping, at present we only account for plant uptake.

The modeling framework was developed and applied to nitrogen cycling in urban systems. Nitrogen was selected for two reasons. First, nitrogen species have been shown to be problematic in both air and water. NOx emissions promote the production of tropospheric ozone through reactions with peroxy radicals in the atmosphere [Seinfeld, 1989], and excessive inputs of nitrate (NO_3^-), ammonia (NH_3), and organic nitrogen compounds to water bodies can lead to harmful algal blooms, eutrophication, anoxia, loss of species diversity, and fundamental changes to ecosystem structure [Novotny and Olem, 1994; Rabalais, 1997; Paerl *et al.*, 2000]. Second, nitrogen compounds transport across media, and emissions to air may adversely impact water bodies following atmospheric reactions and deposition [Paerl, 1997].

The CIT, SWMM and WASP5 models are linked in the following manner. The time-variable deposition amounts calculated by CIT are summed over the simulation period to provide total deposition amounts onto the ground surface. These total loads can then be entered into SWMM as initial loads per subcatchment. Although the link between CIT, SWMM, and WASP5 is seemingly straightforward, the former produces mass fluxes of contaminants to the surface per grid cell while SWMM and WASP5 operate on user defined areas such as a water body element or a subcatchment. The basic unit in CIT is a 25-km^2 grid cell, while SWMM and WASP5 have much smaller unit sizes. SWMM AND WASP5 operational units rely on mass loads to calculate water quality not mass fluxes, so a conversion is required. The grid cells, watershed subcatchments, and water body elements do not, in general, correspond one-to-one. In the model linkage, CIT calculates the dry deposition flux of nitrogen compounds originating from airborne emissions at hourly intervals. The fluxes are summed over the simulated dry weather period to obtain the total load deposited to the land and water surfaces. CIT also calculates concentrations of contaminants in the atmosphere at each time step. At the start of a rainfall event, the concentrations computed at that time step are vertically integrated and used in the calculation of wet deposition. The modified accumulated dry deposition fields are inserted into the SWMM input file as an initial load on the watershed surface. The computed average concentrations of contaminants in the precipitation are then inserted directly into the SWMM input file. SWMM then simulates the washoff and transport of the dry-deposited material and the transport of the material in the precipitation through the stormwater drainage system to the watershed outlet.

Dry Deposition Disaggregation

CIT simulates the major forms of atmospheric nitrogen deposition including nitric acid (HNO_3), ammonium nitrate (NH_4NO_3), ammonia (NH_3), nitrogen oxides (NOx), peroxyacetyl nitrate (PAN), and alkyl nitrate aerosols. A major issue is the fate and transport of these compounds in the water and soil following dry and wet deposition. CIT calculates the dry deposition flux of nitrogen compounds originating from airborne emissions at hourly intervals during dry weather in a set of 25 km^2 grids. At each time step, CIT outputs one dry deposition flux per grid cell, which is the result of an aggregation of the different deposition velocities for each land use within a grid cell. This single value therefore does not accurately represent the dry deposition to individual land uses, land covers, or water surfaces within each grid cell, which limits the estimation of the deposited load to hydrologically effective areas (i.e., impervious surfaces directly connected to the drainage system) of the watershed. Without an accurate estimate of the deposited load on those areas, the fraction of the total deposited load that could potentially be washed off during a rainstorm is unknown. Below we explain our approach to disaggregate the deposited load in each grid cell between pervious and impervious surfaces.

The dry deposition spatial disaggregation approach is based on several assumptions. First, the dry deposition amount for NH_3 is primarily from plant uptake and surface adsorption. To estimate the fraction of dry deposition due to plant uptake we determined the relative fractions of pervious (vegetated) surface, water surface, and impervious (asphalt, concrete, rooftops) surface per grid cell. The deposition velocities to vegetated, impervious, and water surfaces were estimated based on surface resistance assuming unstable meteorological conditions. The fraction of land cover per grid cell and the calculated deposition velocities were used to calculate two weighting factors:

$$WF_w = V_{dw}P_w/[V_{dw}P_w + V_{dg}P_g + V_{du}(1\text{-}P_w\text{-}P_g)] \quad (1)$$

$$WF_g = V_{dg}P_g/[V_{dg}P_g + V_{dw}P_w + V_{du}(1\text{-}P_w\text{-}P_g)] \quad (2)$$

where WF_w is the fraction of material depositing to water in a CIT grid cell, WF_g is the fraction of the deposited material taken up by plants, V_{dg} is the estimated deposition velocity to green surfaces, V_{du} is the estimated deposition velocity to urban impervious surfaces. V_{dw} is the estimated deposition velocity to water surfaces, P_w is the percent of water surface area in the CIT grid cell, and P_g is the percent of green surface area in the CIT grid cell. The water weighting factor times the calculated deposition flux for each grid cell estimates the amount of deposition to water surfaces. The green surface weighting factor is used to estimate the amount of deposition to vegetated surfaces. The load not deposited to water or vegetated surfaces is assumed to be deposited onto impervious surfaces directly connected to the drainage system and potentially available for washoff by subsequent rainfall events.

Gaseous PAN and NOx have high surface resistances and low water solubility and thus have low deposition rates to both land and water surfaces. The PAN and NOx deposition loads that are calculated by CIT are due primarily to plant uptake and unavailable for washoff (i.e., $WF = 1$). Moreover, the dry deposition flux of PAN and NOx are low relative to HNO_3 and NH_4NO_3, so that their contribution to the total nitrogen load is minimal. Studies have shown that the surface resistance to HNO_3 and NH_4NO_3, is relatively small and does not depend on the land cover to a great extent (i.e., $V_{dg} \cong V_{du}$) [Dollard *et al.*, 1987; Hanson and Lindberg, 1991]. Thus, we assumed that the dry deposition of HNO_3 and NH_4NO_3 onto vegetated surfaces and water is strictly proportional to the fraction of each grid cell covered by vegetated surfaces or water.

CIT calculates mass loading rates per unit area in each grid cell for each simulated compound. To link the CIT output with the SWMM model the deposition mass loading rate is summed for a specified dry-weather period prior to a precipitation event and multiplied by the weighting factor accounting for the fraction of dry deposition due to plant uptake. The remaining amounts of deposited material are then translated into the dissolved-phase compounds of interest, nitrate (NO_3^-) and ammonium (NH_4^+) using stoichiometric relationships. The resulting amounts of NO_3^- and NH_4^+ are then prescribed as the pollutant loads from dry deposition available for washoff in the SWMM model at the beginning of the storm event. CIT to WASP5 data transference also occurs by summing the total dry deposition for a specific period (typically a day) and multiplying it by the appropriate weighting factor providing a time series of fluxes. These fluxes are then multiplied by the water surface area resulting in a time series of pollutant loads directly to the water body (kg/day).

Wet Deposition

Wet scavenging may be described in two parts, that occurring within the cloud (rainout) and that occurring below the cloud (washout) [Engelmann, 1971]. The processes of below-cloud scavenging are better understood and more experimental data are available than for in-cloud scavenging processes. Studies of the concentrations of ionic compounds throughout precipitation events also suggest the first part of a storm scavenges contaminants from the atmosphere in an initial washout process. Subsequent rainfall passes through a cleaner atmosphere and concentrations are generally found to decrease [Hendry and Brezonik, 1980]. The simulation of the rainout and washout processes and the time variable atmospheric concentrations is extremely complex. Developing and implementing an algorithm representing the individual scavenging processes is beyond the scope of the present research project. A simpler method is to group the individual processes together into a bulk scavenging ratio to estimate wet-deposition fluxes.

The concept of scavenging ratios is based on the simplified assumption that the concentration of the component in precipitation is related to the concentration of the respective compound in the air [Engelmann, 1971; Kasper-Giebl *et al.*, 1999]. Algorithms have been developed to use the scavenging ratio concept to estimate the amount of contaminant removed during rainstorms given atmospheric concentrations at the beginning of the rainfall event and other characteristics of the rainfall (e.g., raindrop size, intensity) and contaminant (e.g., size, solubility, Henry's Law constant). In our research we used a formulation developed by Slade [1968] to represent the washout of aerosols and gaseous pollutants [Novotny and Olem, 1994]:

$$D_{wet} = C_{air} * H\,(1-e^{-\lambda t}) \tag{3}$$

where D_{wet} is the wet deposition per unit area, C_{air} is the atmospheric concentration before the rain event, λ is the washout coefficient, t is the duration of rainfall, and H is the depth of atmosphere through which the pollutant plume is mixed. We performed the washout calculations in an MS Excel spreadsheet. The value for C_{air} is calculated by CIT for each grid cell by finding the vertically integrated atmospheric concentration over H. The washout coefficient is a function of rainfall intensity and was determined at each time step using relationships from Slade [1968]. More complex rainout and washout algorithms exist, but their implementation in this framework was beyond the scope of this research project.

Nitrogen in precipitation is most often present as ammonium (NH_4^+), nitrate (NO_3^-), and dissolved organic nitrogen (DON) [Halverson *et al.*, 1984; Hendry and Brezonik, 1980; Huff, 1976; Russell *et al.*, 1998]. For our study we are interested in the quantity of NH_4^+ and NO_3^- entering water bodies directly and indirectly from atmospheric deposition. NH_4^+ in precipitation is a result of the dissolution of atmospheric NH_3 gas and the scavenging of NH_4^+ containing aerosols, while NO_3^- in precipitation results mostly from the dissolution of HNO_3. In the CIT-SWMM linkage, we assume the PAN, alkyl nitrates, and NOx contributions to NH_4^+ and NO_3^- concentrations in rainfall are negligible because of their relative insolubility. The other major nitrogen deposition compounds simulated by CIT, namely HNO_3, NH_4NO_3, and NH_3 all dissolve readily in water at pH values commonly observed in precipitation and stormwater runoff. Therefore, we will assume that each contributes to the NH_4^+ and NO_3^- rainfall concentrations.

Contaminant concentrations in rainfall can be represented in SWMM, but the code does not allow for time variant rainfall concentration. To determine the constant concentration in the rainfall, the calculated wet deposition masses of NO_3^- and NH_4^+ were divided by the rainfall event volume. A time variable rainfall concentration would be more accurate, but implementing this formulation in the SWMM code was not possible in the time frame of the project. Although a constant rainfall concentration might not result in accurate representations of within storm runoff concentrations, the overall storm event load should be accurately estimated. In this study of nitrogen loading to estuarine receiving waters, we are interested in the storm event load, not intrastorm concentrations, because the estuarine receiving water quality response time to nutrient inputs is on the order of days to weeks [Donigian and Huber, 1991]. Therefore, the importance of the accuracy of the simulated intrastorm concentration is minimized.

MODEL APPLICATIONS

The modeling framework discussed above can be applied to a variety of problems. At present the urban air to water modeling framework is still modular with the capability of operating each model separately still extant. In order to more accurately simulate the non-linear feedbacks between the different environmental compartments and urban infrastructure, a more thorough integration is required, but the current modular formulation of the framework allows the use of each model separately or in tandem with other models in the framework. In so doing, the cost of implementing the framework can be tailored to specific problems. The particular models used to simulate a given problem will be a function of the nature of the problem or questions to be answered. The framework has been applied to evaluate several environmental problems in Los Angeles, CA [Burian *et al.*, 2001; Burian *et al.*, 2002]. Los Angeles was selected due to its size and air and water quality problems. Air quality in Los Angeles has historically been poor because of the degree of urbanization and the regional meteorological and topographical characteristics. This research focused on the Ballona Creek watershed, which drains portions of downtown Los Angeles, and its downstream receiving waters, the Ballona Creek Estuary and Santa Monica Bay (Figure 2).

The Ballona Creek watershed has been shown to be one of the most significant sources of non-point source pollution to Santa Monica Bay [Wong *et al.*, 1997; Suffet *et al.*, 1999]. The Ballona Creek Estuary is a small near coast estuary that has been severely degraded due to urban development. Santa Monica Bay is a large coastal embayment that is a significant economic resource to coastal Los Angeles [SMBRP, 1994]. The individual models of this system were developed using the best available data for the Los Angeles airshed, Ballona Creek watershed, Ballona Creek Estuary, and Santa Monica Bay.

We linked the CIT photochemical model and the SWMM stormwater model to evaluate the relative importance of dry and wet deposition versus urban runoff

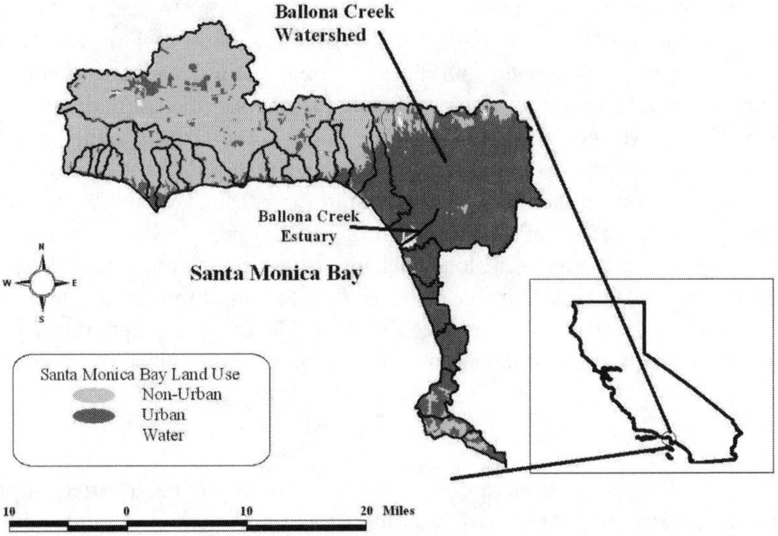

Figure 2. Study Area for Application of the Integrated Environmental Modeling Framework.

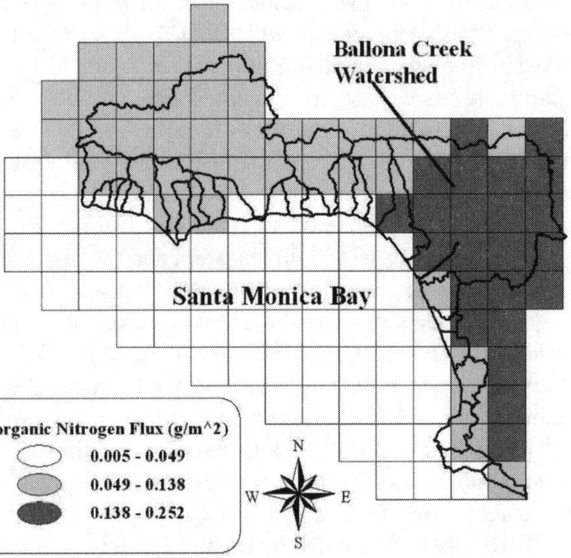

Figure 3. CIT Simulated Dry Deposition Flux (g/m^2) to the Santa Monica Bay watershed.

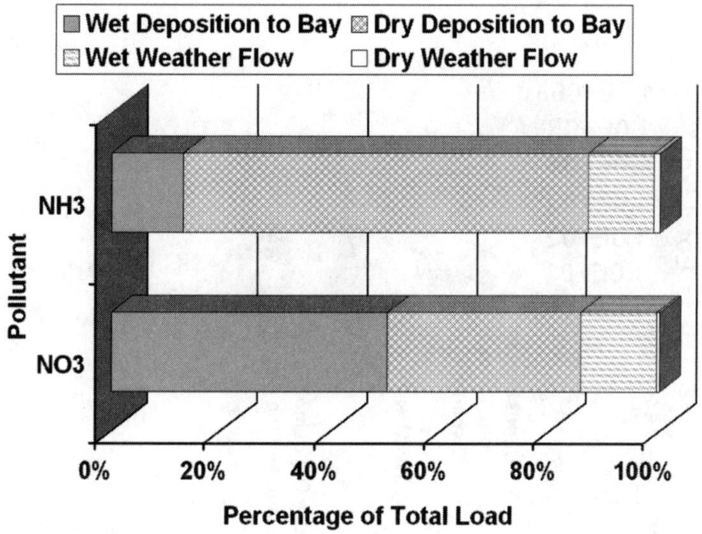

Figure 4. Percentage of Total Non-point Source Load from Deposition and Urban Runoff.

pollutant loads of nitrogen compounds to coastal waterbodies [Burian *et al.*, 2001]. Figure 3 shows the CIT simulated dry deposition flux (g/m^2) of inorganic nitrogen to the Santa Monica Bay watershed. The deposition flux to the watershed and Bay is greatest in the more developed regions of the watershed. The direct deposition flux of inorganic nitrogen to the Bay is generally less than that to the watershed. While the dry deposition flux to the watershed is greater than that to the Bay, a significant quantity of the deposition to the watershed is not discharged in stormwater runoff. Figure 4 shows the percentages of pollutant loads due to wet and dry weather runoff and direct dry and wet deposition to Santa Monica Bay in a 2 week period in the wet season of 1987. Urban runoff from the Ballona Creek watershed accounts for only 10-15% of the total load to the Bay during this time period. Although the load from direct atmospheric deposition is considerably larger than the load from Ballona Creek urban runoff, the atmospheric deposition load is distributed across a much greater area of the Bay than the urban runoff, which is discharged directly to near coast areas. Therefore atmospheric deposition may not be as harmful as Ballona Creek or other coastal watershed urban runoff.

We also applied a linkage between the CIT photochemical model, the SWMM stormwater model and the WASP5 receiving water quality model to assess the importance of atmospheric deposition relative to urban runoff in both Santa Monica Bay and the Ballona Creek estuary [Burian *et al.*, 2002]. This application indicated air quality policy could be important in water quality management if the receiving water of interest is of sufficient size. Figure 5 displays a com-

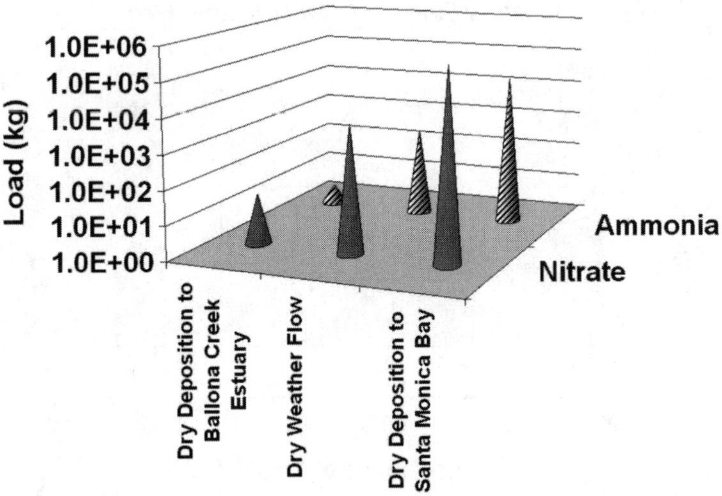

Figure 5. Comparison of the Dry Season Dry Deposition Pollutant Load to two Coastal Receiving Waters to the Load in Dry Weather Urban Runoff.

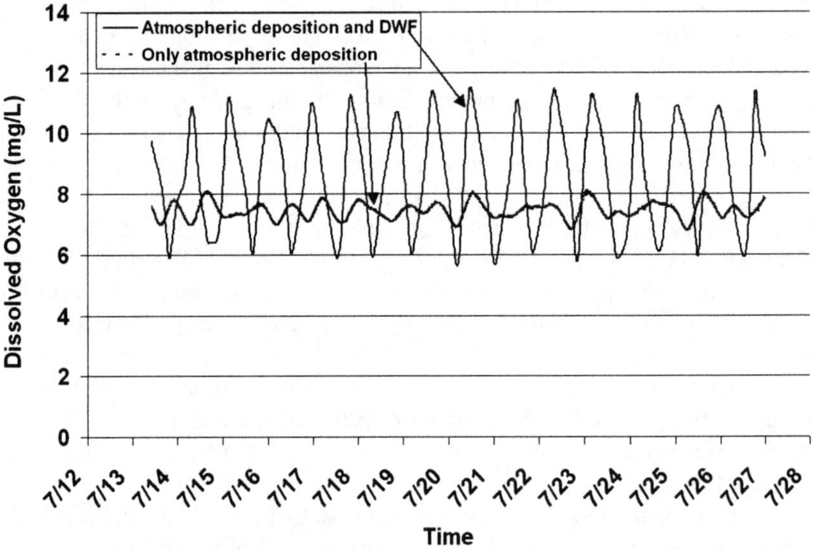

Figure 6. Comparison of the Dissolved Oxygen Dynamics in the Ballona Creek Estuary with and without Dry Weather Flow Pollutant Loading. In each scenario, the system receives atmospheric deposition of nitrogen species.

parison of the dry deposition to Santa Monica Bay and the Ballona Creek Estuary relative to dry weather discharges from the Ballona Creek watershed. In smaller, near coast estuaries, e.g., the Ballona Creek Estuary, the importance of direct dry deposition is minimal compared to the pollutant load in dry weather urban runoff. This is due to the low surface area of the near coast estuary.

When the effects of the dry deposition load on the Ballona Creek Estuary are compared to the effects of the dry weather flow pollutant loading using the WASP5 water quality model, the relative importance of the dry weather flow becomes more apparent. Figure 6 shows the dissolved oxygen pattern in the Ballona Creek Estuary under two scenarios. The first scenario is the dissolved oxygen pattern in the estuary when it receives both the dry weather flow and atmospheric dry deposition nitrogen loads. The second scenario is the dissolved oxygen pattern when the system receives only atmospheric deposition of nitrogen compounds. The dissolved oxygen fluctuations in the system are much less when the estuary receives only atmospheric deposition of nitrogen indicating the greater relative importance of the urban dry weather runoff. The strong daily fluctuation noted when the simulated system receives urban dry weather flow is due to increased algal activity promoted by the elevated nitrogen load. It is important to note that in neither of these cases were any dissolved oxygen water quality standards exceeded.

Results from the application of the modeling framework described herein are similar to those listed in Meyers *et al.* [2000] in which atmospheric deposition loads to major bays or estuaries were assessed using National Atmospheric Deposition Program data and RADM modeling output. Table 1 compares the annual mass flux of nitrogen to Santa Monica Bay from this modeling effort to those in Meyers *et al.* [2000] for other bays and estuaries. This comparison was made assuming the atmospheric deposition to the Bay simulated by the CIT model is representative of average deposition conditions to surface waters. Those values were then adjusted to get an annual flux. The modeling framework can predict annual dry deposition flux to the Santa Monica Bay that is similar to deposition fluxes to other waterbodies. As can be seen, Santa Monica Bay has a greater dry deposition flux then other major west coast bays and estuaries and a similar flux to other bays near major metropolitan centers, e.g. Tampa Bay and Galveston Bay. The dry deposition flux to the watershed surface predicted by this modeling application was not compared to the dry deposition fluxes to the surface of other watersheds listed in Meyers *et al.* [2000]. Such a comparison was avoided because the days simulated in this application were not representative of typical meteorological conditions. The days simulated were known air quality event days with high air concentrations of NOx and therefore higher deposition values. While our simulation limits direct comparison of the dry deposition values from the two studies, it does indicate the importance of air quality in water quality management. If air quality is consistently poor (i.e., consistently exceeds air quality standards) in the Los Angeles airshed, dry deposition to the associat-

Table 1. Comparison of the annual dry deposition flux for Santa Monica Bay (this study) to other U.S. bays and estuaries (Meyers *et al.*, 2000).

Watershed	Dry Deposition to Water Surface (kg/ha-yr)
Chesapeake Bay	9.35
Gardiner's Bay	9.08
Delaware Bay	9.04
NJ Inland Bays	8.69
Hudson River/Raritan Bay	8.55
Long Island Sound	8.18
DE Inland bays	7.87
MD Inland Bays	7.87
Narrangansett Bay	7.19
Waquoit Bay	6.06
Buzzard's Bay	6.06
Massachusetts Bay	5.96
Galveston Bay	5.87
Tampa Bay	5.13
Santa Monica Bay	**5.11**
San Francisco Bay	2.69
Puget Sound	2.01

ed watersheds can be expected to be significant and this can in turn increase pollutant wash-off during storm events.

SUMMARY

Integration of detailed deterministic models into an integrated environmental modeling framework is a developing field. In this work we described the pertinent issues in the field and the development of a linked urban airshed-watershed modeling framework. We described the linkage of HOTMAC and RAMS meteorological codes, CIT air chemistry code, the SWMM urban runoff code, and the WASP5 receiving-water quality code in an effort to simulate the fate and transport of nitrogen compounds through the air and water pathways in an urban environment. In the current applications discussed, the meteorological models are not integrated into the linked modeling framework, but those models will also be integrated providing increased functionality at determining the long term effects of climate change on existing urban systems. The linkage provides a comprehensive modeling framework that can be used to study contaminants that transport through both the urban air and water environments. The integrated modeling system also allows assessment of cross-media affects of environmental policy.

Acknowledgments. This work was sponsored by the Urban Security Initiative, a Laboratory Directed Research and Development (LDRD) project at Los Alamos National Laboratory (LANL). We would like to thank the LDRD program for funding and supporting research in urban geosciences. We would also like to thank Grant Heiken and Greg Valentine of LANL for their leadership and support of the Urban Security Initiative and the laboratories of Mel Suffet and Michael Stenstrom at the University of California, Los Angeles for their assistance in the acquisition of data used to develop and parameterize the watershed and water quality models used in this work.

REFERENCES

Ackerman, B., METROMEX: Wind Fields over the St. Louis Metropolitan Area, *Bulletin of the American Meteorological Society, 55*, 93-95, 1974.

Alexander, R.B., R.A. Smith, G.E. Schwarz, S.D. Preston, J.W. Brakebill, R. Srinivasan, and P.A. Pacheco, Atmospheric Nitrogen Flux from the Watersheds of Major Estuaries of the United States: An Application of the SPARROW Watershed Model, in *Nitrogen Loading in Coastal Water Bodies: An Atmospheric Perspective, American Geophysical Union Coastal and Estuarine Studies No. 57,* edited by: R.A. Valigura, R.B. Alexander, M.S. Castro, T.P. Meyers, H.W. Paerl, P.E. Stacey, and R.E. Turner, American Geophysical Union, Washington D.C., 2000.

Amargit, S., *Air deposition studies: A Review of Air Deposition Literature, Technical Memorandum: RPO-MOD-TM03.00*, Rouge River Wet Weather Demonstration Project, Wayne County, Michigan, 1994.

Ambrose, R.F., T.A. Wool, and J.L. Martin, *The Water Quality Analysis Simulation Program, WASP5, Part A: Model Documentation*, Environmental Research Laboratory, Athens Georgia and AScI Corporation, Athens Georgia, 1993.

Angel, J., D. Pack, C. Dickson, and W. Hoecker, Urban Influence on Nighttime Airflow Estimated from Tetroon flights, *Journal of Applied. Meteorology, 10*, 194-204, 1971.

Athayde, D.N, P.E. Shelley, E.D. Driscoll, D. Gaboury and G. Boyd. *Results of the Nationwide Urban Runoff Program, Executive Summary.* EPA, Office of Water Program Operations, Water Planning Division. Washington DC, 1983.

Atwater, M., Thermal Effects of Urbanization and Industrialization in the Boundary Layer: a Numerical Study, *Boundary Layer Meteorology, 3*, 229-245, 1972.

Bledsoe, B.P. and C.C. Watson, Effects of Urbanization on Channel Instability, *Journal of the American Water Resources Association, 37*, 2, 255-270. 2001.

Bornstein, R., Observations of the Urban Heat Island Effect in New York City, *Journal of Applied Meteorology, 7*, 575-582, 1968.

Bornstein, R., The Two-dimensional URBMET Urban Boundary Layer Model, *Journal of Applied Meteorology, 14*, 1459-1477, 1975.

Brandmeyer, J.E. and H.A. Karimi, Coupling Methodologies for Environmental Models. *Environmental Modelling and Software, 15*, 479-488, 2000.

Burian, S.J., G.E. Streit, T.N. McPherson, M.J. Brown, and H.J. Turin, Modeling the Atmospheric Deposition and Stormwater Washoff of Nitrogen Compounds, *Environmental Modelling and Software, 16,* 467-479, 2001.

Burian, S.J., T.N. McPherson, M.J. Brown, G.E. Streit, and H.J. Turin, Modeling the Effects of Air Quality Policy Changes on Water Quality in Urban Areas, *Environmental Modeling and Assessment*, in press, 2002.

Clarke, J., Nocturnal Urban Bboundary Layer over Cincinnati, Ohio, *Mon. Wea. Rev., 97*, 582-589, 1969.

Costanza, R. R. d'Arge, R. de Groot, S. Farber, M. Grasso, B. Hannon, K. Limburg, S. Naseem, R.V. O'Neill, J. Paruelo, R.G. Raskin, P. Sutton, and M. van den Belt. The Value of the World's Ecosystem Services and Natural Capital, *Nature, 387*, 15, 253-259, 1996.

Dennis, R.L. Using the Regional Acid Deposition Model to Determine the Nitrogen Deposition Airshed of the Chesapeake Bay Watershed. In: Baker, J.E. (Ed.), *Atmospheric Deposition of Contaminants to the Great Lakes and Coastal Waters*, 393-413, 1997.

Dollard, G.J., D.H.F. Atkins, T.J. Davies, and C. Healy, Concentrations and Dry Deposition Velocities of Nitric Acid, *Nature, 326*, 481-483, 1987.

Donigian, A.S. and W.C. Huber. 1991. *Modeling of Non-point Source Water Quality in Urban and Non-urban Areas.* EPA/600/3-91/039. United States Environmental Protection Agency, Office of Research and Development, Washington, D.C.

Engelmann, R.J. Scavenging Prediction using Ratios of Concentrations in Air and Precipitation, *Journal of Applied Meteorology, 10,* 493-497, 1971.

Finkenbine J.K., J.W. Atwater, and D.S. Mavinic. Stream Health after Urbanization. *Journal of the American Water Resources Association, 36*, 5: 1149-1160, 2000.

Franz, T.P., S.J. Eisenreich, and T.M. Holsen. Dry Deposition of Particulate Polychlorinated Biphenyls and Polycyclic Aromatic Hydrocarbons to Lake Michigan, *Environmental Science and Technology, 32*, 23, 3681-3688, 1998.

Greenfelt P. and H. Hultberg, Effects of Nitrogen Deposition on the Acidification of Terrestrial and Aquatic Ecosystems, *Water Air, and Soil Pollution, 30*, 4, 945-963, 1986.

Hanson, P.J. and S.E. Lindberg, Dry Deposition of Reactive Nitrogen Compounds: A Review of Leaf, Canopy and Non-Foliar Measurements, *Atmospheric Environment, 25A*, 1615-1634, 1991.

Harner, T. and D. Mackay. Model of the Long Term Exchange of PCBs Between Soil and the Atmosphere in the Southern U.K., *Environmental Science and Technology, 29*, 5, 1200-1209, 1995.

Halverson, H.G., DeWalle, D.R., Sharpe, W.E., Contribution of Precipitation to Quality of Urban Storm Runoff, *Water Resources Bulletin, 20*, 859-864, 1984.

Hendry, C.D., Brezonik, P.L., Chemistry of Precipitation at Gainesville, Florida. *Environmental Science and Technology, 14*, 843-849, 1980.

Hicks, B.B., Atmospheric Deposition and its Effects on Water Quality. In: Christensen, E.R., O'melia, C.R., (Eds.), Proceedings, Workshop on Research Needs for Coastal Pollution in Urban Areas, 27-36, 1997.

Hicks, B.B., Wind, Water, Earth, and Fire - A Return to an Aristotelian Environment. *Bulletin of the American Meteorological Society, 79*, 1925-1933, 1998.

Hinga, K.R., Keller, A.A., Oviatt, C.A., Atmospheric Deposition and Nitrogen Inputs to Coastal Waters. *Ambio, 20,* 256-260, 1991.

Hosker, R., The Effects of Buildings on Local Dispersion, Modeling the Urban Boundary Layer, *Am. Met. Soc.*, Boston, 1987.

Hornbuckle, K.C., D.R. Achman, S.J. Eisenreich. Over-Water And Over-Land Polychlorinated Biphenyls In Green Bay, Lake Michigan. *Environmental Science and Technology, 27,* 1, 87-98, 1992.

Huff, F.A., Relation Between Atmospheric Pollution, Precipitation, and Streamwater Quality Near a Large Urban-Industrial Complex, *Water Research,* 10, 945-953, 1976.

Huber, W.C. and R.E. Dickinson, *Storm Water Management Model, version 4, part A: User's Manual*, EPA-600/3-88-001a, U.S. Environmental Protection Agency, Athens, Georgia, 1988.

Kasper-Giebl, A., Kalina, M.F., and H. Puxbaum, Scavenging Ratios for Sulfate, Ammonium and Nitrate Determined at Mt. Sonnblick (3106 m a.s.l.). *Atmospheric Environment, 33*, 895-906, 1999.

Linker, L.C. and R.V. Thomann, The Cross-media Models of the Chesapeake Bay: Defining the Boundaries of the Problem, In: *Proceedings, Watershed '96, A Conference on Watershed Management*, June 8-12, 1996, Baltimore, Maryland, 112-114, 1996.

Linker, L.C., C.G. Stigall, C.H. Chang, and A.S. Donigian. Aquatic Accounting: Chesapeake Bay Watershed Model Quantifies Nutrient Loads, *Water Environment and Technology*, 8, 1, 48-52, 1993.

Long Term Ecological Research (LTER) Network. http://lternet.edu/, 2001.

Ludwig, F., Urban Temperature Fields, WMO Publ. Tech. Note 108, 80-107, 1970.

MacKay, D. and S. Paterson, Calculating Fugacity, *Environmental Science and Technology, 15*, 1006-1014, 1981.

MacKay, D. and S. Paterson, Evaluating the Multimedia Fate of Organic-Chemicals: A Level III Fugacity Model, *Environmental Science and Technology, 25,* 427-436, 1991

McElroy, J., A Numerical Study of the Nocturnal Heat Island Over a Medium-Sized Mid-Latitude City (Columbus, Ohio), *Boundary-Layer Meteor., 3*, 442-453, 1973.

McRae, G., Goodin, W., Seinfeld, J., Development of a Second Generation Mathematical Model for Urban Air Pollution: I. Model Formulation. *Atmospheric Environment, 16*, 679-696, 1982.

Meyers, T., Sickles, J., Dennis, R., Russell, K., Galloway, J., and T. Church, Atmospheric Nitrogen Deposition to Coastal Estuaries and Their Watersheds, in *Nitrogen Loading in Coastal Water Bodies: An Atmospheric Perspective, American Geophysical Union Coastal and Estuarine Studies No. 57*, edited by: R.A. Valigura, R.B. Alexander, M.S. Castro, T.P. Meyers, H.W. Paerl, P.E. Stacey, and R.E. Turner, American Geophysical Union, Washington D.C., 2000.

Myrup, L., A Numerical Model of the Urban Heat Island, *J. Appl. Meteorology, 8*, 896-907, 1969.

Nix, S.J., *Urban Stormwater Modeling And Simulation.* Lewis Publishers: Boca Raton, Florida, USA, 1994.

Novotny, V., Olem, H., *Water Quality: Prevention, Identification, and Management of Diffuse Pollution.* Van Nostrand Reinhold: New York, NY, USA, 1994.

Oke, T., The Surface Energy Budgets of Urban Areas, Modeling the Urban Boundary Layer, *Am. Met. Soc.*, Boston, 1987.

Oke, T. & East, C., The Urban Boundary Layer In Montreal, *Boundary-Layer Meteorology*, 1, 411-437, 1971.

Paerl, H.W., Atmospheric Nitrogen Deposition in Coastal Waters. In J.E. Baker (ed.). *Atmospheric Deposition of Contaminants to the Great Lakes and Coastal Waters.* Denver, CO. Society of Environmental Toxicology and Chemistry. 1997.

Paerl, H.W., W.R. Boynton, R.L. Dennis, C.T. Driscoll, H.S. Greening, J.N. Kremer, N.N. Rabalais, and S.P. Seitzinger, Atmospheric Nitrogen Deposition in Coastal Waters: Biogeochemical and Ecological Implications, in *Nitrogen Loading in Coastal Water Bodies: An Atmospheric Perspective, American Geophysical Union Coastal and Estuarine Studies No. 57*, edited by: R.A. Valigura, R.B. Alexander, M.S. Castro, T.P. Meyers, H.W. Paerl, P.E. Stacey, and R.E. Turner, American Geophysical Union, Washington D.C., 2000.

Paraiba, L.C., J.M. Carrasco, and R. Bru, Level IV Fugacity Model by Continuous Time Control System, *Chemosphere*, 38, 8, 1763-1775, 1999.

Pielke, R. W. Cotton, R. Walko, C. Tremback, W. Lyons, L. Grasso, M. Nicholls, M. Moran, D. Wesley, T. Lee, and J. Copeland. A Comprehensive Meteorological Modeling System. *Meteorol. Atmos. Phys. 49*, 69-91, 1992.

Pitt, R.E., Small Storm Urban Flow And Particulate Washoff Contributions To Outfall Discharges. Ph.D. dissertation, University of Wisconsin-Madison, Madison, Wisconsin, USA, 1987.

Roesner, L.A., J.A. Aldrich, and R.E. Dickinson, Storm Water Management Model, version 4, part B: EXTRAN addendum, EPA-600/3-88-001B, U.S. Environmental Protection Agency, Athens, Georgia, 1988.

Rouge River Wet Weather Demonstration Project (RRWWDR), Model Review And Assessment, Technical Memorandum: RPO-MOD-TM-04.04, Wayne County, Michigan, 1993.

Russell, A.G., K. McCue, and G. Cass, Mathematical Modeling of the Formation of Nitrogen-Containing Air Pollutants. 1. Evaluation of an Eulerian Photochemical Model. *Environmental Science and Technology, 22*, 263-271, 1988.

Russell, K.M., J.N. Galloway, S.A. Macko, J.L. Moody, and J.R. Scudlark, Sources of Nitrogen in Wet Deposition to the Chesapeake Bay Region, *Atmospheric Environment, 32*, 2453-2465, 1998.

Sauvajot R.M., M. Buecher, D.A. Kamradt, and C.M. Schonewald, Patterns of Human Disturbance and Response by Small Mammals and Birds in Chaparral near Urban Development, *Urban Ecosystems, 2*, 279-297, 1998.

Santa Monica Bay Restoration Project (SMBRP), State of the Bay 1993: Characterization Study of the Santa Monica Bay Restoration Plan, Santa Monica Bay Restoration Project, 101 Centre Plaza Drive, Monterey Park, CA 91754, 1994.

Seinfeld, J.H., Urban Air Pollution: State of the Science, *Science, 243*, 4892, 745-752, 1989.

Slade, D.H., Meteorology and Atomic Energy. TID24190, U.S. Atomic Energy Commission, Washington, DC., 1968.

Smith, W.S., J.M. Reisner, J.E. Bossert, and J.L. Winterkamp, Tracer Modeling in an Urban Environment, 2nd Annual American Meteorological Society Urban Environmental Conference, Albuquerque, NM, LA-UR-98-3563, 1998.

Smith, W.S., J.M. Reisner, and C.-Y.J. Kao, Simulations of Flow Around a Cubical Building: Comparison with Towing-Tank Data and Assessment of Radiatively Induced Thermal Effects, *Atmospheric Environment, 35,* 3811-3821, 2001.

Suffet, I.H. and M.K. Stenstrom, A Study of Pollutants from the Ballona Creek Watershed During Wet-Weather Flow. Final Report to US Army Corps of Engineers, UCLA, Los Angeles, California, 1999.

Suzuki, N., M. Yasuda, T. Sakurai, J. Nakanishi. Simulation of Long-Term Environmental Dynamics of Polychlorinated Dibenzo-P-Dioxins and Polychlorinated Dibenzofurans Using the Dynamic Multimedia Environmental Fate Model and its Implication to the Time Trend Analysis of Dioxins, *Chemosphere,* 40, 969-976, 2000.

United States Census Bureau, 1999 TIGER/Line Files Technical Documentation 1999, U.S. Government Printing Office, Washington, DC, 1999.

United States Census Bureau, Current Population Reports, Series P23-205, Population Profile of the United States: 1999, U.S. Government Printing Office, Washington, DC, 2001.

U.S. EPA, http://www.epa.gov/glnpo/lmmb. Great Lakes Program, Chicago, Illinois. 2000.

U.S. EPA, Deposition of Air Pollutants to the Great Waters: Second Report to Congress, EPA-453/R-97-011, Research Triangle Park, North Carolina, 1997.

U.S. EPA, Deposition of Air Pollutants to the Great Waters: Third Report to Congress, EPA-453/R-00-005, Research Triangle Park, North Carolina, 2000.

Valigura, R.A., W.T. Luke, R.S. Artz, and B.B. Hicks, Atmospheric Nutrient Inputs to Coastal Areas: Reducing the Uncertainties. National Oceanic and Atmospheric Administration, Air Resources Laboratory, 1996.

Wang, P., L. Linker, and J. Storrick, Chesapeake Bay Watershed Model: Application and Calculation of Nutrient and Sediment Loadings, Report by the Chesapeake Bay Program Modeling Subcommittee, 1997.

Wong, K.M., E.W. Strecker, and M.K. Stenstrom, GIS to Estimate Storm-Water Pollutant Mass Loadings, *Journal of Environmental Engineering, 123,* 737-745, 1997.

World Resources Institute, United Nations Environment Programme, United Nations Development Programme, World Bank, *World Resources 1996-1997: The Urban Environment*, Oxford University Press, NY, 1996.

Yamada T. and J. Kao, A Modeling Study on the Fair Weather Marine Boundary Layer of the GATE, *J. Atm. Sci., 43*, 3186-3199, 1986.

11

Urban Environmental Modeling and Assessment Using Detailed Urban Databases

Steven J. Burian, Timothy N. McPherson, Michael J. Brown, Gerald E. Streit, and H.J. Turin

INTRODUCTION

Throughout history the earth's population has been predominantly agrarian with isolated pockets of population in settled areas. For many of the industrialized countries this distribution changed during the nineteenth century spurred by the need for the workforce to be located near the urban industrial centers. In the United States (U.S.), for example, the population was less than 5% urban in 1820, but by 1860 that percentage had increased to 16%, and by 1880 had risen to 22.5%. The population shift eventually resulted in a transition during the twentieth century from a rural population to a predominantly urban population [Goldfield and Brownell, 1990]. At the beginning of the twentieth century nearly 40% of the U.S. population lived in urban areas. During the decade spanning 1910 to 1920 the U.S. population shifted to more than 50% urban. According to the metropolitan area population estimates based on the 1990 census, more than 80% of the U.S. population now resides in urban areas.

As cities grow they exert a more significant and far-reaching influence over the environment. Cities themselves, or elements of cities, can deleteriously impact the air, water, and terrestrial resources [e.g., EPA, 2000a; 2000b]. One example involves the strong correlation between the amount of impervious area in cities and the health of nearby waterways [Arnold and Gibbons, 1996; Schueler, 1994; Booth and Reinfelt, 1993; Klein, 1979]. A threshold of 10% imperviousness of a watershed has been noted as the point when environmental degradation is likely

Earth Science in the City: A Reader

10.1029/056SP12

to occur in the receiving stream. Imperviousness of 30% or greater has a high likelihood of severe degradation.

A city both depends on and negatively impacts the natural environment. The growth and sustainability of a city currently requires the importation of resources (e.g., raw manufacturing materials, energy), the consumption of natural resources (e.g., air, water, and land), the production of goods and services, and the discharge of wastes. Eventually as the city grows, resources near the city will become inadequate or completely exhausted, which will require the city to reach further for the necessary resources. The historical development of water supplies in several major U.S. cities clearly demonstrates this expanding influence of a city in response to growing demand and degrading local resources. Philadelphia, Boston, and New York City, for example, had to extend their water supply infrastructures significantly in the early nineteenth century because the local sources of water were contaminated and additional uncontaminated supplies were needed to meet the growing demand [Burian, 2001; Melosi, 2000].

In some ways, a city is analogous to a living organism because both consume resources and produce wastes. Cities and living organisms have analogous networks to distribute raw materials and energy and to collect, process, and dispose of wastes. Failures of these distribution and collection networks can create health problems in living organisms and sustainability shortfalls in cities. The infrastructure of living organisms and cities is often invisible or taken for granted by those that rely on the services provided. For example, when the invisible infrastructure of a living organism (e.g., arteries, lungs) is neglected, health problems can result. Similarly, when the invisible infrastructure of cities (e.g., electricity, water distribution, communications, wastewater collection) is neglected, myriad problems can result (e.g., power shortages and outages, public health risks, pollution discharges). Both cities and living organisms must maintain their infrastructure and develop a harmonious relationship with the environment to achieve sustainability. In the U.S., the status of infrastructure was given a grade of D+ in 2001 by the American Society of Civil Engineers (ASCE) [see <http://www.asce.org/reportcard/>]. The relatively low grade for some infrastructure elements is an indicator of existing and potential environmental problems. The continued deterioration of the U.S. infrastructure is a serious problem, and may result in extensive environmental damage, reduced quality of life, and economic and sociological consequences.

Sustainability of cities requires the balancing of growth, economic and sociological development, and environmental preservation and protection. Sustainable cities must provide the basic needs of its citizens while being within the carrying capacity of supporting ecosystems [Bolund, 2002]. Growing cities need to develop with sustainability as the guiding principle. Sustainable development is defined as "development that meets the needs of the present without compromising the ability of future generations to meet their own needs" [World Commission on Environment and Development and Brundtland, 1987]. Urban sustainability is a pressing, multi-disciplinary issue in Europe, the U.S., and other developed countries where the majority of the population resides in urban areas [European Commission, 1996].

In response to the concern about urban sustainability, several government and non-government programs (e.g., the Smart Growth Program of the U.S. Environmental Protection Agency (EPA), the Smart Growth Network, the Sustainable Cities Project) and research efforts (e.g., the U.S. Geological Society (USGS) Urban Dynamics Program and the National Aeronautics & Space Administration (NASA) Land Cover/Land Use Change Program) have been initiated in part to study urban systems and the environmental impacts associated with urban growth. In addition, two urban areas (Phoenix and Baltimore) have recently been added to the National Science Foundation's Long-Term Ecological Research (LTER) Network to study the change of urban ecosystems over time. The accumulation, analysis, and synthesis of information collected during these and other urban research programs will aid in the successful planning, design, and management of sustainable cities.

Clearly, the study of urban systems is interdisciplinary, which suggests the need for an integrated strategy to analyze and manage the interaction between the urban system and the natural environment. The planning, design, development, and implementation of components of an integrated urban environmental management plan require the collection of a significant amount of information. Comprehensive monitoring can provide much of the information, but it is too costly and too time consuming to provide the immediate feedback that city planners and administrators require for decision-making. One cost-effective option to derive information about the interaction between the urban system and the natural environment is through environmental modeling and assessment. Mathematical modeling and data analysis holds promise for the thorough evaluation of environmental responses to changes in urban growth patterns, growth rates, waste discharge characteristics, and other stimuli. Although mathematical models are powerful tools suited to the purpose of urban system analysis, they require data for development, input specification, variable and parameter definition, calibration, and verification.

This paper is an initial effort to define the general characteristics of detailed urban databases for integrated urban environmental modeling and assessment projects. The primary objective of this paper is to identify the types and sources of the spatial data comprising detailed urban databases. The discussion of data types and sources is by no means comprehensive, and much of the technical detail is omitted in order to maintain brevity. However, the cited literature does include numerous references to the general and technical details of the spatial data. The secondary objective is to review the development and application of a detailed urban database for a case study set in Los Angeles, California. The paper concludes with a summary of two applications of the database to study the interaction between the Los Angeles urban system and the adjacent surface water environment.

URBAN DATABASES

The study of an urban system and its interaction with the surrounding environment requires the development of a coordinated database of multi-disciplinary datasets describing the physical, chemical, and biological characteristics of the city. Applications of the database in integrated modeling studies can provide insight into appropriate environmental management strategies, or aid in the composition of comprehensive city plans or natural hazard mitigation plans. The geographic information system (GIS) is becoming the standard database management and analysis tool for urban environmental modeling. Goodchild [1993] summarized the role of GIS in environmental modeling to include:

1. Preprocessing data into a form suitable for analysis (scale, coordinate system, data structure, data models, etc.).
2. Direct support for modeling, so that tasks such as analysis, calibration, and prediction are carried out in the GIS itself.
3. Post-processing data through reformatting, tabulation, mapping, and report generation.

GIS software is now commonly used to study the water resources component of the integrated urban environmental system [Garbrecht et al., 2001; Ogden, et al., 2001; Sample et al., 2001; Wong et al., 1997]. GIS applications are also becoming more common in the study of urban air quality, environmental justice, and other urban environmental issues.

Currently, numerous governmental entities, research groups, and private enterprises are developing comprehensive urban databases in an effort to study one or more aspects of a particular urban system. In addition, most cities in the U.S. are accumulating GIS datasets and computer-aided drafting (CAD) drawings and making them available to the public via the World Wide Web or on CD-ROM. Municipal employees use the GIS datasets and CAD drawings for operations and maintenance, hazard assessment and mitigation planning, growth planning and management, security and anti-terrorism, and a variety of other tasks. The additional public, private, and scientific uses for the datasets are nearly limitless.

Given the interdisciplinary nature of urban systems, urban databases usually contain a wide variety of data types in several different data formats. This presents a daunting data management task. To aid in the development and management of an interdisciplinary urban database, we subdivide the broad urban database concept into six data categories: (1) physical characteristics, (2) infrastructure elements, (3) socioeconomic attributes, (4) natural resources, (5) climate and meteorology, and (6) environmental quality. These six dataset categories, detailed in the next six subsections, contain various degrees of overlap and connectivity, as is illustrated in Figure 1. We focus data collection efforts on these six categories.

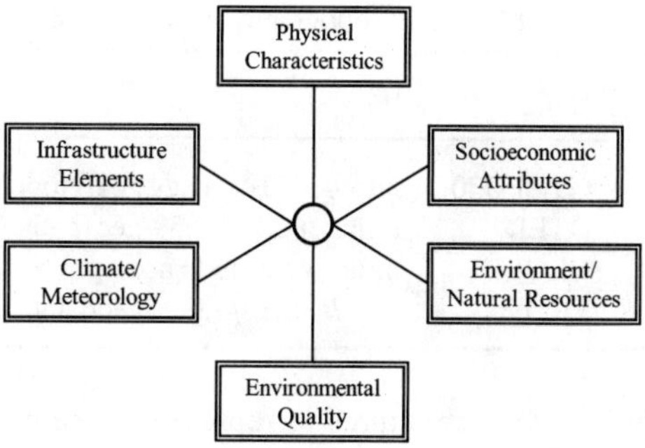

Figure 1. Schematic of primary dataset categories that characterize the urban system for environmental modeling.

Physical Characteristics

Urban environmental modeling requires an accurate description of the physical characteristics of the natural and man-made components of the urban system. One important subset of the physical characteristics category is morphology (e.g., bald-earth terrain, built environment form, vegetative canopy structure). Urban morphology significantly influences the direction and rate of fluid flows through the urban system. Surface topography influences the hydrology in streams, rivers, small lakes, and wetlands during dry weather [Chow et al., 1988]. During wet weather, topography influences the stormwater runoff flow rate, volume, direction, and destination [Wanielista and Yousef, 1993]. Moreover, urban morphology influences the local and regional air flow through drag forces and enhanced turbulent mixing [e.g., Klaiæ, 2002; Fernando et al., 2001; Brown, 1999]. Morphological influence on wind patterns in turn impacts the fate and transport of contaminants in air and water.

Table 1 lists several common sources of global and U.S. digital elevation data. The data structure for most global elevation datasets is a horizontal square-grid mesh with one elevation assigned to each grid cell. The horizontal resolution of the global datasets is approximately 1 km X 1 km, which is not appropriate for detailed urban modeling. Finer horizontal resolution is required to accurately represent the shape and form of the discontinuous structures that make up the urban fabric of the city. In the U.S., the USGS offers a variety of finer resolution digital elevation products that can be downloaded from the web site listed in Table 1. The finest horizontal resolution currently available from the USGS is 10 m X 10 m, but this is currently in production for much of the country.

In most cases, even the 10 m X 10 m resolution USGS data is not appropriate to represent the complex terrain of urban areas, and finer resolution digital elevation data products must be derived or purchased from commercial vendors. Urban digital elevation data products must delineate the precise shape of build-

TABLE 1. Topography data sources.

Scale	Source	Name	Web Site
Global	USGS	GTOPO30	http://edcdaac.usgs.gov/gtopo30/gtopo30.html
	NIMA	DTED	http://164.214.2.59/geospatial/digital_products.htm
	NOAA	GLOBE	http://www.ngdc.noaa.gov/seg/topo/globe.shtml
U.S.	USGS	EROS	http://edc.usgs.gov/geodata/

ings, trees, and other urban structures to represent these elements in modeling and assessment activities. Building morphology datasets can be purchased from commercial vendors in vector and raster data formats. In the vector format, objects (e.g., buildings, trees) are represented by polygons that are precisely positioned in a continuous map space (see Figure 2). On the other hand, raster products divide space into a 2D grid of cells, where each cell contains a value representing the attribute being mapped. Building and tree elevation data products can be derived through analysis of paired stereographic images or collected using airborne laser technology [Ritchie, 1995]. Table 2 lists several vendors of building elevation data products.

TABLE 2. Commercial vendors of building datasets.

Vendor	Web Site
i-cubed	http://www.i3.com
Istar USA	http://www.istar.com
The Gemi Store	http://www.gemistore.com
Urban Data Solutions, Inc.	http://www.u-data.com
Vexcel Corporation	http://www.vexcel.com
Terrapoint (lidar)	http://www.transamerica.com/business_services/real_estate/terrapoint/default.asp

Another physical attribute important for characterizing the urban environment is land use/land cover (LULC). Land use, in general, refers to the specified purpose of land from a human perspective (e.g., high-density residential, commercial & services, industrial), while land cover refers to the state or physical appearance of the land surface (e.g., grass, trees, bare soil, concrete, asphalt). Urban LULC is an indication of human modification and defines characteristics of the urban terrain (e.g., imperviousness, solar reflectivity, heat storage capacity) for environmental modeling applications. The USGS global land cover dataset (http://edcdaac.usgs.gov/glcc/glcc.html) has 1-km horizontal resolution, but this is not sufficient to provide the necessary detail of the urban terrain for most urban environmental modeling activities. In the U.S., finer resolution

Figure 2. 3D building dataset in vector format.

datasets with more urban detail are available from the USGS, the EPA, the National Land Cover Characterization Project (http://landcover.usgs.gov/ nationallandcover.html), and local and regional governmental entities.

Building morphology in a city must be defined accurately for urban wind flow and pollutant dispersion modeling [Hanna et al., 2001; Cionco and Ellefsen, 1998]. However, the complete representation of all building and vegetation elements in a large city for urban dispersion modeling is very difficult because of the high data collection cost, the limitations on computer storage, and the excessive time requirements needed to process the data into a form compatible with the models. Researchers are currently developing computational tools to rapidly process urban morphological data for use in urban dispersion models [Burian et al., 2002a; Ratti et al., 2001]. Work is also underway to define the urban morphological characteristics as a function of urban land use [Brown et al., 2002]. The integration of urban morphology and LULC data is currently being used to parameterize the coupling of urban canopy energy budget models with mesoscale meteorological models for urban air quality studies [e.g., Ching et al., 2002; Dupont et al., 2002].

Socioeconomic Attributes

When conducting a comprehensive urban environmental analysis it is important to understand the socioeconomic attributes of the urban system. The term socioeconomic, by definition, address both social and economic factors. Therefore, population, demographics, spatial distribution of income and wealth, locations of tourist attractions and entertainment districts, locations of cultural centers, and more are potentially important for an integrated urban modeling effort. Socioeconomic factors pertinent to the problem being studied must be identified, assessed, and incorporated into the urban database. Demographic information coupled with infrastructure elements and environmental quality data can be useful for assessing environmental justice issues [EPA, 1995]. In addition, hazard assessment and mitigation planning requires the integration of population, demographics, and infrastructure datasets with hazard simulation [e.g., Heiken et al., 2000]. A digital global population dataset was produced during the Oak Ridge National Laboratory Global Population Project [Dobson et al., 2000]. Demographics datasets are available in electronic format from national census offices, local planning and tax assessment entities, or on the World Wide Web (e.g., the geography network at <http:// www.geographynetwork.com/data/tiger2000/>). Other socioeconomic-related datasets can be obtained from local and regional government entities.

Infrastructure Elements

Infrastructure elements control the rate of movement of raw materials, water, energy, information, waste, and people in a city. Infrastructure datasets are important for a wide range of integrated urban environmental modeling and assessment projects including engineering planning and design, municipal operations and maintenance, hazard assessment, security, and regional planning. The level of detail of the dataset is a primary consideration during database development and each project will have specified needs for infrastructure details. For instance, one project may require only the approximate location of roadways, while another may require information on the precise roadway centerline location, roadway type, roadway and shoulder widths, roadway material, average daily traffic count, the age of the roadbed, the longitudinal slope and cross-slope, and the drainage system characteristics. The size of the dataset will naturally increase with an increase in level of detail, which can present a logistical problem if the project requires high-fidelity datasets.

Historically, urban water infrastructure elements (e.g., water supply, wastewater collection and treatment, and stormwater drainage) have had the most direct influence over public health in cities. In the U.S., improvements to urban water infrastructure had rapid and significant positive impacts to public health. For example, typhoid death rates dropped dramatically from the end of the nineteenth century to the 1920s primarily due to the tapping of uncontaminated supplies of drinking water and the introduction of adequate water distribution infrastructure [Melosi, 2000]. In addition to typhoid, other waterborne diseases were prevalent in the U.S. before adequate urban water treatment and infrastructure elements were constructed [Duffy, 1990]. Waterborne diseases are still a major public health problem in many undeveloped countries with inadequate urban water infrastructure. Even in developed countries the failure of water treatment and distribution systems can have serious public health ramifications (e.g., *cryptosporidium* and *giardia* outbreaks in Milwaukee, USA in 1993 and the scares in Sydney, Australia prior to their hosting of the Olympic Games in 2000).

Inadequacy in wastewater collection and treatment also has severe impacts on the natural environment. Controlled and uncontrolled discharge of wastewater (municipal wastewater, industrial wastewater, wet-weather flow) to water bodies is commonly cited in the U.S. as the reason for impairment of water bodies [EPA, 2000b]. Environmental modeling, along with detailed datasets, can be used to evaluate the performance of urban wastewater infrastructure and specifically evaluate the impacts of system failures on natural resources [Nix, 1997]. Analysis of simulation results can be used to assess public health and aquatic life risk levels associated with urban wastewater discharges [Marr and Freedman, 1997]. From the opposite perspective, integrated environmental modeling can be used to assess the impacts of outside stimuli on urban water quality. For example, integrated urban modeling studies have been performed to assess the influence of air quality on urban runoff and urban water bodies [e.g., Burian et al., 2002b; Burian et al., 2001].

Transportation systems are also of great interest in infrastructure assessments because of their importance for transportation, security, quality of life, and the environment. The air quality impacts of automobile transportation systems have been well studied and numerous investigations have been performed integrating simulations of emissions from transportation networks and air quality [examples of these types of modeling activities can be found at the U.S. EPA Transportation Air Quality Center web site, http://www.epa.gov/omswww/ traq/]. In addition, the potential exposure of drivers and passengers in automobiles to airborne toxic releases has been studied using integrated modeling of the transportation network and urban air quality [e.g., Brown et al., 1997]. In most cities, growth-planning activities must consider transportation systems because of the importance of transportation to the economy, quality of life, and environment.

Energy production and distribution systems are another critical infrastructure element that is important to the sustainability of cities. Fossil fuel combustion provides much of the energy consumed by many cities, but emissions from the combustion process degrade air and water resources. Failure of electrical-power infrastructure can have devastating impacts to the function and productivity of a city. Recent investigations have assessed damages to electrical-power infrastructure caused by natural hazards. For example, Maheshwari and Dowell [1999] used integrated modeling to assess earthquake impacts to the electrical-power infrastructure in the Los Angeles region.

Infrastructure datasets can be collected through field visits with global positioning system units, analysis of aerial photographs and satellite imagery, inspection of construction plans and as-built drawings available at municipal public works departments, or analysis of proprietary datasets. Recently, many U.S. cities have initiated efforts to integrate their infrastructure datasets into coordinated GIS datasets and CAD drawings, and often these datasets are being made available via the World Wide Web or on CD-ROM (see Table 3 for several examples).

TABLE 3. Listing of example U.S. cities with online geodatabases.

City	Web Site
Albuquerque, NM	http://www.cabq.gov/gis/
Chicago, IL	http://w15.cityofchicago.org/mapsites/public/intro.htm
Fayetteville, AR	http://www.faygis.org/
Houston, TX	http://www.jims.hctx.net/jimshome/gis/
Los Angeles, CA	http://gis.lacity.org/
Phoenix, AZ	http://www.ci.phoenix.az.us/GISMETA/theme.html
Portland, OR	http://storefront.metro-region.org/drc/index.cfm
Seattle, WA	http://www.cityofseattle.net/gis/docs/availdata.htm

Climate/Meteorology

A comprehensive urban environmental database also requires information on local and regional climate and meteorology. Important climatic characteristics include monthly, seasonal, and annual averages and trends of temperature, humidity, precipitation, solar radiation, and wind. Climatic characteristics are available in summary datasets or can be derived from meteorological records. Meteorological records provide the necessary time series needed as input datasets to drive environmental models. For example, air quality models require meteorological information on wind fields, precipitation, temperature, solar radiation, and more. Most watershed models require input of precipitation and evaporation time series for prediction of soil storage and runoff over a specified time period. In lieu of meteorological records, mesoscale meteorological models can be used to predict the necessary information for environmental assessment or for input to other environmental models. The use of meteorological models, however, adds complexity and uncertainty to integrated urban environmental modeling [Brown et al., 2000a].

The climate and meteorology can act as the forcing function driving urban systems. One common example is the influence of local meteorology on the energy usage required to heat and cool the inside of buildings. Each summer several heatwaves strike large cities, causing massive consumption of electrical power to cool the buildings. Besides the climate and meteorology forcing the issue, the urban system, on the other hand, can influence the local climate and meteorology. For example, the form and structure of a city can influence the wind flows [e.g., Fernando et al., 2001], the surface energy fluxes and temperature [e.g., Grimmond and Oke, 1999; Voogt and Oke, 1997; Oke, 1987], the precipitation pattern [e.g., Shepherd et al., 2002; Bornstein and Lin, 2000], and other meteorological variables.

Climatic and meteorological datasets can be obtained from federal and local government entities. In the U.S., climatic and meteorological data are archived at the National Oceanic and Atmospheric Administration (NOAA) National Climatic Data Center (NCDC), which can be accessed through the World Wide Web at <http://lwf.ncdc.noaa.gov/oa/ncdc.html>. Radar-rainfall data products can be obtained from NCDC and a number of commercial vendors.

Environment and Natural Resources

Describing the hydrography, soils, vegetation types, and other environmental characteristics is another important task in an environmental modeling and assessment study. The environmental datasets required for a project depend on the project objectives. In most cases, the objective of an urban environmental modeling project will be to assess impacts to the environment from an outside stimulus, such as urban development. In such cases the environmental resources of concern (e.g., a lake or river) must be accurately identified and addressed during data collection efforts. Digital hydrography datasets uniquely identify the stream segments and describe the interconnections between the segments and other surface water elements (e.g., rivers, lakes, wetlands, estuaries, oceans). In the U.S., the EPA maintains a digital dataset of stream and channel reaches called the River Reach Files (http://www.epa.gov/owowwtr1/nps/gis/reach/ html). The Reach Files are distributed in three versions, which have increasing levels of detail.

The soil type influences many environmental phenomena, including the hydrologic cycle. Important soil type characteristics include structure, permeability, water content, and organic matter content. However, when working in urban areas, soils are often disturbed during urban development and this must be factored into modeling efforts. Infiltration rates, for example, are often related to soil type, but if the soils have been compacted then infiltration rates commonly quoted in the literature may not be applicable [Pitt and Lantrip, 2000; Hamilton and Waddington, 1999]. Soil coverage for the U.S. can be obtained from the U.S. Department of Agriculture (USDA) Natural Resources Conservation Service (NRCS) web site (http://www.statlab.iastate.edu/soils/ nsdaf/). The NRCS STATSGO database contains soil maps at the state scale for the conterminous U.S., Hawaii, and Puerto Rico. The STATSGO coverage can be linked to the Soil Interpretations Record (SIR), which contains more than 20 physical and chemical soil properties. The NRCS SSURGO database contains county scale soil maps, but currently data are only available for selected counties throughout the U.S. The SSURGO database can be linked to a Map Unit Interpretations Record (MUIR) attribute database, which contains soil characteristics similar to the SIR.

Surface cover, morphology, soils, and vegetation type are important factors in describing the urban canopy energy budget and its impact on the urban climate. For example, these factors are important in computing the storage heat flux in urban areas. The storage heat flux is the net uptake or release of energy by sensible heat changes in the urban canopy layer, buildings, vegetation, and the ground [Grimmond and Oke, 1999]. Quantifying the storage heat flux is important for modeling evapotranspiration, sensible heat flux, boundary layer growth, and more.

Natural hazards are another important consideration for many urban environmental studies. Information on the location, likelihood, and expected impacts of natural hazards can be used to assess potential impacts to the urban system. Information about natural hazards can be incorporated into the urban database, or the urban modeling effort might be simulating the occurrence of natural hazards. Natural hazards to consider include earthquakes, hurricanes, tornadoes, landslides, flood/drought, volcanic eruption, and forest fires.

Recently, much environmental and natural resource data have been collected using remote sensing. Remote sensing uses measurements of the electromagnetic spectrum to characterize the landscape or infer properties of it [Garbrecht et al., 2001]. The primary advantage of remote sensing is the ability to rapidly characterize the spatial distribution of earth system characteristics. Common variables in integrated urban environmental modeling defined using remote sensing include land use and cover, vegetation indices, soil moisture, surface temperature, precipitation, and snow cover. Remote sensing data can be purchased from commercial vendors (e.g., SPOT Image Corporation <http://www.spot.com> and Space Imaging <http://www.spaceimaging.com>) or obtained from government organizations (e.g., NASA http://daac.gsfc.nasa.gov/).

In general, environmental and natural resources digital datasets are most often obtained free-of-charge, or for a marginal cost, from government agencies responsible for the management of natural resources. The USGS EROS data products center (http://edc.usgs.gov/geodata/) is an especially useful repository of several types of environmental and natural resources digital data.

Environmental Quality

Most integrated environmental assessment or modeling studies require an accurate characterization of the environmental quality of the area [Schnoor, 1996]. Data are needed for analysis and synthesis with other urban datasets and to develop, calibrate, and validate environmental models. Data from previous studies must first be collected and analyzed to identify gaps. Then, additional field monitoring protocols must be established to supplement existing data. In the past, urban environmental modeling and assessment studies have had a single media focus (e.g., air, land, water); consequently, few comprehensive multimedia field datasets exist. Field monitoring activities can be expensive, especially in the case of multimedia, long-term collection efforts. The U.S. EPA BASINS modeling system (http://www.epa.gov/ostwater/BASINS/) contains a large collection of archived environmental quality information organized on a watershed basis.

CASE STUDY: LOS ANGELES

The following case study of Los Angeles (LA) will illustrate the concept of the integrated urban environmental database, explore potential analysis and synthesis activities, and demonstrate two example applications of the database. It was necessary to develop an integrated environmental database for LA because it was selected to be the study area for several integrated modeling projects. Therefore, a coordinated effort to develop the urban database was initiated to eliminate redundant data collection and management costs. Once completed, the urban database was used in integrated modeling studies of the relationship between the LA urban area and local and regional air and water resources.

The City of LA is located along the Pacific Coast in Southern California in the western U.S. (see Figure 3). LA covers approximately 1215 km2 and has an estimated year 2000 population of 3,823,000. Based on the 1990 Census the average population density is approximately 2,855 persons per km2. The LA metropolitan area extends beyond the Los Angeles city limits to include a large part of LA County. LA County is 10,616 km2 and has a 2000 population of nearly 10 million. LA has a highly variable terrain ranging from steep slopes to flat areas nearer the coast. The climate in LA is Mediterranean with the lowest average temperatures occurring in December and January and the highest average temperatures occurring in July and August. Average annual rainfall is approximately 381 mm with nearly 90% occurring during the wet season (October to March).

In the 1980s and 1990s LA commonly exceeded federal and state air quality standards for ozone and respirable particulate matter (PM10). LA County has historically been amongst the counties with the worst air quality in the United States [source: U.S. EPA web site]. However, the air and water resources in and around LA are vital to the economy of the region. Santa Monica Bay, the most significant water resource in the region is a large open embayment located seaward from LA. Santa Monica Bay has a surface area of approximately 691 km2. Approximately 500,000 tourists and local residents visit the beaches annually and nearly 6% of the U.S. population lives near its shore [SMBRP, 1994]. The relationship between the sprawling urban system and Santa Monica Bay must be carefully studied to improve decision-making in the region regarding urban growth, natural resources preservation, and environmental protection.

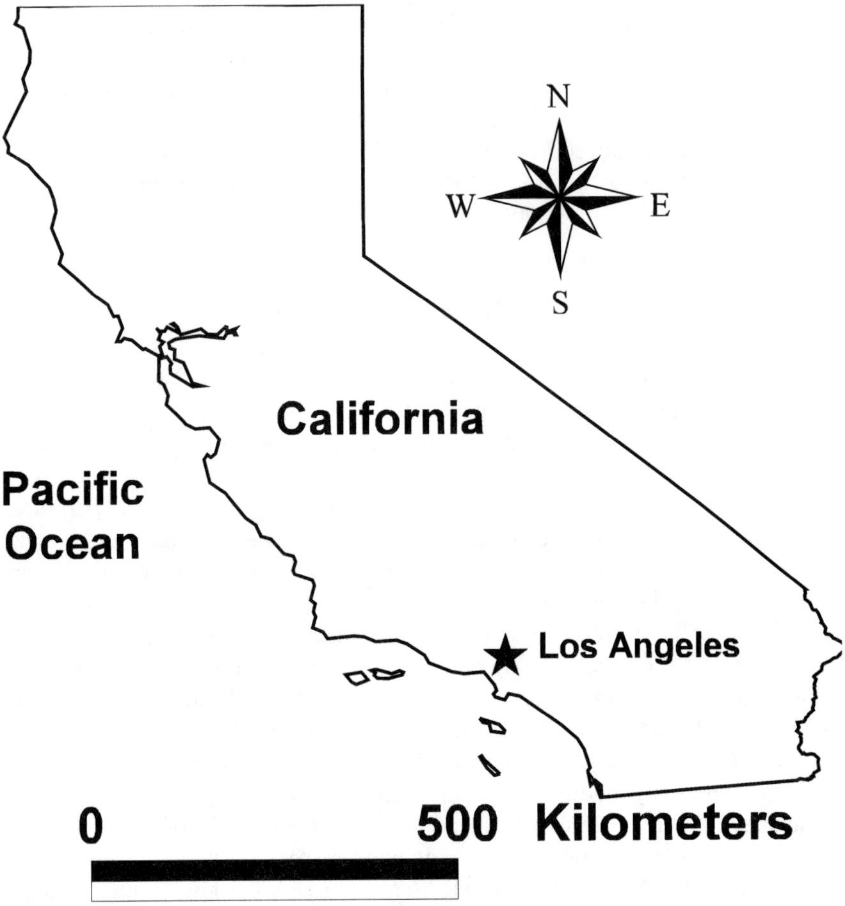

Figure 3. Location of Los Angeles in the State of California.

Urban Database Development

A significant amount of data for LA was readily available from city, county, and national sources. The physical characteristics of the watershed were our initial focus, followed by meteorology and infrastructure, and then environmental quality and natural hazards. After the initial GIS database development, several gaps in the data were identified and field monitoring protocols and additional data collection efforts were planned and performed to fill the gaps. Two noteworthy data collection efforts were (1) the storm drainage infrastructure and (2) the dry weather water quality. Detailed information about the location, type, size, length, invert slope, material, and shape of the underground storm drains and above ground channels were needed to accurately simulate the urban hydrology, hydraulics, and pollutant transport in LA. Project team members obtained the storm drain details from microfiche as-built construction drawings located at the Los Angeles County Department of Public Works (LADPW). The collection of this information required approximately 1.5 person-months of project time. The digitizing of this information into the GIS database required an additional 1.0 person-month of project time. The second noteworthy data collection effort was the dry weather water quality field protocol. During the summer of 1999 approximately 1.5 person-months of project time was used for the collection of dry weather water quality samples from within the case study watershed [Burian and McPherson, 2000]. One of the lessons learned from our data collection activity was that the time and cost for potential data collection efforts must be factored into initial project budget considerations.

The GIS database development required numerous datasets to be digitized into GIS or converted into a format that could be imported into GIS. Once all the datasets were in GIS-compatible formats they had to be converted into a common map projection. A map projection is a mathematical transformation by which the latitude and longitude of each point on the earth's curved surface are converted into corresponding (x,y) projected coordinates in a flat map reference frame [McDonnell, 1991]. For this study we used the Universal Transverse Mercator (UTM) projection, referenced to the North American Datum of 1983 (NAD 83). All datasets were converted to the UTM NAD 83 projection using standard GIS tools. Table 4 lists the primary datasets accumulated for the integrated urban modeling effort. The base level datasets listed in the table were processed to produce derived datasets and additional model input parameters. For example, the digital elevation model (DEM) was processed using standard GIS functions and scripts to derive slopes and elevation contours and to delineate watersheds.

The land use/land cover (LULC) dataset is one of the most important base level datasets in urban environmental modeling because it is needed for watershed and atmospheric modeling activities, as well as general environmental assessment and watershed characterization tasks. LULC is important for modeling activities because it can be used to parameterize the land surface cover for mesoscale meteorological, atmospheric deposition, hydrologic, and pollutant loading models. Figure 4 displays the LULC dataset for the LA region. The region shown is 49% residential, 29% open space, 12% industrial, 7% commercial, with less than 2% water surface. For the modeling efforts the LULC dataset was used to assign many parameters over the modeling domains for the mesoscale meteorological and air chemistry models. In addition, several hydrologic parameters (e.g., percent directly connected impervious area, interception and depression storage) were assigned based on predominant land use/land cover in each subcatchment in the watershed model.

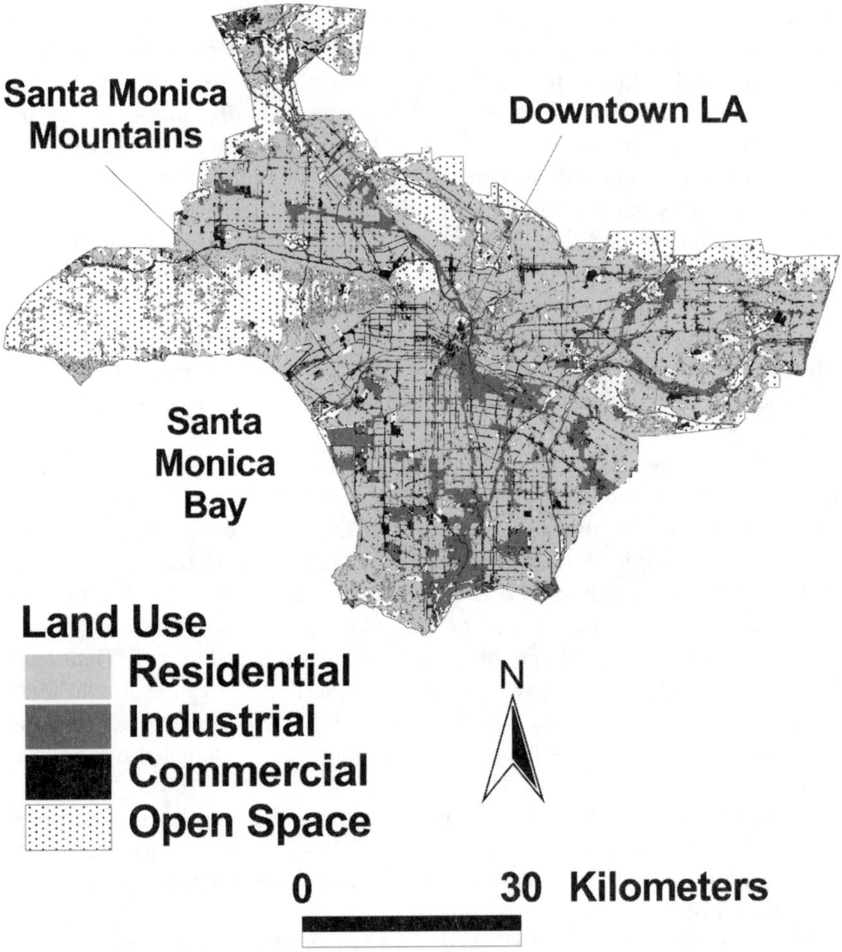

Figure 4. Illustration of land use in Los Angeles and surrounding communities.

Urban Database Analyses and Synthesis Activities

After accumulating digital datasets and organizing them into a coordinated database, the next step in most environmental modeling and assessment projects is to analyze the datasets and derive useful information. Urban databases, in general, can be processed to obtain general information about the study area or to compute model input variables. For one of our projects we had an interest in the morphological characteristics of the study region because LA is a case study location for a national effort to describe the morphological characteristics of

TABLE 4. Primary base level datasets accumulated during the development of the Los Angeles urban environmental database.

Dataset	Source	Notes
Physical Characteristics		
LULC	SCAG	Modified Anderson Level IV; based on 1993 aerial photographs
DEM	USGS	30-m grid cell size
Hydrolography	EPA	Reach File Version 3
Digital Orthophotos	USGS	
Building Footprints	Commercial Vendor	Delivered in vector and raster form and Heights
Infrastructure Elements		
Storm Drainage System	LADPW	Digitized from LADPW as-built drawings
Roads	Tiger Files	Downloaded from ESRI website
Climate/Meteorology		
Rainfall data	NCDC	15-minute and hourly rainfall data for more than 25 rain gauges in LA area
Environmental Quality		
Dry Weather Water Quality	SMBURD/ Monitoring	Database of water quality information obtained from Michael Stenstrom at UCLA. A monitoring effort was performed during the summer of 1999 to supplement data in SMBURD.
Wet Weather Water Quality	LADPW	LADPW Stormwater Monitoring Program
Air Quality	EPA	
Atmospheric Deposition	NADP	Southern California NADP sites

urban land use [Brown et al., 2002]. Specifically, we are integrating three-dimensional building datasets, digital orthophotos, detailed LULC information, bald-earth topography, and roads into a GIS database for analysis of urban morphology for several cities in the U.S. using standard GIS functions, Avenue scripts, and

Fortran programs, we have developed a set of GIS tools to automatically compute numerous urban canopy parameters and aerodynamic roughness parameters [Burian et al., 2002a].

The GIS analysis tools have been used to compute the building height characteristics (e.g., mean height, variance of height, height histograms), as well as other parameters describing the urban morphology of downtown LA including the building plan area fraction (λ_p), building area density ($a_P(z)$), rooftop area density ($a_r(z)$), frontal area index (λ_f), frontal area density ($a_F(z)$), complete aspect ratio (λ_C), building surface area to plan area ratio (λ_B), and the height-to-width ratio (λ_S). In addition, we have used standard morphometric equations to compute the aerodynamic roughness length (z_o) and zero-plane displacement height (z_d) for the entire study area and for each urban land use type. Table 5 shows a summary of building height characteristics for downtown LA as a function of land use type. The building height characteristics for a residential land use class are not shown below because the majority of the residential buildings were less than 8 m. Figure 5 displays the distribution of building heights in the form of histograms for the four urban land use classes. The data analysis suggests that the Industrial and the Industrial & Commercial land use classes are very similar in terms of building height characteristics. The Commercial & Services land use class encompasses the majority of the downtown city center and is characterized by predominantly high-rise buildings.

TABLE 5. Building characteristics for the 17.2 km^2 study area in downtown Los Angeles.

	Commercial & Services	Industrial	Mixed Industrial & Commercial	Mixed Urban
Number	1370	1441	195	319
Mean Height (m)	26.9	6.6	7.7	12.4
Median Height (m)	15	5	5	6
Max Height (m)	331	48	42	52
Stan. Dev. (m)	39.9	4.8	6.0	12.3

In addition to the building height distribution in cities, the plan area fraction (λ_p) of roughness elements (e.g., buildings) is also an important urban canopy parameter for mesoscale meteorological, urban surface energy budget, and urban dispersion models [Brown, 1999]. The building plan area fraction (λ_p) is defined as the ratio of the plan area of buildings to the total surface area of the study region:

$$\lambda_p = \frac{A_p}{A_T} \tag{1}$$

where A_p is the plan area of buildings at ground level, i.e., the footprint area, and A_T is the total plan area of the region of interest, i.e., an arbitrary area that encompasses the buildings. Table 6 shows the plan area fraction as a function of land use type for the downtown LA study site. The results shown in Table 6 indicate a higher building density for Industrial and Mixed Industrial & Commercial land uses. The Residential and Commercial & Services land uses sampled had a significantly lower building density. Figure 6 shows the spatial distribution according to a uniform 100 m X 100 m grid cell mesh of the building plan area fraction for the downtown LA case study location. The grouping of darker grid cells (higher plan area fraction) in the north central part of the site is the high-rise section of the downtown area and the grouping of darker grid cells in the south central area is the industrial region. The more detailed characterization of

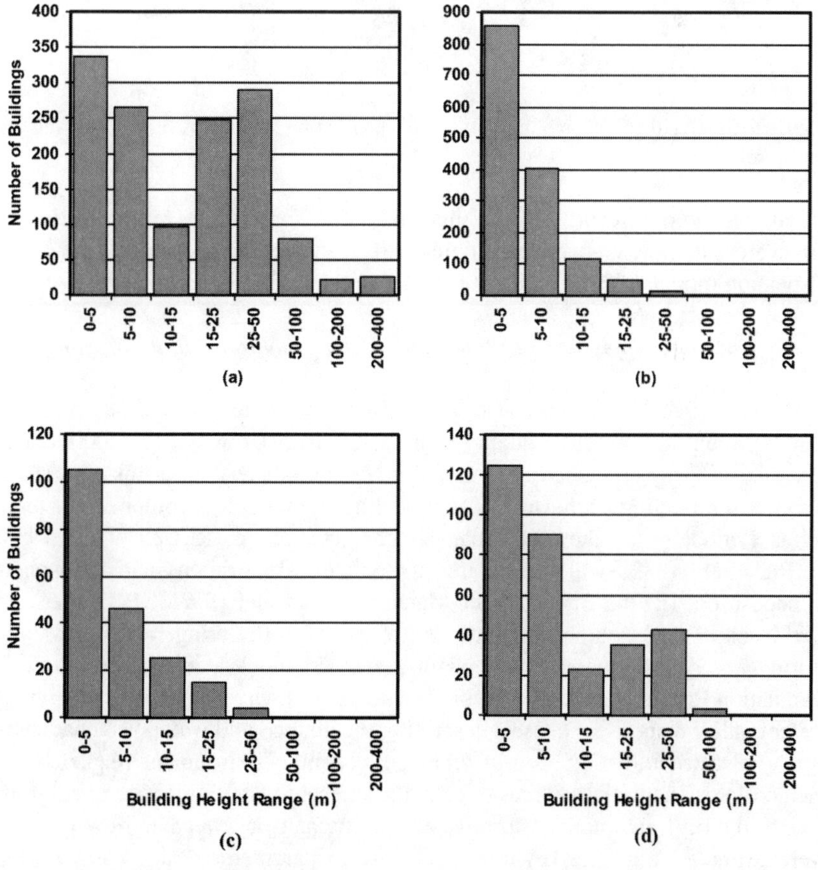

Figure 5. Building height distributions in downtown Los Angeles for (a) Commercial & Services, (b) Industrial, (c) Mixed Industrial & Commercial, and (d) Mixed Urban land uses.

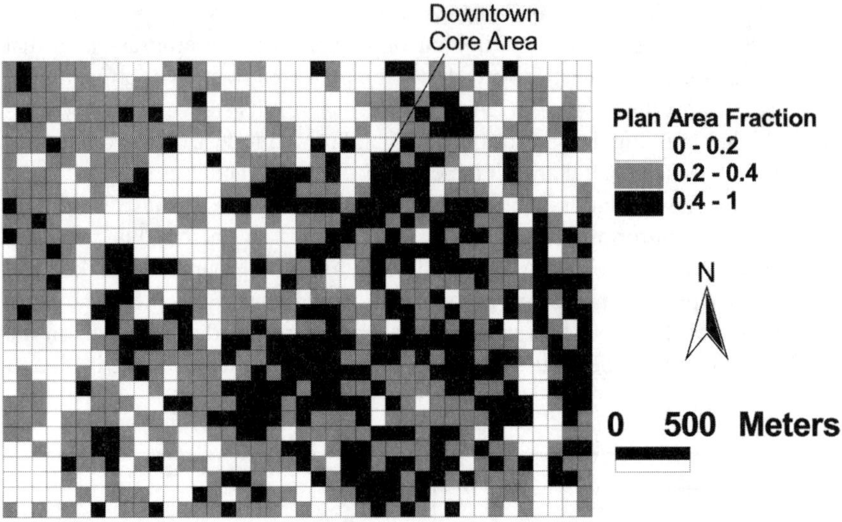

Figure 6. Spatial distribution of building plan area fraction for the downtown Los Angeles study area.

urban morphology resulting from this study will improve the representation of urban areas in mesoscale meteorological, urban surface energy budget, and urban dispersion models [Ching et al., 2002; Brown et al., 2000b; Brown, 1999].

Urban Database Application to Integrated Urban Environmental Modeling

The LA urban database has also been applied to the study of the interrelationship between air and water quality in the LA Basin [Brown et al., 2000a]. The LA database was processed to compute model parameters for a linked airshed-watershed modeling framework. The modeling framework is comprised of three primary models: (1) the CIT urban air chemistry code [Russell et al., 1988; McRae et al., 1982], which simulates photochemical air quality and atmospheric deposition, (2) the Storm Water Management Model (SWMM) [Huber and Dickinson, 1988], which simulates pollutant washoff during rain storms and stormwater runoff quantity and quality, and (3) the Water Quality Analysis Simulation Program (WASP) [Ambrose et al., 1993], which simulates receiving-water quality response. Datasets describing the airshed and watershed characteristics, infrastructure (e.g., storm drainage system), environment (e.g., LULC, meteorology), natural resources (e.g., soils, adjacent water bodies), and environmental quality (air quality, stormwater quality, and dry weather flow quality) were extracted from the LA urban database to parameterize and calibrate the models. Brown et al. [2000a] and McPherson et al. [in this Special Publication] describe the modeling framework in more detail.

TABLE 6. Plan area fraction as a function of land use for the 17.2 km^2 study area in downtown Los Angeles.

Land Use Class	Plan Area Fraction
Residential	0.28
Commercial & Services	0.28
Industrial	0.39
Mixed Industrial & Commercial	0.47
Mixed Urban	0.33

The case study application of the modeling framework and urban datasets was set in the LA airshed, the Ballona Creek watershed (BCW), and the Ballona Creek Estuary (BCE). The total area of simulated airshed is 60,000 km^2. The BCW has an area of 300-km^2 and is composed of more than 85% urban land use (the BCW is within the area shown in Figure 4). Ballona Creek begins as a covered storm drain in downtown LA and increases in size as it travels west towards the coast. Ballona Creek feeds into the BCE, which is tidally influenced by Santa Monica Bay. The CIT model and setup were identical to that in an earlier analysis of the Southern California Air Quality Study (SCAQS) [Harley et al., 1993]. SWMM input parameters were determined from a variety of datasets accumulated during the database development. Surface water transport and reaction of nutrients and algae in the BCE was simulated using the DYNHYD5 hydrodynamics module and the EUTRO5 kinetics module of WASP. The DYNHYD5 hydrodynamic model was calibrated using segment height data collected in the summer of 1999.

The airshed-watershed modeling framework uses CIT to simulate the dry deposition flux of nitrogen compounds (e.g., nitric acid (HNO_3), ammonium nitrate (NH_4NO_3), ammonia (NH_3), nitrogen oxides (NOx), peroxyacetyl nitrate (PAN), and alkyl nitrate aerosols to the land surface and water bodies. These loads are summed over the simulated time period and input to SWMM and WASP. At the beginning of a rainstorm the atmospheric concentrations calculated by CIT are used by an atmospheric washout algorithm to determine the masses of selected nitrogen compounds washed out of the atmosphere during the storm event. The calculated wet deposition loads are also input to SWMM and WASP. During the rainstorm the accumulated dry deposition loads and the calculated wet deposition loads are washed off the watershed using the SWMM first-order washoff algorithm. SWMM routes the runoff and pollutant loads to the receiving water at which point WASP simulates the response to the wet weather loads.

In this paper we present two applications of the linked air-water modeling framework. The first application investigated the nitrogen loading to Santa Monica Bay. The modeling framework was used to quantify the nitrogen load from atmospheric wet and dry deposition directly onto the bay and from wet and dry deposition onto the watershed that is washed into the bay during rainstorms. Figure 7 displays the relative fraction of nitrogen load entering Santa Monica Bay via the four pathways. Figure 7 suggests that most of the nitrogen load enters the bay via direct atmospheric deposition, while the load transported in the stormwater runoff is only a small fraction. Burian et al. [2001] and McPherson et al. [in this Special Publication] provide further details of the nitrogen loading study.

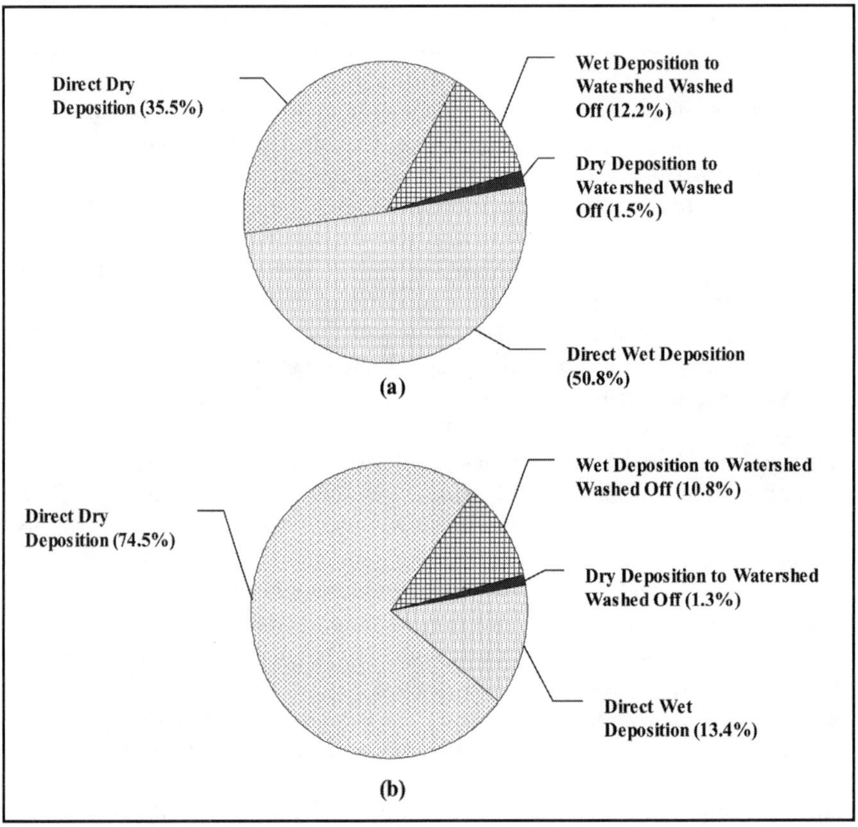

Figure 7. Loads of (a) NO_3^- and (b) NH_4^+ to Santa Monica Bay for the simulated time period of November 18, 1987 to December 4, 1987 [Burian et al., 2001].

The second application of the linked air-water modeling framework studied the impacts of air quality policy implementation on urban environmental quality. Two air emission scenarios were simulated. The first scenario used the 1987 emissions inventory from the South Coast Air Quality Management Plan of 1987 to drive CIT, and the second scenario used an alteration of that inventory based on the air quality goals for 2000 listed in the 1998 Air Quality Management Plan of the South Coast Air Quality Management District (SCAQMD). The reductions in the 1987 inventory are across all source types including point, area, and elevated emissions. NOx emissions were reduced by 32%, volatile organic compound (VOC) emissions were reduced by 51%, and NH_3 emissions were reduced 30%. For both air emission scenarios we simulated a dry season time period and a wet season time period and examined the changes to atmospheric deposition fluxes and stormwater runoff loading caused by the reductions in atmospheric emissions.

The dry season simulation period began on June 1, 1987 and ended on August 31, 1987. The wet season simulation period began on November 18, 1987 and ended on December 4, 1987. Data limitations and time constraints prohibited the simulation of the atmospheric chemistry for the entire dry and wet season simulation periods. Therefore, we chose to simulate August 27, 1987 and December 3, 1987 because they would be representative of the conditions of interest. August 27, 1987 had onshore meteorological conditions, normal for Los Angeles, in which the daylight wind flows are primarily from the coastal areas to the west towards the mountains to the east. December 3, 1987, on the other hand, had increasing offshore wind flow that was insufficient to push the atmospheric contaminants out to sea. Consequently, high aerosol and nitrate levels were observed during the SCAQS stagnating over the western part of the basin on December 3, 1987.

The deposition computed by CIT for August 27, 1987 was representative of the average daily deposition in the LA area [Burian et al., 2002b]. Therefore, the results from the August 27th simulation were used to represent an average meteorological and photochemical pollution day in LA. For the dry season evaluation, the calculated dry deposition for August 27, 1987 was used to represent an average day during the period June 1, 1987 to August 31, 1987. The daily dry deposition load was input to the BCE water quality model.

In the November 18, 1987 to December 4, 1987 wet season evaluation, we estimated the nitrogen dry deposition flux for November 18, 1987 to December 2, 1987 using the results of the August 27, 1987 simulation. The December 3, 1987 simulation provided the results for December 3rd. We included December 3, 1987 in the simulation because of the abnormal meteorological conditions on this date. A small rain event occurred on November 17, 1987 and resulted in 2.5 mm of rainfall. Although that was a relatively small rain event, we assumed all the atmospherically deposited material present on the watershed was removed at that time. The next recorded rainfall event occurred on December 4, 1987 and produced an average of 26 mm of rainfall in the BCW. Summing the December 3rd deposition and 15 times the August 27th deposition (representing the deposition for the 15 days between November 18, 1987 and December 2, 1987) gave us an estimate of the total dry deposition for the 16 days between the November 17th and December 4th rainfall events. For both the dry season and the wet season, the 1987 emission inventory and the emission inventory modified to reflect year 2000 emission goals were used to drive the simulations.

The August 27, 1987 simulation using the 1987 emissions inventory produced a dry deposition flux of 0.174 kg N/ha-day to the BCW, some of which is due to plant uptake and thus is not available for stormwater washoff. This flux is more than twice the historical average daily deposition flux of 0.063 kg N/ha-day reported by Takemoto et al. [1995] for the general Los Angeles area, but our flux includes several more nitrogen compounds. The December 3, 1987 dry deposition flux calculated with the 1987 emissions inventory was 0.27 kg N/ha-day. Our flux is 33% higher than the 0.18 kg N/ha-day dry deposition flux calculated by Russell et al. [1993] using CIT for a location in an experimental forest near Los Angeles. Russell et al. [1993] were simulating August 30-31, 1982, a time period that also had exceptional meteorological conditions producing elevated photochemical pollution levels in the Los Angeles area. Our higher flux is partially explained by the difference in location between the urbanized BCW and the experimental forest. Urbanized sites tend to have higher atmospheric concentrations of certain contaminants (ammonia/ ammonium) than forested sites, which would lead to higher deposition levels.

Table 7 shows the daily dry deposition fluxes to the BCW and Santa Monica Bay for the 1987 emissions levels and the proposed 2000 goals. The August 27, 1987 nitrogen flux to the BCW was reduced 15.0%, while the nitrogen flux to Santa Monica Bay was reduced 21.4% by changing the emissions from 1987 levels to 2000 levels. For the December 3rd simulation, the nitrogen flux to the BCW was reduced 14.7% and the nitrogen flux to Santa Monica Bay was reduced 18.8%. Interestingly, the emission reduction had a greater influence on the reduction of dry deposition nitrogen flux to the Bay than it did to the BCW. Also, regardless of the meteorological conditions the percent reductions in dry deposition nitrogen flux to the BCW and the Bay produced by reducing emissions were similar.

TABLE 7. Daily dry deposition flux of nitrogen to the Ballona Creek watershed and Santa Monica Bay.

	1987 Emissions		2000 Emissions	
Date	Flux to BCW (kg N/ha-day)	Flux to Bay (kg N/ha-day)	Flux to BCW (kg N/ha-day)	Flux to Bay (kg N/ha-day)
8-27-1987	0.174	0.014	0.148	0.011
12-3-1987	0.470	0.048	0.401	0.039

Table 8 shows the total calculated NO_3^- and NH_4^+ dry deposition loads for June 1, 1987 to August 31, 1987 to both the BCE and Santa Monica Bay for the 1987 emissions and the year 2000 goals. The loads include all potential pathways incorporated into the bulk deposition velocities used by CIT (e.g., diffusion, gravitational settling, biological uptake). During the dry season, the air emissions reductions produced a 14.5% reduction in NO_3^- dry deposition load and a 36.5% reduction in NH_4^+ dry deposition load to Santa Monica Bay. The air emissions reductions produced an 18.5% reduction in NO_3^- dry deposition load and a 27.0% reduction in NH_4^+ dry deposition load to the BCE. Note the small loading to the BCE from atmospheric deposition compared to Santa Monica Bay because of the small surface area of the estuary. As a comparison to the loads shown in Table 8, the NO_3^- and NH_4^+ loads from Ballona Creek dry-weather flow (DWF) inputs to the BCE for the same time period are 5,600 kg and 400 kg, respectively. DWF includes all flow in Ballona Creek during dry weather (e.g., nuisance flows, landscape sprinkling runoff, potable water line leaks). Clearly, the DWF load from Ballona Creek is more significant than the dry deposition load to the BCE.

TABLE 8. Dry Season NO_3^- and NH_4^+ loads to the Ballona Creek Estuary and Santa Monica Bay from dry deposition, June 1, 1987 to August 31, 1987.

Receiving Water	Pollutant Load	1987 Emissions	2000 Emissions
Ballona Creek Estuary	Dry Deposition NO_3^- (kg)	30	24
	Dry Deposition NH_4^+ (kg)	4	3
Santa Monica Bay	Dry Deposition NO_3^- (kg)	310,000	265,000
	Dry Deposition NH_4^+ (kg)	20,500	13,000

Table 9 shows the NO_3^- and NH_4^+ dry deposition loads from November 18, 1987 to December 4, 1987 to the BCE and Santa Monica Bay for both the 1987 emissions and the year 2000 emission goals. The altered emissions inventory produced a 14.5% reduction in NO_3^- dry deposition load and a 37.8% reduction in NH_4^+ dry deposition load to Santa Monica Bay. Similar to the dry season simulation, the loads to the BCE are small relative to the loads to Santa Monica Bay because of the small surface area. From these results it can be surmised that the DWF load of nitrogen likely controls the productivity in the BCE.

Figure 8 shows the relative nitrogen loads to the BCW and Santa Monica Bay from November 18, 1987 to December 4, 1987 for both the 1987 emissions and the year 2000 goals. Nitrogen wet deposition to the bay was reduced by 18.6%, dry deposition to the bay was reduced by 15.0%, dry deposition to the watershed was reduced by 15.5%, wet deposition to the watershed was reduced by 16.8%, and stormwater runoff load was reduced by 16.1%. The dry deposition load to the BCW watershed is much higher than the other loads shown, but less than 5% of the load is available for washoff and removed during the simulated runoff event.

The low fraction of dry deposition load washed off the watershed indicates that much of the deposited load is either (1) removed from the watershed before the storm event by the action of any of several processes (e.g., biological uptake, street sweeping, nuisance flows, re-suspension), (2) inhibited from being removed, or (3) fixed to the watershed by soil fixation or chemical reactions. The low fraction could also indicate an underestimate of the fraction of dry deposition due to plant uptake or an incorrect assumption about the form of the dry deposited compounds. These considerations will be factored into our future integrated modeling efforts. More importantly, the percent reductions shown on Figure 8 all indicate the potential of air emissions reductions to reduce loads to water bodies.

TABLE 9. NO_3^- and NH_4^+ loads to the Ballona Creek Estuary and Santa Monica Bay from dry deposition, November 18, 1987 to December 4, 1987.

Receiving Water	Pollutant Load	1987 Emissions	2000 Emissions
Ballona Creek Estuary	Dry Deposition NO_3^- (kg)	5	4
	Dry Deposition NH_4^+ (kg)	1	1
Santa Monica Bay	Dry Deposition NO_3^- (kg)	63,400	54,200
	Dry Deposition NH_4^+ (kg)	3,700	2,300

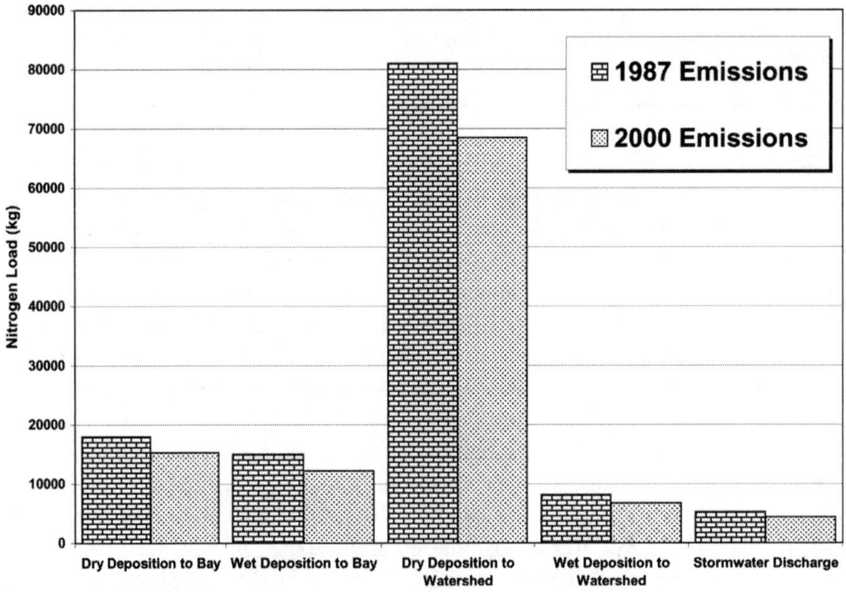

Figure 8. Nitrogen loads from November 18, 1987 to December 4, 1987 for 1987 emissions and year 2000 goals.

SUMMARY

Urban databases are an integral component in the analysis and management of urban environmental systems. It is important for a comprehensive urban database to include multi-disciplinary datasets describing the physical characteristics, economic and sociologic characteristics, infrastructure elements, climate and meteorology, environment and natural resources, and environmental quality of the urban system. A high level of detail and accuracy is currently attainable given the widespread availability of digital datasets from government agencies and commercial vendors. Management, visualization, and analysis of the urban digital datasets are also relatively straightforward using geographic information systems (GIS), database management systems, and visualization tools.

In this paper, we briefly reviewed the types and sources of digital data to include in a detailed urban environmental database. We described the development of a high-fidelity urban geodatabase for the Los Angeles metropolitan area. Similar geodatabases have been developed for other major cities in the U.S. (e.g., Phoenix, Salt Lake City, Portland, Houston, Albuquerque). The databases have been applied in numerous environmental studies and have been provided to government agencies and private entities for other studies and applications. This paper briefly described morphological analyses of the 3D LA building database and reviewed the application of the LA urban database to two interdisciplinary urban environmental modeling activities.

Acknowledgments. Part of this work was supported by the Urban Security Initiative at Los Alamos National Laboratory (LANL). We would like to thank Grant Heiken of LANL for his support, and Eric R. Pardyjak of the Department of Mechanical Engineering at the University of Utah for his helpful comments on an earlier version of this manuscript. Listing of vendors, products, and trade names in this paper is not an endorsement by the writers or their agencies.

REFERENCES

Ambrose, R. F., T. A. Wool, and J. L. Martin, The Water Quality Analysis Simulation Program, WASP5, Part A: Model Documentation. Environmental Research Laboratory, Athens Georgia and AScI Corporation, Athens Georgia, 1993.

Arnold, C. L. and C.J. Gibbons, Impervious Surface Cover: The Emergence of a Key Environmental Indicator, *Journal of the American Planning Association*, 62(2), 243-258, 1996.

Booth, D. B. and L. E. Reinfelt, Consequences of Urbanization on Aquatic Systems – Measured Effects, Degradation Thresholds, and Corrective Strategies, in *Proceedings of the Watershed '93 Conference*, March 1993, 545-550, 1993.

Bornstein, R. and Q. Lin, Urban Heat Islands and Summertime Convective Thunderstorms in Atlanta: Three Case Studies, *Atmospheric Environment*, 34, 507-516, 2000.

Brown, M. J., Urban Parameterizations for Mesoscale Meteorological Models, in *Mesoscale Atmospheric Dispersion*, Edited by Z. Boybeyi, WIT Press, 193-255, 1999.

Brown, M. J., S. J. Burian, S. P. Linger, S. P. Velugubantla, and C. Ratti, An Overview of Building Morphological Characteristics Derived from 3D Building Databases, in *Preprints 4th AMS Urban Environment Symposium*, Norfolk, VA, 2002.

Brown, M. J., C. Muller, and P. Stretz, Exposure Estimates Using Urban Plume Dispersion and Traffic Microsimulation Models, LA-UR-97-3930, in *Preprints, 10th AMS Conference on Air Pollution Meteorology*, Phoenix, AZ, 1997.

Brown, M. J., S. J. Burian, T. N. McPherson, G. E. Streit, K. Costigan, and H. J. Turin, Air and Water Quality Modeling System: Application to the Los Angeles Metropolitan Area, in *Preprint Proceedings of the 11th Joint AMS-AWMA Conference on Applications of Air Pollution Meteorology*, Long Beach, CA, 2000a.

Brown, M. J., S. J. Burian, and C. Müller, Analysis of Urban Databases With Respect to Mesoscale Modeling Requirements, in *Preprints 3rd AMS Urban Environment Symposium*, Davis, CA, 2000b.

Burian, S. J., Developments in Water Supply and Wastewater Management in the United States During the Nineteenth Century, *Water Resources Impact*, 3(5), 14-18, 2001.

Burian, S. J., G. E. Streit, T. N. McPherson, M. J. Brown, and H. J. Turin, Modeling the Atmospheric Deposition and Stormwater Washoff of Nitrogen Compounds, *Environmental Modeling & Software*, 16, 467-479, 2001.

Burian, S. J., M. J. Brown, and S. P. Linger, Morphological Analyses Using 3D Building Databases: Los Angeles, California, LA-UR-02-0781, Los Alamos National Laboratory, Los Alamos, NM, 2002a.

Burian, S. J., T. N. McPherson, M. J. Brown, G. E. Streit, and H. J. Turin, Modeling the Effects of Air Quality Policy Changes on Water Quality in Urban Areas, *Environmental Modeling & Assessment*, in press, 2002b.

Burian, S. J. and T. N. McPherson, Water Quality Modeling of Ballona Creek and the Ballona Creek Estuary, in *Water Quantity and Quality Issues in Coastal Urban Areas*, Proceedings of the American Water Resources Association Annual Water Resources Conference, 6-9 November, 2000, 305-308, 2000.

Ching, J., T. L. Otte, S. Dupont, S. Burian, and A. Lacser, Urban Morphology for Houston to Drive Models-3/CMAQ at Neighborhood Scales, in *Preprints 4th AMS Urban Environment Symposium*, Norfolk, VA, 2002.

Chow, V. T., D. R. Maidment, and L. W. Mays, *Applied hydrology*, McGraw-Hill, Inc., New York, 572 pp, 1988.

Cionco, R. M. and R. Ellefsen, High Resolution Urban Morphology Data for Urban Wind Flow Modeling, *Atmospheric Environment*, 32(1), 7-17, 1998.

Dobson, J. E., E. A. Bright, P. R. Coleman, R. C. Durfee, and B. A. Worley, LandScan: A Global Population Database for Estimating Populations at Risk, *Photogrammetric Engineering & Remote Sensing*, 66(7), 849-857, 2000.

Duffy, J., *The sanitarian*, University of Illinois Press, Urbana, Illinois, 1990.

Dupont, S., I. Calmet, and P. G. Mestayer, Urban Canopy Modeling Influence on Urban Boundary Layer Simulation, in *Preprints 4th AMS Urban Environment Symposium*, Norfolk, VA, 2002.

EPA, Deposition of Air Pollutants to the Great Waters: Third Report to Congress, *EPA-453/R-00-005*, Research Triangle Park, North Carolina, 2000a.

EPA, The Quality of Our Nation's Waters: A Summary of the National Water Quality

Inventory: 1998 Report to Congress, *EPA841-S-00-001*, U.S. Environmental Protection Agency, Washington, DC, 2000b.

EPA, Environmental Justice Action Agenda, *EPA540-R-95-023*, U.S. Environmental Protection Agency, Washington, DC, 1995.

European Commission, *European sustainable cities*. Final report, DG XI Environment, Nuclear Safety and Civil Protection, Brussels, 1996.

Fernando, H. J. S., S. M. Lee, J. Anderson, M. Princevac, E. Pardyjak, and S. Grossman-Clarke, Urban Fluid Mechanics: Air Circulation and Contaminant Dispersion in Cities, *Environmental Fluid Mechanics*, 1, 107-164, 2001.

Garbrecht, J., F. L. Ogden, P. A. DeBarry, and D. R. Maidment, GIS and Distributed Watershed Models, I: Data Coverages and Sources, *Journal of Hydrologic Engineering*, 6(6), 506-514.

Goldfield, D.R. and B. A. Brownell, *Urban America: A History*, 2nd Edition, Houghton Mifflin, Boston, MA, 1990.

Goodchild, M. F., The State of GIS for Environmental Problem-Solving, in *Environmental Modeling with GIS*, Edited by Goodchild, M. F., B. O. Parks, and L. T. Steyaert, Oxford University Press, oxford, 8-15, 1993.

Grimmond, C. S. B. and T. R. Oke, Heat Storage in Urban Areas: Local-Scale Observations and Evaluation of a Simple Model, *Journal of Applied Meteorology*, 38, 922-940.

Hamilton, G. W. and D. V. Waddington, Infiltration Rates on Residential Lawns in Central Pennsylvania, *Journal of Soil and Water Conservation*, third quarter, 564-568, 1999.

Hanna, S. A., R. E. Britter, and P. Franzese, *The effect of roughness obstacles on flow and dispersion at industrial and urban sites*, AIChE/CCPS, 3 Park Ave., New York, NY, 2001.

Harley, R. A., A. G. Russell, G. J. McRae, G. R. Cass, and J. H. Seinfeld, Photochemical Modeling of the Southern California Air Quality Study, *Environmental Science and Technology*, 27, 378-388, 1993.

Heiken, G., G. A. Valentine, M. Brown, S. Rasmussen, D. C. George, R. K. Greene, E. Jones, K. Olsen, C. Andersson, Modeling Cities: The Los Alamos Urban Security Initiative, *Public Works Management & Policy*, 4(3), 198-212, 2000.

Hicks, B. B., Atmospheric Deposition and its Effects on Water Quality, in *Proceedings, Workshop on Research Needs for Coastal Pollution in Urban Areas*, edited by E.R. Christensen and C.R. O'Melia, 27-36, 1997.

Huber, W. C. and R. E. Dickinson, Storm Water Management Model, Version 4, Part A: User's Manual, *EPA-600/3-88-001a*, U.S. Environmental Protection Agency, Athens, Georgia, 1988.

Klaiæ, Z. B., T. Nitis, I. Kos, and N. Moussiopoulos, Modification of the Local Winds due to Hypothetical Urbanization of the Zagreb Surroundings, *Meteorology and Atmospheric Physics*, 79, 1-12, 2002.

Klein, R. D., Urbanization and Stream Quality Impairment, *Water Resources Bulletin*, 15, 4, 948-963, 1979.

Maheshwari, S. and Dowell, L. J., Integrated Modeling of Earthquake Impacts to the Electric-Power Infrastructure: Analysis of an Elysian Park Scenario in the Los Angeles Metropolitan Area, in *Preprints of the URISA 1999 Annual Conference*, Chicago, Illinois, 1999.

Marr, J. K. and P. L. Freedman, Receiving-Water Impacts, in *Control and Treatment of*

Combined Sewer Overflows, 2nd Edition, Edited by P. E. Moffa, Van Nostrand Reinhold, New York, 71-133, 1997.

McDonnell, P. W., *Introduction to map projections*, 2nd Edition, Landmark Enterprises, Rancho Cordova, CA, 1991.

McPherson, T. N., S. J. Burian, M. J. Brown, G. E. Streit, and H. J. Turin, Integrated Environmental Modeling of the Urban Ecosystem, in *Earth Sciences in the Cities* [in this Special Publication], Edited by G. Heiken, R. Fakundidy, and J. Sutter, 2002.

McRae, G., W. Goodin, and J. Seinfeld, Development of a Second Generation Mathematical Model for Urban Air Pollution: I. Model Formulation, *Atmospheric Environment*, 16, 679-696, 1982.

Melosi, M. V., *The sanitary city*, The Johns Hopkins University Press, Baltimore, 2000.

Nix, S. J., Mathematical Modeling of the Combined Sewer System, in *Control and Treatment of Combined Sewer Overflows*, 2nd Edition, Edited by P. E. Moffa, Van Nostrand Reinhold, New York, 27-70, 1997.

Ogden, F. L., J. Garbrecht, P. A. DeBarry, L. E. Johnson, GIS and Distributed Watershed Models, II. Modules, Interfaces, and Models, *Journal of Hydrologic Engineering*, 6(6), 515-523.

Oke, T. R., *Boundary layer climates*, Routledge, 1987.

Pitt, R. and J. Lantrip, Infiltration Through Disturbed Urban Soils, in: *Applied Modeling of Urban Water Systems*, Monograph 8, Edited by W. James, CHI, Guelph, Ontario, 1-22, 2000.

Ratti, C., S. Di Sabatino, R. Britter, M. Brown, F. Caton, and S. Burian, Analysis of 3-D Urban Databases With Respect to Pollution Dispersion for a Number of European and American Cities, in *Proceedings, The Third International Conference on Urban Air Quality: Measurement, Modelling, and Management*, 19-23 March 2001, Loutraki, Greece, 2001.

Ritchie, J. C., Airborne Laser Altimeter Measurements of Landscape Topography, *Remote Sensing of Environment*, 53(2), 85-90.

Russell, A., K. McCue, and G. Cass, Mathematical Modeling of the Formation of Nitrogen-Containing Air Pollutants. 1. Evaluation of an Eulerian Photochemical Model, *Environmental Science and Technology*, 22, 263-271, 1988.

Russell, A., D. A. Winner, R. A. Harley, K. F. McCue, and G. R. Cass, Mathematical Modeling and Control of the Dry Deposition Flux of Nitrogen-Containing Air Pollutants, *Environmental Science and Technology*, 27, 2772-2782, 1993.

Sample, D. J., J. P. Heaney, L. T. Wright, and R. Koustas, Geographic Information Systems, Decision Support Systems, and Urban Storm-Water Management, *Journal of Water Resources Planning and Management*, 127(3), 155-161.

Schnoor, J. L., *Environmental modeling: Fate and transport of pollutants in water, air, and soil*, John Wiley & Sons, Inc., New York, 1996.

Schueler, T. R., The Importance of Imperviousness, *Watershed Protection Techniques*, 1, 3, 100-111, 1994.

Seinfeld, J. H., Urban Air Pollution: State of the Science, *Science*, 243, 745-752, 1989.

Shepherd, J. M., H. Pierce, and A. J. Negri, On Rainfall Modifications by Major Urban Areas – Part I: Observations From Space-Borne Radar on TRMM, *Journal of Applied Meteorology*, 2002.

SMBRP, State of the Bay 1992, Santa Monica Bay Restoration Project, Monterey Park, CA, 1994.

Takemoto, B. K., B. E. Croes, S. M. Brown, N. Motallebi, F. D. Westerdahl, H. G. Margolis, B. T. Cahill, M. D. Mueller, and J. R. Holmes, Acidic Deposition in California: Findings From a Program of Monitoring and Effects Research, *Water, Air and Soil Pollution*, 85, 261-272, 1995.

Voogt, J. A. and T. R. Oke, Complete Urban Surface Temperatures, *Journal of Applied Meteorology*, 36, 1117-1132, 1997.

Wanielista, M. P. and Y. A. Yousef, *Stormwater management*, John Wiley & Sons, Inc., New York, 579 pp, 1993.

Wong, K. M., E. W. Strecker, and M. K. Stenstrom, GIS to Estimate Storm-Water Pollutant Mass Loading, *Journal of Environmental Engineering*, 123(8), 737-745.

World Commission on Environment and Development and G. H. Brundtland, *Our common future*, Oxford University Press, Oxford, 400 pp, 1987.

SECTION IV

THE REMOTELY SENSED CITY

Today's cities, be they in the high-tech San Francisco Bay area of California or the slums of Dhaka, are growing rapidly. As we strive to protect the environment from human misuse, we must also protect people from damage by the environment—both require monitoring on a systematic basis by observational satellites owned by governments and industry. Such satellites are excellent tools for tracking urban change and for cost-effective planning.

12

Mapping the City Landscape From Space: The Advanced Spaceborne Thermal Emission and Reflectance Radiometer (ASTER) Urban Environmental Monitoring Program

Michael S. Ramsey

INTRODUCTION

As the global population expands, concentrating in the large urban centers of the world, the stress placed on these local environments will also magnify. It is estimated that in the next 25 years nearly two-thirds of the global population (over 5 billion) will come to live in cities [WRI, 1996]. Not since the Industrial Revolution has the world experienced such urbanization and human population expansion. Monitoring this growth and the subsequent land-use change can be a fundamental source of information for physical and social scientists intent on understanding the patterns of expansion, the impacts such growth will have on the local environment, and the demands it places on the population. An excellent synoptic means of gathering these data is by using repeat coverage remote sensing. In the past decade numerous new satellite instruments have been launched and many of these are being used to study earth science in urban settings. This paper describes one such satellite instrument and data collection program: the Urban Environmental Monitoring (UEM) project of the Advanced Spaceborne Thermal Emission and Reflectance Radiometer (ASTER) instrument. Detailed here is the algorithm development and testing—first using Landsat Thematic Mapper (TM) and NASA airborne sensor data, the UEM planning and implementation procedure, and the initial results utilizing ASTER.

Earth Science in the City: A Reader

10.1029/056SP13

BACKGROUND

In most countries, including the United States (US), a vast majority of the fastest-growing urban centers are vulnerable to natural hazards and ecological degradation because of their proximity to coastal and semi-arid environments [WRI, 1996; USCB, 2001]. The changes that occur to the urban core as well as the surrounding metropolitan area are significant and commonly detectable even with moderate to low spatial resolution satellite data [Anderson et al., 1976; Haack et al., 1987; Stefanov et al., 2001a]. Monitoring this urban population expansion by extension directly affects the largest percentage of a country's population and resources. Therefore, this activity on a global scale is seen as an important effort over the next two decades. Land cover mapping and monitoring provide input data into Geographical Information System (GIS)-derived models of infrastructure modifications, utility needs, economic development, and the potential vulnerability of the population to natural hazards and environmental damage [Lindgren, 1985; Martin et al., 1988; Treitz, 1992; Lyon et al., 1998; Balmford et al., 2002].

The current urban expansion and subsequent pressure on the fragile resources of highly-populated regions has given rise to new areas of urban science such as ecology, remote sensing, and geology relating to hazard mitigation. For example, the National Science Foundation (NSF) awarded the first ever urban Long-Term Ecological Research (LTER) projects to Phoenix, AZ and Baltimore, MD in 1997 [Grimm et al., 2000; Pickett et al., 2001]. The primary objective of the 21-site LTER network is to monitor and assess long-term ecological change in diverse ecosystems in the United States and elsewhere in the world. Whereas other LTER projects have focused on pristine locations well removed from the myriad effects brought about by extensive human modification and dominance of ecosystems, the two urban LTER programs are providing a unique opportunity to monitor human-induced ecological changes.

U.S. Growth

Urbanization of the semi-arid regions of the southwestern United States is a comparatively recent phenomenon in the history of the country, occurring largely in the last 50 years. The 1990 US Census identified eight of the ten fastest growing cities and six of the fastest growing metropolitan areas as being located in the west and southwest. For example, Arizona has been the second fastest-growing state in the US for the past six years, and the population of the Phoenix metropolitan area has doubled twice in the past 35 years. This growth has pushed the urban fringe into areas formerly occupied by agricultural land and pristine desert. Analysis of the official 2000 Census data show this trend continuing, with the largest increase in population occurring across the southern tier of the country and in the west (Table 1). This expansion is focused on both the central cities within each region as well as the surrounding area (metropolitan-region). The

TABLE 1. Data From The United States Census Showing The Growth Rate Percentage Of U.S. Regions From 1990 To 1999

U.S. Region [a]	Central City Growth	Metro-Region Growth
Northeast	-2.4	4.1
Middle Atlantic	-2.2	4.2
South Atlantic	2.5	20.2
East North Central	-1.1	10.1
East South Central	1.9	18.0
West North Central	2.0	15.0
West South Central	9.5	23.8
Mountain	20.3	35.8
Pacific	8.1	14.9
United States (Total)	4.1	14.2

[a]Northeast: ME, NH, VT, MA, RI, CT; Middle Atlantic: NY, NJ, PA; South-Atlantic: DE, MD, DC, VA, WV, NC, SC, GA, FL; East-North Central: OH, MI, IN, IL, WI; East-South Central: KY, TN, AL, MS; West-North Central: MN, IA, MO, ND, SD, NE, KS; West-South Central: AR, LA, OK, TX; Mountain: MT, WY, CO, NM, ID, UT, AZ, NV; Pacific: WA, OR, CA, AK, HI.

dramatic growth of cities in the Mountain, West-Central and Pacific regions is at the expense of the urban populations in the Northeast, Middle-Atlantic, and East-North Central regions. However, despite many of those regions experiencing slow to negative growth, most still had positive growth in the metropolitan regions, indicating a trend toward suburbanization that began after World War II.

In order to carry out monitoring rapidly and efficiently, the LTER project in Arizona has relied heavily remote sensing. Described by Stefanov et al. [2001b], these products include vegetation, soil, and urban cover types (Figure 1). Because central Arizona is located at the major geographic and climatic transition zones between the Sonoran and Chihuahuan Deserts and the Sierra and Rocky Mountain ranges it has a unique ecosystem and climate. With less than 18 cm of annual rainfall, Phoenix is situated in a semi-arid landscape that provides excellent remote sensing opportunities due to minimal cloud and vegetation cover. However, this climate also produces a strong reliance on surface and groundwater sources, a high moisture evaporation rate, and a continual threat of drought. These same issues, faced by populations living in similar to more extreme environments around the world, makes the science and policy issues examined in Phoenix extremely relevant.

Urban Science From Above

In addition to the NSF, NASA is also currently funding research into natural hazard mitigation within the urban environment using remote sensing, relying on

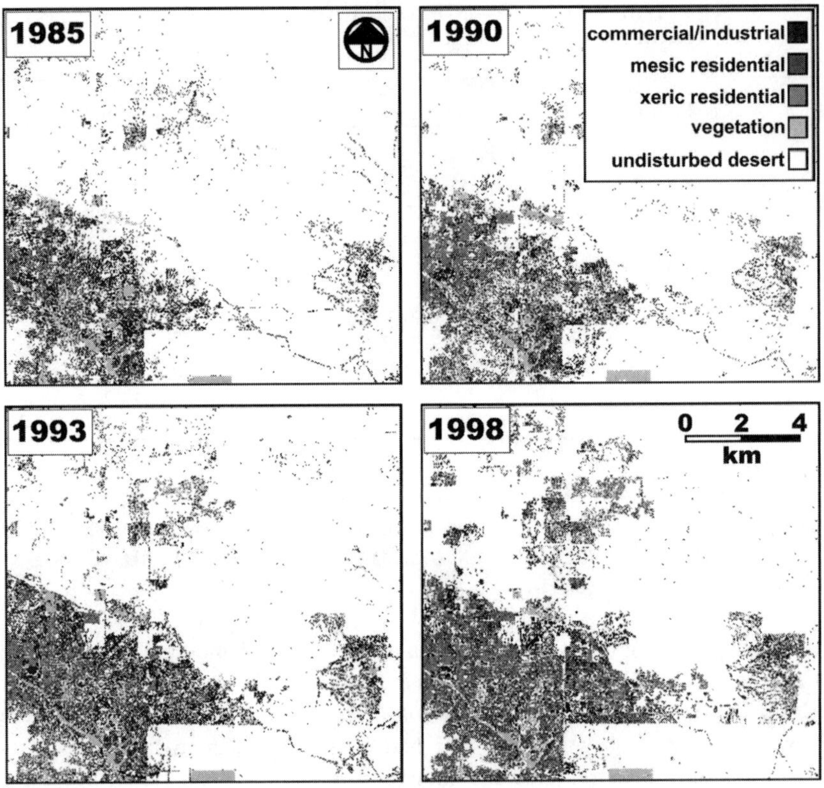

Figure 1. Land use/land cover supervised classification results of Landsat TM data for Scottsdale, AZ from 1985 - 1998. The expert-system derived classification procedure is detailed in Stefanov et al. [2001b]. This time series highlights the spread of urban land cover (residential, commercial/industrial) to the north and northeast, clearly defining the border with the non-developed Salt River Indian Reservation in the southeast. Explosive growth (sprawl) in such short time periods places high stress on the local environment and is a prime target for monitoring using data from ASTER.

both focused studies and global monitoring efforts [Pax-Lenney and Woodcock, 1997; Ridd and Liu, 1998; Quattrochi and Ridd, 1998; Ramsey and Lancaster, 1999; Stefanov et al., 2001b; Zhu and Blumberg, 2002]. However, remote sensing of cities has been limited in the past due to the low spatial resolution of most satellite-based instruments, as well as the lack of demand and use from city officials, planners, and scientists [Townshend, 1981; Harris and Ventura, 1995; Aplin et al., 1999]. This trend has changed with the advent of both innovative processing algorithms and inexpensive, higher spatial resolution data [Gong and Howarth, 1990; Aplin et al., 1997; Stefanov, 2002].

Among these new sensors is the Advanced Spaceborne Thermal Emission and Reflectance Radiometer (ASTER) instrument, launched in December 1999 on the NASA Terra satellite. Its nominal mapping phase, begun on 10 October 2000, is planned to continue until 2006. ASTER was designed by the Japanese Ministry of International Trade and Industry (MITI) and acquires repetitive, high spatial resolution, multi-spectral data. It is the first instrument to ever provide global data of this type in three wavelength regions from the visible/near infrared (VNIR) to the short-wave infrared (SWIR) to the thermal infrared (TIR), and it is the only high-resolution imager of the six instruments on board the Terra satellite [Kahle et al., 1991; Yamaguchi et al., 1998]. The spectral resolution varies between the three subsystems, with three channels in the VNIR, six channels in the SWIR and five channels in the TIR (Table 2). The high spatial resolution, multi-spectral coverage, and the ability to generate digital elevation models (DEMs) make it a critical tool for urban topographic and compositional analyses.

The Terra platform follows a sun-synchronous, polar orbit ~ 30 minutes behind the Landsat satellite providing ASTER with a nominal repeat time of 16 days (Table 2) and local overpass times of ~ 10:15 am/pm. However, with a cross-track pointing capability, the repeat time can be decreased to as low as five days with the added advantage of image collection up to 85° north/south latitude. ASTER has a 60km swath width and a ground instantaneous field of view that increases from 90 meters in the TIR to 30 meters in the SWIR to 15 meters in the VNIR (Figure 2). The instrument also has the ability to perform along-track stereo imaging by way of a 27.6° backward-pointing telescope in channel 3 (0.78 - 0.86μm) (Figure 3). This feature allows high-resolution digital elevation models (DEMs) to be created from the ASTER stereo pairs [Welsh, 1998]. Finally, ASTER data are acquired using one of several dynamic ranges in order to reduce data saturation (over highly reflective targets) and low signal to noise (over minimally reflective targets).

Unlike the previous and current Landsat TM instruments, ASTER is scheduled due to the large data volume it generates. It therefore operates on an 8% average

TABLE 2. ASTER Instrument Design Specifications

	VNIR	SWIR	TIR
Wavelength Range (μm)	0.52 - 0.86	1.60 - 2.43	8.13 - 11.65
Wavelength Channels	3 (+ 1 back-looking)	6	5
Spatial Resolution (m)	15	30	90
IFOV (μrad)	21.3	42.6	127.8
Repeat Time (days) [a]	5	16	16
Pointing Angle (degrees)	± 24	± 8.55	± 8.55

[a]Nominal repeat coverage can be substantially improved with the cross-track pointing capability.

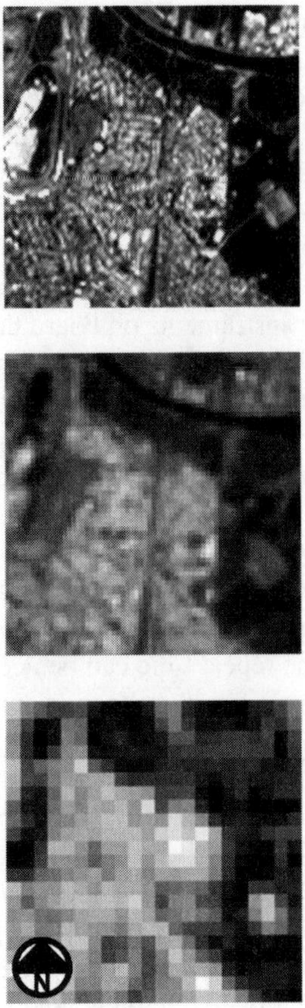

Figure 2. Comparison of the spatial resolution for each of the three ASTER subsystems over a 2 km portion of São Paulo, Brazil. The ASTER L1B scene was acquired on 19 March 2002 (13:23:17 UT). All images have a 2% linear stretch applied. (top) 15 m/pixel VNIR band #2 (0.63-0.69 μm). (middle) 30 m/pixel SWIR band #7 (2.24-2.29 μm). (bottom) 90 m/pixel TIR bands #12 (8.93-9.28 μm).

duty cycle during the lifetime of the Terra mission. Scheduled targets are determined for each orbit from a priority function, which is calculated by including such variables as time of year, resource allocation, cloud coverage, the size of the data request, the presence of a ground campaign, etc. Small and potentially one-time only targets, known as data acquisition requests (DARs), comprise 25% of the total resource allocation of ASTER [Yamaguchi et al., 1998]. The remaining 75% is divided into the global map collection (50% of resource time) and the science team acquisition requests (STARs), which account for the remaining 25%. The global map is a primary goal of the data collection, designed to produce a cloud-free map of the entire land surface of the Earth in all spectral bands by the end of the mis-

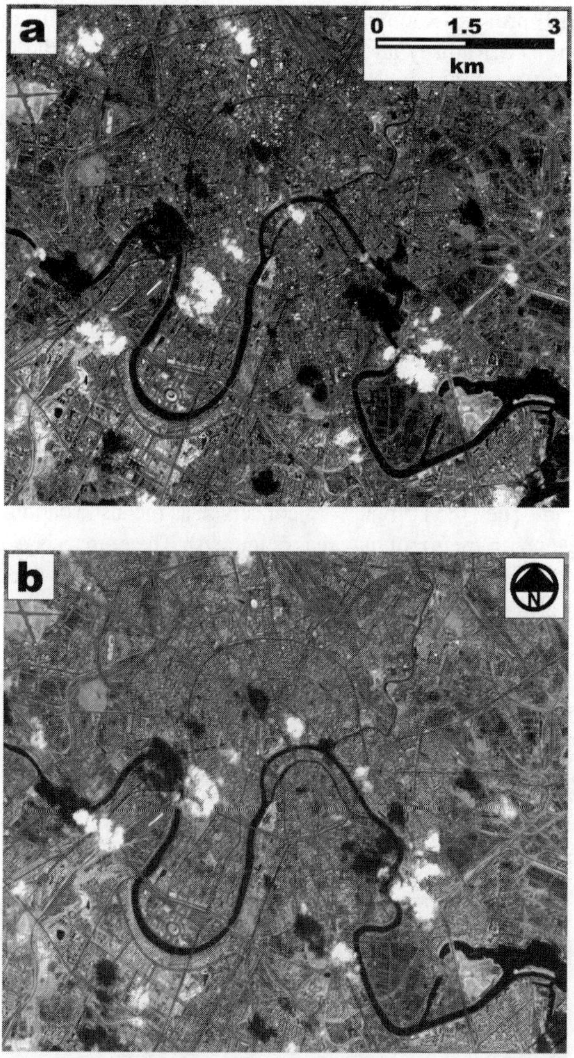

Figure 3. Band #3 (0.76-0.86 μm) ASTER L1A gray-scale images of Moscow, Russia collected on 28 August 2000. (a) Band #3n (nadir-looking telescope). (b) Band #3b (back-looking telescope). Band #3b is acquired with a different viewing geometry (note the cloud positions in each image with respect to their shadows). Images pairs such as this provide the ability to generate along-track digital elevation models (DEMs).

sion. The STARs, on the other hand, are dedicated to large global science objectives that demand larger resources from the instrument than DARs. There are numerous STAR objectives, including for example volcano observations, coral reef mapping, deforestation observations, the Global Land Ice Measurements from

Space (GLIMS) project [Raup et al., 2000, Wessels et al., 2000], the Arid Lands Monitoring project [Ramsey and Lancaster, 1999], as well as the UEM program described here [Ramsey, et al., 1999; Stefanov et al., 2000a]. The capability of ASTER to perform repeated global inventories of land-cover and land-use change from space make it ideal for assessing urban growth and change [Abrams, 2000; Ramsey et al., 1999]. The underlying philosophy of this strategy is to understand the consequences of human-induced change for continued provision of ecological goods and services. The planning, design and logistics for the UEM globally distributed data collection program are described below.

METHODOLOGY

UEM Planning

One of the core STARs of ASTER is the Urban Environmental Monitoring (UEM) program. The UEM project was conceived as means to capture data over the world's largest urban metropolitan areas (Figure 4). The emphasis is on those cities experiencing fast growth, facing potential environmental threats, and those concentrated in semi-arid environments (Figure 5). The data collection effort demands ded-

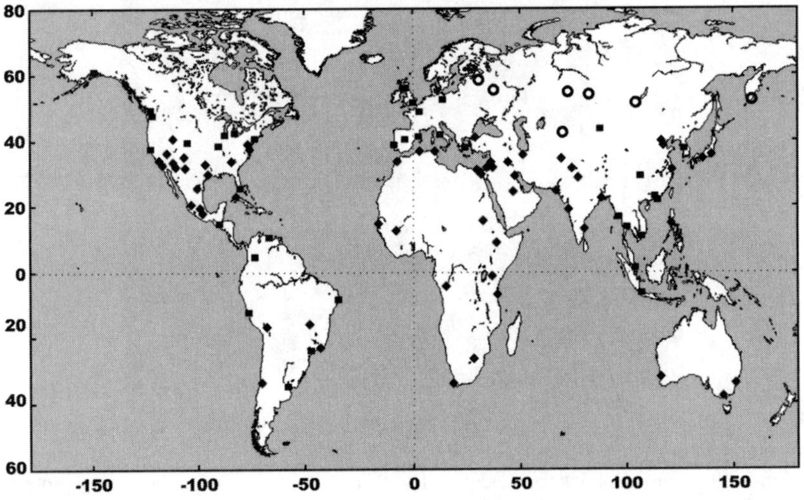

Figure 4. Urban targets of the UEM Science Team Acquisition Request (STAR). This project is divided in to high (solid diamonds) and low (solid squares) priority targets on the basis of specific criteria (see text). High priority targets comprise approximately two-thirds of the data request and are being monitored twice per year during the lifetime of the mission. The remaining low priority targets will imaged at least twice during the mission. The open circles denote the former Soviet Union (FSU) cities included as part of separate "spin-off" monitoring objective. See Table 3 for city names and exact locations.

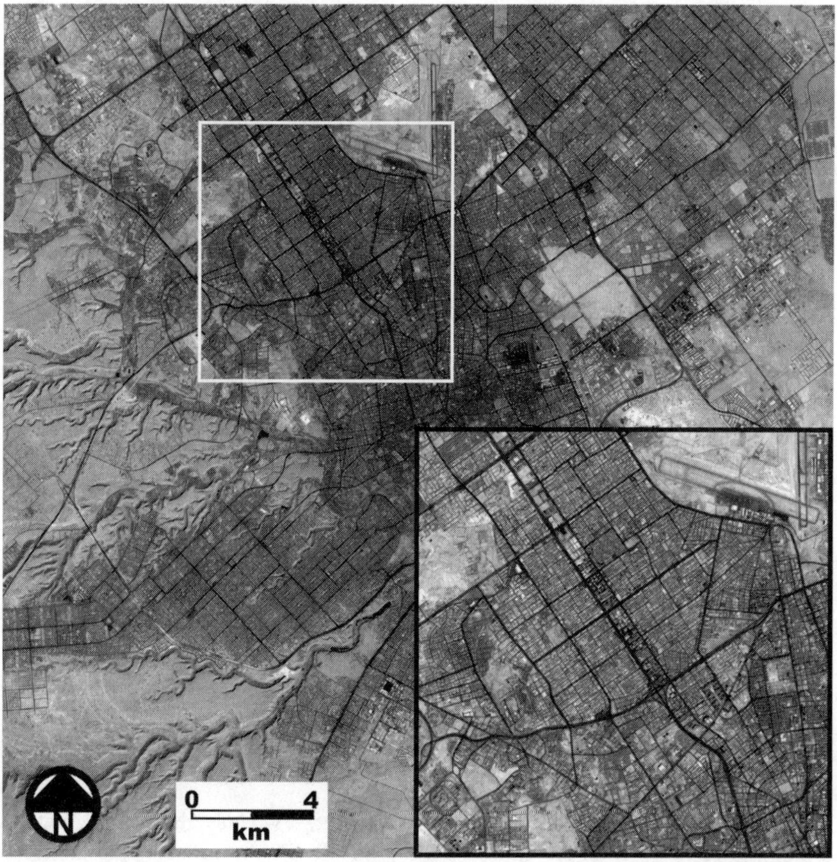

Figure 5. ASTER band #3n (0.76-0.86 μm) image of Riyadh, Saudi Arabia acquired on 23 November 2001 (07:46:23 UT). This scene has been processed to level 1B (L1B) with radiometric and geometric corrections applied. The imagery has a linear 2% stretch that highlights vegetation in white, desert soils in white to gray, and urban regions darker gray. The inset is displayed at full ASTER resolution and covers the central core of the city (denoted by the white rectangle).

icated processing and dissemination to local government officials and scientists. The primary goal of the UEM program is to ensure these data are acquired, processed and made available. The planned products are calibrated and geometrically accurate land use change, material identification, and heat island maps. As mentioned, these products form an integral part of the ecological modeling ongoing at locations such as the urban Long-Term Ecological Research (LTER) sites. In addition, these data form the geospatial context for studies that are examining urban hazard mitigation such as the detection of fire scars and their relationship to localized flooding, slope analyses and the spawning of landslides in connection to development, and soil identification pertaining to industrial "brown field" sites.

The Urban Environmental Monitoring STAR was originally conceived by P. Christensen as an extension of remote sensing pilot studies over portions of the Phoenix metropolitan area from 1992-1995 [P. Christensen, pers. comm., 1995]. The program as originally proposed consisted of 86 urban targets chosen somewhat arbitrarily, but based mainly on their location within semi-arid climates. Because all STAR proposals submitted prior to launch of the Terra spacecraft were reviewed and approved by the ASTER science team, feedback was provided on the UEM project, target selection, and science objectives. This process along with a nearly two-year delay in the launch of Terra resulted in significant revision and expansion of the urban monitoring program [Ramsey et al., 1999; Stefanov et al., 2001a]. The UEM project goals were expanded to include:

1. the infrastructure planning to ensure that all ASTER UEM targets are collected, processed and archived.
2. the production of certain derived data products including, but not limited to: calibrated surface reflectivity/emissivity; land use classification/change; surface heat flux maps.
3. the establishment of local contacts in regions/cities willing to work with the ASTER urban products and serve as points of contact for local research projects.
4. the dissemination of information and data to those contacts upon image acquisition.

Because of the enormous scale of this program, it was originally designed as a collaborative effort, as are most of the ASTER STAR projects. By including investigators worldwide, it is guaranteed that the data are disseminated and used for local science investigations. It also provides a point of contact to the ASTER team, and allows for future feedback, for example to add/delete entries in the database, and make changes to existing data collection parameters. During the launch delay, new collaborators were brought onto the project, the UEM parameters were finalized, and the list of cities was expanded to 100 targets (Table 3) using the following metrics:

1. a current metropolitan population near or in excess of one million people.
2. a predicted rapid growth in the next decade.
3. current or future environmental issues resulting from growth and/or land-use practices.
4. current or future geo-hazard potential due to location, environment and/or population.
5. a roughly even geographical distribution around the world thereby limiting the potential of focusing too heavily on one country/region.

TABLE 3. Global UEM Target List Showing The High And Low Priority Targets

City	Country	UEM Priority
Addis Ababa	Ethiopia	High
Albuquerque	United States	High
Alexandria	Egypt	High
Algiers	Algeria	High
Anchorage	United States	Low
Amman	Jordon	High
Athens	Greece	Low
Atlanta	United States	High
Baghdad	Iraq	High
Baltimore	United States	High
Bangkok	Thailand	Low
Bamako	Mali	High
Barcelona	Spain	Low
Beijing	China	High
Berlin	Germany	Low
Bogota	Colombia	Low
Bombay	India	High
Brasilla	Brazil	High
Buenos Aires	Argentina	Low
Cairo	Egypt	High
Calcutta	India	High
Cape Town	South Africa	High
Caracas	Venezuela	Low
Casablanca	Morocco	High
Chicago	United States	Low
Chongqing	China	Low
Dakar	Senegal	High
Dallas	United States	High
Damascus	Syria	High
Delhi	India	High
Denver	United States	Low
Der Es Salaam	Tanzania	High
Detroit	United States	Low
Edinburgh	Scotland	Low
El Paso	United States	High
Guadalajara	Mexico	High
Guangzhou	China	Low
Guatemala City	Guatemala	Low
Havana	Cuba	High
Ho Chi Minh City	Vietnam	Low
Houston	United States	High
Istanbul	Turkey	High
Jakarta	Indonesia	Low
Johannesburg	South Africa	High
Kabul	Afghanistan	High
Karachi	Pakistan	High
Khartoum	Sudan	High
Kinshasa	Zaire	High
Kuwait City	Kuwait	High
La Paz	Bolivia	High

TABLE 3. Global UEM Target List Showing The High And Low Priority Targets (Continued)

City	Country	UEM Priority
Lahore	Pakistan	High
Las Vegas	United States	High
Lima	Peru	Low
Lisbon	Portugal	Low
London	England	Low
Los Angeles	United States	High
Madras	India	High
Madrid	Spain	Low
Manila	Philippines	High
Melbourne	Australia	High
Mexico City	Mexico	High
Miami	United States	Low
Monterrey	Mexico	High
Moscow	Russia	Low
Nairobi	Kenya	High
New York	United States	Low
Novosibirsk	Russia	Low
Osaka	Japan	High
Paris	France	Low
Perth	Australia	High
Phoenix	United States	High
Puebla	Mexico	High
Rangoon	Myanmar	Low
Recife	Brazil	Low
Rio De Janeiro	Brazil	High
Riyadh	Saudi Arabia	High
Rome	Italy	Low
Salt Lake City	United States	High
San Francisco	United States	Low
San Diego	United States	High
San Paulo	Brazil	Low
Santiago	Chile	High
Seattle	United States	Low
Seoul	South Korea	Low
Shanghai	China	High
Singapore	Malaysia	Low
St. Louis	United States	Low
St. Petersburg	Russia	Low
Sydney	Australia	High
Tashkent	Uzbekistan	High
Tehran	Iran	High
Tel Aviv	Israel	High
Tianjin	China	High
Tokyo	Japan	High
Tucson	United States	High
Tunis	Tunisia	High
Urumqui	China	Low
Vancouver	Canada	Low
Washington D.C.	United States	High
Xianggang	China	Low

The first two criteria in the above list were moderated by the remaining three. In other words, targets were not strictly chosen on the basis of overall population because a vast majority would be concentrated in China and India. Similarly, targets were no longer chosen simply because they were located in arid to semi-arid climates, even though it is clear that most of the fastest-growing urban centers are located in such environments. It was determined that some percentage of the target list would be modified to include: cities with declining or zero-growth populations (to serve as controls and examine land-cover issues unique to such cities), and more widely varying building materials and growth patterns (Tables 3, 4).

In order to conserve instrument resources, the ASTER science team placed further constraints on the individual DAR targets within a STAR request and the total area of each STAR. The new allotment of 400,000 km^2 per STAR included the total cumulative data collection are over the lifetime of the mission. This constraint in particular produced a limitation on data volume that would have made it impossible to meet the twice-yearly objective for each of the 100 cities. A solution to this limitation was derived by dividing the database in to high and low pri-

TABLE 4. United States Census 2000 Data For The US Metropolitan Regions That Are Part Of The UEM Acquisition Plan [a]

City Name	Size Ranking [b]	Urban Core	Metro Area
Albuquerque	62	420,578	678,820
Anchorage	138	257,808	257,808
Atlanta	11	401,726	3,857,097
Baltimore	4	632,681	7,359,044
Chicago	3	2,799,050	8,885,919
Dallas	9	1,076,214	4,909,523
Denver	19	499,775	2,417,908
Detroit	8	965,084	5,469,312
El Paso	60	612,770	701,908
Houston	10	1,845,967	4,493,741
Las Vegas	33	418,658	1,381,086
Los Angeles	2	3,633,591	16,036,587
Miami	12	369,253	3,711,102
New York	1	7,428,162	20,196,649
Phoenix	14	1,211,466	3,013,696
Pittsburgh	20	336,882	2,331,336
Salt Lake City	35	171,151	1,275,076
San Diego	17	1,238,974	2,820,844
San Francisco	5	746,777	6,873,645
Seattle	13	537,150	3,465,760
St. Louis	18	333,960	2,569,029
Tucson	57	466,591	803,618
Washington DC	4	519,000	7,359,044

[a]Census 2000 website.

[b]Rankings are out of the top 276 metropolitan regions in the United States.

ority targets. The high priority cities (~ 65% of the database) satisfy all the original UEM objectives, whereas the low priority cities have a limited coverage of only two observations over the lifetime of ASTER (Table 3).

Study Areas, Algorithm Testing & Development

The governing principle of the UEM program is that remote sensing and image processing techniques developed in other branches of the geosciences can be applied to urban regions in order to provide answers to problems facing their local populations. This foundation was formed during a series of NASA-sponsored pilot project from 1992-1995 in conjunction with the City of Scottsdale, AZ. As conceived, the project goal was to study the applicability of VNIR and TIR airborne data for the purposes of urban scene classification, environmental assessment, and change detection. These projects resulted in the collection of a large volume of data from numerous sources including space-based: Landsat Thematic Mapper (TM) and Shuttle Imaging Radar (SIR-C), as well as airborne: Thermal Infrared Multispectral Scanner (TIMS), airborne TM simulator (NS001) and color VNIR aerial photography. These data sets have been integrated into state and local activities to improve decision-making and planning. For example, the derived data products have been used in surface impermeability studies for storm runoff assessments; development versus preservation surrounding local mountain parks; soil identification to better understand hill slope processes [Stefanov et al., 1998], and brush fire hazards [Ramsey and Arrowsmith, 2001; Misner et al. 2002].

The urban landscape within the Phoenix metropolitan area provides a unique test and excellent ground truth for the validation of surface classification models [Harris and Ventura, 1995; Quattrochi and Ridd, 1998; Stefanov et al., 2001b]. Examination of multi-temporal scenes and identification of land-use patterns clearly show the "urban sprawl" commonly associated with large western US cities [Haack et al., 1987]. An example of this growth pattern monitoring of Phoenix, AZ is shown in Figure 1. Where available, instruments with multi-spectral TIR wavelength bands provide the means to produce very accurate temperature maps from which to study the spatial distribution of heat islands (Plate 1). These data are critical inputs into micro and regional climate models that attempt to predict variations over time with changing land use and urban growth [Stoll and Brazel, 1992; Hafner and Kidder, 1999].

Coinciding with the end of the pilot projects, the LTER program began and provided further resources for the expansion of the remote sensing analysis to the entire metropolitan area. This expansion included the acquisition of historic Landsat Multispectral Scanner (MSS) and TM data (for more complete temporal coverage), the Advanced Very High Resolution Radiometer (AVHRR) data (to examine the effects lower spatial scale), and the acquisition of new NASA data sets including the airborne ASTER simulator (MASTER) in 1999 and 2000.

Phoenix, AZ was chosen as the prime calibration target for the remaining 99 cities within the larger UEM project, because of the presence of the LTER, the man-years already invested, the ease of access to field calibration sites, and the unique growth issues facing the region. Studies are ongoing to monitor urban growth, land use change, impacts on the surrounding environment, and the development of urban heat islands (Figure 1, Plate 1).

As mentioned, a fundamental data set required to monitor these growth patterns and as input into the LTER ecosystem analyses is accurate land use/land cover change. This derived product consists of the major types of land cover and their areal percentages present in the study area. Land cover refers to the physical nature of the surficial materials present in a given area, whereas land use refers to the specific type and pattern of human development [Anderson et al., 1976; Sabins, 1997]. Collection of these data in a large urban environment is obviously very time-consuming and in some cases impossible. A more efficient approach is to use remotely sensed data with field verification to classify land cover types [Anderson et al., 1976; Hixson et al., 1980, Ridd, 1995]. Once the land cover classification is obtained it can be used as an input into a variety of ecological models, and land cover maps can be constructed to aid in planning field-sampling strategy. The land cover types can also be linked to different land use categories to investigate temporal and spatial changes in the urban ecosystem [Stefanov et al., 2001b; Zhu and Blumberg, 2002].

On a global scale, such a data analysis effort can only be accomplished with extensive testing and technique development. However, commonly used remote sensing and image processing approaches suffer from slow analyses techniques that would be impossible at the scale of the UEM project. The need exists for a robust method of identification and classification of the most common land cover types. In order to accomplish this task, an expert system approach to land cover/land use classification has been developed using TM data of Phoenix, AZ. This methodology, fully explained by Stefanov et al. [2001b], relies on well-calibrated data, the derivation of urban texture mapping, and the input of other land use data sets in a GIS hierarchical modeling approach to improve standard supervised classification results. Significant to the results of this study is the increased accuracy over other similar studies achieved with relatively poor resolution data (30m of Landsat TM). The authors were able to produce classifications with an overall accuracy of 85% and maintain twelve distinct land cover/land use classes. A large percentage of the data input into the classification model was derived directly from the data themselves (calibrated reflectance, vegetation indices, urban texture). Clearly, the need for extraneous GIS-based land cover data sets is a limiting factor for many remote cities of the world. However, testing of the model for other cities is now underway using ASTER data. It is being shown that even without the inclusion of non-image derived data sources, accuracy remains high due to the increased spatial and spectral resolution of the ASTER sensor [Stefanov, 2002; Zhu and Blumberg, 2002].

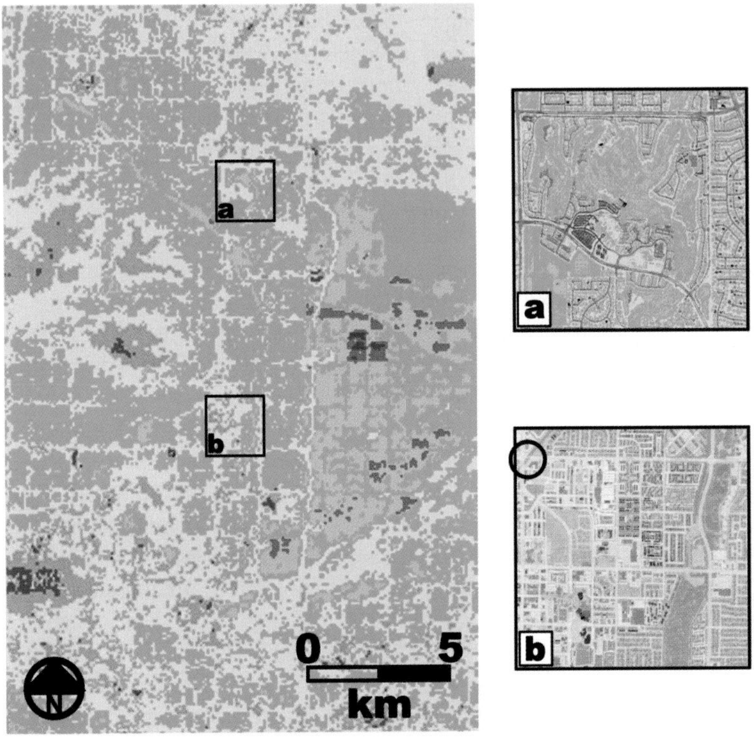

Plate 1. Ground surface brightness temperature images derived from both high spatial resolution airborne data as well as ASTER. The 90m/pixel spatial resolution ASTER TIR night time image (shown on the left) covers a large portion of the eastern Phoenix, AZ Valley. The data were acquired on 7 May 2001 and calibrated to a standard L2 product (atmospherically-corrected ground-leaving radiance). Brightness temperature was derived from the L2 product using an emissivity normalization approach and a maximum emissivity of 0.985. The color scale applied is as follows: 27-30°C (red), 24-27°C (orange), 24-27°C (yellow), 21-24°C (green), 18-21°C (cyan), 15-18°C (magenta), 12-15°C (blue), and < 12°C (black). Non-urbanized land use regions (Salt River Indian Reservation to the north and agricultural fields to the south) show a significantly cooler temperatures compared to the urbanized regions. Insets (a) and (b), denoted by the boxes, show regions covered by 4m/pixel Thermal Infrared Multispectral Scanner (TIMS) airborne data on 14 July 1995 at 02:45 LT. High resolution data such as these are used for calibration of the ASTER TIR products. (a) This region is dominated by commercial/industrial land use, and temperatures vary from 3-40°C, with cooler temperatures shown in darker gray (i.e., the Indian Bend Wash green way and the Central Arizona Project Canal). Note the 5°C cooling of Camelback and Scottsdale Roads as they pass over the canal (circled). (b) Color density slice of another portion of the same TIMS data set showing mesic and xeric residential land use. The cooler overall core of this block is caused by the presence of a golf course. Color scale: 36-40°C (red), 33-36°C (orange), 30-33°C (yellow), 27-30°C (green), 24-27°C (cyan), 21-24°C (magenta), 18-21°C (blue), and < 18°C (black).

This approach is continuing to be refined and updated as it is applied to new ASTER data (Plate 2). In order to produce meaningful land cover maps of all the UEM cities, the classification approach must be tested on numerous cities, which have been constructed in, and are subject to, a variety of conditions different than those of experienced at Phoenix. These include, but are not limited to, local/regional climate, development density, use of native building materials, different urban classes, and transportation patterns. Similar reasoning was the impetus for the selection of two fundamentally different LTER cities (Phoenix, AZ and Baltimore, MD).

Because of the author's relocation to Pittsburgh, PA from Phoenix, AZ, the former has also been selected as a UEM calibration target site and added to the UEM target database. The city of Pittsburgh differs from Phoenix, AZ in many ways and is represented by a declining population, urban decay/infill, denser development, and the presence of a higher percentage of surface water and vegetation. It is also the site of the some of the largest urban renewal projects in the United States producing significant land cover change in a short time period with the construction/demolition of sports stadiums, the growth of river-front retail zones at the sites of former steel mills, and the construction of new highways. Larger proposed projects include major redevelopment of the downtown and north shore of the Ohio River, and a high speed magnetic-levitation (maglev) train connecting the airport to the west and constructed through the urban core. The city therefore has the potential of being a unique test site combining the complications dense urban change in a region with less than ideal weather for remote sensing studies.

The first cloud-free ASTER scene of Pittsburgh, PA was acquired on November 24, 2000 (16:35:04 UT) and since then nine other scenes have been acquired, including a cloud-free summer scene on August 19, 2002 (16:17:23 UT) shown in Plate 2. The data reveal a higher density urban environment with large rivers, and dominated at this time of year by the presence of large amounts actively photosynthesizing vegetation (tree canopy). In contrast to semi-arid cities such as Phoenix, urban land cover in cities like Pittsburgh will change markedly during other seasons with the appearance of snow cover and loss of tree canopies (Plate 2). Such change will clearly impact the classification of natural land cover types, but also has the benefit of revealing more of the urban classes previously masked by vegetation. Application of a modified version of the land cover classification model has been performed on the data. Qualitative estimates of this preliminary analysis show very good agreement to current land cover in the region. More quantitative estimates of the model accuracy are expected to include the amount of change detected from the winter to summer seasons. Modifications in the urban classification model developed at the Phoenix site will be implemented and tested at the other non-desert UEM cities in addition.

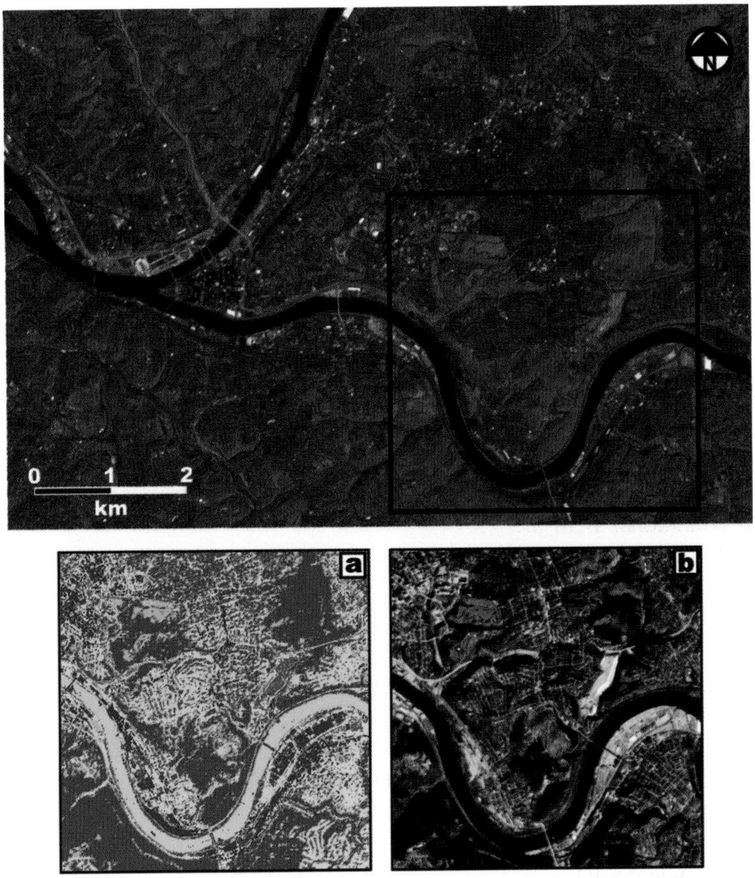

Plate 2. Subset of an ASTER L1B scene of Pittsburgh, PA and surrounding regions acquired on 19 August 2002 (16:17:23 UT). Growth rates, construction materials and patterns, vegetation, and environmental issues in industrial cities of the northern US such as Pittsburgh are dramatically different than those in high growth cities such as Phoenix, AZ. The visible/near infrared (VNIR) color composite (bands 3n, 2, 1 in red, green, blue, respectively) shows vegetation in red and clearly denotes the urban regions in blue-green, with the black box indicating the areas covered by the insets below. (a) Supervised classification of the VNIR and SWIR data highlighting the dominant land use classes: tree canopy (dark green), grass (light green), commercial/industrial (blue) residential (yellow), disturbed surfaces (red) and water (cyan). The large commercial/industrial regions along the river were both sites of former steel mills that have been demolished and replaced by retail and office complexes in the past decade. (b) Difference in the land cover appearance in the winter months (24 November 2000, 16:35:04 UT). Classification results such as these are being compared to similar products derived for other cities in the UEM program.

DISCUSSION

The UEM program has already served as a catalyst for a variety of "spin-off" projects that either use currently available UEM data or seek to expand the list of cities to include new areas. One example of this process is the data analyses ongoing in Pittsburgh, PA and the inclusion of it into the UEM database. Other active projects include the monitoring of the cities of the former Soviet Union (FSU), the study of brush fire and flooding hazards at the urban-wilderness interface, and the impact of rapid urban sprawl on the vulnerability of people living in the mega-city of São Paulo, Brazil.

UEM "Spin-off" Projects

Former Soviet Union cities For a decade, the region that included the former Soviet Union and the states of Central and Eastern Europe has been undergoing fundamental and, at times, tumultuous change. Societies are being transformed, and economic and political systems are being rebuilt under a variety of models and conditions. With the current fiscal condition of Russia and the other FSU countries, both ecological and environmental problems within the urban population centers are commonly overlooked. ASTER data are important in addressing these problems because of the reasons mentioned previously. The principal goals of this particular study are to develop urban scene classifications, environmental assessments and begin a program of change detection of these major urban centers.

In order to accomplish that task, a DAR was submitted to the ASTER science team that augmented the current UEM STAR, which already contained the Russian urban centers of St. Petersburg, Novosibirsk and Moscow (Figure 3). As part of the new request, these cities were increased to high priority (collection of scenes twice/year) and four new urban centers were added (Tashkent in Uzbekistan, and Omsk, Irkutsk, and Petropavlovsk in Russia). Scientific collaborations have already been established with government and academic institutes in England and Moscow. Plans are also underway for field visits and verification of the ASTER data in these cities. This program also has the potential of being both a scientific and political outreach tool over the next several years.

Urban brush fire research The summers of 2000 and 2002 were the worst fire seasons in the past 50 years for the western United States. As of early September, 2000 over 6.5 million acres were burned and the cost of fighting the fires is exceeded $1 billion US dollars. Much of the damage was concentrated in the remote, high elevation pine forests of the western states. However, a large percentage of the Los Alamos, NM fire in May, 2000 and the Rodeo-Chediski fire in July, 2002 attest to hazards of desert brush fires. Where these fires impinge on rural and urban fringe development, the potential cost to lives and property becomes considerable. Further, these burned regions have the potential to facilitate flash flooding and soil erosion during the monsoon rainy seasons over the next few years [Ramsey and Arrowsmith, 2001]. As people expand into these environments and their exposure to hazards increases, the ability to predict and control fires becomes increasingly important. Remote sensing together with detailed field data has been used to characterize areas scarred by past fires with the goal of assessing the risk for burning in the future [Misner et al., 2002].

Remote sensing of the urban environment and surrounding region of several southwestern US cities has revealed the presence of old brush fire scars dating back 30-50 years. Depending on the wavelength region examined, the age and surface properties of the scars can be determined. A NASA-sponsored research study now underway in Phoenix, AZ and Los Angeles, CA, is examining the linkage between the fire scar age, vegetation type/recovery, soil type, and local topography, using data from the Landsat ETM, SIR-C radar, and airborne MASTER [Ramsey and Arrowsmith, 2001; Misner et al., 2002]. Once burned, it is hypothesized that the removal of vegetation may facilitate rapid flood run-off and erosion during intense periods of precipitation. By examining the spatial variability of numerous scars in one location, and given the potential to evaluate their relative ages automatically, it should be possible to establish fire recurrence intervals around any urban area. This can be compared with lightning frequency, climate, vegetation, and terrain characteristics to vastly improve the characterization of hazards associated with semi-arid environment brush fires.

Urban sprawl and vulnerability in São Paulo, Brazil The urban core of São Paulo has experienced some of the fastest growth of any city in the world over the past 75 years. It has expanding from a modest-size agriculturally dominated urban center to become a mega-city with a population in excess of 20 million people. This extreme growth coupled with poor to non-existent laws and enforcement of land-use has produced chaotic urban sprawl conditions that impact on the physical and economic vulnerability of the city's inhabitants. The settlement patterns are a result of direct and indirect public policies that drive populations from the more densely populated urban cores to the less densely settled periphery. Sprawl is a complex socio-economic process and policies to address sprawl are even more complicated and controversial.

The dominant concern about rapid urban growth that results in sprawl is a function of economic costs versus quality of life. It is also more or less of a concern depending on the country in which the urban sprawl is occurring. Prosperous nations have the technology, resources and interest to limit, or at the very least, debate the issue of sprawl. However, the quality of life of people in every large and rapidly expanding city around the world is impacted by this problem. Left unchecked, ecological assets and their services, such as water storage values of forested hillsides or the landscape value of natural hilltops, are either unrecognized or unwittingly sacrificed [Balmford et al., 2002].

The application of ASTER data to the growth issues faced by people of São Paulo and surrounding regions is intended to develop and implement a method for predicting what the economic, quality of life and ecological costs of sprawl processes will be. These predicted costs can be contrasted to alternative development patterns designed to be more conscious of these costs. For example, these data are being used to perform an urban risk analysis, a hydrological/geological assessment of the region, and urban change detection. Included are "point hazards" such as landslides, ground collapse and fire, as well as "distributed hazards" such as flooding, waterway pollution/health, and severe weather impacts. This collaborative project with three universities in the São Paulo region was initiated in the summer of 2002 and discussions and data analyses are in their initial stages.

ASTER data structure and sources

The ASTER science team has developed and tested numerous software packages designed to derive higher-level data products from the calibrated ASTER radiance [Yamaguchi et al., 1998; Abrams, 2000]. The complete description of these products can be found at the ASTER web site [http://asterweb.jpl.nasa.gov/] or within the Algorithm Theoretical Basis Documents (ATBD) located at the Earth Observing System web site [http://eospso.gsfc.nasa.gov/eos_homepage/for_scientists/publications.php]. The products derived from radiance at sensor (Level 1A) or calibrated radiance at sensor (Level 1B) data have the designation of level 2 and include surface emissivity, kinetic temperature, reflectance, DEMs, and several others. Any data that ASTER has acquired are available at the Earth Observing System Data Gateway (EDG) site [http://edcimswww.cr.usgs.gov/pub/imswelcome/], which is coordinated by the USGS land processes Distributed Active Archive Center (DAAC) in Sioux Falls, SD. These data sets (including the UEM targets) will continue to be archived at the DAAC and available at the aforementioned web site. However, a limited number of Level 1A and 1B scenes (mostly those of high priority calibration cities) and all the derived urban data products (such as land cover classifications) for the UEM targets will be available at the project web site [http://elwood.la.asu.edu/grsl/UEM/cities/]. Groups at both the University of Pittsburgh and Arizona State University are monitoring the progress of the UEM collection, examining the data, and refining classification algorithms for the targets already acquired.

ASTER data has began to arrive in earnest as of early 2001. As an example, during the first ten months of data collection (May, 2000 - February, 2001) over 65,000 scenes were processed and made available. Even more amazing is the fact that during most of those ten months, ASTER was engaged in minimal data collection as it underwent calibration and validation tests. Only a small fraction of those scenes comprised urban-focused data. However, since ASTER has been returning data there now exist approximately 650 city scenes in the UEM database. Managing the large volumes of images and meta-data is a challenge where dealing with a globally distributed, multi-temporal program like the UEM. Research groups that are leading such programs must be ready to ingest, process, and disseminate global data sets. Automated routines to produce for example land cover/land use maps are critical. However, the outcome of such an effort provides a valuable resource for urban science as a historical record and as near real-time hazard monitoring tool.

CONCLUSIONS

The primary application of remote sensing data to examine urban regions is to provide a synoptic means for extrapolating local detailed measurements to a regional context. Specifically, multi-spectral image classification can be used to identify land cover types, such as different grasses, crops, trees, soils, man-made materials, water, and native vegetation. Where used with field validation, these data provide accurate identification and estimates of the areal distribution of these different units [Martin, 1988; Treitz, 1992; Stefanov et al., 2001a,b]. These data can then be used to create regional land use thematic maps that depict different processes. For example, urban versus native materials, permeable versus impermeable surfaces, and transportation systems (asphalt and concrete materials) can all be mapped. Over time, the monitoring of surface units allows for the detection of change. Temporal analysis of Landsat TM data has proven critical in identifying ecosystem loss, monitoring growth-related issues, and as input into governmental policy.

During the early stages of the ASTER mission, fine-tuning occurred on the UEM database and the process is still ongoing. In addition, work continues on such items as the dissemination of data sets to the investigators in near-real time, ensuring they have the tools to analyze those data, creating a rapid search tool by way of the world wide web, and continuing to test and refine land cover mapping algorithms. However, it is expected that these issues will not present any major obstacles to the overall success of the program. The field of urban remote sensing is ever expanding and many of the tools used by remote sensing geologists, ecologists and social scientists are directly applicable to these types of analyses. The ASTER UEM project provides important new data for many cities around the world, and the future of the project depends on the availability of calibrated ASTER data and the continued collaboration with investigators worldwide.

Acknowledgments. Research funding for this program has been provided by NASA through the ASTER science project, and the Solid Earth and Natural Hazards (SENH) Research Program. The planning and preparation of the UEM program took place while the author was at Arizona State University (ASU). Current oversight and daily operation of this project is now the responsibility of Dr. William Stefanov, who also contributed greatly to the Phoenix land cover analysis. The author would also like to acknowledge Dr. Philip Christensen at ASU for the original UEM concept, Dr. William Harbert at the University of Pittsburgh for his determination in bringing the former Soviet Union (FSU) city project to life, and Jeff Mihalik for his help with figure preparation and São Paulo, Brazil research.

REFERENCES

Abrams, M., The Advanced Spaceborne Thermal Emission And Reflectance Radiometer (ASTER): Data products for the high spatial resolution imager on NASA's Terra platform, *Int. J. Rem. Sens., 21,* 847_859, 2000.

Anderson, J. R., E. Hardy, J. Roach, and R. Witmer, A land use and land cover classification system for use with remote sensor data, *U.S.G.S. Prof. Paper, 964,* 1976.

Alpin, P., P. M. Atkinson, and P. J. Curran, Fine spatial resolution satellite sensors for the next decade, *Int. J. Rem. Sens., 18,* 3873_3881, 1997.

Alpin, P., P. M. Atkinson, and P. J. Curran, Fine spatial resolution simulated satellite sensor imagery for land cover mapping in the United Kingdom, *Rem. Sens. Environ., 68,* 206_216, 1999.

Balmford, A., A. Bruner, P. Cooper, R. Costanza, S. Farber, R. E. Green, M. Jenkins, P. Jefferiss, V. Jessamy, J. Madden, K. Munro, N. Myers, S. Naeem, J. Paavola, M. Rayment, S. Rosendo, J. Roughgarden, K. Trumper, and R. K. Turner, Economic reasons for conserving wild nature, *Science, 297,* 950_953, 2002.

Gong, P., and P. J. Howarth, The use of structural information for improving land-cover classification accuracies at the rural-urban fringe, *Photogramm. Eng. Rem. Sens., 56,* 67_73, 1990.

Grimm, N. B., J. M. Grove, C. L. Redman and S. T. A. Pickett, Integrated approaches to long-term studies of urban ecological systems, *BioSci., 70,* 571_584, 2000.

Haack, B., N. Bryant, and S. Adams, An assessment of Landsat MSS and TM data for urban and near-urban land-cover digital classification, *Rem. Sens. Environ., 21,* 201_213, 1987.

Harris, P. M., and S. J. Ventura, The integration of geographic data with remotely sensed imagery to improve classification in an urban area, *Photogramm. Eng. Rem. Sens., 61,* 993_998, 1995.

Hafner, J., and S. Q. Kidder, Urban heat island modeling in conjunction with satellite-derived surface/soil parameters, *J. Appl. Met., 38,* 448_465, 1999.

Hixson, M., D. Scholz, N. Fuhs, and T. Akiyana, Evaluation of several schemes for classification of remotely sensed data, *Photogramm. Eng. Rem. Sens., 66,* 1547_1553, 1980.

Kahle, A. B., F. D. Palluconi, S. J. Hook, V. J. Realmuto, and G. Bothwell, The Advanced Spaceborne Thermal Emission And Reflectance Radiometer (ASTER), *Int. J. Imaging Syst. and Tech. 3,* 144_156, 1991.

Lindgren, D. T., *Land Use Planning and Remote Sensing,* Martinus Nijhhoff, Inc., Boston, 1985.

Lyon, J. G., D. Yuan, R. S. Lunetta, and C. D. Elvidge, A change detection experiment using vegetation indices, *Photogramm. Eng. Rem. Sens., 64,* 143_150, 1998.

Martin, L. R. G., P. J. Howarth, and G. Holder, Multispectral classification of land use at the rural-urban fringe using SPOT data, *Canada J. Rem. Sens., 14,* 72_79, 1988.

Misner, T., M. S. Ramsey, and J. R. Arrowsmith, Analysis of brush fire scars in semi-arid urban environments: Implications for future fire and flood hazards using field and satellite data (abs), *Eos Trans. AGU, 83,* B61C-0740, 2002.

Pax-Lenney, M., and C. E. Woodcock, The effect of spatial resolution on the ability to monitor the status of agricultural lands, *Rem. Sens. Environ., 61,* 210_220, 1997.

Pickett, S. T. A., M. L. Cadenasso, J. M. Grove, C. H. Nilon, R. V. Pouyat, W. C. Zipperer, and R. Costanza, Urban ecological systems: Linking terrestrial ecology, physical, and socio-economic components of metropolitan areas, *An. Rev. Ecol. Syst., 32,* 127_157, 2001.

Quattrochi, D. A., and M. K. Ridd, Analysis of vegetation within a semi-arid urban environment using high spatial resolution airborne thermal infrared remote sensing data, *Atmos. Environ., 32,* 19_33, 1998.

Ramsey, M. S., W. L. Stefanov, and P. R. Christensen, Monitoring world-wide urban land cover changes using ASTER: Preliminary results from the Phoenix, AZ LTER site, *Proc. 13th Inter. Conf., Appl. Geol. Rem Sens., 2,* 237_244, 1999.

Ramsey, M. S., and N. Lancaster, Using remote sensing to derive sediment mixing patterns in arid environments: Future global possibilities with the ASTER instrument (abs), *Geol. Soc. Am. Abst. with Progs., 30,* A360, 1999.

Ramsey, M. S., and J.R. Arrowsmith, New images of fire scars may help to mitigate future natural hazards (abs), *EOS, Trans. Amer. Geophys. Union, 82:36,* pp. 393_398, 4 Sept. 2001.

Raup, B. H., H. H. Kieffer, T. M. Hare, and J. S. Kargel, Generation of data acquisition requests for the ASTER satellite instrument for monitoring a globally distributed target: Glaciers, *IEEE Trans. Geosci. Rem. Sens., 38,* 1105_1112, 2000.

Ridd, M. K., Exploring a V-I-S (vegetation-impervious surface-soil) model for urban ecosystem analysis through remote sensing: Comparative anatomy for cities, *Int. J. Rem. Sens., 16,* 2165_2185, 1995.

Ridd, M. K., and J. Liu, A comparison of four algorithms for change detection in an urban environment, *Rem. Sens. Environ, 63,* 95_100, 1998.

Sabins, F. F., Land Use and Land Cover: Geographic Information Systems, in *Remote Sensing: Principles and Interpretations,* pp. 387_416, W.H. Freeman and Company, New York, N. Y., 1997.

Stefanov, W. L., P. R. Christensen, and M. S. Ramsey, Mineralogic analysis of soils using linear deconvolution of mid-infrared spectra (abs), *Geo. Soc. Amer. Abs. Progs. 30,* A138, 1998.

Stefanov, W. L., P. R. Christensen, and M. S. Ramsey, Remote Sensing of Urban Ecology at Regional and Global Scales: Results from the Central Arizona-Phoenix LTER Site and ASTER Urban Environmental Monitoring Program, *Regensberger Geographische Schriften, 35,* 313_321, 2001a.

Stefanov, W. L., M. S. Ramsey, and P. R. Christensen, Monitoring the urban environment: An expert system approach to land cover classification of semiarid to arid urban centers, *Rem. Sens. Environ., 77,* 173_185, 2001b.

Stefanov, W. L., Assessment of landscape fragmentation associated with urban centers using ASTER data (abs), *Eos Trans. AGU, 83,* B61C-0739, 2002.

Stoll, M. J. and A. J. Brazel, Surface/air temperature relationships in the urban environment, *Phys. Geog., 2,* 160_179, 1992.

Townshend, J. G., The spatial resolving power of Earth resources satellites, *Prog. Phys. Geogr., 5,* 32_55, 1981.

Treitz, P. M., Application of satellite and GIS technologies for land-cover and land-use mapping at the rural-urban fringe: A case study, *Photogramm. Eng. Rem. Sens., 58,* 439_448, 1992.

United States Census Bureau (USCB), Metropolitan Area Population Estimates, (http://eire.census.gov/popest/data/metro.php), Population Division, Washington, D.C., 2001.

Welsh, R., T. Jordan, H. Lang, H. Murakami, ASTER as a source for topographic data in the late 1990s, *IEEE Trans. Geosci. Rem. Sens., 36,* 1282_1289, 1998.

Wessels, R., J. S. Kargel, H. H. Kieffer, R. Barry, M. Bishop, D. MacKinnon, K. Mullins, B. Raup, G. Scharfen, and J. Shroder, Initial glacier images from ASTER and test analysis for GLIMS (abs), *Eos Trans. AGU, 81,* 2000.

World Resources Institute (WRI), *World Resources: The Urban Environment 1996-1997,* pp. 1_30, Oxford University Press, New York, N.Y., 1996.

Yamaguchi, Y., A. B. Kahle, H. Tsu, T. Kawakami, and M. Pniel, Overview of the Advanced Spaceborne Thermal Emission And Reflectance Radiometer (ASTER), *IEEE Trans. Geosci. Rem. Sens., 36,* 1062_1071, 1998.

Zhu, G., and D. G. Blumberg, Classification using ASTER data and SVM algorithms; The case study of Beer Sheva, Israel, *Rem. Sens. Environ., 80,* 233_240, 2002.

13

Airborne Laser Topographic Mapping: Applications to Hurricane Storm Surge Hazards

Dean Whitman, Keqi Zhang, Stephen P. Leatherman, and William Robertson

INTRODUCTION

In the United States, the population and urbanization of the coastal zone is rapidly increasing. Currently, it is estimated that the population in the U.S. coastal zone increases on average by over 3600 people each day [Cullinton, 1998]. Cities along the Southeast and Gulf coast of the United States are particularly vulnerable to the hazards of hurricanes. The dramatic increases in the cost of hurricane damage experienced in recent decades can be directly attributed to increases in the population and wealth of these communities [Pielke and Landsea, 1998].

One of the greatest hazards posed by a hurricane is the storm surge. A storm surge is the abnormal rise of water levels along a coast caused by wind and pressure forces of an approaching hurricane or other intense storms. Historically, the storm surge has caused 90% of all hurricane related deaths, mostly from drowning [Simpson and Riehl, 1981; Elsner and Kara, 1999]. Flooding caused by storm surges is also a major cause of property damage.

Accurate topographic information is essential for predicting storm surge damage and flooding. These data are an integral component in the construction of evacuation maps based on numerical storm surge models such as the NOAA SLOSH model [Jelesnianski, et al., 1992]. In the U.S., the best existing topographic data usually consist of U.S. Geological Survey (USGS) contour maps produced at 5 to 10 foot (1.5 and 3 m) contour intervals. The absolute vertical accuracy of these maps is limited due to poor sampling and the analog techniques used to produce the contours. In low relief coastal plains, this poor accuracy and resolution can result in large errors in predicted flooding.

Earth Science in the City: A Reader

10.1029/056SP14

Airborne LIDAR (acronym for LIght Detection and Ranging) is an emerging technology which can accurately and inexpensively map topography over large areas. We present results of an Airborne Laser Terrain Mapping (ALTM) survey of eastern Broward County in southeast Florida and demonstrate how these data can be used to better predict the extent of storm surge flooding.

LASER TOPOGRAPHIC MAPPING

Airborne Laser Topographic Mapping (ALTM) is a subset of an active remote sensing technology known as LIDAR. LIDAR systems direct pulses of laser light toward the ground and detect the return times of reflected or back-scattered pulses in order to determine ranges to the reflecting surface. The use of LIDAR for airborne topographic mapping began in the late 1970's, but early systems suffered because of poor determination in the aircraft position and orientation. By the early 1990s, advances in navigation technology, electronic miniaturization and laser technology lead to the development of the first commercial ALTM systems. Other common acronyms used for ALTM include ALSM (Airborne Laser Swath Mapping) and ALS (Airborne Laser Surveying). A comprehensive review of current LIDAR mapping systems is given in Baltsavias [1999a] and Wehr and Lohr [1999].

Most ALTM systems consist of four basic components (Figure 1): the laser range finder, the scanner, the Inertial Measurement Unit (IMU), and a kinematic Global Position System (GPS). Data are recorded in flight and are later post processed to return X, Y, Z coordinates of the ground surface. Additional data analysis and filtering allows separation of non-surface features from the terrain surface. Finally, irregularly spaced points are usually interpolated onto a regularly spaced grid to produce a digital elevation model (DEM).

The LIDAR sensor detects the range from aircraft to ground by recording the time difference between laser pulses sent out and reflected back. Pulse repetition rates of most ALTM systems range between 5 and 25 kHz [Baltsavias, 1999a]. In addition, many systems allow the recording of multiple returns and the return intensity for each laser pulse. A scanner allows measurements to cover a wide swath beneath the flight path. In most systems, an oscillating mirror allows the laser to scan back and forth. This oscillation of the scanner mirror in combination with forward motion of the aircraft typically results in a zigzag scan pattern beneath the flight path (Figure 1).

Aircraft positioning and orientation are provided by the GPS and IMU systems. GPS receivers mounted in the aircraft and at one or more known ground positions continuously record GPS carrier phase data at sample rates of 1 Hz or higher. Post flight, differential GPS techniques compute a precise aircraft trajectory from the aircraft and ground station carrier phase data [Mader, 1986; Krabill and Martin, 1987]. The IMU consists of a set of gyroscopes and accelerometers that continu-

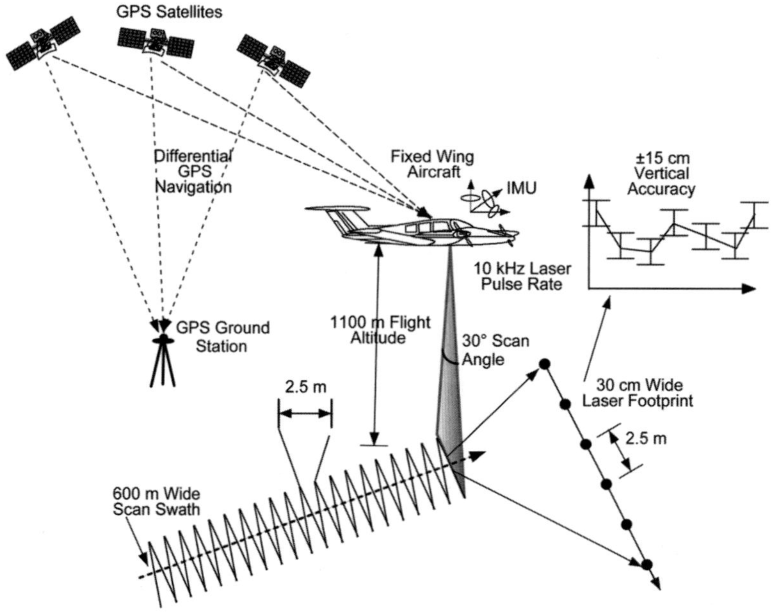

Figure 1. Schematic diagram showing the components of an ALTM system along with data acquisition parameters used for the Broward County, FL survey.

ously measure the roll, pitch, and heading of the aircraft from 10 to 100 times per second. After the flight, the aircraft trajectory is combined with the laser range data, the scanner mirror angle, and the IMU measurements to determine the precise horizontal coordinates and vertical elevations of each laser reflection.

TOPOGRAPHIC DATA

Broward County lies in a low relief coastal plain with elevations ranging between 0 and 8 m (Plate 1). With a population of over 1.5 million people (1999), Broward is Florida's second most populous county and includes the municipalities of Ft. Lauderdale and Hollywood. The most prominent topographic feature in Broward County is the Atlantic Coastal Ridge. Before development, the Atlantic Coastal Ridge formed the eastern rim of the Everglades. Early urbanization in southeast Florida was confined to the higher elevation ridge because these areas were less susceptible to flooding. Starting in the 1930s canals were cut through the ridge to drain water from the Everglades and provide more land for farming and urbanization. In recent decades, urbanization has spread westward into the low-lying wetlands of the Everglades and eastward onto the coastal lowlands and barrier islands. This urban growth increasingly has placed popula-

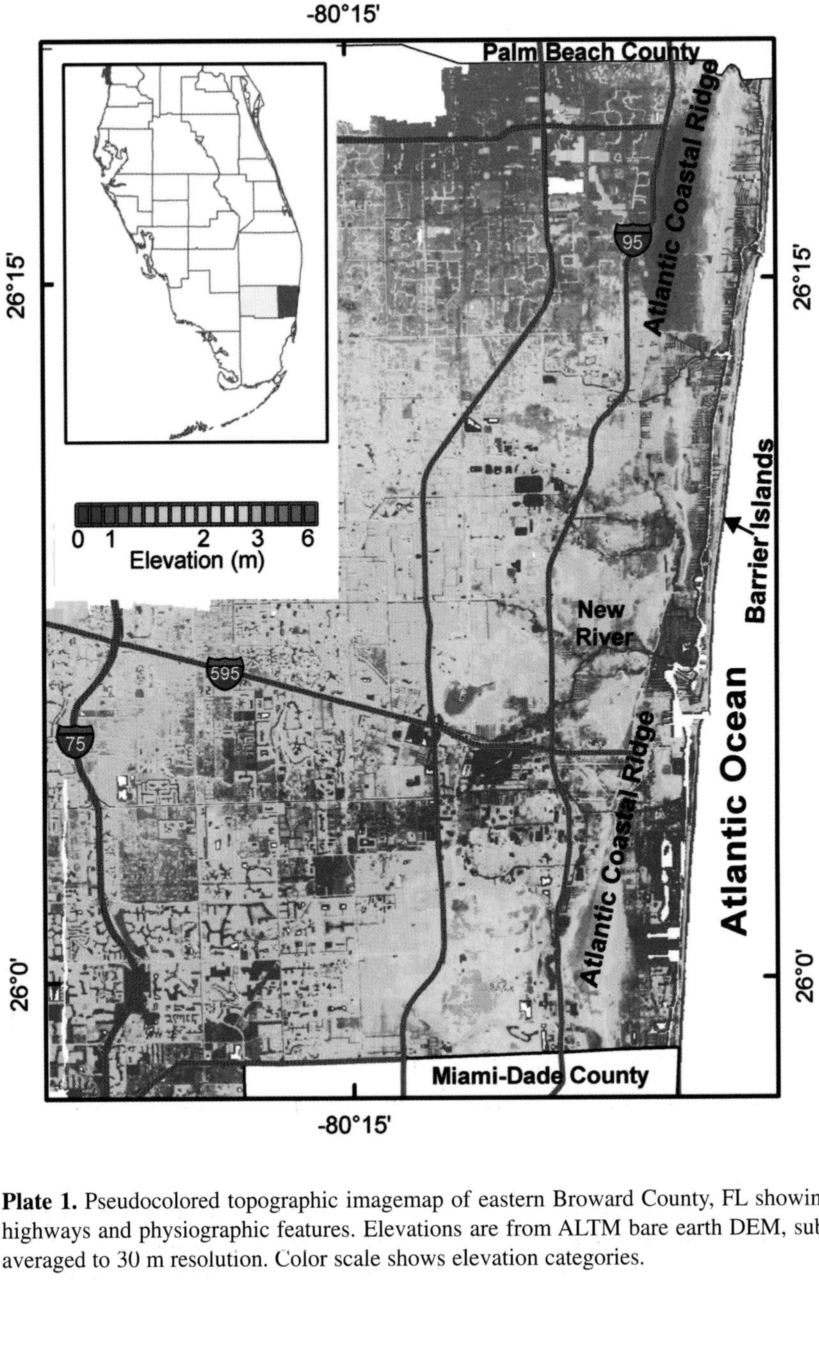

Plate 1. Pseudocolored topographic imagemap of eastern Broward County, FL showing highways and physiographic features. Elevations are from ALTM bare earth DEM, sub-averaged to 30 m resolution. Color scale shows elevation categories.

tions in areas susceptible to both inland flooding due to rainfall and to storm surge [Finkl, 1994; 2000].

In 2000-2001, Florida International University (FIU) collected ALTM measurements in eastern Broward and Palm Beach Counties to assist emergency management personnel in revising their hurricane evacuation maps. Elevations were collected with an Optech ALTM 1210 LIDAR mapping system jointly owned and operated by FIU and the University of Florida [Gutelius et al., 1998; Shrestha, et al., 2000]. Data were collected as a series of 600-meter-wide swaths consisting of points spaced approximately every 2.5 m beneath the flight path. Flight lines were spaced 500 m apart to allow sufficient overlap in order to avoid data gaps and to assess measurement repeatability. Each deployment typically took 4-5 hours during which GPS data were continuously recorded on both the aircraft and on the ground. In total, 160 separate swaths were collected.

Data from overlapping swaths were checked for internal consistency, combined and subdivided into smaller and more manageable sized portions. These consisted of 1.5-km^2 tiles, each containing 1 – 2 million points. In total, the project measured over 700 million irregularly spaced ground elevations and covered over 1300 km^2. Additional technical details of the data acquisition for this project are found in the report by Whitman [2000].

The ALTM system returns a 3-dimensional cloud of points corresponding to laser reflections off various objects (Plate 2). In order to model and visualize variations in the ground surface, reflections from non-ground features such as buildings, vegetation, and vehicles must be classified and removed [Kraus and Pfeifer, 1998; Shrestha et al., 1999]. Since a given DEM pixel can often contain both ground and non-ground surface reflections, terrain classification is best performed on the raw, irregularly spaced laser points rather than on gridded data. After classification, the remaining ground surface points are then gridded to produce a "bare earth" DEM.

A simple approach for removing non-ground points is to estimate a minimum ground surface envelope and classify the reflections based on their proximity to that envelope. An iterative algorithm, which utilizes expanding search windows and proximity thresholds, was used to classify the points. First, points outside a specified vertical range were excluded. Each tile was then subdivided into a series of overlapping 1 m square blocks and all points except the minimum elevation in each block were discarded. For the next iteration the blocks were doubled in size and the minimum elevation in each block was determined. Then, all points with elevations greater than a threshold above the minimum were discarded. The process was repeated with the block widths and classification thresholds doubling in size until the block size was 128 m or no points were discarded from the previous iteration.

After filtering, data for each tile were gridded into a 2 m resolution DEM. Because terrain filtering often produces large data gaps in areas covered by buildings and vegetation, elevations were interpolated using kriging with a

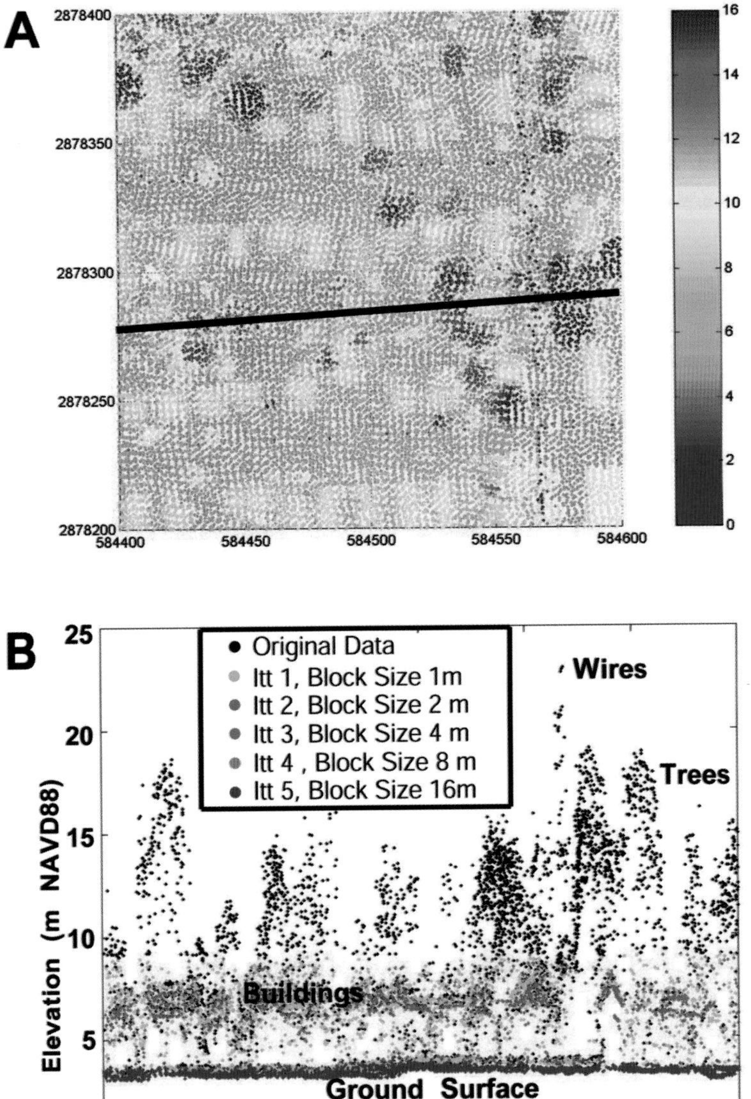

Plate 2. Example of raw ALTM data before terrain filtering and gridding. A) Color coded point elevations (in meters, NAVD88) of irregularly spaced ALTM. Black line denotes position of section in B. Horizontal coordinates are in UTM 17 meters. B) Cross profile showing points remaining after each iteration of terrain filter. Elevations were projected from a 75 m-wide swath into the section shown in A (black line). After 5 iterations, only ground surface returns remain (blue dots).

search radius of 50 m. Grid cells outside the 50 m search radius were assigned a value of NODATA. An example of a tile gridded after terrain filtering is shown in Plate 3. The color shaded relief image clearly shows the Atlantic Coastal Ridge running through the center of the tile. Roads appear as lower elevations cut into the background topography. Even subtle drainage features such as the elevated road crowns can be resolved. The footprints of buildings removed by terrain filtering appear as raised platforms, which presumably correspond to the ground elevations at the base of the buildings.

Like all remote measurements, airborne LIDAR measurements are subject to error. Errors arise from three main sources: laser range, aircraft trajectory, and INS measurements (Baltsavias, 1999b). Comparison of the LIDAR data with an independent dataset of higher accuracy is necessary in order to estimate absolute uncertainties in the elevations. Verification of the data is also necessary in order to ensure against systematic errors or offsets in the data caused by instrument malfunctions or processing blunders.

Accuracy was assessed by comparing the bare earth DEMs with an independent dataset consisting of approximately 321 GPS control points provided by the Broward County Engineering Department. These control points usually consist of survey tacks placed in the pavement of road intersections and are spaced approximately every 800 m and have vertical and horizontal accuracies of 1-2 cm. At each control point, the DEM elevations were calculated by bilinear interpolation and were compared with the control point elevations. This analysis returned a vertical root mean squared error (RMSE) of 0.12 m.

FLOOD MODELS

In the U. S., the most widely used numerical storm surge model is the National Weather Service SLOSH (sea, lake, and overland surges form hurricanes) model [Jelesnianski et al, 1992]. The SLOSH model computes water height above mean sea level at a network of grid points in a pie-shaped geographical area known as a basin. SLOSH uses a hyperbolic coordinate system and the model cells vary in size. For a typical basin, the size of each grid cell varies from 0.5 km near the center or pole of the basin to over 7 km at the outer boundaries of the basin. Typically, a basin is oriented such that the highest density of points is over land where surge heights are of greatest interests. Bathymetry or topography relative to sea level is specified at each grid point. The model can also incorporate subgrid cell features such as barriers, levees, rivers, and channels. A series of overlapping basins provide coverage for most of the Gulf and Atlantic coastlines.

Output from a composite of numerous SLOSH runs are used to define flood prone areas for evacuation planning. Strength is modeled using central pressure and storm eye size parameterized by the five categories of storm intensity developed by Saffir and Simpson [Simpson and Riehl, 1981]. For each category of

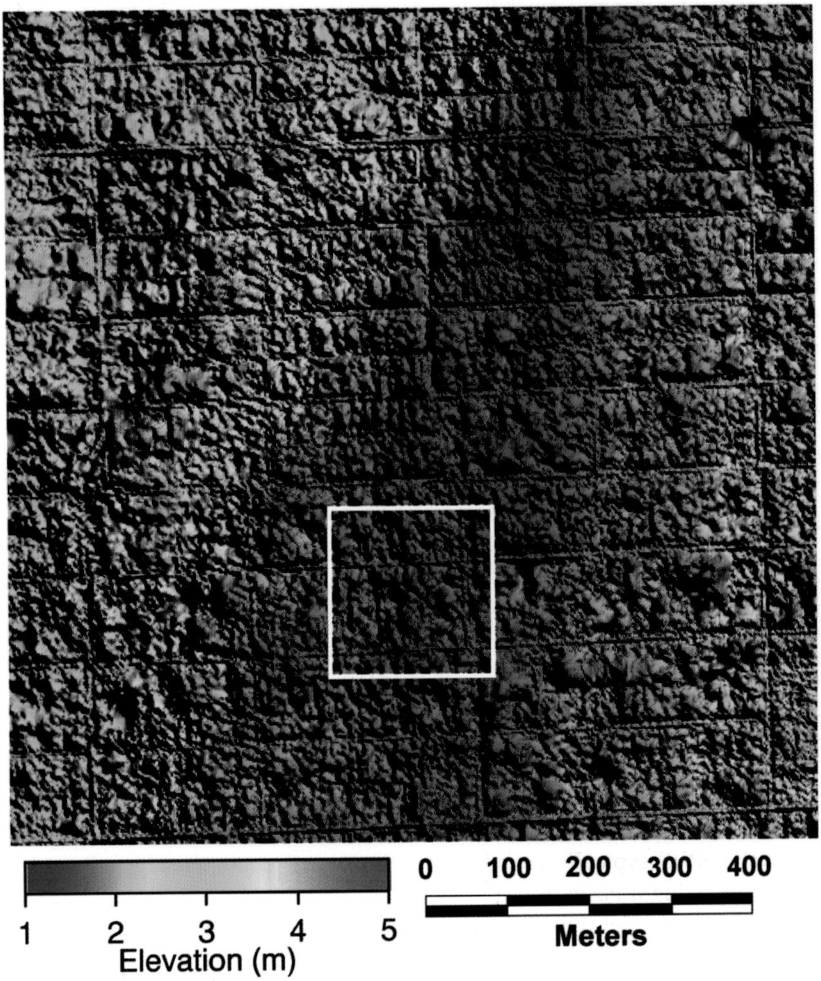

Plate 3. Color shaded relief map of bare earth DEM gridded from filtered point elevations in a 1 km^2 tile in a residential neighborhood of Hollywood, FL. Over 200,000 irregularly spaced measurements were gridded to produce this DEM. Linear features in this image are road crowns. The elevated feature in the center of the tile is the Atlantic Coastal Ridge. The white box shows location of the data in Plate 2A.

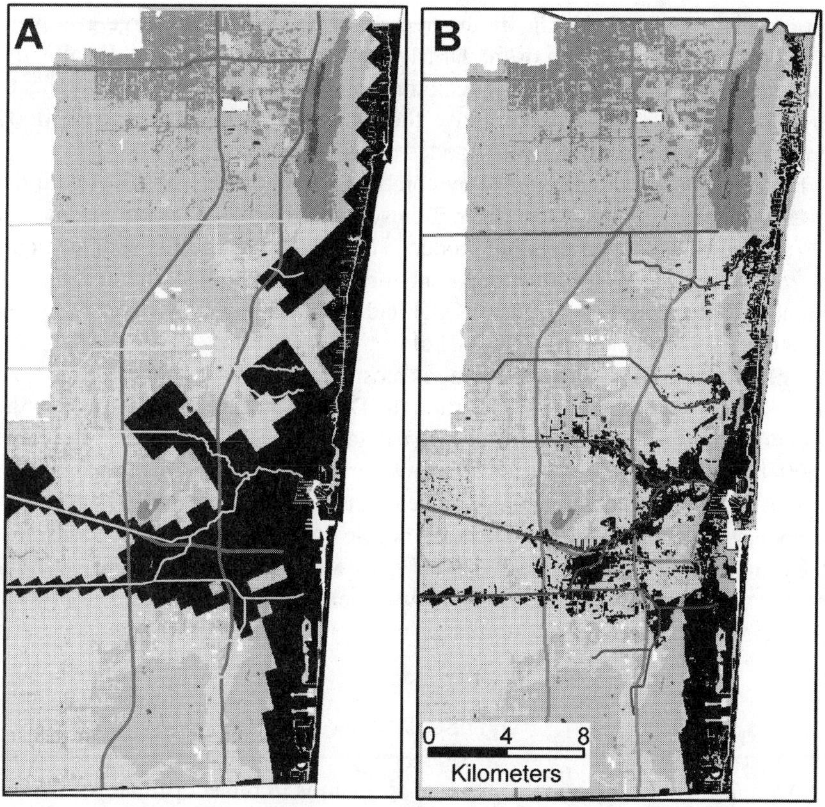

Figure 2. Maximum predicted flooding from category 5 hurricane. A) Flooding predicted from SLOSH model only. B) Predicted flooding from combined SLOSH and ALTM 30m DEM. Background topography is from ALTM 30m DEM shown in Plate 1.

storm, the NWS typically calculates surge heights for 200 - 300 hypothetical storms impacting a basin at various locations and from various directions. The results of these model runs are combined, and the maximum storm surge height at each grid cell is selected in order to construct a map of maximum potential storm surge height for each Saffir-Simpson category. These maximum of maximum (MOM) storm surge maps indicate all areas that could potentially flood for a given storm strength.

The SLOSH model is not sensitive to topographic features of small dimension and tends to overestimate the flooded area because of its relatively coarse resolution (> 500m) (Figure 2A). In order to simulate the effect of higher resolution topography on predicted flooding, we use a GIS to combine the MOM output with higher resolution DEMs. First, a lower resolution topographic dataset was produced by mosaicing the tiles and subaveraging the 2 m pixels to 30 m resolution (Plate 1). Then, heights in the hyperbolic SLOSH storm surge grid were resampled to the resolution of the DEM with bilinear interpolation. Finally, the DEM elevations were overlaid and subtracted from the SLOSH flood heights to produce a map of flood depth above the ground surface. Regions with flood depths greater than zero indicate flooded areas.

Flooding maps were calculated by combining both the ALTM DEM with the MOM maps computed for Safir-Simpson category 1-5 storm scenarios. Differences in predicted flooding produced by combining SLOSH with the higher resolution DEMs are most apparent for the category 5 storms (Figure 2). Inundated areas predicted from SLOSH and ALTM topography cover 54% less area than that predicted from SLOSH alone.

The reduction in predicted flooding is most apparent along E-W trending tidal water bodies such as the New River and the Dania Cutoff Canal that traverse the Atlantic Coastal Ridge (Figure 2). This is largely a consequence of the large (600–800 m wide) SLOSH cells. These water bodies are considerably narrower than a SLOSH model cell and are modeled in SLOSH as sub-grid cell features. Flooding does not occur throughout the whole cell because high elevations near the canal confine flooding to a relatively narrow strip. A similar effect is seen for portions of the coastal barrier islands where elevations as high as 5 m prevent flooding even for category 5 storms.

DISCUSSION

An important goal in emergency management is to protect people who are in potential danger while minimizing the overall impact and disruption to society. When a hurricane warning is issued by the NHC, local authorities usually request the evacuation of residents living in predetermined evacuation zones susceptible to storm surge. Evacuation of people constitutes a significant expense in any hurricane emergency, with estimates as high as one million dollars per mile of coastline evacuated. These costs remain even for cases of "false alarms" where warnings are issued, but the hurricane does not strike. In addition, persons who are unnecessarily evacuated are often placed in harm's way. Often, the safest place for people to reside during a hurricane is at home, unless the residence is subject to storm surge flooding or is of a type vulnerable to wind damage (e.g. mobile homes). For this reason, it is important that the best possible information be used when determining whom to evacuate.

Broward County used SLOSH model output combined with results of this ALTM study to revise their hurricane evacuation zones in 2000 (Figure 3). In designing these zones, the Broward County emergency managers also considered other information such as road access and population demographics. For practical purposes, well-known cultural features such as major roads were used for the zone boundaries. The revised maps significantly reduced the evacuation areas for

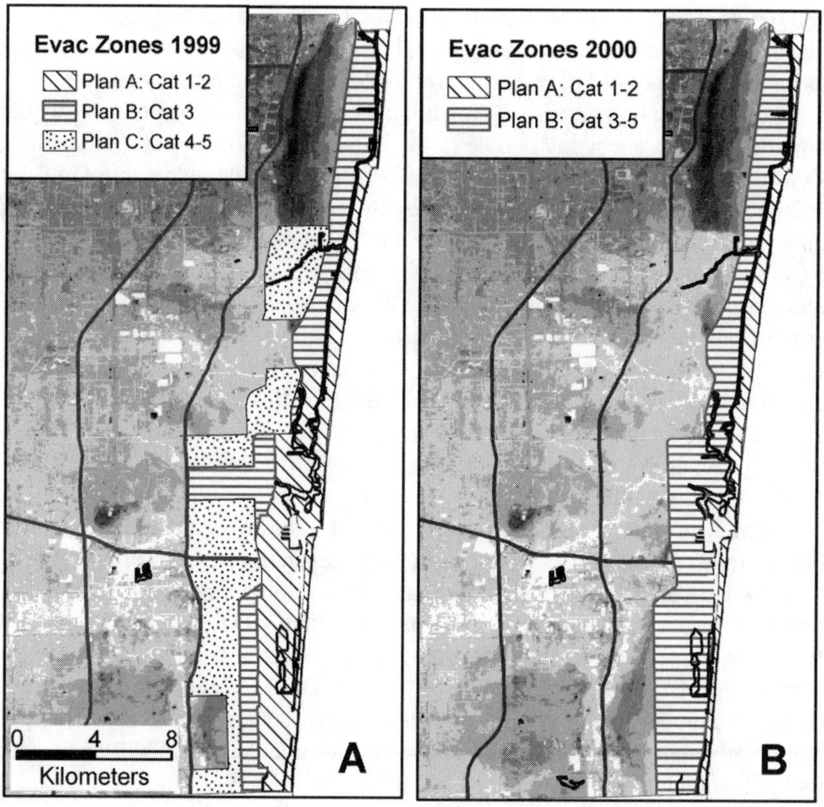

Figure 3. Revision of Broward County evacuation zones. A) Evacuation plan for the 1999 hurricane season. B) Evacuation Plan for the 2000 hurricane season [Source: Broward County Department of Emergency Management]. Background topography is from the ALTM 30m DEM shown in Plate 1.

all Saffir-Simpson categories. Category 1-2 evacuations (Figure 3, Plan A) are reduced in area by over 68% and are confined only to the barrier islands to the east of Atlantic Intercoastal Waterway. For major hurricanes (Saffir-Simpson categories 3-5), the revised evacuation zone area decreased by over 45% and in general, only includes regions to the east of the Atlantic Coastal Ridge. Over 175,000 less people will need to evacuate in the event of a major hurricane impacting Broward County.

One surprising result of this study was that much of the coastal barrier islands do not appear to be flooded even for the largest storms. However, the revised evacuation plan still requires evacuation of these islands for all hurricanes. This decision was prudent because access to the bridges and causeways to these islands will likely be threatened in any hurricane. In addition, SLOSH does not model wave set-up and run-up, which could be significant along the Broward County coastline. The identification of potentially dry areas on the islands will be useful in determining locations of refuges of last resort and for deployment of emergency vehicles.

ALTM data combined with 3-dimentional computer visualization can play an important role in educating people to the hazards of storm surges. Studies have shown that 85% of the current residents of the hurricane-vulnerable Gulf and Atlantic coasts of the US have never experienced the effects of a direct hit by a major hurricane [Jarrell et. al, 1992]. In the US, most people are well informed by the media about whether they live in an evacuation zone. In spite of this knowledge, many choose not to evacuate. One reason that some ignore hurricane evacuations may be that they do not relate the 2-dimensional maps of the evacuation zones to their 3-dimensional life experience. Three-dimensional computer graphics afford the opportunity to substitute for this lack of personal experience. Digital images taken from the air and from ground-level photography would provide the real-life skin to be put over the wire-mesh of elevations obtained by the airborne laser.

ALTM data have many other applications to storm hazard mitigation. The intense rainfall of tropical cyclones can cause severe flooding even for relatively weak storms. For example, flooding from a minimal category 1 hurricane, Irene, caused over $600 million of damage in Miami-Dade, Broward, and Palm Beach Counties in October, 1999. Many of the affected areas were not even in mapped FEMA flood zones. High resolution DEMs combined with surface hydrologic models and rainfall estimates from weather radar will allow the prediction of flooding in urban areas on a street-by-street basis and will be a useful tool in updating flood insurance maps.

Tropical cyclones and other coastal storms are also a major cause of coastal erosion. Recent studies have demonstrated the utility of airborne laser altimetry in mapping shoreline position and topography [Carter et al., 1998; Sallenger, et al., 1999; Krabill et al., 2000, Zhang et al., 2000]. Shoreline surveys can be scheduled seasonally or after the passage of major storms to assess quantitatively the amount of beach erosion, dune scarping, and overwash deposition. This new technology will fill a major void in terms of providing timely and accurate data for scientific assessments and coastal management programs.

This paper has focused on the application of high-resolution airborne laser altimetry to the hazards caused by hurricanes and other coastal storms. Obviously, this technology will find a wide range of other applications in urban settings. In hazard mitigation, applications include flood plain mapping, identification of ground subsidence related to groundwater withdrawals, and detection of active fault scarps for determining seismic risk. For urban planning, ALTM data may be used for mapping infrastructure, planning construction, and cataloging vegetation. For public information purposes, these data can be combined with advanced computer graphics and database technology to create realistic 3-D renderings of the urban landscape.

CONCLUSIONS

High-resolution topography collected by airborne laser topographic mapping systems is a useful tool in delineating hazard zones due to hurricane storm surge. When combined with the output from lower resolution numerical storm surge models, these data can provide more accurate predictions and can reveal the extent of flooding on the scale of individual streets. In Broward County, southeast Florida, topography from an ALTM survey combined with output from the SLOSH storm surge model reduces the areas of predicted flooding by over 50% compared with that predicted from SLOSH alone. In Broward County, the results of this survey may prevent over 175,000 people from being unnecessarily evacuated in the event of a major hurricane.

Acknowledgments. This study was made possible by many people. R. Shrestha, W Carter, and M Sartori and their graduate students in the UF Geomatics Program provided system development, consulting and software training on this project. D. Bloomquist piloted the aircraft. G. Mader computed the GPS aircraft trajectories. S Baig and B Jarvinen of the National Hurricane Center provided SLOSH model output. C. Finkl provided helpful comments on the manuscript. This project was supported by grants from FEMA, and the Broward County, FL, Department of Emergency Management and the FIU QIP program.

REFERENCES

Baltsavias, E.P., Airborne laser scanning: existing systems and firms and other resources, *Journal of Photogrammetry & Remote Sensing,* 54, 164-198, 1999a.

Baltsavias, E.P., Airborne laser scanning: basic relations and formulas, *Journal of Photogrammetry & Remote Sensing,* 54, 199-214, 1999b.

Carter, W. E., R.L. Shrestha, and S. P. Leatherman, Airborne Laser Swath Mapping: Applications to Shoreline Mapping, in: Proceedings International Symposium on Marine Position, INSMAP, 30 November to 4 December, 1998, Patrich L Fell, Muneendra Kumar, George A. Maul and Gunter Seeber, editors, pp, 323 - 333, Melbourne, Florida, 1998.

Cullinton, T. J., Population: distribution, density and growth, NOAA State of the Coast Report, Silver Spring, MD, 1998.

Elsner, J. B., and A. B Kara, *Hurricanes of the North Atlantic: Climate and Society,* Oxford University Press, New York, 1999.

Finkl, C.W., Identification of unseen flood hazard impacts in southeast Florida through integration of remote sensing and geographic information system techniques, *Environmental Geosciences*, 7, 119-136, 2000.

Finkl, C.W., Disaster Mitigation in the South Atlantic Coastal Zone (SACZ): A Prodrome for Mapping Hazards and Coastal Land Systems Using the Example of Urban Subtropical Southeastern Florida, *Journal of Coastal Research Special Issue No. 12: Coastal Hazards*, p 339-366, 1994.

Gutelius, G, W.E. Carter, R.L. Shrestha, E. Medvedev, R. Gutierez, and J.G. Gibeaut, Engineering Applications of Airborne Scanning Lasers: Reports from the Field, The

Journal of American Society for Photogrammetry and Remote Sensing, LXIV(4), 246-253, 1998.

Jarrell, J.D., P.J. Helbert, and M. Mayfield, Hurricane Experience Levels of Coastal County Populations from Texas to Maine, NOAA Tech. Memo, NWS NHC-46, 1992.

Jelesnianski, C.P, J. Chen, and W.A. Shaffer, SLOSH: Sea, Lake, and Overland Surges from Hurricanes, NOAA Technical Report NWS 48, National Oceanic and Atmospheric Administration, U.S. Department of Commerce, 1992.

Krabill, W.B., and C.F. Martin, Aircraft positioning using global positioning carrier phase data, *Navigation,* 34, 1-21, 1987.

Krabill, W.B., C. W. Thomas, R.N. Swift, E.B. Fredrick, S.S. Manizade, J.K. Yungel, C.F. Martin, J.G. Sonntag, M. Duffy, W. Hulslander, and J.C. Brock, Airborne Laser Mapping of Assateague National Seashore Beach, *Photogrammetric Engineering and Remote Sensing*, 66, 65-71, 2000.

Kraus, K., and N. Pfeifer, Determination of terrain models in wooded areas with airborne laser scanner data, *Journal of Photogrammetry & Remote Sensing,* 53, 193-203, 1998.

Mader, G.L., Dynamic Positioning Using GPS Carrier Phase Measurements, Manuscripta *Geodaetica,* 11(4), 272-277, 1986.

Pielke, Jr., R.A., and C.W. Landsea, Normalized hurricane damages in the United States: 1925-95, *American Meteorological Society,* 13, 621-631, 1998.

Sallenger Jr., A.H., W. Krabill, J. Brock, R. Swift, M. Jansen, S. Manizade, B. Richmond, M. Hampton, and D. Eslinger, Airborne Laser Study Quantifies El Nino-induced Coastal Change, *EOS Transactions,* 80, 90-93, 1999.

Shrestha, R.L., W.E. Carter, M. Lee, P. Finer and M. Sartori, Airborne Laser Swath Mapping: Accuracy Assessment for Surveying and Mapping Applications, *Journal of American Congress on Surveying and Mapping*, Vol. 59, p. 83-94, 1999.

Shrestha, R.L., W.E. Carter, and M. Sartori, Airborne laser swath mapping research at the University of Florida, *The Florida Surveyor,* April, 10-12, 2000.

Simpson, R.H., and H. Riehl, *The hurricane and its impact,* Louisiana State University Press, Baton Rouge, LA, 1981.

Wehr, A., and U. Lohr, Airborne laser scanning – an introduction and overview, *Journal of Photogrammetry & Remote Sensing,* 54, 68-82, 1999.

Whitman, D.,1999-2000 ALTM data collected in eastern Broward County, Florida, Report to Broward County Commission, International Hurricane Center, Miami, FL, 27 pages.

Zhang, K., D. Whitman, S. Leatherman, W. Robertson, R. Shrestha, and W. Carter, Airborne lidar applications to quantification of coastal morphology, *Carbonate Beaches 2000*, Key Largo, FL, 2000.

SECTION V

INTEGRATED EARTH SCIENCES AND URBAN DEVELOPMENT AND SUSTAINABILITY

Urban planning and management is now much more than the work of architects and engineers designing a new subdivision. True long-term planning requires that integrated geological data, along with information on infrastructure and dwellings are entered into Geographic Information System databases for every city worldwide. Here the geological and atmospheric sciences work with representatives of all aspects of city government, ranging from highway engineers to the police. The interconnectivities between the components of a city are many and will require large databases and modeling; modern computing systems can support this goal for the future.

14

Integrating Geological Information into Urban Planning and Management: Approaches for the 21st Century

B. R. Marker, J. J. Pereira, and E. F. J. de Mulder

INTRODUCTION

This contribution examines difficulties experienced by many geoscientists in making a proper contribution to urban planning and management. It examines some key aspects of urbanisation in relation to geoscience issues that affect safe, cost efficient development and conservation of the cultural and natural heritage. It refers to lack of awareness amongst many urban managers of the range and significance of these factors. Much progress was made in the second half of the 20th century towards better compilation and provision of relevant information and this will be illustrated by two examples. One is an urban area undergoing regeneration in the United Kingdom and the other is a part of Malaysia that is in the process of rapid urbanisation. However, a great deal remains to be done to promote recognition of key issues by planners through dissemination of results and constructive dialogue.

Subsequent sections consider current activities and trends that might help. For instance, consideration of sustainable urban management may provide the necessary link between planners, managers and geoscientists. Within the broad principles of sustainable development, a number of new ways of considering the environment are developing. These include the identification of indicators of environmental change; new approaches to the appraisal of, and identification of the consequences of actions on, the environment; and better, more flexible delivery of information using information technology. In addition, international scientific organisations are undertaking initiatives to raise awareness of relevant issues and to secure better consideration of environmental aspects of geoscience in education.

Earth Science in the City: A Reader

10.1029/056SP15

These initiatives may help to improve dialogue between urban managers and geoscientists, but geoscientists also need to collaborate more fully with other environmental scientists to provide information on the environment as a functional whole. Such improvements can provide a better basis for decisions on urban development and conservation

URBANISATION AND GEOSCIENCE

Urban Population

The current rapid growth in world population, estimated by the United Nations at 1.2% per annum, is well known. The current population of about 6.1 billion could reach about 9.3 billion by 2050. It is less widely appreciated that the rate of growth of urban population is more rapid than the overall rate of population increase (McGraw 1990). By 2005, half of world population is likely to be living in cities because of higher rates of infant survival in urban areas, compared with the countryside, and movements of rural people to urban centres in search of a better life (Gugler 1996). Currently, the percentages of urbanised population are highest in Europe and North America but that is changing. The rate of urbanisation is highest in Africa and southern Asia. Of 13 cities, which have populations above 10 million people, 7 are in Asia. While planning for urban development is even more urgent in these continents than elsewhere, long-developed areas are also subject to increasing urbanisation (UNESCO 1997). In Europe, for instance, about 74 percent of people live in towns and cities but this is expected to increase to about 83 percent by 2025. It is a global issue.

Characteristics of Urban Areas

Many of the World's great cities have had a very long period of occupation and industry. This has given rise to an important historical, architectural and archaeological legacy. Historic buildings give character and provide a stimulus to tourism. It is important to preserve or conserve these, as far as possible, for both cultural and economic reasons but old cities also have less prosperous areas with outworn buildings and infrastructure, vacant sites and derelict land. These may become very undesirable places in which to live. Derelict sites are often industrially contaminated and areas that have been mined for minerals may be potentially unstable. Such features often occur close to city centres but may also be found in the suburbs. Strategies for urban regeneration and investment (Roberts and Sykes 2000) are required to rectify these problems but care is needed to avoid sacrificing the historic fabric of the city (Moseley 1998; Oxley 1998).

Patterns of development are driven mainly by social and economic circumstances but are also influenced, to some extent, by current theories in planning. For instance, functional theories arising in the 1920s and 1930s in Europe led to the idea of segregation of industrial activity from residential areas. This was partly a reaction to a legacy of 19th century pollution from factories intermingled with residential streets (e.g. Forshaw and Abercrombie 1944). The result was that more people needed to travel further to work and, ultimately, this led to greater car use and increased environmental effects of traffic emissions. It is now possible, although not always achieved, to undertake a great deal of industrial activity with relatively limited pollution. These, and the greatly expanding service industries, can be integrated more readily with residential land uses. This provides more options for locating homes near places of employment to reduce travel to work whilst improving the quality of urban life in commercial and residential centres. Therefore there is a growing emphasis on neighbourhood issues and public participation in land use planning. The increasing use of new technology also assists in the dispersal of work in the service sector and is beginning to change patterns of population dispersal and movements.

A crucial factor, which reflects economic activity, is the rate of growth (O'Sullivan 2000). Where growth is measured and modest it is possible to plan ahead. Where, however, development and population increase is rapid plans may be outstripped and urbanisation may outrun provision of infrastructure, such as provisions for wastewater and sewage treatment. Ultimately, this may lead to an urban sprawl poorly designed for the needs of the people and poorly adjusted to the environment. Additional problems arise if development becomes uncontrolled, for instance where rural dwellers flock to the city and, without resources, build temporary shelters wherever they can. Such communities are often located on land, which has been avoided because it is potentially unstable, or flood prone, so that individuals who are least able to cope may be placed at the greatest risk (Gugler op cit.; Walmsley and Lewis, 1993).

While the early stages of city development tend to adjust to local ground conditions, the growing city gains an economic dynamic of its own (O'Sullivan op cit). Because of the higher value of land close to the city centre, peripheral development often expands into areas that have been avoided previously because of problems such as poor bearing strength, slope instability or potential for flooding. There may also be extensive remodelling of the landscape in order to accommodate more development. This may conceal geological factors that may subsequently affect future development and redevelopment. However, planned reshaping of the land can provide new opportunities and, especially in some geographically constrained urban areas, such as Hong Kong and Singapore, may be inescapable (Marker 2001). Where there are alternative options, however, cheaper and more sustainable solutions may be possible.

Preferred patterns of active development vary from place to place. Inner city regeneration may be important in old industrial towns but, even in these, much development takes place peripherally leading to suburban sprawls, and, ultimately, conurbation or the growth of megacities. Alternatively, rapid building and industrialisation, linked to population growth or increasing prosperity, especially in the developing countries, may lead to the creation of new cities. However, overall, buildings tend to become taller, and the amount of underground construction increases. Greater loads are placed on geologically less suitable ground, because the best sites have been occupied already. When 3.5 billion people live in less than 1% of the land area of the Earth, with all their demands for earth materials, the delicate balance between the natural environment and human occupation becomes severely disrupted. Major impacts on the physical environment in, and around, urban centres become inevitable.

Geological Constraints to Development

There are a wide variety of geological and related factors that may constrain development or the way in which it is undertaken. Some relate to disastrous, rapid geological hazards such as earthquakes, volcanic eruptions, flooding and landslides. Some result from slower processes including some types of land subsidence, coastal erosion or less obvious processes like shrinkage or swelling of clay beneath foundations (McCall 1996a).

The average world economic loss due to natural hazards in 1993 was almost $45 billion US, or 0.23 % of the world GNP. The Munich Reinsurance Company (1982) estimated that $100 billion would be lost in 2000, $270 billion in 2010 and about $700 billion in 2020. Because of concentration of capital in cities, a strong relation exists between economic losses and urbanisation. The larger the city, the higher the potential risk and greater damage that is likely. However it would be wrong to dismiss less spectacular constraints to development as less important. Damage arising from shrinking and swelling of clay, for instance, cost as much as $150 million per annum in the UK alone. Given the magnitude of such costs, one might expect reduction of natural hazards to be at the forefront of prudent urban planning, but that is not so. Many urban managers think of major disasters as "Acts of God" that cannot be prevented. Events that occur infrequently quickly fade from memory. Less dramatic processes are seldom considered at all. A common reaction is that these matters should be left to engineers at the site investigation stage and to building control officers when the development is being undertaken, on the assumption that most problems can be predicted and prevented or, at least, minimised. As a result, many developments are poorly located, suffer delays when works are commenced, and involve large preventive and remedial costs. These could be avoided, or reduced, if more consideration is given to ground conditions at the planning stage (Marker 1996).

Many urban planners do not appreciate the diversity of issues that may affect the area that they deal with or that all cities have a variety of geological problems. For example, the geological surveys within Europe have jointly formed an Urban Topic Network known as GEURBAN (Annells 1998). One task of this Network was to check the problems and potential of selected urban centres in 16 European countries in terms of 15 key issues. Several of these were related directly to geological and topographical conditions. Five more, triggered primarily by human intervention, were linked indirectly such as pollution and contamination. Three issues related to the depletion of natural resources, such as groundwater, arable land and building materials in and around cities (Table 1). While this survey was not comprehensive, it did give a reasonable indication of the variety of issues that need to be taken into account. It was concluded that cities in Europe's Mediterranean countries suffer most from geological and topographical risks and, except for Sweden, cities in the Nordic countries least. This comes as no surprise to a geoscientist since it reflects the characteristics of a relatively young and dynamic fold mountain belt in southern Europe and generally more stable geological environment in most of Scandinavia. However that is not at all obvious to someone untrained in the geosciences. The European results are probably broadly similar to those that could be expected in any fairly large and geographically varied area of the world. Adequate geoscience information is needed, therefore, to reduce the tendency to overlook ground conditions in the drive towards necessary development.

Natural Resources

Urban areas require large amounts of minerals, energy, water, food, and other materials to support construction, infrastructure, industry, services and the population. The larger the city the greater, in general, its requirements and the wider the area it secures these essentials from. Most cities call, at least, on a regional or national area for supplies. Megacities are fully international in their effects. To satisfy the basic needs of citizens enormous flows of earth resources are required: approximately 20 tonnes per head per year. Because of transport costs most must come from as close as possible. The result is depletion of nearby resources (Bobrowsky 1998). For example, the Urban Topic Network study (Table 1) illustrated, for Europe, that depletion of natural resources was strongest in selected cities in Luxembourg, Portugal, Austria, Spain, France, Italy and the Netherlands. The increasing need for long distance transportation has its own environmental penalties such as increased air pollution and use of energy.

In addition, the exploitation of natural resources affects the delicate balance in the earth systems. Water may come from surface or underground sources but urbanisation can put the quality of both a risk through potential pollution. Once polluted these may be difficult to restore (Foster et al 1998; Lawrence and Chaney 1996). Exploitation may also store up problems for the future. Thus, groundwater extraction may cause land subsidence or flooding. Mining for minerals and outworn industry may leave land derelict (McCall and Marker 1996; Mather et al 1996). Building on unstable slopes may cause landslides, and underground mining can cause subsidence (McCall 1996a). Little thought is given to soils during urbanisation even though these are a valuable resource that

Table 1: Relative relevance of various geological factors in selected cities within the European Economic Area from a survey undertaken by the Eurogeosurveys Geurban Topic Network.

Key issues		Country															
		A	B	D	DK	E	F	FI	G	IR	IT	L	NL	N	P	S	UK
Geological and topographical	Failure of foundations and under-ground structures	2	3	1	1	3	3	3	1	1	2	1	2	1	2	3	3
	Ground collapse	2	3	1	1	2	3	0	2	2	3	2	2	0	2	3	3
	Landslides and rockfalls	3	2	1	1	3	2	1	3	1	2	3	0	3	2	2	1
	Land loss, erosion, siltation	1	2	1	2	2	1	1	2	3	1	1	2	0	3	1	0
	Earthquakesa nd volcanic eruptions	1	1	1	0	3	2	0	3	0	3	0	0	0	3	0	0
	Natural radiation	1	2	1	2	0	1	2	1	2	2	2	0	1	2	3	1
	Flooding and coastal inundation	1	2	3	2	3	3	3	2	3	3	2	3	3	3	2	1
	Sub-total	11	15	9	9	16	15	10	14	12	16	11	9	8	17	14	9
Human interference	Contamin-ated land	2	2	2	2	3	3	2	3	2	3	2	3	3	2	3	3
	Derelict land	1	2	2	0	1	3	2	2	1	2	0	0	2	2	2	2
	Ground-water pollution	3	2	2	3	2	3	3	3	3	2	3	3	3	3	3	3
	Surface water pollution	2	2	2	3	2	3	3	3	3	2	3	2	3	3	3	1
	Urban waste disposal	3	2	2	3	3	3	2	3	2	2	2	1	1	3	3	3
	Sub-total	11	10	10	11	11	15	12	14	11	11	10	9	12	13	14	12
Natural resources	Shortage of natural construction materials	3	0	2	1	3	2	1	2	1	2	3	2	0	1	3	2
	Shortage of water resources	2	1	1	2	2	2	1	2	2	2	2	1	0	3	1	1
	Depletion of green and arable land	1	1	2	0	1	2	1	1	2	2	2	3	3	3	1	1
	Sub-total	6	2	5	3	6	6	3	5	5	6	7	6	3	7	5	4
Total		28	27	24	23	33	36	25	33	28	33	28	24	23	37	33	25

Key to countries: A= Austria: B= Belgium: D= Germany: DK= Denmark:
E= Spain: F= France: FI= Finland: G= Greece: IR= Irish Republic: IT= Italy:
L= Luxemburg: NL= Netherlands: N= Norway: P= Portugal: S= Sweden:
UK= United Kingdom
Relevance of hazard: 0 = almost none: 1= low: 2= medium: 3=high

is difficult to replace. It makes sense to recover these for use elsewhere rather removing them and wasting them or, simply, building over them (Simpson 1996). Care is also needed to avoid contamination of soils elsewhere in the urban area so that these remain as resources for improving the present quality of life and for future regeneration (Pratt 1993; Simpson op cit).

Urban growth can also overrun areas of land, which could otherwise be used for agriculture, livestock farming, forestry, or wildlife conservation (Barker 1996). It can also grow across mineral resources thus sterilising them unless care is taken to safeguard these from development. This destruction or sterilisation of resources can be minimised through soundly based planning. Adequate geoscience information is needed, therefore, both to secure raw materials, such as groundwater resources and construction materials, and to protect these, and soils, from the effects of built development and to secure regeneration that is in balance with the natural environment, in general, and earth surface processes in particular (Hurley 1996; Ward et al 1994).

Pollution and Waste

Cities give rise to pollution of soils and surface and underground water. The Urban Topic Network study (Table 1) demonstrated that, in selected European cities, the impact of contamination and pollution showed a less distinct geographical distribution than either geohazards or depletion of natural resources thus certain cities in Greece, France, Portugal were considered to be affected strongly, as well as others in Sweden, UK and Finland. This is to be expected because some cities show the effects of past industrial processes, others experience contemporary problems, and some have to deal with both. Air or water pollution can travel long distances (Geernaert and Jensen 1997; Mather et al op cit), as well as having effects on nearby people and wildlife, or causing local contamination of soils (Bolton and Evans 1997). Particular problems may arise if rapid growth exceeds the capacity of infrastructure, such as wastewater and sewage disposal arrangements are inadequate. Conversely there may be opportunities for developing more sustainable urban drainage systems (CIRIA 2001).

Cities also produce massive amounts of waste, on the average some 500 kg per person per year. OECD has predicted a 3% increase per annum. Even with increasing levels of recycling in many countries, landfill remains the most widespread form of waste management. Problems occur if urban expansion takes place across former landfills giving rise to risks from landfill gas emissions or ground with poor bearing properties (Eyles and Boyce 1997). Thus geoscience is required for assessment of the extent of contamination and pollution, pathways, likely effects and risks, the design of remediation strategies (Ecotec Research and Consulting 1987; Pratt op cit), and for the development of well-contained landfills (Proske et al in press).

The Hinterland

Each urban area affects, and is affected by, its surroundings and is vulnerable to processes, which may commence far away. For example, deforestation of upland parts of a river catchment and poor farming practices may lead to soil erosion and excessive run-off of precipitation which, when collected together downstream may lead to silting up of channels and ultimately over bank flooding of an urban area (UNESCO op cit page 66). Transport of resources from the hinterland and of wastes and products out of the city may give rise to excessive traffic emissions unless, for instance, rail or water transport can be used. There is a need to consider the relationship of the city to the resources to establish whether provision be made close enough to sustain the development and whether, for instance, it is likely to place an excessive burden on scarce local water supplies. Regional geoscience information is needed for such assessments.

Despite an emphasis on spatial issues amongst planners (Bradford and Kent 1993) a little attention is paid generally to geohazards and constraints to development, natural resources, pollution and contamination in urban planning. Cities are only rarely considered within the wider environmental context of the areas in which they exist. This arises partly because planners seldom have training in geoscience but also because few geoscientists are trained in urban planning and management. Much information on the hinterland of cities is available in geological surveys but a great deal more is often widely dispersed. A great deal is technical and needs expert interpretation. There is a need to collate and simplify information to aid communication.

LAND USE PLANNING TO ACHIEVE SUSTAINABLE DEVELOPMENT

Planning of Land use

In most countries, development and conservation is implemented through a land use planning system. It is difficult to generalise too much about this topic since approaches vary from country to country: some are fairly unrestrictive and loosely enforced whilst others are fairly tightly drawn and rigorously applied. However the main aims are generally to secure:

- necessary development that is appropriate to the location and circumstances of the area; and
- the conservation of the cultural and natural heritage.

Most planning systems consist of two main stages—the preparation of planning policies for specified areas of land and the determination of applications for permission to develop the land (Marker 1998). The identification of planning policies for development in a specific administrative area (municipality, district, county, region, or combination of these) is usually undertaken in development plan documents. These may contain general policies on, for example, what mixture of facilities may be need during the plan period and criteria for where these might be acceptable. Many development plans identify areas in which particular types of development might be particularly appropriate or might be resisted (a process referred to as allocation or zoning). In general, the smaller area that is covered by the plan, the more likely it is to identify specific sites on maps.

If a development plan is to be prepared or reviewed and updated, it is necessary for the responsible authority to collate information on the principal relevant characteristics of the area. It has long been recognised that "All good planning commences with a survey of the actual resources" of the area and any survey should be "complete, accurate, well organised, and relevant" (Mumford 1961). The preparation of any strategic plan for an area needs to be proceeded by a survey of the principal social, economic, physical, and ecological characteristics of the area concerned. This is the basis for identification of resources, which may be exploited, features which merit preservation or conservation, constraints, which may prevent development or, at least, make it more expensive, and opportunities for development. Incomplete or incorrect information can lead to wrong decisions.

Since land use planning operates at several levels—national, regional and local—good information is needed at appropriate levels of detail for all of the issues that are relevant, including, where appropriate, geoscience data. A key step is the identification of criteria for deciding how to take specific characteristics into account. Commonly, there is strong emphasis on the social and economic features and, increasingly, on ecological aspects but the geological and geomorphological environment tends to be neglected. There is a need to rectify this by convincing planners that geoscience is important because it saves money and prevents risks to urban society.

Site Appraisal

A planning application submitted by a prospective developer will be considered by the planning authority and will either be permitted or refused. If permission is given then it is usually subject to planning conditions that are intended to protect people and the wider environment during and after development. When a planning application is prepared, data is required on the specific site. It is usually the developer who is charged with collecting this information or with hiring consultants to do so either through site investigation or, for major developments, through environmental impact assessment (EIA). The information is necessarily much more detailed than that collected for preparation of a development plan. However the main focus is within the boundaries of the proposed development site (DETR 2000). Even EIA, that should take wider issues fully into account, may be relatively superficial if insufficient regional data on the environment can be readily obtained. Whilst geoscience is firmly embedded in both site investigation and EIA, some relevant factors may not be taken into account. For instance, many investigations in old urban areas do not, as a matter of course, include checks on whether land is contaminated unless information on past uses of the site suggests that this might be a significant issue. Ready availability of good information is the key to improved design and content of site investigations and appraisals (Cendrero et al 2001; Tucker et al 1998).

Ultimately, the test of any good development plan is whether it is capable of full implementation and delivers the intended outcomes and benefits to the community through appropriate decisions on individual planning applications. However planning systems are tested to the full in urban areas where many social, economic and environmental problems coincide, the pace of change may be great, and quick decisions are often needed.

Whilst land use planning has always included consideration of natural resources and constraints to development these have often been poorly integrated into the land use planning process as a whole, particularly in urban areas. There is a need to secure a better link between the social and economic factors and environmental factors on the other, and the need is becoming more urgent because of the rate of urbanisation. Over the past two decades, the concept of sustainable development has offered a potential way of bridging the gap. This requires proper account to be taken of the environmental, social and economic implications both for now and the future and, thus, a change from the narrow, short-term considerations that have led to many unfortunate decisions in the past (CEC 1996). Short-term profits may vanish in the face of longer-term costs.

However the best balance of environmental, social and economic costs may be perceived very differently from place to place. In a relatively poor country the need for development may be seen as overriding whilst in a relatively rich country it is easier to choose to constrain development. Even so, the challenge is to achieve the best balance for both people and the environment. However, that requires good information on the environment, a sound understanding of how it works, and wide awareness of the issues.

Making Links

There are some issues that can be used to explain the links between land use planning and geoscience. To take a few examples:

- How can the best use be made of the landscape? Consideration of how land uses can be harmonised with topography, imaginative use can be made of perspectives and viewpoints, gradations between uses can be built in requires an understanding of the landforms, and, crucially, the processes that mould and maintain them. The skills of the landscape architect need to be blended with those of the engineering geomorphologist and engineering geologist (Simonds 1978). In addition, the water engineer needs to be brought in to consider, for instance, how gravity flow and storm sewers should be designed and constructed to maximise efficiency and minimise the costs of construction, maintenance and potential for flooding.

- What should be done to reduce risks in areas with potential ground problems? Potential hazards are not regarded as good news. These may affect land prices and investment so some planners and managers might prefer not to know about them. This is short sighted since, ultimately, these will give rise to damage to property or even loss of life and the effects of this might cause more blight than early knowledge followed by sensible planning. Areas with ground problems may also provide opportunities thus potentially unstable slopes could, for example, be used as areas for urban wildlife or, if the hazard is not too great, as public open space (McCall 1996a). Areas prone to flooding could have similar functions or could bear types of developments, which are specially designed or not particularly vulnerable to the hazard. In these examples engineering geoscientists need to collaborate with ecologists and architects respectively.
- How can we best preserve or conserve interesting landscape features and natural land cover and make opportunities for wildlife? It may be possible to leave landscape features such as high quality agricultural land, wildlife habitats and interesting topography undeveloped. Alternatively, there might be opportunities for reclaiming derelict land into pleasant open areas and for making provision for wildlife on buildings as well as around them (Barker op cit). The landscape architect and ecologist are likely to take the lead in this but they will need to draw on the expertise of geoscientists, especially soil scientists.
- How can wastes from the urban area best be managed? Even with trends towards waste minimisation, recycling and incineration, a great deal of waste will continue to go to landfill with potential problems for the environment if sites are not well located, well engineered and managed properly using natural (geological) barriers (Eyles and Boyce op cit).
- Where can supplies of natural resources, including minerals, be secured? It makes sense in terms financial and environmental of costs of bulk transportation to secure resources from as nearby as possible. This requires the avoidance of sterilisation of nearby resources by built development (McCall and Marker op cit).
- What use can be made of underground infrastructure and development? There is a tendency to think about planning in terms of surface development, even where cities are crowded already. There may be much potential for development of underground facilities but this can be constrained by choices of foundations, underground transport tunnels, and sewage infrastructure. Such developments may help to improve the quality of life at the surface.

There is a need to consider the ground in three dimensions to identify and fully realise its real potential (Paul et al 2002).

These, and many other, questions need to be addressed during the process of defining land use options to deliver the maximum social, economic and environmental benefits at the least costs (Marker 1996). There is a need to accommodate cultural and political variations. It is also being realised that it is prudent to involve since the general public in urban design whether through the land use planning process or specific environmental initiatives such as those developed, for instance, under Local Agenda 21 (ICLEI 1996).

There is, therefore, a need for better communication and collaboration:

- between environmental scientists (including geoscientists) on the one hand and urban planners and managers, and the general public, on the other; and
- among various specialists in different aspects of environmental science and other disciplines such as engineering and architecture. This would help to secure more sustainable approaches to dealing with natural resources, waste management and natural hazards in the urban environment (Thompson et al 1998).

The next section reviews briefly some progress made during the second half of the 20th century to addressing the links between urban planning and management and the use of geoscience information. The subsequent section makes some remarks on necessary steps for the first half of the 21st century.

LINKING URBAN PLANNING AND GEOSCIENCES: CASES

A rapid increase in interest in urban geoscience in the second half of the 20th Century grew from preparation of engineering geology and, later, environmental geology maps. Legget (1973), working in Canada, emphasised the need for better, readily accessible engineering geology information and improved communication between geoscientists and civic authorities. His comments remain relevant today, a reflection on the slow overall progress made during the subsequent 30 years.

In the 1960's and early 1970's, engineering geology maps were being prepared and improved most actively in eastern Europe, notably the Czech and Slovak Republics, and in North America. Environmental geology maps, addressing a wider range of issues, were developed in the USA during the 1970's. While many of these maps addressed rural areas, a proportion considered urban areas. The potential connection with planning of land use was recognised early on and was the main motivation behind many of the initiatives. However, most of these maps were strongly geological in content and, essentially, intended for use by geoscientists. Most were of limited value in communicating essential geoscience information to people untrained in the subject. It was difficult to interest planners and developers in these results that thus remained interesting but were not fully exploited (see reviews in Commission of Engineering Geology Maps of the IAEG 1976; Brook and Marker 1987; Marker and McCall 1990; Robinson and Speiker, 1978).

Programmes of work were undertaken in a number of countries to explore better approaches to communicating results (see, for example, work reviewed in McCall and Marker 1989 and Smith and Ellison 1999). In the UK, for example, attempts began to develop more appropriate keys to mapped information. By the late 1990's these had progressed to maps specifically designed for land use planning and which included planners in their preparation (Marker 2001). These gave the basis for an initiative to increase awareness of the use of environmental geology information in planning, including specific guidance to planners (Thompson et al 1998).

It is instructive to select two examples of contrasting areas to illustrate the general level that recent investigations have reached. The first deals with a long urbanised area in the UK that is undergoing regeneration. The second example concerns an area of rapid urbanisation in Malaysia.

A Long Urbanised Area: Wigan, UK

Old urban areas have both advantages and problems. They benefit from a legacy of architecture, historical and archaeological remains, which might merit conservation but also suffer from difficult ground due to contamination, pollution and potential instability.

The Metropolitan Borough of Wigan can be taken as an example based on work by Forster and others (1995). Wigan is located in north-west England close to Liverpool and Manchester. It has an area of about 200 km^2 that contains a population of about 306,000. The topography is gently undulating at about 20-150m above sea level. The geological succession (Jones et al 1938) is summarised in Table 2.

Wigan was a small market town in the 12th C. By the 16th C coal extraction from surface outcrops and bell pits, local smelting and fabrication of pewter were going on, alongside a wool industry. In the 18th century iron foundries were established, fuelled from drift coalmines in the early part of century, and shaft mines towards end of century. Industrial activity and coal extraction peaked in the late 19th C after which the town became overshadowed by Manchester and Liverpool and went into industrial decline. This persisted through the first half of the 20th C. but, from 1950 onwards, efforts were made to regenerate the area and, in particular, to reclaim extensive derelict and damaged land left by closed industries and to replace out-worn housing (Hannavey 1990). The industrial heritage is now being developed to stimulate tourism. However extensive ground problems remain.

TABLE 2: Geological succession in the area around Wigan, UK

Period	Geological units	
Quaternary	Glacial tills, aeolian sands, peat, fluvial sand and gravel cover most of the area These infill palaeo-valleys and tend to even out the topography. Much of the area has been affected by human activity thus made, excavated and disturbed ground mantles most of the area	
Permo-Trias	Manchester Marls	Soft to weakly lithified siltstones
	Sherwood Sandstone	Weakly cemented porous sandstones, a major aquifer
Carboniferous	Coal Measures	Rhythmic sediments – shales, sandstones and coal, including economic resources of coal

The British Geological Survey with planning consultants Roger Tym and Partners was commissioned by the then Department of the Environment, Transport and the Regions to prepare maps of the area drawing on its archives, site investigation reports from engineering consultancies, records from the water industry, coal mine plans and historical records, as well as field survey (Forster et al 1995). Some of the findings, relating to contamination, pollution, and potentially unstable land, are referred to here.

Land contamination and water pollution. Patterns of land use since late 19th C are known fairly well and can be used to indicate sites where soils may have become contaminated e.g. foundries, gas works, chemical works. However contamination is not confined to these sites because:

- locations of earlier contaminating uses are not well recorded;
- emissions from chimneys may have been carried downwind into land used for other purposes, including agricultural land; and
- methane gas from old coalmines migrates through mined ground and other voids.

Therefore problems may be encountered more widely than expected but, even so, available data can help in designing reclamation strategies.

Water pollution is an important issue because the main source of supply is the Sherwood sandstone aquifer. Water quality may be affected by pollution from industry and landfill sites, or from acidic drainage from coalmines. Caution is needed to avoid unsuitable development close to abstraction points and to safeguard these when the ground is disturbed during reclamation. The aquifer is protected by relatively impermeable tills in parts of the area but is very vulnerable to pollution where it outcrops or is beneath permeable surface deposits. It is important to know where that is the case (Taylor 1957).

Land stability problems. Coal mining in the Wigan area included:

- early, shallow excavations at or near the outcrop and now backfilled or concealed, and drift mines including pillar and stall workings, which have remained unfilled, both of which can cause surface instability;
- mining from deep shafts—collapse of deep voids seldom cause surface subsidence because of bulking-up of debris but some shafts, which were not capped, can give rise to serious subsidence incidents; and
- longwall mines in which controlled subsidence took place as mining proceeded and few residual problems remain.

There are reasonably good records from the 1870's to the end of coal extraction in 1992. But earlier workings are largely unrecorded. It was necessary, therefore, to classify the ground into areas where:

- mining is unlikely to have taken place—essentially areas that are not underlain by coal-bearing strata;
- mining is likely to have taken place at depths greater than 30m—areas where collapse of voids is only rarely likely to cause subsidence of the surface; and
- mining is likely to have taken place less than 30m from the surface—areas where subsidence potential may exist.
- Access shafts can occur throughout the areas underlain by Coal Measures area and differential subsidence may occur where the strata are cut by faults so the locations of these should also be taken into account.

These (and other) significant factors for planning and development were depicted in simplified maps and special report for planners and developers (as well as a technical report), which alerted them to:

- the significance of each issue for planning and development;
- key indicators of particular ground problems;
- locations where each factor might be particularly important;
- appropriate action which should be taken;
- background reading material; and
- organisations, which could be contracted for, further information.

This enabled land use planners, many of whom had no geoscience training, to appreciate the significance of, and to take account of, these issues when preparing development plans and considering planning applications, and in obtaining professional advice quickly when necessary.

An Area of Rapid Urbanisation: the Langat Basin, Malaysia.

The Langat Basin is a rapidly urbanizing area with an area of nearly 3000 km^2. It is located in the central western portion of Peninsular Malaysia, between the latitudes of 2° 35' and 31° 07', and the longitudes of 101° 12' and 102° 00'. The basin is drained by the Langat, Labu and Semenyih Rivers, which flow west into the Straits of Melaka. The highest peak in the Eastern Highlands of the Basin, at 1493m, is Nuang Mountain.

The geology of the Langat Basin is relatively simple (Figure 1). Metamorphosed sediments, consisting of quartzite, phyllite and shale, outcrop in the west and central parts of the basin while an extension of the Main Range Granite is mainly confined to the upper reaches. The bedrock has been affected by deep tropical weathering. Much of the area has a cover of alluvium, and a broad alluvial plain overlies metasediments near the coast. The Eastern Highlands are covered with primary forests, including a number of forest reserves. These serve as the catchment for the Langat and Semenyih Dams. The central portion of the basin is contains most of the urbanized areas. The western part of the basin is agricultural land, with tracts of oil palm and rubber plantations. Mangrove swamps are restricted to a narrow zone along the coast.

Before the early 1990s only a small part of the basin had been built over. There was subsequent rapid expansion of housing and industry until the rate decreased due to an economic downturn in 1997. Renewed expansion is, however, likely in the future. Development in the basin requires adequate supplies of construction aggregates. There is also a danger that new development might be exposed to geohazards. Studies of the Langat Basin were, therefore, undertaken. This summary is based on work undertaken at the Institute for Environment and Development (LESTARI) at the Universiti Kebangsaan Malaysia.

Sterilisation of construction aggregates. The urbanisation of parts of the Langat Basin and of nearby areas in the southern development corridor, creates demand for construction aggregates (Pereira and Komoo 2003). The intensity of development is reflected by a per capita consumption in the Langat basin area of about 37 tonnes compared with the Malaysian per capita annual consumption of aggregates of only about 4.7 tonnes. The resulting expansion of quarrying of granite, at 21 reported quarries, is reflected by an annual increase in production of 5% per annum between 1980 and 1989 rising to 30% between 1990 and 1996 (Figure 2). This fell by 60% in 1997 but is expected to rise again.

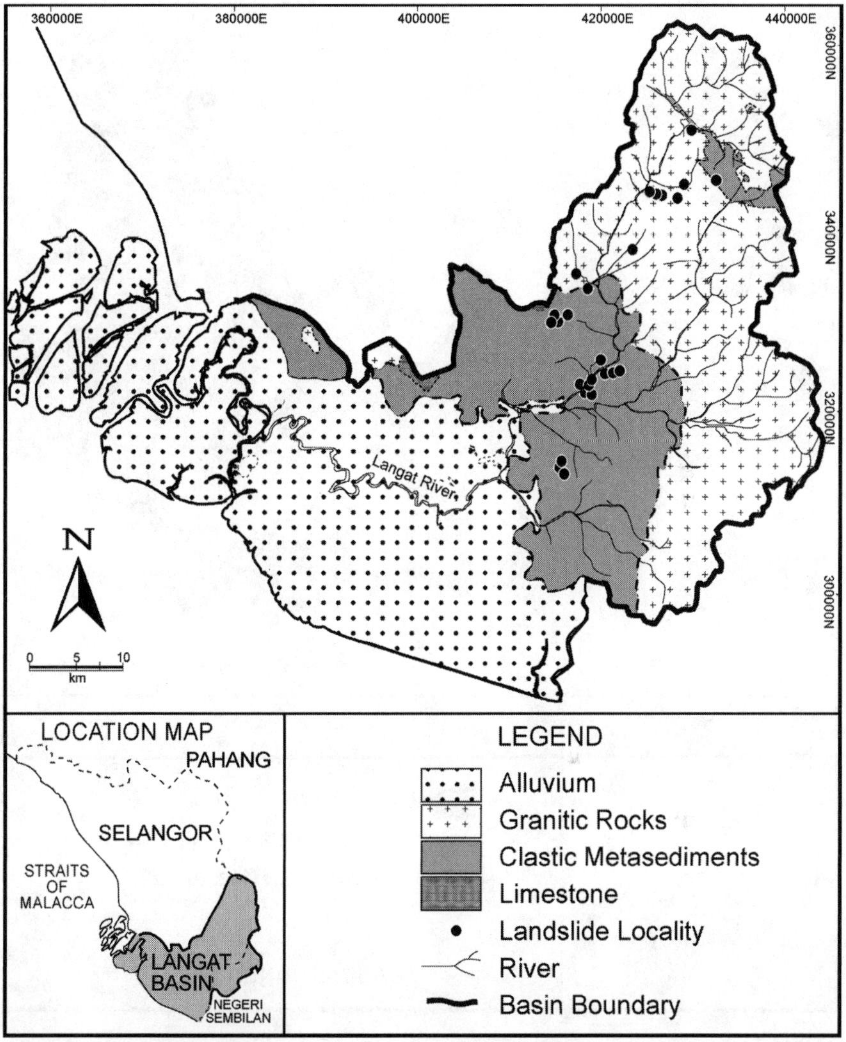

Figure 1. Simplified geology, and landslide locations, Langat Basin, Malaysia.

Most of the urbanisation to date has been at the expense of agricultural land. However it is only a matter of time before pressure is felt on forested areas, including those worthy of conservation. In addition, residential and industrial areas are encroaching on the existing quarries leading to increasing complaints about dust, noise, vibration, traffic, and adverse effects on water quality. There has been little reclamation or rehabilitation of quarried ground. Whilst environmental and health and safety regulations exist the capacity to enforce these is limited.

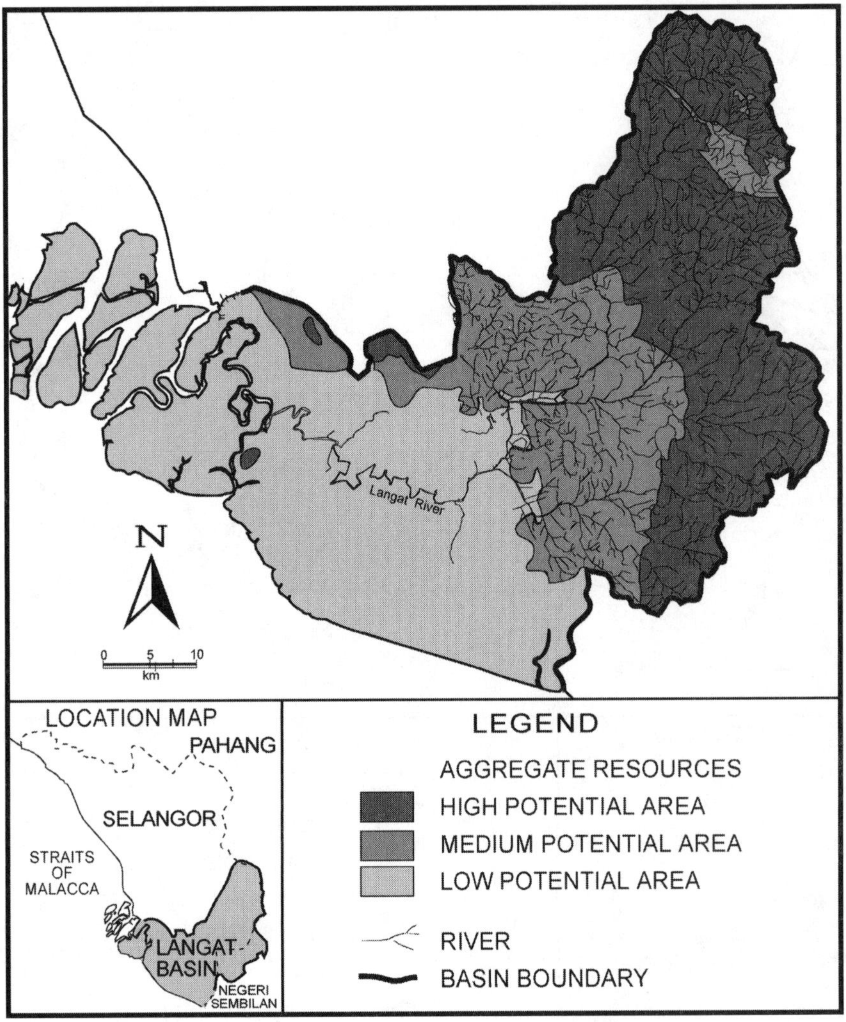

Figure 2. Distribution of aggregate resources in the Langat Basin, Malaysia.

About 35% of the Basin is thought to be underlain by aggregate resources. However about 15% of this is already sterilised by built development and 54% is covered by forest reserves and water. This leaves only 31% available for identification of suitable sites for mineral extraction. There was insignificant sterilisation before 1981 but the annual rate since then is about 10%. This implies that there could be total sterilisation of the unconstrained resources within about 15 years after, which there would be pressure to quarry the forest reserves. The results provided a basis for safeguarding unconstrained mineral resources and identification of buffer zones to prevent encroachment of built development.

Occurrences of landslides. Geohazard studies have been conducted in the Langat Basin only since the mid-1990's. The Minerals and Geoscience Department of Malaysia prepared terrain classification maps for parts of the Basin that had been earmarked for national development projects. A broad scale study of the entire basin was initiated only recently, as part of the effort to determine the overall state of ecosystem health in the Langat Basin. The geohazards that have been reported are landslides, land and coastal erosion, floods, including flash floods, as well as ground settlement. However landslides and land erosion are the most widespread geohazards that occur within the Basin and are referred to here (for a full account see Komoo et al 2003; Wan Mohd Muhiyaddin et al 2000).

The study of landslides focused on the upper sub-basin of the Langat Basin because it was representative of the overall geology and patterns of land use (Figure 1). The research involved the preparation of a spatial inventory of existing landslides based on information from air-photo interpretation, field survey and literature review. During the field survey, information was collected on:

- locations, types and sizes of landslides;
- amounts of displaced material;
- geology;
- topography; and
- land use of the site.

It was not possible to establish ages of individual landslides from the field survey. A lack of annual systematic mapping and the limited temporal air-photo coverage prevented identification of the annual incidence of landslides. It was not possible, therefore, to establish trends regarding the annual distribution and intensity of landslide occurrences in the context of the different land uses. Continuous annual assessment for at least three to four years into the future is required before such trends can be identified.

The combination of heavy rainfall and removal of vegetation caused landslides in 70 localities within the main sub-basin of the Langat Basin, between 1995 and 1998. So far, only two deaths due to landslides have been reported in the urban centre of the Langat Basin. Economic losses associated with damage to infrastructure such as roads and buildings are unknown, but significant. The landslides in the Langat Basin have occurred due to a combination of natural causes and human intervention. Six types of slope failures are observed and these are earth slump, earth fall, rock fall, earth slide, complex landslide and debris flow. The two most common types of slope failure are earth fall and earth slump, which have been recorded at 26 and 25 localities, respectively, out of the total of 70. Earth fall often occurs in slopes with steepness more than 60° while earth slump is common in slopes with steepness less than 45°. The other types of failures have occurred in less than five localities within the sub-basin.

The average size of slope failure was calculated by dividing the total volume of displaced material with the number of landslides in a given area for each category of topography, geology and land cover within the sub-basin. For the category of topography, the volume of displaced material per landslide within the flood plains is relatively smaller compared to the hilly areas. This indicates that the size of the slope failure increases with the ruggedness of the topography. In the case of geology, it was observed that the slope failures over granitic terrain are larger than those within metasediments. As the granitic terrain forms a significant proportion of the rugged areas, this result is consistent with the earlier observation for the various types of topography.

The larger slope failures occur within agricultural and built-up areas, while slope failures in cleared areas and forests are smaller in comparison. The intensity of landslides was determined from the number of landslide occurrences per unit area (per 1000 ha). The displaced material is obtained by dividing the total volume of material associated with all slope failures in a given area. Generally, the landslide intensity and displaced material are greater in the flood plains and undulating lowland terrain compared to the hilly area. The intensity of landslides is nearly 60 times greater, while the displaced material is six times greater within the metasediments, compared to the granitic terrain. In terms of land cover, the lowest intensity of landslides and displaced material occurs in forests. The intensity of landslides is the highest in cleared area, followed by built-up and agricultural areas. However, more material is displaced per unit area within the agriculture and built-up areas compared to the cleared area. This means that cleared areas have a greater number of small slope failures. The results were used in conjunction with land cover and landform analyses to establish limits for clearing land in the Langat Basin.

Soil erosion due to land clearing. Land erosion in the study area relates to detachment and removal of soil and rock by the action of raindrops, running water, subsurface water and mass wasting (Selby 1993). The conventional approach to assess levels of soil loss involves the mapping of those factors that contribute towards soil erosion. The assessment of soil loss can now be conducted using a combination of GIS and remote sensing techniques, based on a universal soil loss equation model (Omakupt 1987 cited in Komoo et al 2003). TM Landsat images of the whole of the Langat Basin for three periods—1993, 1996 and 1998—were used to classify the area into five categories: forests, agricultural land, water bodies, built-up areas and cleared land. An erosion risk map produced by the Agriculture Department in 1996 was then overlaid onto this. Images for 3 periods allowed examination of the situation before, during and after construction. Soil loss was then calculated based on appropriately weighted factors such as land cover, the level of risk and the ruggedness of the topography.

Between 1993 and 1998, the forest coverage in the Langat Basin was reduced annually by about 10%. During the same period, the cleared, built-up and agricultural areas increased annually by 15%, 10% and 3%, respectively. The highest amount of soil loss, over the same period, was generally associated with undulating lowlands and hilly terrain, followed by mountainous areas and the flood plains. It was generally very high for cleared areas (152 tonne/ha/year). In comparison, soil losses in the forest and agricultural areas were insignificant (less than 4 tonne/ha/year). Between 1993 and 1998, the soil loss from cleared areas fluctuated between 4% and 13% annually but no significant changes were observed for soil loss in the forest and agriculture areas. These changes in rates of soil loss could be attributed to the location of the cleared land, whether it is on flood plains, undulating lowlands, and hilly or mountainous areas. Clearing of land makes it susceptible to gully, sheet and splash erosion as a result of the heavy tropical rainfall. The increase in soil loss has been correlated to the increased levels of suspended sediment in the automatic monitoring stations of rivers draining the Langat Basin, during the period between 1993 and 1998. The largest amounts of suspended sediments were also recorded at monitoring stations located near areas being cleared, during the phase construction. The levels observed normally exceeded the standards set for aquatic life, drinking water and irrigation by the Department of Environment, Malaysia. The concentration of suspended sediments normally dropped to its original levels after the construction phase. Thus the resulting sediment generally finds its way to nearby channels, and causes siltation of the rivers. The trends for soil erosion can be established more readily from the manipulation of remote sensing and GIS modeling.

In general, therefore, the Langat Basin can be divided into three types of areas with different levels of soil erosion and landsliding. The forested areas in the mountainous terrain, located in the eastern part of the Langat Basin, have soil loss less than 2 tonne/ha/year and landslides do not exceed overall natural levels in term of size, intensity or displacement of material. The undulating lowland and hilly areas, in the central part of the basin, where development is focused, have soil loss generally between 12-13 tonne/ha/year. In areas with land clearing activities for major construction of projects such as the Kuala Lumpur International Airport and Putra Jaya, the soil loss is as much as 152 tonne/ha/year. The highest intensity of landsliding and greatest amounts of displaced materials within this setting. The coastal and floodplains located in the western part of the Langat Basin show intermediate characteristics. Soil losses in these mainly agricultural areas are generally less than 3.6 tonne/ha/year. The intensity of landslide occurrences is quite high but the size of slope failures and volumes of displaced material are relatively low.

Soil loss in the forest and agriculture areas of the Langat Basin have not changed significantly (less than 0.2 and 3.7 tonne/ha/year, respectively) between 1993 and 1998. Thus, these sectors have not contributed significantly to the deterioration of ecosystem health with respect to soil loss. However, taking 1993 soil erosion levels as a baseline, the soil loss associated with development areas increased annually by about 20% up to 1998. This situation is not sustainable. Management measures should be enforced to arrest the problem of excessive soil erosion during the construction phase of development projects in the Langat Basin. It was concluded, therefore, that land clearing should not exceed 60 km^2 (or 2% of the total basin) in any one year in order maintain basin conditions at 1993 levels. This limit of land clearing would also ensure that landslide occurrences do not exceed background levels.

Discussion

The main issues in Wigan were ground instability from past underground mining and contamination and pollution from previous industry. There were voluminous existing records on the area but these had to be collected and analysed, and supplemented by some field investigation, before they could be used readily. The resulting database could not be regarded as complete but was a good basis for classifying the area in terms of potential constraints to development. Land use planners were involved in the process of developing simplified keys to the maps. The keys were designed mainly to alert planners to issues that might need to be considered and whom they should contact to get more detailed professional advice. The results were used by planners in the review of the development plan soon after work was completed. However there is evidence that, after only a few years, the material is now being forgotten. This indicates a need for repeated dissemination of results rather than just at the completion of the study.

The main issues in the Langat Basin were safeguarding mineral resources and potential hazards from soil erosion and landslides. Here the main emphasis was on field survey and the use of remote sensing data. Consideration of the mineral resources allowed identification of key areas that should be safeguarded for future extraction and where buffer zones should be employed to prevent encroachment of built development. Study of the geohazards led to simple management rules about the maximum amount of land that should be cleared annually in the development zone so that soil erosion and landsliding could be minimised.

Whilst one study concerned an urbanised area in need of regeneration and the other concerned an area experiencing urbanisation for the first time, both show the importance of developing a simple basis for planning and management of the land. One provided the basis for a strategy to reclaim derelict and despoiled land. This could then be set alongside planning considerations and, ultimately urban renewal, which is now progressing well. The other identified where key pressure points for urbanisation would arise as a foundation for a rational development strategy provided the planning framework and decisions are enforceable. If this is not achieved, there is a risk that the Langat Basin will become derelict and despoiled in the future, just as the area around Wigan did in the past.

To avoid this, there is a need to take careful account of geoscience information during development, to record and archive information to support future decisions, and to ensure that pollution and contamination are minimised so that the need for remediation does not arise. Where problems do occur, as they inevitably will, then these should be carefully recorded to assist future generations.

THE FIRST HALF OF THE 21ST CENTURY: FURTHER PROGRESS TOWARDS SUSTIANABLE DEVELOPMENT?

A number of developments and initiatives over the past few years have provided a basis, which, if exploited, might help to secure better progress in the coming years. These relate to:

- recognition of the need to assess progress towards sustainable development and measurement of environmental change;
- development of new approaches towards description and assessment of land and the environment;
- development of approaches to considering risk in addition to hazards;
- improved approaches to compilation and delivery of information;
- the potential for improving awareness and education; and
- better recognition of the need to collaborate on environmental issues at all levels.

Progress Towards Sustainable Development and Measurement of Environmental Change

There are widespread attempts to adopt indicators of more sustainable practices. However "indicators" are often crude and are frequently based on categories of information that are available already, rather than by developing a framework for monitoring and assessment of policies (Petersen et al 1999). Even so, attempts to develop these have conveyed widely the concept of indicators and awareness of the need for monitoring.

This awareness has been reinforced by the need to establish and to monitor indicators of environmental change including phenomena associated with climate change. A wide range of parties, regardless of background, is receptive to this overall concept. This potential link needs to be exploited to ensure that funding agencies support the collection and retention of the right information. To that end, the Commission on Geological Sciences for Environmental Planning (Cogeoenvironment) of the International Union of Geological Sciences (IUGS) launched a geoindicators initiative that has led to increasing activity through successful workshop meetings (Cogeoenvironment Working Group on Geoindicators 1996; see also www.lgt.lt/geoin/ and McCall 1996b).

Description and Assessment of Land and the Environment

Academic description of land and the environment has often been discipline based except, perhaps, in the case of ecology where a wide range of contextual factors must be taken into account. However, until recently, most ecological work has avoided urban settings. Applied work commissioned in relation to development concentrates on the possible implications of the development and is very much focussed on the proposed development site. The result is a patchwork of information on different topics to different levels of detail, available from a variety of sources or, sometimes, not very readily retrievable at all. It is only where there has been conscious investment in strategic environmental monitoring or in specific regional work that broad background information is readily available from databases or publications.

However there are signs of change. There is increasing monitoring as a basis for assessing environmental performance, widespread interest in a variety of new ways of considering environmental issues, and steps in some quarters to extend the principles of Environmental Impact Assessment (EIA) from development proposals to broader assessment of the implications of plans and policies.

New approaches to considering the environment include consideration of:

- environmental capacity of an area—the extent to which environmental change might be accommodated before causing irreparable damage to the landscape, ecosystems and to the quality of life (Entec UK 1997);
- density characteristics of urban occupation with the aim of achieving the best balance of population against urban environmental improvement in order to improve the quality of life (Urban Task Force 1999);
- life-cycle analysis to determine the full range of environmental implications of present practices and of options for change (Ravetz 2001).
- the environmental footprint—the area that is affected by the functioning of a specific system compared with that which would be affected if more sustainable practices are introduced (Imperial College of Science, Technology and Medicine 2000); or
- ecosystem health—the identification, analysis, and monitoring of factors that help to determine whether an area is degrading, stable, or improving in environmental terms (Aguilar 1999; King and Hood 1999).

All of these require information of the right types from a number of disciplines and at the right levels of detail. A major problem has been the cost of survey and data collection. Recent advances in remote sensing are leading to broad data sets for some purposes quickly and at relatively limited costs compared with ground survey. Improvements to resolution and discrimination are still required for some purposes but these systems are developing quite rapidly.

The European Union has recently adopted a Directive (law) on the assessment of plans and programmes and prepared by public authorities. This will require all draft plans and programmes to undergo systematic environmental evaluation—"strategic environmental assessment". This will reinforce the need for a wide variety of regional data sets to be readily available. Whilst the Directive would apply only to those States already within, or likely soon to enter, the EC the principles may spread more widely. There is already a United Nations Economic Commission for Europe proposal for a protocol on strategic environmental assessment within the Espoo Convention on EIA in a transboundary context. (www.unece.org/env/documents/2001/EIA/ac1/mp.eia.ac.1.2001.3.e.pdf)

Risk and hazard

A great deal of work on geoscience in support of planning and management of land use has been concerned with evaluation of the nature and extent of hazards to development, rather than risk. However it is an appreciation of risk that allows urban planners and managers to evaluate whether action is needed; to set priorities for dealing with problems; and to decide whether the introduction of warning and civil emergency procedures is justified. The risk assessments that have been undertaken tend to relate to specific problems that arise. There has been limited material to date that aims to communicate the general approaches to dealing with generic risk assessment in urban areas, in respect of geohazards. Much of that which is available relates to landslides (see, for instance: Dai et al 2001; Fell 1994). There is also a need for better communication of the perception of risks, especially relative risks, to the general public. Unlikely events often cause as much, or even more, concern than more likely ones.

Improved Compilation and Delivery of Information

In recent years there has been an enormous increase in the capacity and flexibility of databases, the power of Geographic Information Systems, and the ability to export large data sets from one system to another. This has set the scene for much better storage and delivery of information (Nathanial and Symonds, 2001).

Much information used by geoscientists and by land use planners is map-based, which should give a basis for mutual exchange and use of information. Major constraints have been how much information and map can show before it becomes excessively complicated and how quickly the map is outdated by changes to the area and collection of new data. Updating of databases and GIS layers has now brought the flexibility that is needed to keep information up to date and to ease the analysis of information by manipulating it in logical steps. This brings new possibilities to work similar to the case studies described earlier, including the development of customised interfaces for various users of the information. With that in mind, for example, the UK Office of the Deputy Prime

Minister and the Natural Environment Research Council have jointly commissioned the British Geological Survey with the Centre for Ecology and Hydrology and the University of Nottingham to develop an environmental information system for urban planning and management. The aim of the work is to develop a software "gateway" to a major database of environmental information customised for use by planners, and to be developed and tested, in practice, in the offices of selected local planning authorities. This work is not limited to geoscience but, rather, takes full account of soils/geology, water, ecological and atmospheric sciences and the interfaces between these. It is recognised, however, that a flexible product that can be easily used by non-specialists should not be allowed to give the impression that specialist interpretation is not needed. Rather it is intended that there should be several levels, or interfaces. These will help the planner or developer to identify relevant issues and to frame the right questions and then to guide them to the people or organisations that can provide expert information advice. At a deeper level the system will contain the detail that is required by the specialists in their day-to-day work.

On a broader scale, the partners in GEURBAN have submitted a project proposal called SECURIS to the European Union. The proposed aim is to examine geological problems in terms of their significance for urban society using standardised environmental risk assessment procedures. All available geoscientific and social information will be combined to produce geohazard risk management systems for (European) cities, intended for eventual use by lifeline companies, investors, planners, developers, insurance companies and citizens. In addition to Eurogeosurveys, other scientific and technological partners have been identified and six cities have been identified as potential users of the results. If approved by the EU, SECURIS will eventually make Europe's cities safer and prevent loss of property for the citizens and the urban industries both by increasing awareness, particularly amongst city planners and managers, of the risks, which threaten cities from beneath.

EuroGeoSurveys has also developed a metadatabase on geo-data for 16 European countries (GEIXS). In the Netherlands a very large database (DINO), eventually to contain some 120 subsidiary geological databases, covering the whole country, was opened for public use in April 2001. Similar activities are now well advanced in many of the developed countries.

Rapid developments in information technology have led to information systems, based on facilities developed for the oil industry, in which the subsurface conditions of cities can be plotted and which can be used for simulating the effects of human activities on the subsurface conditions and vice versa. Such systems are operational now and can be applied for assisting urban management in coping with geo-problems. This is leading to a revolution in the ways in which some geological surveys work. For example, both the British Geological Survey and the Illinois State Geological Survey are working towards digital geoscience spatial models as a basis for future three-dimensional mapping of the ground.

Each is putting particular emphasis on the shallow subsurface for the purposes of engineering geology in support of development (Central Great Lakes Geologic Mapping Coalition 1999; Walton and Lee 2001).

Improving Awareness and Education

The use of the Internet as a source of information is increasing rapidly. However it is often difficult to find the information that is needed quickly and there are no guarantees of quality when it has been located. While the specialist is in a position to seek the right information and to evaluate it, the non-specialist may not know what to look for or whether it is accurate. For this reason, there is an increasing role for gateway sites that help the user by providing some background information and by linking to other useful sites. There are a number of these now, or shortly to be, available in respect of urban geoscience and planning including, for instance, the websites of:

- Cogeoenvironment (www.sgu.se/hotell/cogeo) containing information on environmental geology;
- the Urban Geology Working Group (www.pro.ukm.my/urbangeology/index.asp) for both geoscientists and for planners and managers in urban areas and designed specifically to give different levels of detail and approaches for these audiences;
- the IUGG Georisk Commission (www.rusartknife.urbanet.ru/georisk) containing comprehensive information on geohazards in the extended sense of the term including, for instance, material on storms and wildfire; and
- a European Commission research programme providing advice on geotechnical issues for urban areas (www.bygg.ntnu.no/fakadm/eng/).

These could simply duplicate effort but it makes more sense for them to be complimentary. It is encouraging, therefore, that there are current contacts between these initiatives with the aim of securing added value.

The delivery of information is important, but it is crucial that the intended audiences should be aware that they should take notice. That is a matter for education, in its widest sense, and collaboration. There is a need, therefore, for;

- properly funded and organised initiatives to prepare books, guidance, papers and software targeted on specific audiences such as planners, developers, financial institutions and the general public as well as schools and universities (Robinson and McCall 1996).
- conferences and workshops at which different interest groups meet to discuss common problems involving, for example, both geoscience and planning organisations. All too often conferences consist of one group of specialists discuss issues with their subject peers (Worth 1987);

- articles for the popular press and presentations for TV and radio to raise public awareness of the issues; and
- imaginative facilities for people to visit whether based in museums or consisting of city walks, for example, the Isle of Wight County Council in the UK has developed a visitor centre in the town of Ventnor that explains the effects of landsliding in the area and provides leaflets for walks round the town to see the resulting features and so, to some extent, turning a problem into an attraction (Lee et al 1991).

Collaboration on Environmental Issues

There is a clear and increasing need for collaboration between specialists from many different disciplines in addressing problems. Already there is a trend towards various types of environmental scientists working together to resolve problems that has been reinforced by the need to collaborate on environmental appraisals of plans and development proposals. However, so far there have been only limited attempts to bridge the divide between these and social and economic scientists, let alone to involve practitioners in urban planning and management directly. Funding agencies need to recognise that more work across traditional subject boundaries is important .

CONCLUSIONS

The latter part of the 20th century has seen many major advances in securing and presenting geoscience information to support sound urban planning and management. However this is still under-used and the remains a gulf of awareness between geoscientists and those who should respond to their findings. This is due partly to under-representation of geoscience in many educational programmes but other important influences are the pace of change in urban areas and the social and economic imperatives that drive the process. Most planners and managers have too little time to increase their awareness and may not recognise the need to do so. That results partly from geoscientists presenting information at the wrong level of detail and in language that is too technical.

A great deal has been done over the last 50 years to address these issues. Work has progressed from traditional geological mapping, through engineering and environmental geology mapping, to simplified maps designed for planners and developers. However these, and the databases from which they have been compiled, are available only for scattered trial areas and general publications tend to cover scattered examples from around the world. This makes it difficult to demonstrate the local circumstances that are most meaningful to managers of specific towns and cities.

The cited case studies demonstrate that different approaches are needed in old developed urban areas, where existing data plays and important part, and areas ripe for new development where primary survey and remote sensing provide most of the data. In both cases, however, it has been necessary to summarise and generalise results so that they can be communicated properly to the target audiences.

A common language and approach is needed to overcome these barriers. Since the planning of land use is essentially a map based discipline there is a clear basis if mapped information can be provided in the right forms. The current rapid development of database software, GIS, decision aid systems and three-dimensional graphics offers opportunities for joint development of facilities. New ways of considering the environment in terms of indicators of progress and of change and different, more holistic approaches to environmental appraisal and the assessment of the environmental, as well as economic, implications of development provide a battery of tools that can be employed if they are developed by urban managers in liaison with environmental scientists from all relevant sub-disciplines.

In addition, a great deal of effort is needed to increase the level of awareness of relevant environmental issues at all levels, from school education to communication with professional planners and developers. In addition, links need to be made with the general public through the popular press and activities such as those promoted under Local Agenda 21. International geoscience organisations are recognising the need to collaborate on this to achieve medium and longer-term improvements. However the work will only achieve it full potential if the collaboration is extended to organisations representing planners and developers and other scientific disciplines. This is a major challenge but, given the rate of urbanisation and of environmental change, it is imperative that adequate effort should be devoted to addressing these issues.

Acknowledgments. Work on the Wigan area was undertaken by the British Geological Survey under contract to the then Department of the Environment, Transport and the Regions (now the Office of the Deputy Prime Minister). The research on the Langat Basin was undertaken by LESTARI with funding through MATREM CP/5220-97-02, a project of the United Nations Environment Programme Network for Environmental Training at the Tertiary Level in Asia and the Pacific (NETLAPP).

REFERENCES

Aguilar, B. J., 1999 Application of Ecosystem Health for the Sustainability of Managed Systems in Costa Rica, *Ecosystem Health* 5, 36-48.

Annells R., 1998, EuroGeoSurveys: Joint Actions by the Geological Surveys of Europe *Earthwise,* Issue 11, British Geological Survey (Keyworth) page 8.

Barker, G. M. A. 1996 Sites in Urban Areas: lessons from wildlife conservation In: Bennett, M. R; Doyle, P.; Larwood, J. C.; and Prosser, C. D., *Geology on Your*

Doorstep—the Role of Urban Geology in Earth Heritage Conservation. Geol Soc (London) 270pp.

Berger, A R and Iams, W [Eds] 1996 *Geoindicators—Assessing Rapid Environmental Changes in Earth Syst*ems. 480pp Balkema (Rotterdam)

Bobrowsky, P T [Ed] 1998 *Aggregates Resources—a Global Perspective.* 480pp Balkema (Rotterdam)

Bobrowsky P T [Ed] 2001 *Geoenvironmental Mapping—Method, Theory and* Practice Balkema (Rotterdam) 500pp

Bolton K and Evans L J 1997 *Contaminant Geochemistry of Urban Sediments and Soils* In: Eyles N [Editor] pages373-382 [separately referenced]

Bradford M and Kent A 1993 *Understanding Human Geography—People and Their Changing Environments.* Oxford University Press (Oxford) 309 pp

Brook D and Marker B R 1987 *Thematic Geological Mapping as an Essential Tool in Land Use* Planning In: Culshaw M G et al 211-4 [separately referenced]

CEC 1996 *European Sustainable Cities: Report by the Expert Group on the Urban Environment* European Commission (Luxembourg) 108pp

Cendrero A, Panizza M, and Marchett M [Eds] 2001 *Geomorphology and EIA* Balkema (Rotterdam) 220pp

Central Great Lakes Geologic Mapping Coalition 1999 *Sustainable Growth in America's Heartland: 3D Geologic Maps as the Foundation.* USGS Circular 1190, 17pp

CIRIA (Construction Industry Research and Information Association) 2001 *SUDS Best Practice Manual.* CIRIA Report C523. CIRIA (London)

Cogeoenvironment Working Group on Geoindicators 1996 *Tools for Assessing Rapid Environmental Changes—the 1995 Geoindicators Check List.* ITC Publication 46 (Enschede) iii+102pp

Commission of Engineering Geology Maps of the IAEG 1976 *Engineering Geology Maps.* Earth Sciences No. 15. UNESCO (Paris) 79pp

Corfield M, Hinton P, Nixon T and Pollard M [Eds] 1998 *Preserving Archaeological Remains in Situ.* Proc. Conference 1-3 April 1996. Museum of London Archaeological Service (London) 189pp

Dai, F C; Lee, C F and Ngai, Y Y 2002 Landslide Risk Assessment and Management: an overview. *Engineering Geology* 64, 65-87

Department of the Environment, Transport and the Regions (DETR) 2000 *EIA—a Guide to Procedures.* Thomas Telford (London) 111pp

Ecotec Research and Consulting Ltd 1987 *Greening City Sites: Case Studies of Good Practice in Urban Regeneration* HMSO (London) 127pp

Entec UK 1997 *The Application of Environmental Capacity to Land Use Planning* Department of the Environment, Transport and the Regions (London) 86pp

Eyles N [Ed] 1997 *Environmental Geology of Urban Areas.* Geological Association of Canada Text 3 Geological Association of Canada (St Johns) 590pp

Eyles N and Boyce J I 1997 *Geology and Urban Waste Management in Southern Ontario* In: Eyles, N [Editor] pages297-322 [referenced separately]

Fell R 1994 Landslide risk assessment and acceptable risk. *Canadian Geotech Jl* 31, 261-272

Forshaw J H and Abercrombie P 1944 *County of London Plan.* Macmillan (London) xii+188pp

Forster A, Arrick A, Culshaw M G and Johnson M 1995 *A Geological Background for*

Planning and Development in Wigan. Vol. 1 A Geological Foundation for Planning 89 pp 9 maps at 1:25000 scale. *Vol 2 A Users Guide to Wigan's Ground Conditions.* BGS Technical Report WN 95/3 British Geological Survey (Keyworth) 42pp

Foster S, Lawrence A, Morris B 1998 *Groundwater in Urban Development—Assessing Management Needs and Formulating Policy Strategies.* World Bank Technical Paper 390 xiv+55pp (1-8)

Geernaert G L and Jensen S S 1997 Health and Economics of Urban Air Pollution Management. In: Inececik S, Ekinci E, Yardim F and Bagram A (Eds.) *Environmental Research Forum Trans Tech Pubn* 7 (Zurich), 502-7

Gugler J 1996 *The Urban Transformation of the Developing World.* Oxford University Press (Oxford) xviii+327

Hannavey, J 1990 *Historic Wigan, 2000 Years of History.* Carnegie Publishing Ltd (Preston) 167pp

Hurley D 1996 Urban Geology and Natural Conservation In : Bennett, M R et al [Editors] pages155-162 [separately referenced]

ICLEI 1996 *The Local Agenda 21 Planning Guide—an Introduction to Sustainable Development Planning* ICLEI, IDRC and UNEP (Toronto)

Imperial College of Science, Technology and Medicine *2000 Island State—an Ecological Footprint and Analysis of the Isle of Wight.* The Future Centre (Oxford) 54pp www.bestfootforward.com

Jones R C B, Tonks, L H and Wright E B 1938 *Wigan District (One Inch Geological Sheet 84 New Series)* Mem Geol Survey G B. HMSO (London)

King, L A and Hood V L 1999 Ecosystem Health and Sustainable Communities: North and South. *Ecosystem Health* 5, 49-57

Komoo, I; Pereira J J and Muhiyuddin Ibrahim, W M 2003 Diagnosing ecosystem health of the Langat basin in the context of geohazards. In: In: Rapport, D J; Lasley, W L; Rolston, D E, Nielsen, N O; Qualset, C O and Damania, A B [Eds] *Managing for healthy ecosystems.* Lewis Publishers (Boca Raton, Florida) 1385-1419.

Lawrence AR and Chaney C 1996 *Urban Groundwater.* In: McCall et al [Editors] pages 61-80 [separately referenced]

Lee E M, Doornkamp, J C, Brunsden D and Noton N H 1991 *Ground Movement in Ventnor, Isle of Wight.* Geomorphological Services Ltd (Newport Pagnell) vi+65pp

Legett R F 1973 *Cities and Geology.* McGraw Hill (New York) xii+624

Marker B R 1996 Urban development—Identifying Opportunities and Dealing with Problems. In: McCall et al [Eds] 181-214 [separately referenced]

Marker B R 1998 Incorporating Information on Geohazards into the Planning Process. In: Maund J C and Eddleston M [Eds] *Geohazards in Engineering Geology.* Geological Society Lond. Special Pub. 15, 385-9

Marker B R 2001 Encouraging Better use of Geological Information by Planners and Developers in England. In: Bobrowsky, P T [Ed] 429-450 [separately referenced]

Marker B R and McCall G J H 1989 Environmental Geology Mapping 201 -234 In: McCall G J H and Marker B R [separately referenced]

Marker B R and McCall 1990 Applied Earth Science Mapping: the Planners' Requirement. *Engineering Geology* 29, 403-11

Mather J D, Spence I M and Brown MJ 1996 Man Made Hazards. In McCall et al [Editors] pages 127-162 [separately referenced]

McCall G J H 1996a Natural Hazards In McCall et al [Editors] pages 81-126 [separately

referenced]
McCall, G J H 1996b Geoindicators of Rapid Environmental Change—the Urban Setting In Berger A R and Iams W [Editors] pages 311-18 [separately referenced]
McCall G J H, de Mulder E F J and Marker B R [Eds] 1996 *Urban Geoscience*. AGID Special Pub. 20 Balkema (Rotterdam) viii+273
McCall G J H and Marker B R [Eds] 1989 *Earth Science Mapping for Planning, Development and Conservation*. Graham and Trotman (London) 268pp
McCall G J H and Marker B R 1996 Mineral Resources. In McCall et al [Editors] pages13-34 [separately referenced]
McGraw, E 1990 *Population: the Human Race* Bishopgate Press (London) 141pp
Moseley H 1998 Archaeology and Development. In: Corfield et al [Eds] 47-50 [separately referenced]
Mumford, L 1961 *The City in History*. Harcourt Brace (New Jersey) 365pp
Munich Reinsurance Co. 1982 *Loss Adjustment After Natural Disasters*. Munich Re-Insurance Co (Munchen) 33pp
Nathanail P and Symonds A [Eds] 2001 Geographical Information Systems. In: Griffiths, J S [Ed] *Land Surface Evaluation for Engineering Practice*. Geol Soc Engineer Gp Spec Pubn 18, 57-58
O'Sullivan A 2000 *Urban Economics* Irwin McGraw Hill (Boston) 4th Edn. xxvii+740pp
Oxley, J 1998 Planning and the Conservation of Archaeological Deposits. In: Corfield et al [Eds] 51-54 [separately referenced]
Paul T, Chow F and Kjeksted, O 2002 *Hidden Aspects of Urban Planning—Surface and Underground Development*. Thomas Telford (London) 85pp
Pereira, J J and Komoo, I 2003 Addressing Gaps in Ecosystem Health Management: the Case of Mineral Resources in the Langat Basin. In: Rapport, D J; Lasley, W L; Rolston, D E, Nielsen, N O; Qualset, C O and Damania, A B [Eds] *Managing for healthy ecosystems*. Lewis Publishers (Boca Raton, Florida) 905-916
Peterson P J, Sani S and Nordin M 1999 *Indicators of Sustainable Development in Industrialising Countries: Vol. 3. Key Indicators for Tropical Countries*. Institute for Environment and Development (LESTARI), Universiti Kebangsaan Malaysia (Bangi) 107pp
Pratt M 1993 *Remedial Processes for Contaminated Land* Institute of Chemical Engineers (Rugby) viii+140
Proske H, Vlcko, J and Rosenbaum M (in press) Report on Special Maps for Waste *Disposal Bull. Int. Assoc. Engineer. Geol and Environ.*
Ravetz, J 2001 *City Region 2020: Integrated Planning for Sustainable Development* Earthscan Publications Ltd (London) 307pp
Ravishankar H M 1994 Watershed Prioritisation Through the Universal Soil Loss Equation Using Digital Satellite Data, an Integrated Approach. *Asia-Pacific Remote Sensing Journal* 6, 56-60
Roberts P and Sykes H 2000 *Urban Regeneration: a Handbook*. Sage Publications (London) xvi+320pp
Robinson J E and McCall G J H 1996 Geoscience Education in the Urban Setting In: McCall G J H et al [Editors] pages 235-252 [separately referenced]
Robinson G D and Speiker, A M [Eds]1978 *Nature to be Commanded—Earth Science Maps Applied to Land and Water Management*. USGS Professional Paper 950. USGS (Washington DC) iv+95pp

Selby M J 1993 *Hillslope Materials and Processes.* Oxford University Press (Oxford) 286pp

Simonds J O 1978 *Earthscape—a Manual of Environmental Planning and Design.* Van Nostrand Reinhold (New York) 340pp

Simpson T G 1996 Urban Soils In: McCall, G J H et al [Editors] pages 35-60 [separately referenced]

Smith A and Ellison R A 1999 Applied Geology Maps for Planning and Development—a Review of Examples from England and Wales 1983-1996. *Quarterly Journal of Engineering Geology* 32 (supplement) S1-S44

Taylor, B J 1957 *Report on Underground Water Resources of the Permo-Triassic Sandstone Areas of South Lancashire.* BGS Technical Report WD/57/8

Thompson A, Hine P D, Poole J and Greig J R 1998 *Environmental Geology in Land Use Planning—a Guide to Good Practices* Symonds Group (East Grinstead) 80pp + 14 case study leaflets

Tucker P, Ferguson C and Tzilivakis J 1998 Integration of Environmental Assessment Indicators into Site Assessment Procedures In: Lerner D N and Walton N R G *Contaminated Land and Groundwater—Future Directions* Geol. Soc. Lond. Eng Group Spec Pub 14, 127-133

UNESCO 1997 *Geology for Sustainable Development* Bulletin II Urban Geology Royal Museum for Central Africa (Tervuren, Belgium) xiii+153pp

Urban Task Force 1999 *Towards an Urban Renaissance* E and F N Spon (London) 328pp

Walmsley D J and Lewis G J 1993 *People and Environment—Behaviour Approaches in Human Geography.* 2nd Edition. Longman (Harlow) xii+290pp

Walton G and Lee M K 2001 *Geology for our Diverse Economy.* British Geological Survey (Keyworth) 99pp

Wan Mohd Muhiyaddin, Komoo I and Pereira J J 2000 Land Degradation in Langat Basin: a Case Study of the Ulu Sungi Langat Sub-basin. *Proc Symp on the Langat Basin, Shah Alam.* Institute for Environment and Planning (Lestari), Universiti Kebangsaan Malaysia (Bangi) 126-134

Ward D, Holmes, N and Jose P 1994 *The New Rivers and Wildlife Handbook.* Royal Society for the Protection of Birds, National Rivers Authority and Royal Society for Nature Conservation (Sandy) xvi+320pp

Worth D H 1987 Planning for Engineering Geologists. In Culshaw M G, Bell F G, Cripps J C and O'Hara M [Eds] *Planning and Engineering Geology.* Geol. Soc. Lond. Eng. Gp. Spec Pub 4, 39-46

15

Greater Phoenix 2100: Building a National Urban Environmental Research Agenda

Jonathan Fink, Frederick Steiner, Nancy B. Grimm, and
Charles L. Redman

INTRODUCTION

At the start of the twenty-first century, scientists of all types are becoming increasingly bold in their willingness to examine complex systems. Multidisciplinary teams using the latest information technology are now attacking problems of societal relevance that had, until recently, been considered too difficult for the available tools. Nowhere is this transformation better illustrated than in the growing attention ecologists, geologists, and other natural scientists are paying to the study of cities and their metropolitan regions. The multidimensional challenge of addressing science problems in urban settings is compounded by intense interest from policy makers and the general public in their solutions, especially in cities that are experiencing the pressures associated with rapid growth.

One of the places where this new kind of exploration is being most aggressively carried out is metropolitan Phoenix, Arizona, which has been among the fastest-growing urban areas in the United States over the past decade. A confluence of federally funded scientific research, public interest in quality-of-life issues, and academic aspiration of the region's only research university have positioned Greater Phoenix to be an ideal national laboratory for interdisciplinary study of the complex interactions between an urban population and its physical, biological, and social environment. A project called Greater Phoenix 2100 was launched in April 2001 by Arizona State University (ASU), in conjunction with various state and federal agencies, to capitalize on these research opportunities, and to use them to better inform public policy debates about regional

Earth Science in the City: A Reader

10.1029/056SP16

growth. This chapter describes the background, goals, and proposed implementation strategy for the Greater Phoenix 2100 project.

BACKGROUND: DRIVERS OF CHANGE

Cities present important challenges for scientific inquiry. Rapidly growing urban regions pose especially complex issues for science, policy, and our futures. Most natural scientists in the United States have eschewed the study of such human-dominated ecosystems as cities in favor of learning about the interactions of more purportedly pristine environments. American geophysical and biological scientists have discovered urban places only relatively recently as field laboratories for their work [e.g., Botkin and Beveridge, 1997; Pickett et al., 1997; Collins et al., 2000; Grimm et al., 2000; Valentine and Heiken, 2000; Fagan et al., 2001; Fernando et al., 2001]. The consequences of this discovery are potentially wide ranging. If we can better understand urban processes, then we might be able to create healthier metropolitan regions, or "city-regions" [e.g., Calthorpe and Fulton, 2001].

Perhaps because of the change in the millennium, several scholars and writers have urged society to look ahead more critically. Jonathan Weiner [1990] speculates about planet Earth over the next one hundred years. Stewart Brand [1999] takes an even longer view—10,000 years. Robert Costanza [2000] advocates envisioning as a tool for policy analysis. Such queries into the future help us begin to identify drivers of change that are likely to influence the fate of the planet in this new century and millennium. The U.S. Environmental Protection Agency's Science Advisory Board defines such drivers as "the large social, economic, and technological forces that [are] likely to drive future changes in environmental conditions" [1995, p. 4]. Further, they observe that such drivers "can generate environmental stressors…that cause adverse effects on specific human health and ecological endpoints" [1995, p. 4]. Some possible drivers of change that might affect metropolitan regions include: (1) population dynamics and consumption; (2) urbanization; (3) connectivity and networks; (4) technology, economics, and politics; (5) culture and the arts; (6) education and human services; and (7) global and regional environmental processes. In addition, metropolitan regions themselves contribute substantially to all types of global environmental change; thus, we must look to cities and their social and ecological dynamics to understand the origin and possible solution to global environmental change.

Population growth and migration include those factors that will change the global demographic structure. The Earth currently has about six billion inhabitants. The United Nations projects the world's population to plateau at 9.4 billion by the year 2050 and then slowly rise to 10.4 billion by 2100 [Barrett and Odum, 2000]. This translates into some 12.6 billion additional individuals appearing on the planet over the next century [Brand, 1999]. Half of the world's population will soon live in cities, and the number of these urban inhabitants is expected to double by 2030 [United Nations Development Programme et al., 2000]. By 2050, two-thirds of the people in the world will be living in urban regions.

Population growth drives change because everyone requires water, food, shelter, clothing, and energy. However, the levels of consumption vary widely. The United Nations notes that globalization tends to separate the costs from the benefits because "consumers derive goods and services from ecosystems around the world...This [appropriation] tends to hide the environmental costs of increased consumption from those doing the consuming" [United Nations Development Programme et al., 2000, p. 23]. Society's desires to consume the basics and the amenities of life affect the level of resources necessary to fulfill those demands.

Human populations consume natural resources but also produce wastes as byproducts of this consumption. These wastes, such as excess CO_2 from fossil-fuel burning, NO_x from automobile exhaust and nitrogenous solutes from fertilizer, must be assimilated by natural or designed ecosystems or they will accumulate in the environment. That accumulation does occur is obvious from the alteration of biogeochemical cycles and the CO_2-induced warming that is already well documented [e.g., Vitousek, 1994]; what is perhaps less well known is generation of much of this waste can be traced directly to cities.

Population changes caused by growth and migration, and the associated consumption, are closely linked to urbanization. The movement of people to cities and metropolitan regions involves the transformation of land use from rural and natural to urban and suburban, the urbanization of the wild, the abandonment of the rural, and the recovery of the core city and older suburban neighborhoods. Some key questions related to both population growth and urbanization are: (1) Why do people choose to live where they do? (2) What policies direct/affect growth and development? (3) What are the long-term impacts of these policies, both for ecological systems and for further development?

Connectivity involves the ways that new networks and information systems will alter communities, knowledge transfer, time, social relationships, and education. Connecting technologies such as the automobile and the Internet may also divide. While we can assume that connectivity will continue to transform human society, the details remain obscure. Uncertainties surround such questions as: (1) What will communities look like when people no longer need to be next to each other for commercial reasons? (2) How will business, educational, and public institutions be affected? (3) How will connectivity affect use, knowledge, experience, and perception of place? (4) What will be the ecological consequences of a reduced need for transportation?

Connectivity and networks from new technologies are likely to drive global changes. Technological change is often linked to politics. Examples of technological and political linkages include: war, energy policy, and scientific advances, such as space exploration and biotechnology. Changes in the gross national and domestic products, extractive enterprises, industry and manufacturing, food and fiber, tourism, and transportation drive economics too.

The culture and the arts also drive change. Recreation and entertainment affect our aspirations and expectations. The Beatles, for example, helped define the youth culture of the late twentieth century. The past also helps shape the future. As a result, understanding the history and prehistory of nations, regions, and communities can help society anticipate possible changes.

Historic innovations in education such as universal public, primary and secondary schools and the GI Bill in the United States have resulted in dramatic transformations. Future alterations in education and other human services are likely to have similar impacts. For example, how health care is delivered to an aging population will no doubt drive major changes.

The global environment is also likely to see major changes in the future [Harrison and Pearce, 2000]. Global warming trends are well documented. Species are becoming extinct at a dramatically accelerating rate, from rain forests to coral atolls to deserts. Meanwhile, local climate changes as a result of urban heat island (or heat archipelago) effects are becoming better known. For example, summer nighttime average temperatures in the Phoenix metropolitan region increased by 2.2°C between 1970 and 1990 [Brazel et al., 2000]. Additional environmental drivers of change influencing the global commons and, to varying degrees, specific regions and landscapes, include: natural disasters, the nitrogen cycle, energy uses and greenhouse effects.

GREATER PHOENIX 2100

Against this global backdrop, ASU launched a new research initiative called Greater Phoenix 2100. It will combine a regional perspective with the latest scientific information to address several socially-relevant questions, including: (1) What are the desirable characteristics of Greater Phoenix that today's citizens want to preserve or create for their great-great-grandchildren? (2) How do we best describe the metropolitan Phoenix region as it exists today? (3) How do we characterize explosive urban growth? (4) What tools can help citizens and policy-makers make educated forecasts about the region's future? (5) How will trends in science and technology affect the development of metropolitan regions like Greater Phoenix? (6) How will social and economic changes drive metropolitan policies during the next 100 years? The project also will serve as a focal point for coordinating federal, state, and academic information programs relating to the environment of the region, and will be linked with similar studies in other metropolitan areas. Greater Phoenix 2100 will build state-of-the-art forecasting and decision tools and theories. Coupled with the National Science Foundation-supported Central Arizona-Phoenix Long-Term Ecological Research Project (CAP LTER) Greater Phoenix 2100 has the potential to launch a network of similar undertakings nationally and internationally.

Greater Phoenix 2100 has been influenced by several other national efforts, such as the Los Alamos National Laboratory's Urban Security and Sustainability

Project, the U.S. Geological Survey's (USGS) Urban Dynamics Project, the National Science Foundation's (NSF) Urban Research Initiative, various university institutes, and state "smart growth" initiatives. In the mid-1990s, Los Alamos Laboratory developed a multidisciplinary urban modeling approach [Heiken et al., 2000]. Their team used their modeling capabilities to develop urban security and sustainability scenarios, focused on a few cities, including Los Angeles, Dallas, Portland (Oregon), and Albuquerque. In 2000, Phoenix was added to the list of cities to be modeled. However, that year the budget for the program was greatly reduced.

The USGS Urban Dynamics Research program supports studies of the landscape transformations that result from the growth of metropolitan regions over time. By combining a variety of data sources including historic maps, Landsat satellite data, and aerial photography, the project's scientists document past effects of urbanization on landscapes, and model land-use changes under alternative growth scenarios. Among the metropolitan regions that have been analyzed are Portland (Oregon), Chicago, Baltimore–Washington, San Francisco, and New York. The program is a partnership among the USGS, University of California at Santa Barbara, NASA, and others.

The Urban Research Initiative (URI) was an interdisciplinary program coordinated by the National Science Foundation's Social, Behavioral, and Economic Sciences Directorate (http://www.nsf.gov/pubs/1998/nsf9898/nsf9898.txt). Its goal was to support projects studying processes that determine or constrain the nature and direction of change in urban environments. This research was intended to facilitate development of a predictive understanding of the complex interactions among people, the natural environment, and the physical settings of urban environments. Originally launched as a multi-year program in 1998, the URI became incorporated into NSF's "Biocomplexity" initiative in 1999.

Several universities have projects with goals similar to those of Greater Phoenix 2100. For example, the University of Texas at Austin's Urban Issues Program focuses research on topics like the built environment and housing, community development, demographic change, economics, the natural environment, planning, urban form, and design. Portland State University has established an Institute of Portland Metropolitan Studies. This institute serves the Portland, Oregon region by providing new access to its higher education resources. The institute is attempting to create a shared understanding of the region, its issues and prospects by providing a neutral forum for the study of metropolitan policy issues. The University of Texas at Austin and Portland State University exemplify the trend among urban universities to contribute to the knowledge base of their region.

In concert with the American Planning Association's Growing Smarter Project, several states have adopted new statutes to modernize their approach to growth management [Meck et al., 1999]. For example, Maryland enacted a Smart Growth program in 1997. By revamping Maryland's laws, the state's leaders expressed their conviction that improved growth management laws would be one of the most important ways to reduce pollution and to stimulate economic development. In 1998, Arizona adopted a Growing Smarter law, followed in 2000 by Growing Smarter Plus. The Arizona law requires counties and cities to adopt comprehensive plans to guide their growth. Maryland and Arizona typify numerous examples of the growing nationwide interest in the nature of urban growth.

In spite of these notable beginnings by governments and universities, few activities are systematically coordinated over long periods of time or across complex bureaucratic divides. Several conditions discourage the needed cooperation. Federal/state/private/academic collaborations are fragmented. The social, biological, and physical sciences are disconnected from one another as well as from the humanities, law, engineering, public policy, and the environmental design arts. Growth debates lack a scientific foundation and tools for forecasting the environmental and social consequences of growth are limited. With the framework provided by Greater Phoenix 2100, we believe that the Phoenix metropolitan region can address and move beyond these current constraints.

WHY STUDY METROPOLITAN PHOENIX?

The Phoenix region possesses three characteristics that make it an ideal urban laboratory: it is geographically delimited; it is a national leader in growth; and it typifies the arid urban west. Water and energy availability impose sharp resource constraints that define the physical boundaries of the metropolitan region. From both a modeling and a political perspective, the boundary conditions of the region are relatively simple: it is encircled by public land and most of the urban, built up area is contained in a single county with 27 jurisdictions (fewer than most other large U.S. regions). As a result, until relatively recently, few "leap-frog" developments have occurred beyond a well-proscribed, contiguous metroplex.

Throughout the twentieth century, the Phoenix region grew consistently and rapidly, especially after the second world war (Figures 1 to 5). The population of the Phoenix metropolitan region has been growing at some of the most rapid rates in the U.S. Maricopa County grew 44.8 percent between 1990 and 2000, increasing from 2,122,101 to 3,072,149 people. This made it the fastest growing and fourth most populous county in the country. During the 1990s, the City of Phoenix topped a million people and became the sixth largest city in the nation. Its spatial expanse has eclipsed that of the city of Los Angeles. According to City of Phoenix Planning Department data, the region is growing by about 63,000 residents per year (more than 2000 per week), who require about 23,000 new housing units. The number of people in Arizona is expected to double in the next 20 years [Gammage, 1999; Morrison Institute, 2000].

Much of the expansion has been in the form of suburban sprawl. Recently, 9,000 acres of land per year (one acre per hour) have been developed, resulting in loss of both natural desert environments and productive irrigated farmlands

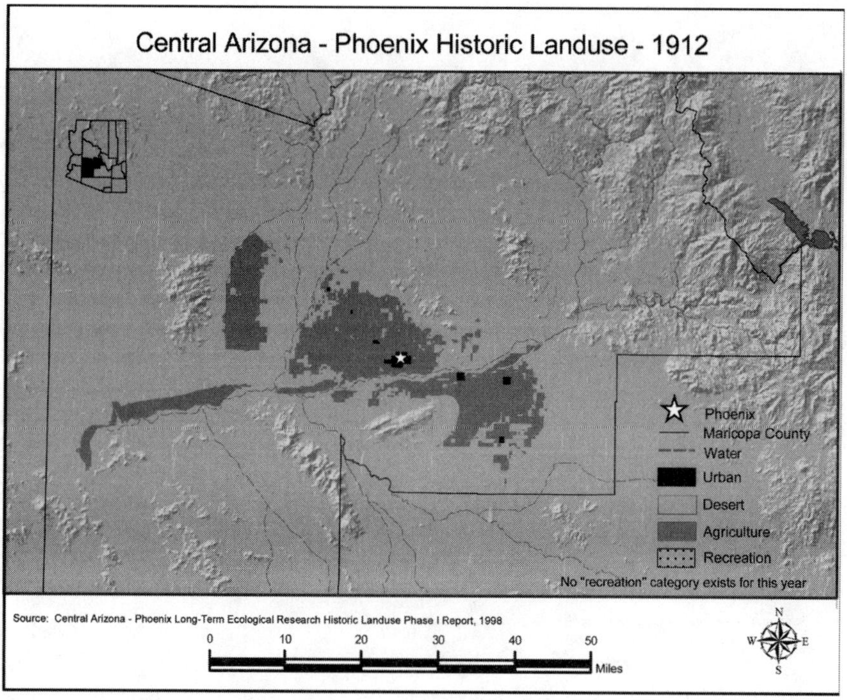

Figure 1. Central Arizona Phoenix Historic Landuse – 1912.

[Gammage, 1999; Morrison Institute, 2000]. Air pollution has increased to alarming levels, with Maricopa County failing to meet federal standards for two air pollutants (particulates and ozone) at various times in the past decade. Other signs of environmental degradation abound, ranging from visual clutter to water diversion from natural stream courses, to contaminated ground water plumes, to noise and traffic problems [Gammage, 1999; Morrison Institute, 2000]. As these trends continue, an area once well known for its scenic beauty and health benefits is developing a more negative image and reality. This transformation has serious implications for the state's important tourism, retirement, and film industries.

Phoenix also typifies other rapidly growing cities in the arid and semi-arid American West. The economy has a strong and growing high tech sector, spurred by business-friendly tax policies and an abundance of relatively cheap land. However, mass transit options are limited, social welfare needs are underfunded, and the continuing influx of new residents is nearly balanced by a relatively large exodus of existing citizens. From a global perspective, Phoenix shares environmental similarities with many of the most rapidly urbanizing regions around the world, adding to the relevance of projects like Greater Phoenix 2100.

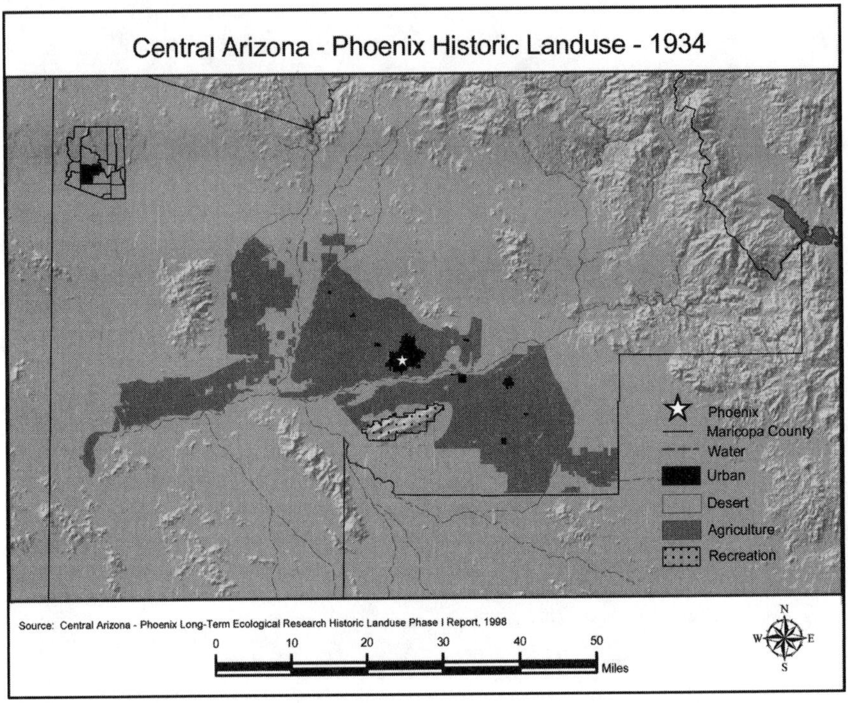

Figure 2. Central Arizona Phoenix Historic Landuse – 1934.

WHY HAVE A UNIVERSITY-LED PROJECT?

Besides the above advantages that metropolitan Phoenix offers for an urban environmental science and policy initiative, the region's one major university (ASU) is well-positioned to play a leadership role. ASU contains several high-profile environmental research projects and teams, it has the institutional persistence to engage in a long-term project of this type, and it is largely detached from the economic and political interests with a stake in possible conclusions reached by studies of growth impacts on the urban and surrounding environment.

The flagship of ASU's environmental research portfolio is the Central Arizona–Phoenix Long-Term Ecological Research (CAP LTER) project, which was selected in 1997 by the National Science Foundation to be one of two urban sites in the LTER network [Grimm et al., 2000]. The aim of CAP LTER is to understand the changing urban fabric of the Phoenix region's arid ecosystem, through an understanding of how land-use change and other human activities alter ecological conditions in and around the metropolis, and, conversely, how these ecological changes feed back to affect further human decisions, behavior and activity. CAP LTER is a multidecade-scale monitoring project, involving 48

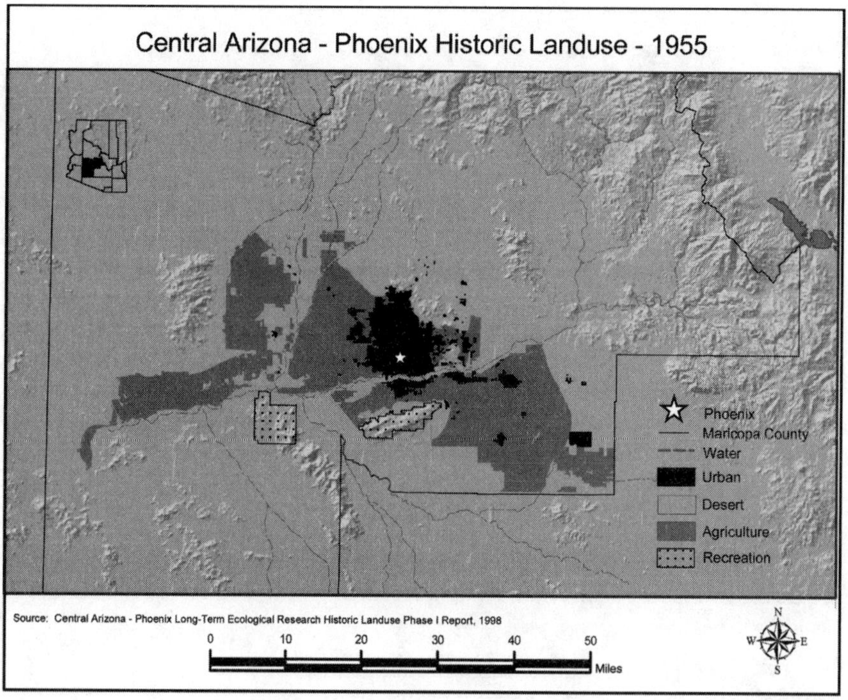

Figure 3. Central Arizona Phoenix Historic Landuse – 1955.

co-investigators from 14 ASU departments, and partnerships with numerous state, federal, and city agencies. The other urban LTER project is in Baltimore, which differs from Phoenix in age, growth rate, environmental setting, and politics [Grimm et al., 2000].

In addition to CAP LTER, ASU houses several other relevant research projects funded by federal, state, and local agencies. As part of NSF's Urban Research Initiative, ASU researchers have been studying the growth and distribution of a carbon dioxide "dome" within the metropolitan region. NSF also recently awarded ASU one of their prestigious IGERT (Integrated Graduate Education and Research Training) grants, for urban ecology. NASA has supported a collaborative study between ASU planetary scientists and the City of Scottsdale to develop terrestrial remote sensing tools for urban resource management. A spin-off of this project, funded by NASA's Mission to Planet Earth, is a comparative remote sensing study of 100 rapidly growing cities around the globe, using ASTER (Advanced Spaceborne Thermal Emission and Reflection Radiometer) and other remote sensing instruments on the Terra platform. ASU's Environmental Fluid Dynamics Program has extensive research partnerships with the EPA's Office of Atmospheric Research and with the Arizona Department of Environmental Quality for modeling airflow within the Phoenix metropolitan area.

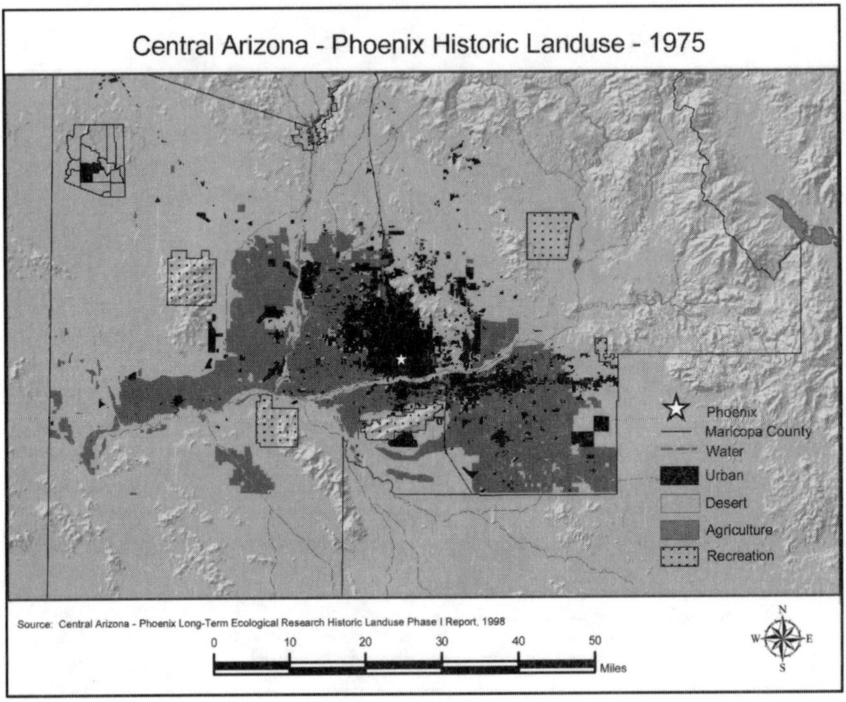

Figure 4. Central Arizona Phoenix Historic Landuse – 1975.

Besides the above federally funded projects, ASU has also worked with planners in the City of Phoenix to develop a GIS database for the city's 134-square-mile, largely undeveloped, North Area. Multiple data layers including geology, soils, drainage, vegetation, land ownership, and land use were developed for this parcel, which constitutes approximately 20 percent of the city's area. In addition, ASU researchers conducted original research of specific geological features using NASA data and specific vegetation and wildlife patterns based on field research. The GIS maps and other studies were used to conduct land suitability analyses, the identification of environmentally sensitive areas, as well as forecasts of potential environmental consequences of possible future developments. The ASU work led to approximately one third of the area being set aside by the city and state as a Sonoran Preserve. The city also changed its planning and development requirements for the area [Steiner, 2000].

PLOTTING POSSIBLE FUTURES FOR METROPOLITAN PHOENIX

In concert with the above university-based activities, Greater Phoenix 2100 represents a focal point to coordinate a variety of federal, state, municipal and academic research efforts, ultimately linking with similar studies in other cities.

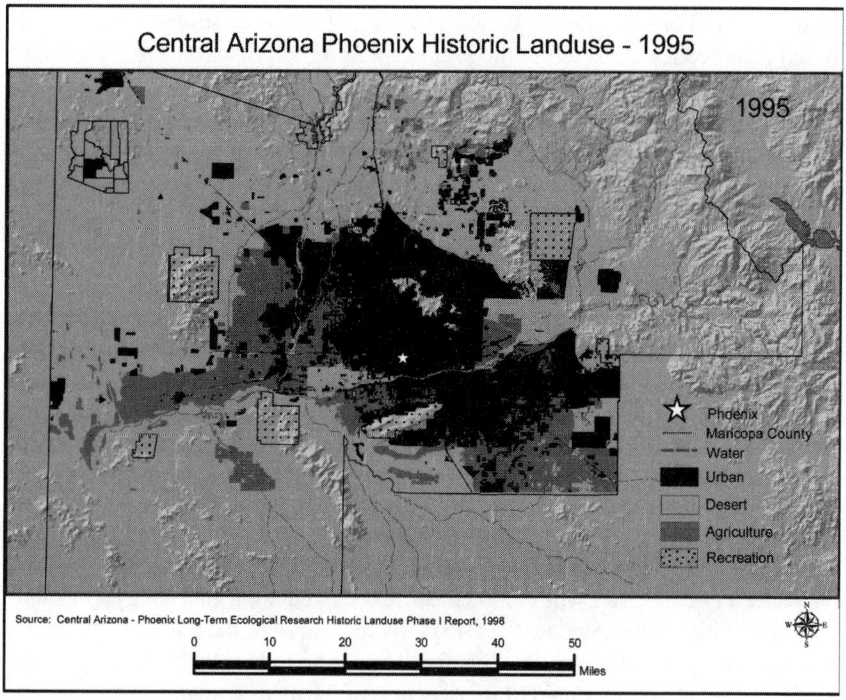

Figure 5. Central Arizona Phoenix Historic Landuse – 1995.

However, the first goal of Greater Phoenix 2100 is to answer environmental policy questions that people in the region care about by providing objective, scientifically based information. To present these results most effectively, Greater Phoenix 2100 will build state-of-the-art forecasting and decision tools and theories.

ASU and various government entities possess a significant storehouse of information about greater Phoenix. ASU faculty and students study and analyze practically every important aspect of central Arizona from its underlying geological structure to daily real estate transactions. Significant data exist concerning climatic variations, the flora and fauna of the Sonoran Desert biome, regional history and economic trends, and health and education of the population. An important goal of Greater Phoenix 2100 is to make this information available in ways that will enable wise, knowledge-based decision-making that can shape the region during the next 100 years.

The 100-year timeframe presents a purposefully longer-term view of the metropolitan region than has previously been developed. While short-term visioning is limited by immediate considerations, a century-long perspective requires the incorporation of multigenerational concerns and changes in technology. A 100-year timeframe also allows for evaluation of impacts of such geologically common but societal rare events as droughts, major floods, and gradual climate changes. In short, Greater Phoenix 2100 will provide a strong, scientifically based resource for considering the region's long-term prospects and for creating the kind of future its residents want.

Several linked products are envisioned to flow from Greater Phoenix 2100. First, existing data can be coalesced into a dynamic warehouse of continuously updated regional information. Such a data repository can be presented to the public through an Urban eAtlas, which will be made available in electronic and more conventional forms. The Urban eAtlas will provide an important resource for the documentation of existing conditions and will be designed to enable the construction of future scenarios. The digital version will be available on-line so that it may be continuously accessed and updated. The data warehouse and Urban eAtlas will contribute to a third major product: a Decision Theater where local leaders, citizens, students, and researchers can explore future options for the region. The Decision Theater will be an immersive physical space, in which scientific data, group dynamics, and interactive computer technology can be used to develop simulations of the region's futures and considerations of their consequences. The simulations and their representations will evolve with new computational and representational technologies as well as with new scientific information.

Greater Phoenix 2100 will complement and augment existing long-term monitoring activities being conducted at ASU, such as those of the CAP LTER project. A long-term goal of Greater Phoenix 2100 might be to launch a satellite in cooperation with NASA to regularly take the pulse of the metropolitan Phoenix area (as well as other cities, depending on orbital characteristics). One option would be for this "Phoenix-Sat" to have a highly elliptical orbit, passing low over the region twice daily, enabling diurnal measurements of such dynamic parameters as traffic, air quality, soil moisture, and construction activity. Tools such as the Urban eAtlas, the Decision Theater, and Phoenix-Sat will enable scholars and decision-makers alike to probe the major issues that metropolitan areas like Phoenix will face in the coming 100 years. As a result, problems may be anticipated and avoided while societally desirable opportunities may be pursued with vigor.

SUMMARY

With the launching of Greater Phoenix 2100, a community of scientists and public stakeholders seeks to tackle some of the most complex scientific and policy issues of the twenty-first century. Through a combination of advanced visualization tools and robust community involvement, Greater Phoenix 2100 will be a clearinghouse for fundamental questions about the impacts of urbanization on natural environments and social function. Results will be provided to inform debates among policy makers and citizens about future options. The hundred year time frame allows unique exploration of issues that might not otherwise be possible. Because the problems and opportunities faced by Phoenix typify those of many of the most rapidly growing urban areas in the United States and the world, the results of Greater Phoenix 2100 can potentially be relevant to hundreds of millions of people.

REFERENCES

Barrett, Gary W. and Eugene P. Odum, The twenty-first century: The world at carrying capacity, *BioScience, 50*(4), 363_368, 2000.

Botkin, D. B. and C. E. Beveridge, Cities as environments, *Urban Ecosystems, 1*, 3_19, 1997.

Brand, Stewart, *The Clock of the Long Now: Time and Responsibility*, Basic Books, New York, 1999.

Brazel, Anthony, Nancy Selover, Russell Vose, and Gordon Heisler, The tale of two climates–Baltimore and Phoenix LTER sites, *Climate Research, 15*, 123_135, 2000.

Calthorpe, Peter and William Fulton, *The City Region*, Island Press, Washington, D.C., 2001.

Collins, James P., Ann Kinzig, Nancy B. Grimm, William F. Fagan, Diane Hope, Jianguo Wu, and Elizabeth T. Borer, A new urban ecology, *American Scientist, 88*(September–October), 416_425, 2000.

Costanza, R., Visions of alternative (unpredictable) futures and their use in policy analysis, *Conservation Ecology, 4*(1), 5, 2000.
[online] URL: http://www.consecol.org/vol4/iss1/art5

Fagan, W. F., E. Meir, S. S. Carroll, and J. Wu, The ecology of urban landscapes. Modeling housing starts as a density-dependent colonization process, *Landscape Ecology 16*(1), 33_39, 2001.

Fernando, H. J. S., S. M. Lee, J. Anderson, M. Princevac, E. Pardyjak, and S. Grossman-Clarke, Urban fluid mechanics: Air circulation and contaminant dispersion in cities, *Environmental Fluid Mechanics, 1*, 107_164, 2001.

Gammage, Jr., Grady, *Phoenix in Perspective: Reflections on Developing the Desert*, Herberger Center for Design Excellence, Arizona State University, Tempe, 1999.

Grimm, Nancy B., J. Morgan Grove, Steward T. A. Pickett, and Charles L. Redman, Integrated approaches to long-term studies of urban ecological systems, *BioScience, 50*(7, July), 571_584, 2000.

Harrison, Paul and Fred Pearce, *AAA Atlas of Population & Environment*, University of California Press, Berkeley, 2000.

Heiken, G., G. A. Valentine, M. Brown, S. Rasmussen, D. C. George, R. K. Greene, E. Jones, K. Olsen, and C. Andersson, Modeling cities: The Los Alamos urban security initiative, *Public Works Management Policy 4*, 198_212, 2000.

Meck, Stuart, Rodney Cobb, Karen Finucan, Denny Johnson, and Patricia E. Salkin, *Planning Communities for the 21st Century*, American Planning Association, Washington, D.C., 1999.

Morrison Institute for Public Policy, *Hits and Misses: Fast Growth in Metropolitan Phoenix*, Arizona State University, Tempe, 2000.

Pickett, S., W. R. Burch, and S. E. Dalton, Integrated urban ecosystem research, *Urban Ecosystems, 1*, 183_184, 1997.

Science Advisory Board, *Beyond the Horizon: Using Foresight to Protect the Environmental Future*, U.S. Environmental Protection Agency, Washington, D.C., 1995.

Steiner, Frederick, *The Living Landscape* (Second edition), McGraw-Hill, New York, 2000.

United Nations Development Programme, United Nations Environment Program, World Bank, and World Resources Institute, *World Resources 2000-2001, People and Ecosystems, The Fraying Web of Life*. Elsevier, Amsterdam, 2000.

Valentine, Greg A. and Grant Heiken, The need for a new look at cities, *Environmental Science & Policy, 3*, 231_234, 2000.

Vitousek, Peter M., Beyond global warming: Ecology and global change, *Ecology, 75*(7, October), 1861_1876, 1994.

Weiner, Jonathan, *The Next One Hundred Years: Shaping the Fate of Our Living Earth*, Bantam Books, New York, 1990.

16

Modeling Cities—The Los Alamos Urban Security Initiative

Grant Heiken, Greg A. Valentine, Michael Brown, Steen Rasmussen, Jonathan Dowell, Sudha Maheshwari, and Denise C. George

INTRODUCTION

"When planning starts, for urban community or region, the area to be developed is not the equivalent of a piece of blank paper ready for the free materialization of the ideas of the designer, but it is rather an environment that has been exposed for a very long period to the effects of many natural modifying factors. The present-day surface of the Earth is the product of most complicated geological, hydrogeological, climatic and other processes, knowledge of which assists in recognition of probable trends of future terrain changes. Development of new communities and the charting of regional development must, therefore, take account of this fundamental organic and dynamic character of Nature so that the works of man may fit as harmoniously as possible into the environment and not disturb the biological equilibrium any more than is essential. To conserve the soundness and productive power of a region for the use of future generations is a basic requirement of overall planning and one that should be observed as a guide in the prosecuting of all engineering works."

Robert F. Leggett, "Cities and Geology," 1973.

Within the United States there are 47 metropolitan areas with populations in excess of a million people and 198 with populations of 100,000 to 500,000. As city populations increase, so do the issues of environmental, economic, and political stability that contribute to the quality of life. These issues may have technical solutions if those solutions are integrated into the more traditional approaches that cities use to manage municipal infrastructure maintenance, emergency

Earth Science in the City: A Reader

10.1029/056SP17

response, public health, environmental sustainability, and planning. As was described so well above by Robert K. Legget, foremost expert on urban geology in the 20th Century, the natural setting of a city is the underpinning of its infrastructure and health. In the past, urban planning decisions were made with little or no regard for the role of the environment in the long-term health and stability of a city. Today, as the Earth's population shifts from rural to urban settings, we must adopt a new way of understanding and managing cities—integrating the linked systems and environments that make up a city.

To support a "systems" view of cities, it is necessary to integrate across such diverse areas as environmental studies, geosciences, atmospheric sciences, architecture, engineering, transportation systems, utilities infrastructures, sociology, economics, public health, demography, and public security. Integrated approaches will produce quantitative city models that can be used by the governing bodies of the world's metropolitan areas to reduce vulnerabilities to natural disasters, overcrowding, and bad planning decisions. As urban planning is done now, managers of the many systems that comprise a city are commonly isolated from one another. For example, the mutual dependence of water, electrical, and sewage systems is not evident until there is a disaster such as an earthquake or wildfire. It is important that cities be run without the management barriers that block an understanding of all natural and man-made systems. Because of the interdependencies of subsystems, an integrated approach is needed: from the data-collection level all the way up to the decision-making level. Integrated urban science will play a vital role in planning and running sustainable cities of the future.

EARLY INTEGRATED SCIENTIFIC MODELS OF CITIES

Thirty-five years ago, the use of computers to create models of urban transportation systems and for planning was initiated with an optimistic attitude that the complex systems that make up cities could be understood (see review by Batty, 1994). This optimism changed to pessimism within the decade, leading to a paper by Lee (1973) on "a requiem for large scale models." Despite Lee's requiem, work on urban models continued, but it was focused on subsystems such as transportation, economics, and housing. In addition, there was a void separating scientific modelers from city-growth theorists and from urban-planning professionals.

Throughout the 1990s, city modeling research continued in many parts of the world, building heavily on GIS and visualization. As discussed in the paper by Burian et al. (this volume), one of the most difficult aspects of such modeling is to find and handle very large data sets, some of which are of questionable quality. To establish usable models, cities must have data standards for everything from electrical infrastructure to air quality. Even now there are no true models of a city, although research on the utility of such models continues. A scientific understanding of how a city functions and evolves is one of today's greatest challenges.

Wegener (1994) has published a thoughtful review of urban modeling activities across the world. At that time, most of the research had been done in 20 cities located in North and South America, Western Europe, Australia, and industrial Asia. These models focused on land use, transportation, housing, and energy consumption. None of the models were comprehensive and none included the geological and hydrological framework, atmospheric processes, or effects of natural hazards. Wegener (1994) puts forth his view of an ideal "model of urban models," for which he includes subsystems involving:

- slow changes: networks and land use
- medium-paced changes: workplaces and housing
- fast changes: employment and population
- immediate changes: goods transport and travel, and
- the urban environment

We believe a 21st Century focus on urban systems, via integrated modeling, is a new challenge for the scientific community, especially professionals in the natural sciences. Science-based integrated modeling of cities is moving forward in metropolitan Phoenix and Baltimore (Grimm et al., 2000; Collins et al., 2000), representing "long-term ecological research" projects. Another attempt to model cities has been pursued at the Los Alamos National Laboratory with its "Urban Security" project.

THE LOS ALAMOS URBAN SECURITY PROJECT

"Mankind's future will unfold largely in urban settings" (Fuchs et al., 1994) was the quote that stimulated a team at the Los Alamos National Laboratory to pursue the elusive goal of city modeling. With the availability of high-performance computing facilities, access to experts in many disciplines, and existing models for air pollution and transportation, Los Alamos assembled a team that included environmental engineers, geologists, software designers, natural hazards experts, mathematicians, hydrologists, civil engineers, atmospheric scientists, chemists, geographic information system specialists, and transportation experts. In collaboration with urban planners and environmental scientists from academia, work began with the immediate goal of developing integrated studies of cities. The greater purpose was to evaluate the vulnerability and response of urban systems to changes in physical environment, socio-political setting, and the economy. The entire urban "system of systems" has yet to be modeled, but progress was made in the six subsets (Heiken et al., 2000) described below.

- A computing framework prototype that enables computer models of urban components to produce simulations on a variety of different machines and in different computer languages (George and George, 1999). The goal is to allow multiple users to log onto the system simultaneously and work independently without disturbing the tasks of others.

- Pilot studies of city drivers exposed to a toxic-gas plume (Brown et al., 1997). Transport and dispersion models for plumes were run in conjunction with a transportation simulation of 100,000 vehicles in north Dallas. Exposed drivers spread the contaminant over an area far larger than the plume itself; simulation scenarios reveal the widely dispersed final locations of vehicles with high exposure. The scenario fundamentals are being abstracted to run on lap-top computers; ultimately, the technology would be used in realtime by emergency-response personnel.
- Integrated models of natural disasters (e.g., earthquakes), in which environmental and infrastructure models were linked (Maheshwari and Dowell, 1999; also discussed later in this chapter).
- Models simulating the growth of an urban center after a major disaster, and exploration of how such models can be used to aid in planning. (Andersson et al., in press and Yamins et al., in press). Simulations of the "re-growth" of an area are based on what existed before the disaster, zoning, physical suitability, transportation access, and land use in adjacent neighborhoods. The models address urban dynamics on a timescale of about a month to several years.
- Links between air pollution, surface water, and storm drain water quality (important when linking infrastructure with environmental considerations for sustainable development) (See papers in this issue by Burian et al. and McPherson, et al.)
- Novel, computer-based concept for achieving consensus among large numbers of stakeholders when making difficult urban-planning decisions (Johnson et al., 1998; Keating et al., 2002; Rasmussen and Goldstein, 2002). A web site is used to establish a consensus among organizations in response to a disaster; for example, determining the best places to store relief supplies in anticipation of an earthquake.

COMPUTING A FRAMEWORK PROTOTYPE

Cities are composed of a wide range of subsystems, including such areas as transportation, construction, energy distribution, communication, water, meteorological phenomena, geology, ecological, solid waste, food and water distribution, economic zones, and demographics. The Los Alamos team worked toward linking the many components of an urban system, rooting the framework in physical models, and abstracting essential physical, social, and economic interactions for decision-makers. (Figure 1). One particularly challenging aspect of this research effort is understanding the relationships between subsystems (right-hand side, Figure 1). These subsystems are linked and their interaction produces the collective and often non-intuitive behavior of the urban system. In the next two subsections we discuss modeling plume releases and the effects of earthquakes on infrastructure.

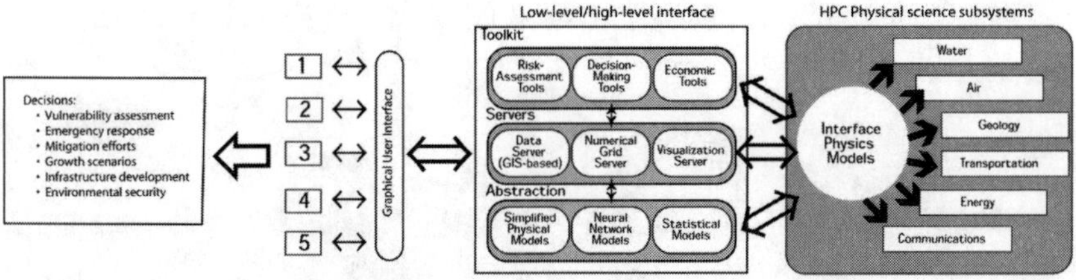

Figure 1. Proposed system architecture for an urban security "system of systems." Numerical models from across disciplines and covering a broad range of scales will be integrated through interface physics modules.

Airborne Toxic Release and Vehicle Exposure

If a tanker truck carrying hazardous chemicals overturns in a city, could emergency response crews estimate the exposure to vehicles that unwittingly drove through the poisonous cloud? Or, if a terrorist releases a chemical or biological agent in the downtown area, could first responders determine how the toxic agent spread, where it ended up, and how much contamination was transported away from the scene by moving vehicles? The ability to accurately simulate meteorology and traffic behavior with computer models could help answer these questions and better prepare emergency response personnel to deal with these types of dangerous situations.

As part of Los Alamos' Urban Security Project, atmospheric and vehicle transportation models were used both to study the movement of a plume in an urban setting and to compute the exposures to vehicles that drove through the plume (Brown et al., 1997). The HOTMAC mesoscale atmospheric model (Williams and Yamada, 1990) was used to simulate the weather and winds over a several-hundred-square-kilometer area centered in Dallas, Texas. A gas source was located at ground level between two buildings; the near-source transport and dispersion of the contaminant cloud was simulated using the GASFLOW computational fluid dynamics code (Travis et al., 1994).

To objectively evaluate the impact of the buildings on the larger scale plume motion and spread, one simulation accounted for the buildings and one did not. We found what we considered a counter-intuitive result: the plume actually traveled farther in the same amount of time when the influence of buildings was calculated (see Figure 2). Initially, we had incorrectly guessed that the plume would travel more slowly in the presence of buildings because of plume entrapment between the buildings. However, although trapping does occur, swirling air between buildings raised parts of the plume high above the ground. At these higher levels, a portion of the plume was rapidly blown downwind by the faster winds aloft. The study concluded that buildings are important in determining large-scale transport; therefore, they were included in all subsequent model simulations.

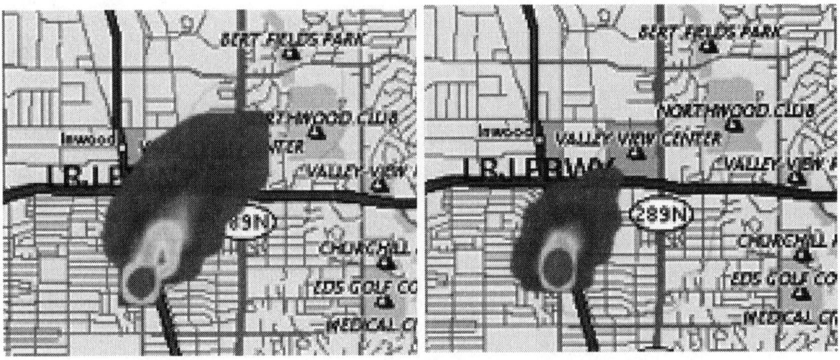

Figure 2. A comparison of plume transport and dispersion (ground-level concentrations) computed (a) with and (b) without explicit modeling of building-scale effects. The plume travels faster when buildings are included because the plume is lofted into the faster moving winds found at higher altitudes.

The traffic flow for over 100,000 vehicles in North Dallas was performed by the Los Alamos Transportation Simulation team, using the TRANSIMS model (Smith et al., 1995; Rickert and Nagel, 1997). TRANSIMS represents a new approach in traffic modeling, in which the second-by-second movement of individual cars is computed. By simulating the interactions between individual cars on the street network, the model more accurately depicts the patterns that occur in common everyday traffic.

Using vehicle routes computed by TRANSIMS and predictions of plume motion and dilution by GASFLOW and HOTMAC, the team estimated exposures to over 36,000 vehicles traveling through the hypothetical contaminant cloud in the Dallas-Ft. Worth area. Vehicle exposure data clearly delineate the major thoroughfares (the North Dallas Tollway and the LBJ freeway) and show that the contaminated vehicles carry the toxic agent away from the source location more quickly than the winds can. In addition, the exposed drivers and vehicles are dispersed over a much larger area than that covered by the plume. Moreover, the final locations of vehicles with high exposure rates are surprisingly and dramatically depicted by the simulation scenario (Figure 3).

Using a simulation tool similar to this model, emergency response personnel could determine the impact zones, optimal routes for response teams, locations of likely casualties, and probable dispersion patterns for the agent. The efforts of clean-up crews and medical teams would be enhanced, as well, through knowledge and predictions of final locations and levels of exposure. There are additional applications for a broad range of cross-disciplinary research programs, including urban air pollution, traffic simulation, counter-proliferation (countering domestic terrorist attacks), and emergency response (mitigation of accidental

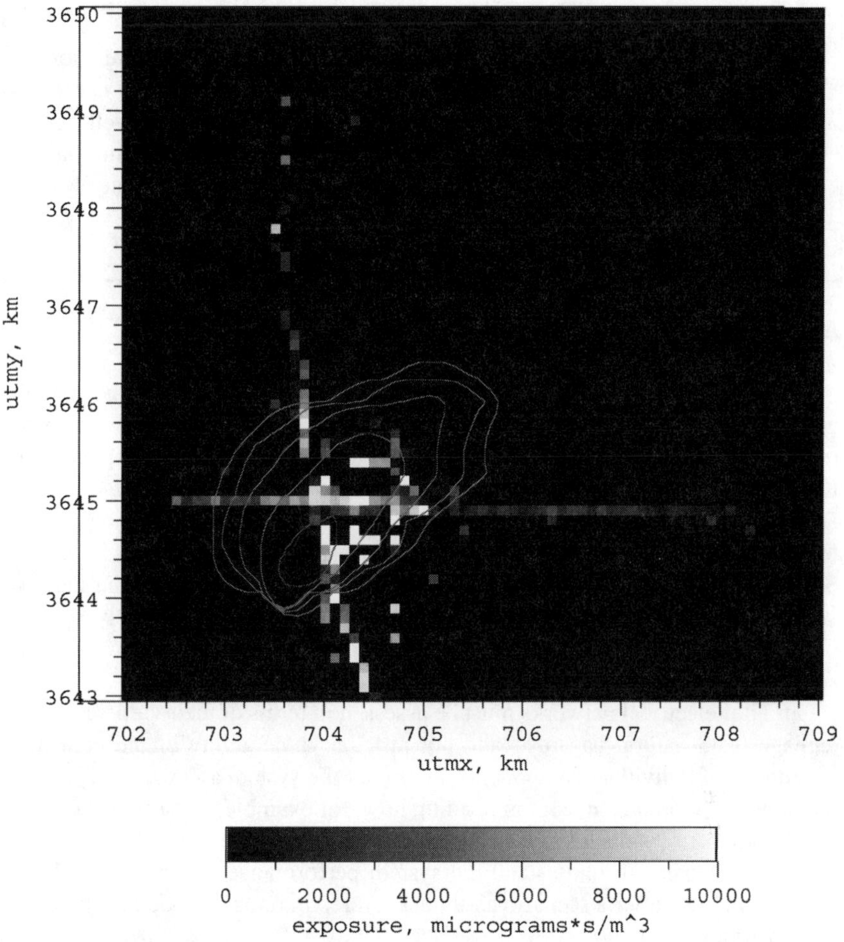

Figure 3. Vehicle exposure as a function of final destination.

chemical spills). Further research efforts are underway at Los Alamos to refine estimates of multiple buildings on the behavior of a toxic plume. We are continuing research on transportation simulation and will include efforts to abstract the fundamental vehicle behaviors so that simulations can be run on personal computers in the field, as needed.

Note: For the modeling done here, GASFLOW and HOTMAC simulations were performed on a Sun workstation but could easily be run on current PCs. The TRANSIMS simulations were performed on Suns as well, and studies involving larger numbers of vehicles are currently being run on a PC Linux cluster. To simulate plume dispersion across a whole city with resolved buildings, a cluster also would be necessary.

Earthquakes and Urban Infrastructure

Earthquake damage to an electrical network is not restricted to the specific geographical location of the quake, but has far-reaching implications for other regions as well. For example, the damage of electrical network components in Los Angeles by the Northridge Earthquake caused power outages as far away as British Columbia, Montana, Wyoming, Idaho, Oregon, and Washington; the longest power interruption, in Southern Idaho, lasted 3 hours (Schiff et al., 1995). Such disruption can be either first-order, direct damage to the generating stations, substations, and transmission and distribution networks or, alternatively, second-order, indirect damage caused by changes in load as a result of fluctuating consumer demand.

Our vital electrical networks must be assessed in terms of their vulnerability to damage from natural hazards—and not just the vulnerability of the network's multitudes of individual components, but rather the system as a whole. This is of the utmost importance in comprehending how, for example, an earthquake in the metropolitan Los Angeles area will likely affect the entire Western one-half of the United States. To understand the system performance of the electrical network during and after a scenario earthquake, three important modeling strategies must be integrated:

- generating the best possible ground-motion parameters (peak ground acceleration, peak ground displacement, or response spectrum) for the scenario earthquake,
- combining the ground motions with component fragility to compute the damage state for each of the power system's components, and
- using the damage-state probabilities to undertake systems-engineering analysis and assess the performance of the electrical system.

Although each of the above models has been developed and used singularly, the need to integrate these models is crucial to any kind of damage assessment; exchange of input and output values from one model to another must be accomplished in a way that the results of one model can be used by another.

Modeling scenario earthquake ground motions. One of the most important aspects of earthquake studies is the ability to predict with a high degree of reliability the ground motions that result from an earthquake (Maheshwari and Dowell, 1999). The distribution of ground motion during earthquakes is often heterogeneous, resulting in nonuniform damage distribution. This tendency could be seen in the Loma Prieta earthquake in the Bay Area of California and again in the Northridge earthquake in Los Angeles. The three-dimensional nature of velocity structures in deep sedimentary basins makes it very difficult to predict the ground motions in Los Angeles (Olsen and Archuleta, 1996). This study included 3D-modeling of velocity structure and a propagating rupture along a finite fault area, a method developed by Olsen and Archuleta (1996). The model employs first-principle simulations of the earthquake ground motions and takes into account subsurface geology, particularly in basin settings. The ground motions used in this study incorporate high (3-Hz) frequencies because the significant energy content at this level or greater is largely responsible for damage to components of urban infrastructure systems. The finite area modeled using this method incorporated a 75 x 75 grid mesh at regular spacing of 2 km and covered the greater Los Angeles area. These modeled ground motions were considered much more satisfactory than the ground motions generated by commonly used earthquake damage estimation models such as HAZUS™ (FEMA, 1997) (Figure 4).

Our integrated approach to modeling is based on an earthquake (of 6.75 magnitude on the Richter scale) along the Elysian Park fault, which is close to downtown Los Angeles. We used results from this simulation both to estimate damage to electrical substations in Los Angeles and the percentage of consumers affected by damage to each of those substations. Some remote agricultural and forested areas northeast of Los Angeles were excluded from the estimate because of uncertainties in identifying the substations serving these low-population areas (these omissions had no significant effect on the analysis).

For the simulation, we used HAZUS™ damage-state probability calculations and a single Monte Carlo sample of damage states for representative substations in the study area. (Substation damage ranged from none to complete.) As can be expected from Monte Carlo simulation, some low-probability results were selected. The HAZUS™ results found 43 substations with a probability of extreme or complete damage; that is, exceeding 50%.

The Monte Carlo-selected scenario provided a geographic correlation between those of the substations that experienced complete failure and the location of the Elysian Park fault (Figure 5). Blackout areas in Los Angeles were caused by either first-order isolation that resulted when a substation failed and cut off the flow of power to the distribution system, or second-order isolation that occurred when substation failures removed all transmission paths to substations downstream (Figure 6).

According to data from the CaISO (California Independent System Operator) 1999 summer-peak-load database, first- and second-order isolation removes

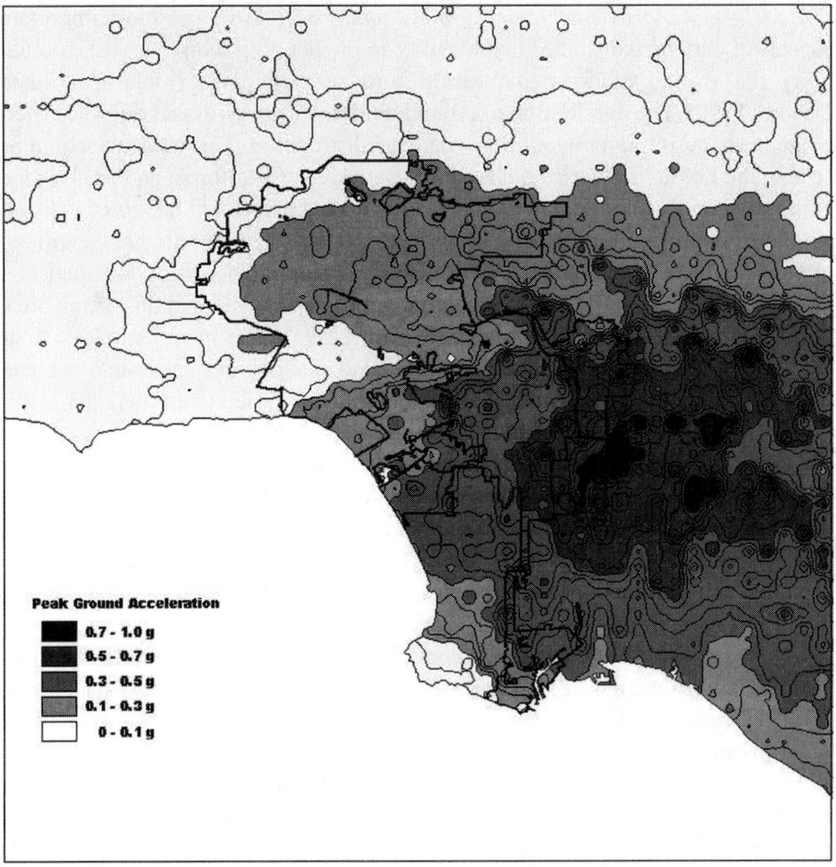

Figure 4. Peak Ground Acceleration for Elysian park Fault Scenario (Olsen and Archuleta, 1996 and Maheshwari and Dowell, 2000)

11,448 MW of consumers' loads and 4,400 MW of scheduled generation from the electric-power infrastructure. The load removed by the Elysian Park earthquake scenario is 8.9% of the entire WSCC EPI (Western States Coordinating Council, Electric-Power Infrastructure) load. These perturbations leave WSCC with a significant excess of scheduled power generation, and consequently, will produce higher-than-nominal voltages and increased system frequency. Manual and automatic generation control must reduce generation to match the post-earthquake load to prevent voltage and frequency problems.

As an academic assumption, we reduced the surviving scheduled generation uniformly (to 95% of the scheduled value) across the WSCC. The consequent power flow revealed surprisingly few problems for the WSCC EPI, but two types of incidents occur. First, power flow is shifted by line failures and changed load

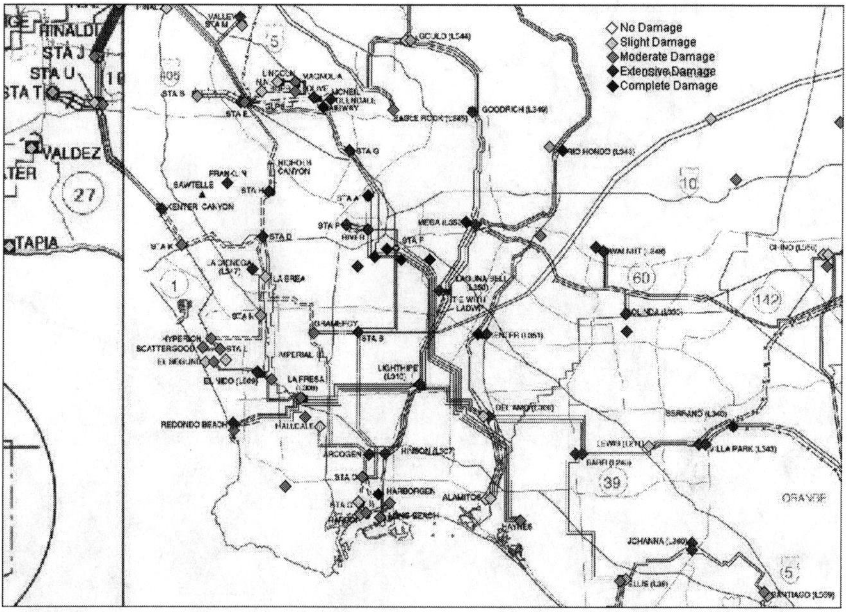

Figure 5. Monte Carlo selected Scenario damage states for each substation in the Los Angeles study area.

and generation schedules, producing thermal overloads (flows that exceed the thermal capacities of transmission lines). Such overloads are small (a few tens of megawatts on only eight transmission lines) and could be mitigated by changing the post-earthquake generation schedule. Second, the failures produce changes in reactive power flow that lead to abnormal voltages at substations. These problems are more serious: 129 substations had voltages 10% or more above normal and 1 substation experienced voltage 20% above normal. These are large excursions from the normal and could exceed the voltage-control apparatus' capability of mitigating abnormal power transmission. The abnormal voltages occurred throughout WSCC. There could be serious consequences if anomalous voltages occur at even several of these substations, particularly those in the high-voltage backbone through northern California, Washington, and Oregon (Figure 7). Abnormal voltages occur at the Malin, Captain Jack, and Grizzly substations in Oregon; Hanford, Ashe, Lower Monumental, Little Goose, and Lower Granite substations in Washington; Round Mountain and Olinda substations in California; and at substations as far away as Colorado and British Colombia. We need more information in order to predict how the WSCC EPI's relays and other devices would respond to these abnormal voltages and to determine if there could be cascading failures that would result in blackouts of areas far removed from the scenario's Los Angeles Elysian Park earthquake.

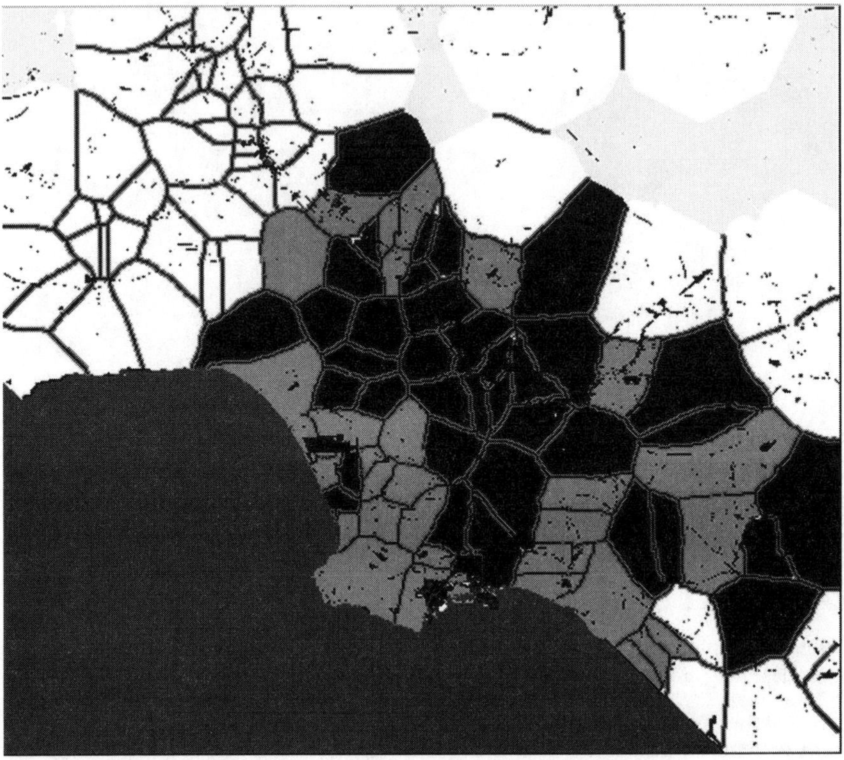

Figure 6. Areas of blackout and "brownout" for Los Angeles following a scenario earthquake on the Elysian Hills fault.

THE FUTURE OF MODELING CITIES

Scientists can supply models and validate data to be used within an integrated urban planning system that has both predictive and responsive capabilities. A great deal of work remains to be done to identify and extract the most important components of the detailed subsystem models. At the same time, we must design and implement a common data repository to ensure consistency in model results. Properly designed, a high-level model could be used by decision makers for managing urban areas in the coming decades.

Cities need integrated teams to collect data and make observations that can be used in quantitative models of the city—models that will help reduce vulnerabilities to natural disasters, terrorist attacks, and poor planning decisions. These teams and their resulting models must be included in existing urban management structures: not to replace, but to enhance them. Cities must employ more scientists, especially from the earth and atmospheric sciences, who can provide expertise in vital areas such as water sources and quality; air quality and circula-

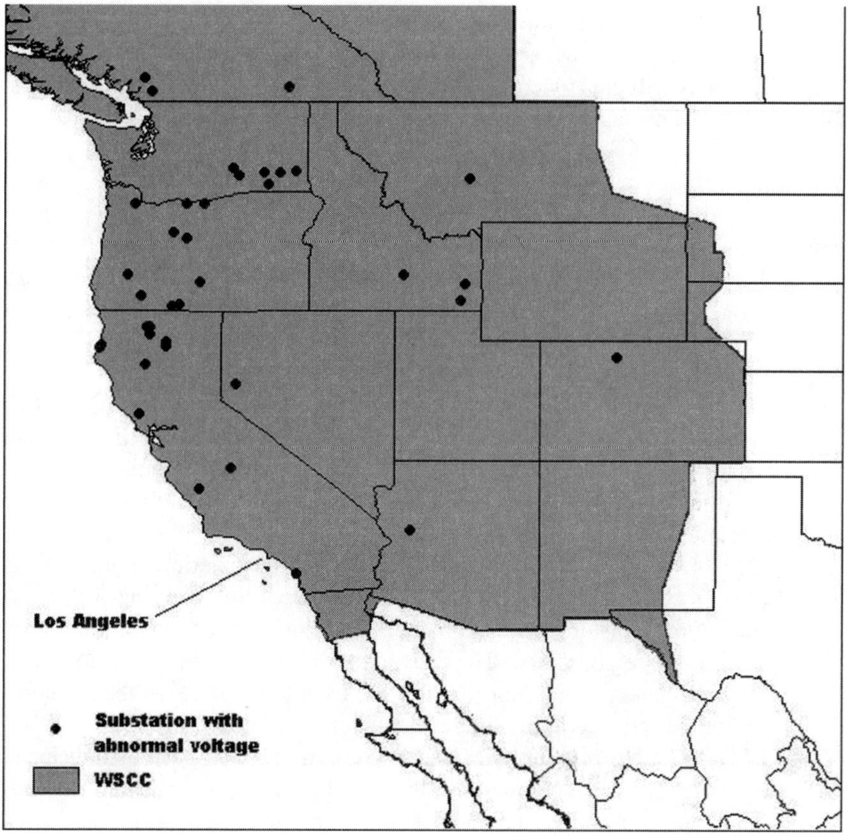

Figure 7. Cities on the western U.S. power grid near an earthquake in Los Angeles (Elysian Hills Scenario).

tion; energy resources and building materials; the fate and transport of chemical species from pollution or chemical/biological attacks; natural hazards mitigation (by hazards mapping, zoning, and predicting hazards scenarios); public health after a natural disaster; the establishment of greenbelts and preservation of urban agriculture; the effects of sea-level rise; thermal extremes and health effects; and understanding urban microenvironments as incubators of disease.

A focus on integrated urban systems also presents a new challenge for the scientific community, especially those of us in the natural sciences who, like many city managers and infrastructure managers, need to step outside conventional work and research roles. Integrated science is becoming a common approach to understanding natural (and man-made) systems; this trend is reflected by the reorganization of some traditional, discipline-oriented university departments into interdisciplinary institutes or divisions. University graduates must be encouraged to explore uncharted roles in urban research; cities must be convinced that they will gain by employing these pioneers—together they can develop new ways of managing a city.

ACRONYMS

GASFLOW—A computer program developed at the Los Alamos National Laboratory that predicts the 3D time-varying flow fields around buildings and other obstacles. It uses mathematical routines optimized for the computer to solve conservation equations for mass, momentum, and energy (Travis et al., 1994).

HAZUS—The Federal Emergency Management Agency, under a cooperative agreement with the National Institute of Building Sciences, has developed a standardized, nationally applicable earthquake loss estimation methodology. This methodology is implemented through PC-based Geographic Information System (GIS) software called HAZUS.

HOTMAC —Higher-Order Turbulence Model for Atmospheric Circulation. A computer code developed at LANL that predicts time-varying 3-d wind, temperature, moisture, and turbulence fields over mountains and valleys, cities, coastal regions, etc. The code uses conservation equations for mass, momentum, energy, and moisture that are used to predict how the meteorological fields change in the future (Williams and Yamada, 1990).

TRANSIMS—TRANsportation SIMulation System. A Los-Alamos-developed modeling system that computes traffic flow based on hundreds of thousands of interactions of individual travelers on the roadway network.

Acknowledgments. This work was funded by the Los Alamos National Laboratory as a Laboratory-Directed Research and Development grant. The authors greatly appreciate the reviews by Jill Singer and Ellis Krinitzky. Los Alamos National Laboratory is operated by the University of California for the National Nuclear Security Administration of the US Department of Energy.

REFERENCES

Anderson, C., S. Rasmussen, and R. White, Urban settlement transitions, *Environment and Planning B,* 2002.

Batty, M., A chronicle of scientific planning—The Anglo-American modeling experience. *Jour. Amer. Planning Assoc.,* 60, 7-15, 1994.

Brown, M., C. Muller, and P. Stretz, Exposure estimates using urban plume dispersion and traffic microsimulation models. *10th Joint AMS/AWMA Conf. On Applic. Of Air Poll.* Meteor., 140, 1997.

Collins, J. P., A. Kinzig, N. B. Grimm, W. F. Fagan, D. Hope, J. Wu, and E. Borer, A new urban ecology: modeling human communities as integral parts of ecosystems poses special problems for the development and testing of ecological theory. *American Scientist*, 88, 416_425, 2000.

FEMA (Federal Emergency Management Administration, *Earthquake Loss Estimation Methodology HAZUS^{TM} User's Manual*, 1997.

Fuchs, R. J., Introduction, in *Mega-city Growth and the Future,* edited by R. J. Fuchs, E. Brennan, J. Chamie, F. C. Lo, and J. I. Uitto, pp. 1-13, United Nations University Press, Tokyo, 1994.

George, J. E. and D. C George, L2F—Legacy to the Future Framework, *ACM 1999 Java Grande Conference Workshop for High-performance Network Computing*, June 12-14, 1999, San Francisco, CA, 1999.

Grimm, N. B., J. M. Grove, C. L. Redman, and S. T. A. Pickett, Integrated approaches to long-term studies of urban ecological systems, *BioScience*, 50, 571-584, 2000.

Heiken, G., G. A. Valentine, M. Brown, S. Rasmussen, D. George, E. Jones, K. Olsen, *and C. Andersson,. Modeling cities—The Los Alamos Urban Security Initiative. Public Works Management and Policy,* 4, 198-212, 2000.

Johnson, N. S. Rasmussen, C. Josly, L. Rocha, S. Smith, and M. Kantor, Symbiotic intelligence: Self-organizing knowledge on distributed networks driven by human interactions, in *Artificial Life VI, Proceedings of the Sixth International Conference on Artificial Life*, edited by C. Adami, R. K. Belew, H. Kitano, and C. Taylor, pp. 403-407, MIT Press, Cambridge, 1998.

Keating, G., S. Rasmussen, M. Raven, E. Tso, J. Cocq, and P. Dotson, Use of web-based consensus building and conflict clarification process for the Navajo Nation Governmental Efficiency Study, Navajo Council, Fall 2001, *(Los Alamos Report LA-UR-01-6207)*, 2001.

Lee, D. B., Jr., A requiem for large-scale models, *Jour. of the Amer. Inst. Of Planners*, 39: 163-178, 1973.

Leggett, R. F., *Cities and Geology.* McGraw-Hill, New York, 1973.

Maheshwari, S. and L. J. Dowell, Integrated modeling of earthquake impacts to the electric-power infrastructure: Analyses of an Elysian Park scenario in the Los Angeles Metropolitan area,*URISA 1999 Annual Conference,*1999.

Olsen, K. B. and R. J. Archuleta, Three-dimensional simulation of earthquakes on the Los Angeles fault system, *Bull.Seis. Soc.Amer.,* 86, 575-596, 1996.

Rasmussen, S. and N. Goldstein, Use of distributed information technology in disasters, submitted to *Policy in Science and Technology*, 2002.

Rickert, M. and K. Nagel, Experiences with a simplified microsimulation for the Dallas/Fort Worth area, *Int. J. Mod. Phys. C*, 8, 483, 1997.

Schiff, A. J., Power Systems, in Northridge Earthquake—Lifeline performance and post-earthquake response. *Tech. Council on Lifeline Earthquake Engineering Monograph 8*, edited by A. J. Schiff, Amer. Soc. Civil Engineers, New York, 1995.

Smith, L., R. Beckman, D. Anson, K. Nagel, and M. Williams, TRANSIMS: TRansportation ANalysis and SIMulation System, *Proc. 5th Natl. Trans. Plan. Meth. Appl. Conf.,* Seattle, WA, 1995.

Travis, J., K. Lam, and T. Wilson, GASFLOW: Theory and computational model, volume 1. *Los Alamos National Laboratory unclassified report* LA-UR-94-2270. Los Alamos, NM, 1994.

Valentine, G. A. and G. Heiken, G., The need for a new look at cities. *Environmental Science and Policy,* 3, 231-234, 2000.

Wegener, M., Operational urban models—state of the art. *Jour. Amer. Planning Assoc.,* 60, 17-29, 1994.

Williams, M., and T. Yamada, A microcomputer-based forecasting model: potential applications for emergency response plans and air quality studies, *Journal of Air & Waste Management Association*, 40, 1266-1274, 1990.

Yamins, D., S. Rasmussen, and D. Fogel, Growing urban roads, *Networks and Spatial Economics,* in press, 2002.

Contributors

Charles Baskerville
Central Connecticut State University
Dept. of Physics and Earth Sciences
1615 Stanley Street
New Britain, CT 06050-4010
Phone: (860) 832-3188/2930
Fax: (860) 832-2946
Baskerville@ccsu.edu

George Bugliarello
Polytechnic University
Six Metrotech Center
Brooklyn, NY 11201
Phone: (718) 260-3330
Fax: (718) 260-3974
gbugliar@poly.edu

Steven J. Burian
Department of Civil Engineering
University of Arkansas
4190 Bell Engineering Center
Fayeteville, AR 72701
Phone: (479) 575 4182
Fax: (479) 575-7168
sburian@engr.uark.edu

Marjorie A. Chan
The University of Utah
Department of Geology and Geophysics
135 S. 1460 E.
Room 719
Salt Lake City, UT 84112-0111
Phone: (801) 581-6551/6553
Fax: (801) 581-7065
machan@mines.utah.edu

Robert Fakundiny
New York State Geological Survey
3140 Cultural Education Center
Empire State Plaza
Albany, NY 12230
Phone: (518) 474-5816
rfakundi@mail.nysed.gov

Jonathan Fink
Arizona State University
Department of Geological Sciences
Tempe, AZ 85287-1404
Phone: (480) 965-3195
Fax: (480) 965-8293
jon.fink@asu.edu

Gerald T. Hebenstreit
Science Applications International Corporation
1710 Goodridge Drive
McLean, VA 22102
Phone: (703) 676-4975
GERALD.T.HEBENSTREIT@saic.com

Grant Heiken
Los Alamos National Laboratory
EES- IGPP, MS C305
Institute of Geophysics & Planetary Physics
Los Alamos, NM 87545
Phone: (505) 667-847
Fax: (505) 665-3687
heiken@lanl.gov

Brian Marker
Minerals and Waste Planning Division
Department of the Environment
Zone 4/A2
Eland House
Bressenden Place
London SW1E 5DU, UK
Phone: 44-00-20-7944-3851
Fax: 44-00-20-7944-3859
brian_marker@detr.gsi.gov.uk

Timothy N. McPherson
Los Alamos National Laboratory
Energy & Environmental Analysis
Group D-4,MS F604
Los Alamos, NM 87545
Phone: 505 665 8521
Tmac@lanl.gov

Giovanni Orsi
Osservatorio Vesuviano
Via Diocleziano no. 328
Napoli, 80124, Italy
Phone: 39-081-6108343
Fax: 39-081-6108344
Orsi@osve.unina.it or orsi@ov.ingv.it

Richard J. Pike
US Geological Survey
345 Middlefield Road, MS 975
Menlo Park, CA 94025
Phone: (650) 329-4947
Fax: (650) 329-4936
rpike@usgs.gov

Michael Ramsey
Department of Geology & Planetary Science
200 SRCC Building
University of Pittsburgh
Pittsburgh, PA 15260-3332
Phone (Office): (412) 624-8772;
Phone (Lab): (412) 624-8773,
Fax: (412) 624-3914
ramsey@ivis.eps.pitt.edu

John M. Sharp, Jr.
The University of Texas
College of Natural Sciences
Department of Geological Sciences
118 Geology Building
Mail Code: C1100
Austin, TX 78712-1101
Phone: (512) 471-5172/3317
Fax: (512) 471-9425
jmsharp@mail.utexas.edu

John Sutter
US Geological Survey
520 N. Park Ave. #355
Tucson, AZ 85719
Phone: (520) 670-5588
jsutter@usgs.gov

Greg A. Valentine
Los Alamos National Laboratory
Hydrology, Geochemistry & Geology Group EES-6
MS D462
Los Alamos, NM 87545
Phone: (505) 665-0259
Fax: (505) 665-3285
gav@lanl.gov

Dean Whitman
Florida International University
International Hurricane Center
University Park, PC344
11200 SW 8th Street
Miami, FL 33199
Phone: (305) 348-3089
Fax: (305) 348-3877
whitmand@fiu.edu